Ethics, Information, and Technology

Ethics, Information, and Technology

Kip Currier

BLOOMSBURY LIBRARIES UNLIMITED
NEW YORK • LONDON • OXFORD • NEW DELHI • SYDNEY

Online resources to accompany this book are available at: https://bloomsbury.pub/ethics-info-tech. If you experience any problems, please contact Bloomsbury at: onlineresources@bloomsbury.com

BLOOMSBURY LIBRARIES UNLIMITED
Bloomsbury Publishing Inc, 1359 Broadway, New York, NY 10018, USA
Bloomsbury Publishing Plc, 50 Bedford Square, London, WC1B 3DP, UK
Bloomsbury Publishing Ireland, 29 Earlsfort Terrace, Dublin 2, D02 AY28, Ireland

First published in the United States of America 2026

Cover Designer: Jess Stevens
Cover image: Istock © cyano66 & Adobe Stock © James Thew

Library of Congress Cataloging-in-Publication Data
Names: Currier, James (Kip) author.
Title: Ethics, information, and technology / James "Kip" Currier.
Description: New York: Bloomsbury Libraries Unlimited, 2026.
Identifiers: LCCN 2024050895 (print) | LCCN 2024050896 (ebook) | ISBN 9781440856662 (PB) | ISBN 9798216186113 (HC) | ISBN 9798216171386 (eBook) | ISBN 9781440856679 (ePDF)
Subjects: LCSH: Information technology–Moral and ethical aspects. | Information science–Moral and ethical aspects.
Classification: LCC T58.5.C876 2025 (print) | LCC T58.5 (ebook) | DDC 004–dc23/eng/20250304
LC record available at https://lccn.loc.gov/2024050895
LC ebook record available at https://lccn.loc.gov/2024050896

ISBN: HB: 979-8-216-18611-3
PB: 978-1-4408-5666-2
ePDF: 978-1-4408-5667-9
eBook: 979-8-216-17138-6

Typeset by Integra Software Services Pvt. Ltd.
Printed and bound in the United States of America

For product safety related questions contact productsafety@bloomsbury.com.

To find out more about our authors and books visit www.bloomsbury.com and sign up for our newsletters.

To my parents, James Hughes Currier and Marilyn Wible Currier.
For countless loving acts, great and small, whose positive
influences are with me every day.

Copyright Acknowledgments

The author and publisher are grateful for permission to reproduce material from the following sources:

American Library Association. (2021). ALA code of ethics. https://www.ala.org/tools/ethics. Used with permission from the American Library Association.

The Ethics Institute Ethical Decision-Making Model. © 2022 Kent Place School. All rights reserved. Used with permission.

A Framework for Ethical Decision Making. Used with permission from the Markkula Center of Applied Ethics, Santa Clara University, www.scu.edu/ethics.

Big Nate comic. © 2017. Used with permission from Andrews McMeel Syndication.

International Federation of Library Associations and Institutions. How to spot fake news. Used with permission from IFLA under Creative Commons License CC-BY 4.0. Icons available individually under FreePik licence.

Bizarro comic. © 2014 Bizarro Studios, distributed by King Features Syndicate, Inc. Used with permission.

Non Sequitur comic. © 2018. Used with permission from Andrews McMeel Syndication.

Contents

Preface

Roughly five weeks after the 2024 US presidential election, I sit at my desk, pondering the contours of an update to this book's preface. Donald Trump defeated Kamala Harris on November 5, 2024. All indications are that the 2024 election results will have *seismic* consequences for our nation and world. Moreover, fiercely contested ethics issues and clashes are likely to be omnipresent, with as-yet-unknown implications for information and technologies.

The focal points of this textbook *are* in fact ethics, information, and technology, as evinced by the book's title; the word *politics* is notably absent. However, even a cursory view of the past three presidential election cycles reveals that politics is inextricably linked to this book's foci of inquiry and critical examination. Take the issue of net neutrality: the Trump administration undid the Obama administration's pro–net neutrality policy position, and the subsequent Biden administration reverted back to supporting net neutrality. Elections, hence, can and often do have significant if not monumental impacts on issues relating to ethics, information, and technology. Political considerations and decisions influence ethics, and ethical choices and decision-making influence politics. The same is true for information and politics (as seen in ongoing culture wars over the information available via books in public and school libraries), as well as for technology and the political world. Consider the ways that Twitter/X owner Elon Musk—the world's richest person—attempted to influence the 2024 election through social media, and how he continues to inject himself into the new Trump administration landscape, earning him the media moniker "shadow vice president."[1]

Ethical issues during the 2016, 2020, and 2024 campaigns (as well as throughout the in-between years) have been ever present and inescapable:

- ethics, information, and data
- ethics and truth
- ethics and media coverage
- ethics and conflicts of interest
- ethics and transparency
- ethics and privacy
- ethics, national security, and classified information
- ethics and social media
- ethics and civil discourse
- ethics and integrity
- ethics and leadership

One of the most memorable, talked-about lines from the 2016 campaign came from then first lady Michelle Obama, while delivering her speech at the 2016 Democratic National Convention in Philadelphia: "When they go low, we go high."[2] We learned that this seven-word tenet was invoked by the first lady's mother when facing the vicissitudes of life with strength and grace. Many subsequently adopted the phrase as a rallying cry for decency and taking the higher road, a moral touchstone for how we *can* or *should* choose to behave and act. As a credo, the words also speak to the *individual* choices people can make about how to respond to negativity, how to treat others, how to uplift rather than tear down. For me, the phrase also evokes President Abraham Lincoln's first inaugural address on March 4, 1861. That historic speech was a call for unity and understanding, before the start of four years of devastating civil strife, pitting US citizens against one another:

> We are not enemies, but friends. Though passion may have strained, it must not break our bonds of affection. The mystic chords of memory, stretching from every battlefield and patriot grave, to every living heart and hearthstone, all over this broad land, will yet swell the chorus of the Union, when again touched, as surely they will be, by the better angels of our nature.[3]

Ultimately, Donald Trump won the 2016 election, securing the US Electoral College with 304 votes, well over the 270 votes required. Hillary Clinton would eventually be declared the winner of the popular vote by nearly three million votes when the counting concluded several weeks later. Though Trump had won the election, Clinton's popular vote count was a fact that was contested by the president-elect, as was the size of the crowd attending the new president's inauguration on the Washington Mall in January 2017. These disparate perceptions of events presaged ongoing debates and disputes that persist today about the very nature of facts, truth, and information.

In the ensuing years since the 2016, 2020 and 2024 presidential elections, ethics issues of all kinds have rocketed to prominence. Many of these issues have clear information- and technology-related nexuses:

- access to information and technologies
- privacy
- surveillance
- data collection and use
- intellectual freedom
- free speech
- censorship
- free and independent presses
- accuracy of information, misinformation, and disinformation
- social media
- intellectual property rights

Like politics, *character* is inseparably connected to ethics, information, and technology. Decisions that touch the everyday lives of all living organisms on this planet are made about information and technology that involve ethics and character. What is "character"? One of the best definitions comes from a 1947 *American College Dictionary* used by my late grandfather Stanley N. Currier, a teacher of ethics and government and, later, a high school principal. Character, it states, is "the aggregate of qualities that distinguishes one person or thing from others." Other aspects include "moral constitution," "reputation," and "good repute." Perhaps, though, the connotation that best captures the varying hues of character is this one: "An account of the qualities or peculiarities of a person or thing."[4]

Theodore Roosevelt said, "Character, in the long run, is the decisive factor in the life of an individual and of nations alike."[5] Ethics issues linked to character—sometimes called professional ethics—are all around us every day, visible in every newspaper and newscast. Sometimes ethical issues involving character upend lives and derail careers; other times, they are ignored and deemed nonissues. Many character issues fall within the categories below, which, among others, are examined later in this book:

- integrity
- misrepresentation
- plagiarism
- research misconduct
- transparency
- conflicts of interest

Into this mix, too, enter the newest disruptive emerging technologies: the 800-pound gorillas of artificial intelligence (AI) and generative AI, along with the Internet of Things (IoT), drones, robots, self-driving vehicles, and immersive technologies. Complicating all of this is the intersectionality and contestability of these areas and issues.

Amid this seemingly ungovernable mélange of subjects and interests, information centers—libraries, archives, museums—still stand as institutions of great public trust, serving the common good and public interest. Has the role of the information center ever been more vital to the planet and its peoples, considering the responsibilities and challenges faced by the information professionals who inhabit those spaces and guide and aid societies and communities with issues of this nature? Yet information professionals find themselves challenged, too, as never before by inconstant societal and technological forces, political winds, and complex ethics questions:

- How *should* librarians, archivists, museum staffs, and other information professionals navigate the manifold ethics issues that envelop them and their communities?
- What tools, people, and organizations can they turn to for advice, counsel, inspiration, and enlightenment?
- How will they determine what the *right* (ethical) thing to do is, or is not?

This book examines the foundations, fundamentals, big questions, and ethical issues involving these and other ethics-related topics. We explore these topics and issues in relation to the convergent and divergent contexts—digital-age technology, legal, ethical, policy, and sociopolitical—within which these dynamic issues are situated. The political context is particularly salient, as leadership changes in the makeup of executive, legal, and judicial branches of the federal government, as well as agencies and departments at all levels, can have substantial effects on information- and data-related issues. The past three decades, for example, have demonstrated a continuum of status quo, moderate, or dramatic shifts in the ways that local, state, and federal entities have addressed or not addressed privacy, data collection and use, digital divides, intellectual freedom, net neutrality, cyberbullying, social media, intellectual property, and more.

We also rigorously examine the roles and responsibilities of a diverse array of information professionals positioned within the contexts identified. An overarching goal of this book is to promote identification and critical examination of a diverse range of stakeholders and ethics issues in the internet technology era. This work highlights some of the key voices that are contributing to discourse, action research, and law, ethics, and policy frameworks in these research areas.

A guiding principle of this book is to present information from a variety of perspectives and positions and to empower readers to come to their own conclusions on where they stand on issues and topics. The "Teaching Philosophy Statement" that I crafted and have included in my course syllabi for the past decade and a half crystallizes this approach to my teaching and the writing of this work:

> As the course instructor, my teaching goals are to share information and to guide students to seek, find, and critically engage with information that is relevant to this course, their programs of study, and professional development, and to promote their individual construction of knowledge and understanding. I am committed to fostering and sustaining an inclusive learning community that is interactive *and* reflective and which encourages students to be active participants in their diverse exploration, discovery, learning, and developmental processes.

What one reader may see as recitation or descriptive explanation, another reader may see as advocacy and pontificating. One of my favorite stories for illustrating the subjectivity of human experience and understanding is a true story I've coined "Solomon's Piano." A gifted young pianist was playing a complex classical music work at a rehearsal. Two esteemed senior performers were to listen and render their judgments on the pianist's interpretation and skills. At the end of the piece, as the pianist's fingers rose from the keys, Performer 1 exclaimed, "Too fast." Performer 2 retorted, "Too slow." Indeed, as this story comically yet pragmatically illustrates, we all see and experience the world—and information—through our own individual lenses. Another objective of this book, therefore, is to identify a variety of stakeholder perspectives on the issues that are examined, in order to present a diversity of viewpoints and arguments.

The topics that are included in this book represent the canon of "traditional" ethics-related areas—intellectual freedom, censorship, freedom of expression, free and independent presses, access, privacy, surveillance, accuracy—as represented in many library science and information studies' ethics courses. These topics are also reflected in the ethics codes and core values statements of information-related professional associations, like the American Library Association (ALA) and the Society of American Archivists (SAA). Many of these topics are also seen as fundamental to healthy and well-functioning democracies and are codified in the US Constitution and Bill of Rights.

Additionally, this book addresses emerging topics—like AI and data ethics—that are increasingly included in the curricula and course offerings of library, information, data, and computing programs. A significant amount of material is included on intellectual property (IP) as this area of the law, particularly copyright law, is arguably the area that most impacts the day-to-day work of information professionals. IP also intersects heavily with existing and emerging technologies. This book is neither intended to serve nor capable of fulfilling the role of a multivolume set of encyclopedic works on each and every ethics topic with a relationship to the information professions. Hence, selection decisions have necessarily been made regarding what to include and what to exclude.

A "self-empowerment" and democratizing aim of this book is to emphasize the roles that anyone and everyone can take on in engaging with activities and shaping discourse that involves ethics. Information professionals, too, have vital leadership roles to step into and carry forward, stemming from the great responsibilities that come with the powers that information can confer. Information centers and professions, like our societies, are ever evolving. Similarly, ethical inquiry, reasoning, and decision-making must continue to be revisited and adapted, as warranted, to reflect the changing needs of people and times. Ethics codes and codes of conduct must also be periodically reexamined and updated, when and where necessary. Dr. Shaun Sellers, a dentist who wrote about ethics, dentistry, and technology over the course of five years, epitomized these ethics aspirations via his final December 2024 *British Dental Journal* piece, "Ethics Revisited": "The ethical landscape of dentistry will continue evolving, shaped by societal shifts, technological advances, and our commitment to doing better. If there's one lesson I've learned, it's that ethics isn't static—it's a living, breathing part of what we do. It challenges us to reflect, adapt, and, above all, remain human in our approach."[6]

Information professionals, like dentists and people in all vocations, must remain human-centered, ethically mindful, and proactively nimble amid twenty-first-century technological changes that present potential perils and advancements for individuals and communities.

Connections Between Ethics and Legal Education/ Experience

My legal education and juris doctor degree from the University of Pittsburgh School of Law (1985–1988), coupled with subsequent experiences as a public interest attorney, have influenced my teaching, research, and writing interests in a number of ways. Moreover, my legal background and orientation have also influenced my approach to writing and organizing this book.

The case study approach is often used in law school courses, as well as in other disciplines and research methodologies. Focusing on a real-world case (or a *hypothetical* fact pattern, as is the wont of law school professors) can highlight the relevance of issues that are being examined and critiqued. The case study approach affords practitioners the opportunity to hone their competence in spotting issues and identifying the laws, ethical principles, and policies that are applicable to the facts and issues in the case. It's also the quintessential method of inquiry for law school tests and bar exams.

The basic premise is that becoming more proficient in analyzing issues and then being able to connect those issues to pertinent laws, ethics, and policies through course exercises can aid students in being better prepared to respond to issues that arise in real life for themselves and others (e.g., clients, customers, patrons). Being able to see issues in new ways and from different perspectives, and to devise novel solutions to problems and challenges, is key not only for lawyers but also for information professionals. The case study method is useful for facilitating these kinds of crucial skill sets. This book's chapters present real-world case studies to delve into at a granular level. At the same time, the case study method allows one also to zoom out and see how the particular facts and issues of the case at hand can yield insights and enable the making of findings and development of conclusions at a macro level.

Another strength of the law school education approach is gaining experience with crafting and arguing varying positions, even positions and viewpoints that one may not agree with. One of the fundamental tenets of the US legal system is that everyone is entitled to legal representation. This precept applies equally to defendants, no matter what their alleged civil or criminal crimes may be. An ethical imperative for all licensed attorneys is that they zealously represent their clients. Hence, being able to compose legal positions and effectively and competently advocate for and assist their clients is requisite, in order for attorneys to remain in good standing with their licensing organizations. I am a licensed attorney in the Commonwealth of Pennsylvania.

Ethics also factors heavily in the education of law school students and in the professional responsibilities of licensed attorneys. In order for attorneys to maintain their licenses with their respective licensing entities, they are required to complete a prescribed number of continuing legal education (CLE) courses and earn a specified number of credits from those courses within each calendar year. These licensing groups also set out the number of credits from these CLE courses that must be ethics-related credits. The undergirding purpose of these CLE requirements is to ensure that licensed attorneys are engaging in ongoing professional development, maintaining competency in the law, and adapting to legal and ethical developments and societal changes. The ethics-related credit requirements also ensure that licensed attorneys are keeping abreast of ethical principles and rules. I have gained experience in working with and staying current on legal and ethics-related subject matter over the course of several decades.

My education and experience as an attorney have informed the courses I have developed and teach at the University of Pittsburgh. In 2011, I relaunched the course "Information Ethics" (which had been created and taught by two prior colleagues, Dr. Toni Carbo and Father R. Stephen Almagno) and have been teaching that class continuously to the present. Around 2017, I revised the name of the course to "Ethics of Data, Information, and Emerging Technologies." Since 2009, I've developed and have been teaching the course "Copyright and Fair Use"; the name was updated to "Intellectual Property and Open Movements" about 2015. The "Managing and Leading Information Services" course I've been teaching since 2007 includes units on managing legal issues and ethics in information centers. Since 2020, the "Information Professional in Communities" course that I developed has incorporated ethics-oriented case studies. Case study examples include the following:

- Ethical and legal issues stemming from a multimillion-dollar rare books theft at the Carnegie Library of Pittsburgh
- A photographic exhibition controversy at the National Archives
- Ethics issues surrounding the provision of life-saving anti-overdose medications at libraries

Legal work with marginalized communities has provided me with experience in working with and on behalf of individuals and groups from historically disadvantaged and underserved populations. These include Native Americans, mental health consumers, persons experiencing housing insecurity, incarcerated individuals, and immigrants. While addressing all members of societies, this book also endeavors to examine ways that issues like surveillance, data collection, the right to be forgotten, and AI algorithms can disproportionately impact some of the most vulnerable members of societies and those at the lower end of the economic spectrum. These kinds of disparities implicate ethical inequities. This book looks at ways that such disparities can be ameliorated and remedied.

My legal and ethics background facilitates an additional skill set: insights on risk management. One of the elemental jobs lawyers perform is to advise their clients about risks:

- How to identify actual and potential risks (i.e., contingencies)
- How to prepare for and protect against risks *before* they happen
- How to confront and manage risks that can turn into negative problems and undesirable outcomes when and if they occur

Perhaps the most common task that lawyers perform for clients is the preparation of wills and other testamentary instruments. The very aim of a *last will and testament* is to avoid the risk of dying *intestate* (i.e., without a will) and thereby forfeiting the power to proactively direct the disposition of one's assets. Therefore, another aim of this book is to assist information professionals in thinking about and being as ready and empowered as possible to foresee and mitigate risks and *proactively* and effectively respond to challenges and possible pitfalls involving ethics, information, and technologies.

Acknowledgments

I am deeply grateful for the people who have helped and encouraged me throughout this research and writing odyssey. A number of individuals especially stand out:

- **Dr. Blanche Woolls** initially contacted me about writing this book and beneficially secured the contract for me, for which I will be forever appreciative; she also supportively served as my editor for a number of years during this project.
- **Jessica Gribble** has more recently provided invaluable editorial guidance and support.

- **Bob Moore** has shared indispensable insights on trusting one's voice, taking one's place at the table, pursuing what's important, and "considering the source."
- This last sage observation was also made often by my late grandmother, **Esther Elizabeth Hughes Currier** (1898–1993), when assessing the intentions, legitimacy, and qualities of a person. Of course, "considering the source" is also a useful metric for evaluating the accuracy, authoritativeness, and objectives of an information source.
- **Stanley Nathan Currier** (1897–1971), my late grandfather, taught me the core value that service to others is the most important work we can do in this life.
- **Dick Marshall** is a helper to his core and has repeatedly reinforced for me the cardinal importance of kindness, empathy, advocacy, and risk-taking.
- **Father Geoffrey Wild** and **Cheryl Wild** offered healing clarity when it was needed most and asked the right questions at the right time.
- **Dr. Nikki Nordenberg** suggested librarianship to me in 1995 (which I ultimately decided to pursue), while I was exploring career retooling with her.
- During my twelve-month-long University of Pittsburgh Master of Library and Information Science (MLIS) Program in 1996, **Dr. Sara Fine**'s required MLIS core course "Behavioral Perspectives" had (and continues to have) a lasting impact on my outlook on information services and what it means to serve people in need. Dr. Fine encouraged her students to actively listen to and engage with users in meaningful conversation, in order to better assist people in uncovering and satisfying their respective information needs.

The importance in our lives of mentors and counselors—those who advise, guide, and see potential paths for us that we may not yet recognize (i.e., think the *X-Men*'s Professor Xavier, *Star Wars*' Yoda, or *Batman*'s Alfred Pennyworth)—can't be overstated. Both Drs. Fine and Nordenberg periodically encouraged me to pursue a PhD during the 1997–2003 years when I was working in librarianship in California, Ohio, and Pennsylvania. Indeed, whenever I updated Dr. Fine, she would acknowledge the librarianship work I was doing, while also adding "but you can do more" by getting a PhD. From those germinating seeds, in early 2003 I met with my former MLIS professor **Dr. Margaret "Maggie" Kimmel** (whose required MLIS core course "Librarianship in Society" was also deeply impactful), as well as with cataloguing and information organizing authority **Dr. Arlene Taylor**, to learn more about earning a PhD from Pitt. I decided to pursue the degree, with the encouragement of many of those previously noted, as well as my family and friends. I left my position as the manager of a branch of the Carnegie Library of Pittsburgh (CLP) to embark on full-time PhD studies at Pitt in August 2003. Right before my departure, a CLP marketing person asked, "Do you *really* think you'll finish [the degree]?" During my PhD program years, I channeled that question as a motivating touchstone for keeping my eye on the prize, whenever tiring times inevitably arose. I received my doctoral degree in the spring of 2007 and have been teaching continuously at my alma mater since that time.

One additional note drawing on my PhD studies: during my first term in the PhD program, I took **Dr. Toni's Carbo**'s 2003 "Information Policy" class (producing a course paper on the potential privacy and security implications of electronic voting machines, an emerging technology at the time!). Dr. Carbo is a former dean of Pitt's School of Information Sciences (SIS), along with numerous other career accomplishments, and a pioneer in information ethics, along with the late **Father R. Stephen Almagno**. Three ethics-grounded principles that Dr. Carbo spoke of during that 2003 course have particularly resonated with me over the past decades: the importance of (1) civil discourse, (2) intellectual curiosity, and (3) intellectual humility. This book is informed by those core values in the ways that I endeavor to steward the topics and issues that are critically investigated and reflected on by readers, while acknowledging the inherent limitations of any research work.

Others have continued to positively influence my professional development in the ensuing years. As one noteworthy example, while working on this book, in February 2017, a few months after the 2016 election, I was doing some self-questioning about what approach I should use for investigating and writing about the ethics topics I'd selected. This period of introspection came after suggestions by a few academic colleagues at a departmental ethics "brown bag" lunch discussion about the approaches that they felt my book *might* take and the kind of ethics book that they would write if they were writing it. One colleague suggested a *Great Philosophers* approach, looking at, say, what Kant might inform us about privacy, Nietzsche about free speech, and so on. That approach did not "call" to me.

Some weeks later, during a lunch conversation with my former PhD dissertation advisor and now-retired University of Pittsburgh colleague **Dr. Ellen Detlefsen**, I referenced the group's comments about the kinds of ethics books they had said they would write if they were doing the book. Dr. Detlefsen reminded me that the book's publishers, knowing my legal background as well as having approved my pitch for the book and the table of contents' topics, were giving me the green light to pursue *my* approach for the book. At the end of our conversation, with a wry grin, she also gave me what became a clarifying touchstone for following my "scholarly journalism" case study vision for this book. With no offense intended to philosophers, scholars, or aficionados of Kant (especially in light of several recent articles regarding the enduring relevance of Kant, e.g., "Why the World Still Needs Immanuel Kant"[7]), *we* don't need another Kantian book on ethics.

Employing a case study approach with law- and ethics-centered lenses, this book aims to serve as a constituent part of a broader, multidisciplinary, global community of ethics-minded thinkers, writers, and other societal actors, in promoting fulsome and crucial conversations about the present and future of ethics, information, and existing and emerging technologies. To that end, the discussion questions at the end of each chapter are intended to foster reflection and exchanges of ideas. It has been said that democracy is messy, imperfect, and subject to such variables as the need for rigorous debate, principled reason, a spirit of compromise, and, perhaps above all, good faith. Within that democratic tradition, this ethics-centered book is similarly offered to contribute to the often messy but necessary work of exploring the ethical dimensions of our shared humanity and singular identities.

Notes

1. Robins-Early, N. (2024, November 9). How Elon Musk became Donald Trump's shadow vice-president. *Guardian*. https://www.theguardian.com/technology/2024/nov/09/elon-musk-trump-administration
2. Obama, M. (2016, July 26). *Remarks by the first lady at the Democratic National Convention*. White House. https://obamawhitehouse.archives.gov/the-press-office/2016/07/25/remarks-first-lady-democratic-national-convention
3. US National Park Service. (n.d.). Lincoln's first inauguration. Retrieved February 12, 2025, from https://www.nps.gov/articles/000/lincoln-s-first-inauguration.htm
4. Barnhart, C. L. (1947). *The American college dictionary*. Random House.
5. American Museum of Natural History. (n.d.). Theodore Roosevelt Rotunda. Retrieved February 12, 2025, from https://www.amnh.org/exhibitions/permanent/theodore-roosevelt-memorial/roosevelt-rotunda
6. Sellars, S. (2024). Ethics revisited. *British Dental Journal,* 237(12), 895–895. https://doi.org/10.1038/s41415-024-8235-5
7. Neiman, S. (2024, April 17). Why the world still needs Immanuel Kant. *New York Times*. https://www.nytimes.com/2024/04/17/arts/immanuel-kant-300-anniversary.html

Part 1

Foundations

Chapter 1

Introduction and Ethical Foundations

Introduction: Ethics, Law, and Policy

Ethics is *the* focal point of this book. It's one of those terms that most people have heard of or encountered at some point in their life. Having heard the term and clearly understanding it, though, are two different things. Like many terms of art, ethics can be difficult to define. Moreover, getting people to come to consensus on what is and isn't considered "ethical" can vary wildly, just as opinions vary on how to define "information" and "data."

What then is ethics? Let's look at a few dictionary definitions to get a clearer sense. The Legal Information Institute (LII) is a useful resource based at Cornell Law School. It provides etymological, definitional, and contextual explanation for the term:

> The word "ethics" is derived from the Greek word ethos (character), and from the Latin word mores (customs). In the legal context, ethics defines how individuals choose to interact with one another. In philosophy, ethics defines what is good for the individual and for society and establishes the nature of duties that people owe themselves and one another.[1]

Santa Clara University's Markkula Center for Applied Ethics defines ethics as "Well-founded standards of right and wrong that prescribe what humans ought to do, usually in terms of rights, obligations, benefits to society, fairness, or specific virtues."[2] One way to get a better handle on what ethics is and isn't is to look at ethics in comparison to "the law." LII states that "Law refers to a system of rules that regulate the conduct of a community, and is often enforced by a controlling authority through penalties."[3]

In light of these definitions, how do ethics and law differ? Laws denote *required conduct*. In contrast, compliance with ethics is, generally, more optional in nature. Put more simply:

- Law tells us what we must do or must not do.
- Ethics tells us what we ought to do or ought not to do.

Driving a car is one way to illustrate distinctions between law and ethics. The law in all 50 US states and territories specifies that anyone wanting to legally drive a motor vehicle must have a valid driver's license. Valid auto insurance is required, too. Laws also spell out things that drivers must refrain from doing while driving:

- not drinking alcohol beyond the legal limit
- not texting

- not going over the speed limit
- not passing when there's a double line
- not holding a phone while driving (a law in some US jurisdictions but not all)

As LII's definition of law indicates, legal rules are "often enforced… through penalties." Not being in compliance with the law—that is, breaking the law—can result in civil and criminal consequences and penalties for the kinds of action or inaction illustrated by these examples. Staying with the driving example, laws also govern vehicular operation issues, such as

- slowing down in construction areas
- moving to the side of the road for emergency vehicles
- having approved car safety seats for children
- adhering to the use of seat belts
- not changing lanes inside tunnels
- turning car lights on when it's raining

A complicating factor in understanding rules that govern our behavior is that not all places have the same laws, as noted with regard to holding phones while driving. Take this example: injuries and deaths have resulted from snow and ice that flew off of the roofs of cars, vans, and trucks and landed on the vehicles of nearby drivers. Advocacy in the wake of such injuries and deaths prompted the Commonwealth of Pennsylvania to enact the Pennsylvania Removal of Snow and Ice From a Motor Vehicle or Motor Carrier Vehicle Law in 2006. This law (codified in Title 75, Section 3721) requires

> a driver of a motor vehicle or motor carrier vehicle being operated on a highway of this Commonwealth shall make reasonable efforts to remove accumulated ice or snow from the motor vehicle or motor carrier vehicle, including the hood, trunk and roof of the motor vehicle or motor carrier vehicle, within 24 hours after the cessation of the falling snow or ice.[4]

A driver can be fined $50 for each offense "regardless of whether any snow or ice is dislodged from the vehicle." If "dislodged or fallen snow or ice" should cause serious bodily injury or death, a fine of $200 to $1,500 can be ordered for each offense.

The Role of Policy

In addition to Section 3721 being a law, there is a persuasive ethical and public policy argument that people *should* remove snow and ice from the roofs of their vehicles, so they don't cause accidents that can harm and kill others. Indeed, a tragic circumstance led to the passage of this law: "The law was proposed by Senator Lisa Boscola (D-Lehigh/Northampton) as 'Christine's Law' in honor of Christine Lambert of Palmer Township. Lambert died after a large piece of ice dislodged from a passing box truck and crashed through her windshield on Christmas Day in 2005."[5] LII defines a policy as "a guiding principle that leads a government or agency to make laws or to govern effectively. From a narrow angle, policy also refers to the rules and regulations made by an administrative department."[6] A guiding principle that led to the passage of Christine's Law was the Pennsylvania legislature's desire, expressed in the form of a law, to promote public safety.

Ethics, in most cases, does not have the weight of legal enforcement and penalties behind it. (Note, though, that there are some situations in which ethics-related laws govern, for example, the activities of elected officials. In such instances, compliance with ethics is *not* optional.) As the LII definition of ethics states, ethics involves decisions about how people choose to interface with each other. Without the law's systems of rules, enforcement authorities, and the possibility of civil and criminal penalties, ethics can often feel less defined and less clear. In the absence of systemic rules, ethics codes of conduct, like policies, can provide guidelines that help people and organizations determine what ethical courses of action to take. We examine ethics codes in the next chapter.

"Legal" but Unethical

Can something be legal and yet unethical? The answer is an emphatic yes. One of the axiomatic truths is that a law may be held by courts to be valid and enforceable at the time of its legislative promulgation and judicial upholding while still being unethical and unjust. This moral paradox is especially made clear when norms and mores change and unjust laws are overturned. Perhaps the most enduring and egregious example of this disconnect between law and ethics is seen in the US Supreme Court's 1857 decision that slavery was legal (often called the Dred Scott Decision):

> On March 6, 1857, Chief Justice Roger B. Taney read the majority opinion of the Court, which stated that enslaved people were not citizens of the United States and, therefore, could not expect any protection from the federal government or the courts. The opinion also stated that Congress had no authority to ban slavery from a federal territory. This decision moved the nation a step closer to the Civil War.
>
> The decision of *Scott v. Sandford*, considered by many legal scholars to be the worst ever rendered by the Supreme Court, was overturned by the 13th and 14th amendments to the Constitution, which abolished slavery and declared all persons born in the United States to be citizens of the United States.[7]

Similarly, world history is replete with examples of laws that have been enacted by leaders and legislative bodies, despite minorities—or even majorities—of people believing such laws are deeply unjust and violative of natural laws and human rights. As Ellen Drew said in the 1945 film *Isle of the Dead*, "Laws can be wrong and laws can be cruel, and the people who live only by the law are both wrong and cruel."[8] The 1950s and 1960s civil rights crusader Dr. Martin Luther King Jr., assassinated in 1968, continues to inspire new generations of social justice activists and human rights advocates to oppose unethical laws, as exemplified by one of his most famous declarations, an idea he attributed to theologian Theodore Parker: "The arc of the moral universe is long, but it bends towards justice."[9]

Ethics Foundations

Textual Records and Oral Traditions

In addition to ethics codes, religious teachings and traditions, philosophical approaches, and other sources and tools can aid individuals in making ethics decisions. Religious texts, cultural artifacts, and archival records and literature are primary source materials that can shed light on the varying ways that human beings have sought answers to questions of how they should or should not act. The world's major religions possess sacred historical texts that have been consulted over the course of millennia (or more recently with regard to newer religions) for ethical guidance and teachings as well as moral prescriptions and proscriptions. Examples include

- Judaism's Torah
- Islam's Koran
- the Vedas in Hinduism
- Buddhism's Tibetan Book of the Dead
- Christianity's Holy Bible
- the Church of Jesus Christ of Latter-day Saints' Book of Mormon

Information technologies—ancient stone tablets, papyrus scrolls, parchment, paper and analog books, and today's digital-era bits and bytes—have enabled the preservation and passing down of ethical and moral precepts from generation to generation. Ancient and existing cultures, many of them indigenous, have also used oral lore to preserve and pass on beliefs and traditions from past ages to the present. Physical examples of these beliefs are evident in extant carvings of the Mayan temples and ruins of Mexico and Central America, as well as on the sarcophagi of ancient Greece, Rome, and Egypt.

Philosophies

Throughout recorded human history, in addition to religious teaching and traditions, people have thought and written about life's "big questions":

- What are the origins of life?
- What is the nature of good and evil?
- What is the meaning of suffering?
- Why does God (or the gods) let "bad things" happen?
- What, if anything, happens after we die?
- Are there consequences after death for people who do bad acts?

For millennia, thought leaders from many of the world's cultures and time periods have pondered such questions, giving rise to the discipline we know as "philosophy": "Classical philosophy studies the fundamental problems concerning human existence through the eyes of our intellectual ancestors. Many thinkers from Classical times were pioneers of our modern philosophical and scientific ideas."[10] When does the historical record tell us that philosophy began? "The earliest beginnings of philosophy are traced back to the sixth century B.C.E., when the first scientists of Western history, the Pre-Socratics—among them Thales, Heraclitus and Parmenides—advanced revolutionary theories concerning the natural world, human knowledge and humans' relationship with the gods."[11]

Since that time, numerous philosophers, philosophies, and ethical theories and approaches have arisen, including but not limited to the following:

- rule utilitarianism
- act utilitarianism
- Kantianism
- subjective relativism
- cultural relativism
- divine command theory
- social contract theory
- teleological ethics
- deontological ethics
- feminist ethics (also referred to as feminist moral theory)
- existentialist ethics (also referred to as existential ethics)
- animal rights theory
- Buddhist ethics
- secular ethics

Later in this chapter, we examine some foundational sources for ethical analysis and decision-making.

Ethics in Everyday Life

Since 2011, I've taught an ethics-related graduate level course ("Information Ethics," retitled and revamped in later years as "Ethics of Data, Information, and Emerging Technologies"). One of the course's learning outcomes is for students to gain greater awareness of how ethical issues and decision-making manifest in their everyday lives. Other disciplines incorporate and investigate the role of day-to-day activities in people's lives. For example, the discipline of human information behavior (HIB), a subarea of the information sciences, includes the work of Reijo Savolainen, a Finnish researcher who developed the concept of *everyday life information seeking* (ELIS).[12] It can be illuminating to identify the many ways, often unnoticed, that ethics is present in our quotidian existence. This in turn reinforces the relevance of ethics to people's lives and the fabric of societies.

A central facet of ethics is that people frequently seek help and guidance from others when they are faced with ethics issues and dilemmas. Since at least the late 1600s, advice givers have utilized the technologies of the times to dispense ethics words of advice and admonition to the uncertain, the angst-ridden, and the lovelorn. A British periodical called the *Athenian Mercury*, which ran for six years (1691–1697), is the earliest known source of the advice column. A 2015 article in the *Atlantic* uncovered some examples of the kinds of queries sent to the *Athenian Mercury*:

- Is the soul born with the body?
- Why can an owl see better by night than by day?
- Which is most noble, to love or to be beloved?
- Is it proper that women should be learned?[13]

For centuries, pseudonymous female "agony aunts," as well as male "agony uncles" occasionally, have delivered a mix of comfort, admonitions, and scolds to supplicants seeking answers. As a 2009 piece in the *Guardian* titled "A Brief History of Agony Aunts" indicates,

> In 1691 a 32-year-old man called John Dunton was having an affair and realised there was no one he could ask for advice about it without revealing his identity. Most of us would have shrugged and struggled on, but in Dunton, a printer and bookseller, the entrepreneurial as well as adulterous spirit was strong. Realising his dilemma could not be unique, he launched the *Athenian Gazette* and opened its pages to the readers. Thus the first agony column—and interactive magazine—was born.[14]

The post–World War II era brought new agony aunts with the rise of advice dispensers like "Dear Abby" and "Ann Landers"—twin sisters penning advice columns that became reader staples in rival newspapers throughout the United States. The growing influence of television in American life over the next decades ushered in televised agony aunts and uncles, as talk show hosts Phil Donahue, Sally Jessy Raphael, and Oprah Winfrey beamed their own brands of advice into American living rooms. In the digital age, newspapers and magazines continue to dedicate space in print and digital formats to columnists supplying ethically oriented advice. For example, "The Ethicist" is a decades-running weekly column in the *New York Times*, covering a wide range of ethics-related topics and questions, currently posed to New York University School of Law philosopher Kwame Anthony Appiah.

How will human advice givers compare over time with advice that is provided through technologies like artificial intelligence (AI)? A July 2024 *Scientific American* article, "AI Chatbots Seen as Ethical as a *New York Times* Advice Columnist,"[15] reported on the findings of a 2023 AI ethics advice study conducted by researchers at the Wharton Business School at the University of Pennsylvania:

> Drawing from a pool of ethical dilemmas published in the *New York Times* column "The Ethicist," we compared the ethical advice given by the human expert and author of the column, Dr. Kwame Anthony Appiah, with AI-generated advice. The comparison is done by evaluating

> the perceived usefulness of the ethical advice across three distinct groups: random subjects recruited from an online platform, Wharton MBA students, and a panel of ethical decision-making experts comprising academics and clergy. Our findings revealed no significant difference in the perceived value of the advice between human generated ethical advice and AI-generated ethical advice.[16]

Today's most popular creative entertainment, much of it streaming content referred to as "prestige television," employs storylines and characters that are steeped in ethics questions and dilemmas. Ethics-challenged antihero protagonists and supporting casts have become a staple of these shows. These characters are grayer and more nuanced in nature than classic black-and-white, good-or-bad, one-dimensional characters of years past. Antihero-driven shows like *The Sopranos*, *The Wire*, *Breaking Bad*, *Nurse Jackie*, *Mad Men*, and more transport viewers on redemption-seeking narrative arcs filled with ethics-tinged perils and character-testing labors.

These "peak TV" shows owe much to Ancient Greek works such as *The Odyssey*, *The Iliad*, *Medea*, and *Oedipus Rex*. The mythological characters in those works frequently face timeless questions of character: the nature of heroism, loyalty, betrayal, sacrifice, vengeance, forgiveness, hubris, and other human frailties and virtues. Joseph Campbell, a Sarah Lawrence College professor of literature, is best known for his print books and talks that were broadcast on public television. Campbell popularized exploration of comparative mythology in the late twentieth century before his death in 1987. His most famous works, such as *The Power of Myth* and *The Hero's Journey*, examine the ways that myths and mythic characters help us to make sense of the world and uncover life lessons about the kinds of ethical choices human beings confront throughout their lives.

Many of literature's canon of Great Books, too, feature protagonists who come face-to-face with existential questions that have ethical dimensions. Shakespeare's Renaissance-era characters are repeatedly beset by moral questions and ethical conundrums: Romeo and Juliet, Macbeth, *The Merchant of Venice*'s Shylock, King Lear, and Othello, just to note a few. Perhaps no Shakespearean line better exemplifies the essence of the ethical dilemma than Hamlet's (1603) life-or-death question to himself: "To be, or not to be." Other classic examples include the following:

- In Victor Hugo's *Les Miserables* (1862), heroic Jean Valjean steals bread to feed the starving and then is mercilessly pursued by the law-enforcing Inspector Javert.
- Charles Dickens's *A Tale of Two Cities* (1859) features the lovable rogue-cum-sacrificial martyr Sydney Carton, and *A Christmas Carol*'s (1843) curmudgeonly misanthrope Ebenezer Scrooge is given a second chance to be kind and generous to those in need.
- Nathaniel Hawthorne's *The Scarlet Letter* (1850) explores themes of shame, deception, forgiveness, and grace through the characters of Hester Prynne and Rev. Arthur Dimmesdale.

Comic book superheroes also serially tangle with ethics and moral dilemmas, in addition to sparring with their nemeses. The poster child of "ethical superheroes" is Marvel Comics' Spider-Man, beginning with his very first issue in 1962's Amazing Fantasy #15. After gaining his eponymous powers from the bite of an irradiated spider during a high school science class field trip, Spider-Man's alter ego Peter Parker succumbs to arrogance and indifference that contributes to the escape of a thief who later murders Peter's beloved Uncle Ben. In consequence, the guilt-bearing hero is forever inspired to live by the words Uncle Ben taught him as a boy: "With great power comes great responsibility."[17]

Marvel's X-Men superheroes, introduced in 1963, employ "mutants" as metaphors for the real-world's persecuted, disenfranchised, and historically disadvantaged minorities—namely, "the other." Characters like Wolverine, Jean Grey, Cyclops, Storm, Nightcrawler, Rogue, Professor Xavier, and Magneto wrestle time and again with the consequences of their actions and their character flaws, fears, traumas, aspirations, limitations, and quests for redemption. The religious precept of "loving one's enemies" is a recurrent X-Men theme and ethical imperative for these "othered" heroes, exemplified by their credo: "Sworn to protect a world that hates and fears them."[18]

Other superheroes, like DC Comics' iconic Superman, Batman, and the Flash, frequently contend with ethics issues, while saving the world and their respective cities of Metropolis, Gotham, and Central City. Fans endlessly debate, for example, whether Batman should end the life of archnemesis the Joker, rather than continually apprehending and confining him to Arkham Asylum until his next inevitable escape. To that point, some Superman fans were shocked when the Henry Cavill version of the character in 2013's *Man of Steel* movie killed the Kryptonian superpowered foe Zod before he could murder a group of innocents with his heat vision. Heated ethical arguments erupted on social media and in the mainstream media over whether Superman had violated his core values in fatally stopping a would-be killer or whether the taking of Zod's life was justified under a "greater good" rationale. The Flash, the Speed Force–channeling hero, has also grappled with the ethics—and the larger multiverse-impacting consequences—of using his super speed to travel back in time to, as one example, save his mother from dying at the hands of a villain when the speedster was a boy not yet with superpowers.

So popular are comic book heroes that they're even used in some college-level courses. As the BBC reported in a 2010 article, "Teaching Philosophy With Spider-Man," "Now, philosophy professors are finding superheroes and comic books to be exceptionally useful tools in helping students think about the complex moral and ethical debates that have occupied philosophers for centuries. Moreover, superheroes are attracting students to a discipline often perceived as overrun by musty books, suede elbow patches and bow ties."[19]

The blog *Law and the Multiverse: Superheroes, Supervillains, and the Law* was started by a couple of comics-loving attorneys as a way to ponder the legal and ethical ramifications of superpowered do-gooders and their respective rogues' galleries. The bloggers subsequently published their book *The Law of Superheroes*, featuring legal analysis of questions like

- Would mutants have civil rights?
- Would the Second Amendment apply to Wolverine's claws?
- Could Batman patent the Batmobile?[20]

This sampling of ethics-challenged characters from TV shows, classic literature, and comic books could easily be expanded to highlight many more instances from their respective genres.

Other genres like opera, film, dramatic works, and musical compositions are similarly replete with ethics questions and issues. Indeed, this book highlights numerous ethics-encompassing examples from movies, plays, and songs in upcoming chapters. Since the dawn of recorded history, audiences have been knowingly *and* unknowingly engaging with creative works that involve ethics and moral dilemmas. These cultural works can serve dual functions of both entertaining *and* educating/informing their user audiences and consumers. Often, too, these works may raise ethical questions that users are left to ponder long after a book's last page has been read or a movie has faded to black.

Ethical Analysis and Decision-Making

Actor Claude Rains's character in the 1941 movie *The Wolf Man* observes that "To some people, life is very simple. They decide that this is good, that is bad, this is wrong, that's right."[21] However, many of life's questions aren't simple or a binary decision. Figuring out whether, when, and how to address ethics issues and dilemmas is often complex. Clear-cut answers can be elusive. In this section, we look at some strategies and tools that can be used to analyze ethics questions, decide what to do, and then assess how the ethical decision-making turned out. Ethics frameworks and decision-making models are an important resource that can help in making decisions and learning from decisions and outcomes.

Santa Clara University's Markkula Center for Applied Ethics

Santa Clara University's Markkula Center for Applied Ethics, located in the heart of Northern California's Silicon Valley, is a leading institutional voice and resource center on all things relating to ethics and technology. Markkula Center regularly hosts thought-provoking webinars and conferences on a wide range of topics; in 2022 alone, I virtually attended the following three webinars:

- "The Character of AI—a Technology Ethics Conference"[22]
- "Space Ethics"[23]
- "Virtual Reality, Real Virtues, and Augmented Norms and Laws"[24]

Hearing practitioners and scholars talk on all kinds of ethics-related topics can expand interdisciplinary awareness of new perspectives and ways of approaching ethical dilemmas and issues.

Sources of Ethical Approaches and Standards

Before we examine a useful ethical decision-making model created by the Markkula Center, knowing the sources of ethical approaches that inform ethical analysis and decision-making is important. The Markkula Center provides information and resources on six approaches to ethical decision-making.

1 The Utilitarian Approach

The utilitarian approach is a balancing test; namely, it balances positives that can arise from a particular action against the negatives that can happen. The Markkula Center explains it as "the ethical action… that provides the most good or does the least harm, or, to put it another way, produces the greatest balance of good over harm."[25] Markkula also points out that this is the ethical standard that "deals with consequences." The utilitarian approach weighs the needs (or the interests) of the one versus the needs of the few or the many. Identifying the needs of the interested parties can be of help in making an eventual decision. The balancing test of this approach was a key plot point in the big-screen films *Star Trek II: The Wrath of Khan* (1982) and *Star Trek III: The Search for Spock* (1984).

2 The Rights Approach

The Markkula Center describes this ethical standard's focus as an action "that best protects and respects the moral rights of those affected."[26] This approach also places priority on the inherent worth and dignity of the individual. Arguments for the ethical treatment of animals, also referred to as nonhumans, and the conferring of legal rights on nonhumans, are anchored in the rights approach. For example, high-profile cases have asked courts to consider whether specific nonhumans within the categories of apes[27] and elephants[28] should be accorded protected legal status. The nonprofit organization Nonhuman Rights Project (NhRP) is challenging long-standing views of animals/nonhumans, as their mission and values statement explains:

> The NhRP is the only civil rights organization in the United States dedicated solely to securing rights for nonhuman animals.
>
> Our groundbreaking work challenges an archaic, unjust legal status quo that views and treats all nonhuman animals as legal "things" with no rights. As with human rights, nonhuman rights are based on fundamental values and principles of justice such as liberty, autonomy, equality, and fairness.[29]

3 The Fairness Approach

The fairness approach draws from the views of ancient philosophers "that all equals should be treated equally," notwithstanding slavery's systemic role in the history of humankind and its tacit acceptance during the times in which many early philosophers lived. An example of this ethical standard is the 2011 Occupy Wall Street movement's position that CEO salaries should not be hundreds of times larger than the pay of others. Data is key in helping to illuminate this issue: a March 2021 *Nonprofit Quarterly* article studying issues of CEO pay disparities observed that "Women and people of color make up a disproportionately large share of today's low-wage workers and a distressingly tiny share of corporate leaders."[30] Analysis of CEO salary ethics example might probe "whether the huge disparity is based on a defensible standard or whether it is the result of an imbalance of power and hence is unfair."[31]

4 The Common Good Approach

This approach sees "life in community" as a public good. Societal relationships are "interlocking" and inform "ethical reasoning." The common good standard prioritizes values like "respect and compassion" and protection of society's most at-risk persons. As Hindu religious leader Mahatma Gandhi famously stated, "The true measure of any society can be found in how it treats its most vulnerable members." We see the common good approach embodied by first responders, libraries and cultural heritage organizations, public education, and public, state, and national parks. COVID vaccine availability and recommendations for people to get vaccinated during the COVID-19 pandemic to mitigate infecting others are examples of ethics issues implicating the common good.

5 The Virtue Approach

The Markkula Center explains that this approach takes the position that "ethical actions ought to be consistent with certain ideal virtues that provide for the full development of our humanity."[32] Oxford Reference defines virtue as a "quality considered morally good or desirable in a person."[33] Examples of virtues include

- courage
- empathy
- temperance
- forgiveness
- justice
- wisdom
- respect
- charity

The virtue approach, like others highlighted here, derives from ancient thinkers and is aligned with many religious faiths. As the *Stanford Encyclopedia of Philosophy* states, "In the West, virtue ethics' founding fathers are Plato and Aristotle, and in the East it can be traced back to Mencius and Confucius."[34]

6 The Care Ethics Approach

The care ethics standard is a newer approach that draws on the universal value of care for others. The Markkula Center explains that "care ethics emphasizes the value of people's relationships, the universality of human dependence on others, the significance of emotions and the body, and the context-sensitive nature of ethical deliberation that does not merely follow abstract moral rules."[35] Additionally,

> care ethics also pays attention to what is required for relationships to be healthy, reciprocal, and caring, rather than abusive, exploitative, or unfair. Furthermore, care ethics is not limited just to considering personal relationships that are especially intimate or close, but can also extend

to professional relationships, political relationships with fellow citizens, and even global relationships with distant others across the world.[36]

A Framework for Ethical Decision-Making

In addition to description of the chief ethical approaches, one of the Markkula Center's most useful contributions is its model for ethical thinking and analysis, "A Framework for Ethical Decision Making." The framework's ten components are articulated in plainspoken language. They also mirror real-life aspects of decision-making, such as identifying ethical issues, gathering relevant facts, considering possible courses of action, and asking someone what they might do when facing an ethical quandary. The framework's steps are reproduced here:

Identify the Ethical Issues

1 Could this decision or situation be damaging to someone or to some group, or unevenly beneficial to people? Does this decision involve a choice between a good and bad alternative, or perhaps between two "goods" or between two "bads"?

2 Is this issue about more than solely what is legal or what is most efficient? If so, how?

Get the Facts

3 What are the relevant facts of the case? What facts are not known? Can I learn more about the situation? Do I know enough to make a decision?

4 What individuals and groups have an important stake in the outcome? Are the concerns of some of those individuals or groups more important? Why?

5 What are the options for acting? Have all the relevant persons and groups been consulted? Have I identified creative options?

Evaluate Alternative Actions

6 Evaluate the options by asking the following questions:

- Which option best respects the rights of all who have a stake? (The Rights Lens)
- Which option treats people fairly, giving them each what they are due? (The Justice Lens)
- Which option will produce the most good and do the least harm for as many stakeholders as possible? (The Utilitarian Lens)
- Which option best serves the community as a whole, not just some members? (The Common Good Lens)
- Which option leads me to act as the sort of person I want to be? (The Virtue Lens)
- Which option appropriately takes into account the relationships, concerns, and feelings of all stakeholders? (The Care Ethics Lens)

Choose an Option for Action and Test It

7 After an evaluation using all of these lenses, which option best addresses the situation?

8 If I told someone I respect (or a public audience) which option I have chosen, what would they say?

9 How can my decision be implemented with the greatest care and attention to the concerns of all stakeholders?

Implement Your Decision and Reflect on the Outcome

10 How did my decision turn out, and what have I learned from this specific situation? What (if any) follow-up actions should I take?[37]

The first step of the framework—"Identify the Ethical Issues"—is, in my view, the most important takeaway. Newspapers, airwaves, and social media sites are brimming every day with examples of individuals and organizations who did not recognize or ignored an ethical issue. Being able to spot *potential* ethics issues before they manifest into *actual* ethics issues is a valuable skill set. Many corporations and organizations have staff whose job responsibilities include ensuring *compliance* with ethics and legal requirements.

What Would *You* Do?

The Markkula Center's ethical decision-making framework has many pluses. The steps and language are straightforward and understandable. It's a model, too, that incorporates diverse skill sets and competencies, such as

- issue spotting
- information gathering
- comparative analysis
- reasoning
- contemplation
- the possibility of conferring with others and seeking counsel
- action
- reflection
- active learning

In using and discussing this model with graduate-level students for more than a decade, I have found that students recognize some of these steps as ones that they already engage in when faced with an ethical question. For example, one of the steps that most of us are probably already using at times is: "8. If I told someone I respect (or a public audience) which option I have chosen, what would they say?"[38]

This step is more often represented and recognizable as the "What would X do?" question. Persons of faith, for example, may consider what the sacred head of their respective religion would do in a particular situation. Internet searches I have done for lectures over the years have yielded additional examples that use this WWXD structure, including

- What would **Frodo** Do?
- What would **Superman** do?
- What would **Katniss Everdeen** do?
- What would **Captain Picard** do?
- (And even a book called) *What Would* ***Wonder Woman*** *Do?*

Aside from imagining how a religious or fictional person might approach a question or challenge, many people also turn to real-life individuals they can ask for insights and guidance. Some persons also may reflect on how a deceased person they admire and trust might advise them in a situation, such as a parent, grandparent, or sibling. Some of the types of people one might consult or think back on include

- family members
- friends
- former or present teachers
- former or present work colleagues
- former or present work supervisors
- coaches
- mentors
- career advisors or guidance counselors
- health providers
- legal advisors/attorneys
- financial advisors

Big Nate is a syndicated comic strip that appears in many newspapers. Written and illustrated by Lincoln Peirce, it follows sixth-grader Nate Wright and the people around him. In May 2017, I discovered a roughly weeklong series of strips in which Big Nate faces the kinds of ethical dilemmas and decision-making that people face in the real world. In one example (see Figure 1.1), Big Nate is asked by his friend who owns a comic book shop to briefly watch over the store while the owner steps out. After the shop owner leaves, Big Nate wonders whether a kid in the store whom he knows has shoplifted a comic book. He weighs what to do and the potential consequences, including harm to himself. Echoing step 8 of the model at issue, Big Nate turns to a life-size cardboard cutout of Marvel Comics' Captain America and says, "What would *you* do?"

Figure 1.1 *Big Nate* comic. (© 2017. Used with permission from Andrews McMeel Syndication)

The Ethics Institute at Kent Place School (EIKPS)

As discussed, the Markkula Center's ethical decision-making model has a number of positive features. However, it's important to note that there are other ethics models from different organizations that are worthwhile to examine, consider, and use. Also, ethics models aren't just for adults: one model that school-age children are actively working with is from Kent Place School (KPS), a private preschool through 12th grade girls' school in Summit, New Jersey, near New York City. The Ethics Institute at Kent Place School (EIKPS) has operated as a part of KPS since 2007. The KPS website explains that it is "a first-of-its-kind institute at the primary and secondary school level. We believe that promoting the process and practice of ethical thinking and decision-making prepares people of all ages to be effective leaders and compassionate citizens."[39]

A brief video produced by the school begins with an administrator candidly stating that "in the beginning, nobody really even understood what the word ethics meant."[40] A 2019 KQED story, "What Students Gain From Learning Ethics in School," reveals that students at KPS are "introduced to ethics in fifth grade, during what would otherwise be a health and wellness class."[41] The students explore "simple case studies" and "consider the various points of view." Dr. Karen Rezach, the EIKPS director and ethics course instructor, "also acquaints them with the concept of right vs. right—the idea that ethical dilemmas often involve a contest between valid but conflicting values."[42]

The ethics curriculum becomes more "structured and challenging" for the middle and high school students.[43] These older students engage with nuanced case studies that invariably require them to "weigh the benefit and the harm" presented by the fact patterns in the scenarios. Underpinning the KPS ethics pedagogy is an ethical decision-making framework that is paired with a set of values. The Ethics Institute Ethical Decision-Making Method consists of the following steps:

1. Study and understand the situation.
2. Identify the values that are involved on all sides.
3. Identify the ethical dilemma—right versus right.
4. Identify the values that influence your position.

5 Communicate effectively and respectfully a decision that demonstrates the ability to apply the facts, identify values that inform your position, and can refer to other sources or stakeholders.[44]

Examples of the kinds of workshops offered at Kent Place School include

- "Be Well, Lead Well"
- "Developing Ethical Leadership Skills in Girls, Ethics and Social Media"
- "Developing Ethical Decision: Making [*sic*] Skills in 21st Century Learners"[45]

AI Apps and Chatbots

Artificial intelligence (AI) apps and chatbots (which we examine in detail later in this book) are increasingly emerging as tools for ethical reasoning and decision-making. Perplexity is an AI tool developed by Google. A user can pose questions to Perplexity on all kinds of subjects and topics and receive answers. What is an example of an ethics-related question for which Perplexity can provide information? On January 22, 2025, I typed the following question into Perplexity's search box: "Give me an example of an ethical dilemma." In a matter of seconds, Perplexity offered the following response:

> An ethical dilemma occurs when two or more moral principles conflict, forcing an individual to choose between them. Here's an example: A physician discovers that their patient intends to harm another person. The doctor now faces a conflict between maintaining patient confidentiality and preventing potential harm to someone else. This situation pits the ethical principle of respecting patient privacy against the moral obligation to protect others from harm. In this scenario, the physician must weigh several factors:
>
> **1** The duty to maintain patient confidentiality, which is a cornerstone of medical ethics.
>
> **2** The potential consequences of not disclosing the information, which could result in harm to an innocent person.
>
> **3** The impact on the patient-doctor relationship if confidentiality is breached.
>
> **4** Legal implications of either decision
>
> This example illustrates the complexity of ethical dilemmas, where there is often no clear right or wrong answer, and the decision-maker must carefully consider the potential outcomes and ethical implications of their choice.[46]

It's important to note that, at the current time, AI apps sometimes provide inaccurate information and responses (we discuss this at length later in the book, too). So it's critical to always verify any information that an AI app provides to ensure its veracity and correctness.

Summary

The term *ethics* derives from the Greek word for character and the Latin word for customs. Law and ethics serve different but cooperative functions in societies. Law tells us what we must or must not do. Ethics guides us in deciding what we should or should not do. Humans have sought guidance from trusted others and philosophical frameworks like religion for much of recorded history. Ethics principles and issues are present in many areas of our everyday lives, such as the culture we consume. Ethical analysis is beneficial for considering the various dimensions of ethical dilemmas and then making well-reasoned decisions. Ethical decision-making models include steps like gathering relevant facts, considering the options, identifying key stakeholders and their respective interests, weighing ethical approaches that are anchored in philosophical and moral principles, consulting with trusted persons or considering how someone admired might handle an ethical question, and assessing decisions after they have been made. In the digital and AI ages, ethics-related tools, including AI apps and chatbots, are increasingly available to aid people with ethical challenges and decision-making.

Questions for Reflection, Exploration, Action, Discussion (READ)

1. Reflect on your ethical thinking.

a. If and when you are faced with ethical dilemmas, what do you rely on or turn to in order to help you decide what to do or not do?
b. Are there principles or standards or guidelines that you draw on to make a decision clearer?
c. Do you have persons or organizations in your life that you ask for advice or counsel when faced with an ethical situation you're not sure how to handle?

2. Explore the 10-step framework for ethical decision-making created by the Markkula Center for Applied Ethics at Santa Clara University.

a. Do you see yourself already using some of the steps in the 10-step framework? Discuss.
b. If applicable, talk about which steps you recognize you are using and ways those steps are useful to you in addressing ethical challenges.
c. Are there other strategies you use—that aren't in the 10-step framework—when faced with ethical decisions? Explain.

3. Ethics in everyday life: Cultural content.

a. Think about the cultural content you engage with, such as books, TV shows, movies, anime, manga, comic books, videogames, dramatic arts, music, and so on. What ethics issues are present or addressed within that content? Identify the ethics issues.
b. Think again about the cultural content you engage with. What ethics issues are *not* present or addressed within that content but, in your view, could or should be present or addressed? Identify the ethics issues.

Notes

1. Legal Information Institute. (n.d.). *Ethics*. Cornell Law School. Retrieved February 12, 2025, from https://www.law.cornell.edu/wex/ethics
2. Velasquez, M., Andre, C., Shanks, T., & Myer, M. J. (2010, January 1). *What is ethics?* Markkula Center for Applied Ethics, Santa Clara University. https://www.scu.edu/ethics/ethics-resources/ethical-decision-making/what-is-ethics
3. Legal Information Institute. (n.d.). *Law*. Cornell Law School. Retrieved February 12, 2025, from https://www.law.cornell.edu/wex/law
4. Pennsylvania General Assembly. (2022). *Title 75*. https://www.legis.state.pa.us/cfdocs/legis/LI/consCheck.cfm?txtType=HTM&ttl=75&div=0&chpt=37&sctn=21&subsctn=0
5. Shannon, B. (2022, November 15). *Pennsylvania law now requires you to remove snow, ice from cars*. PAHomepage. https://www.pahomepage.com/news/pennsylvania-law-now-requires-you-to-remove-snow-ice-from-cars/
6. Legal Information Institute. (n.d.). *Policy*. Cornell Law School. Retrieved February 12, 2025, from https://www.law.cornell.edu/wex/policy
7. *Dred Scott v. Sandford.* (1857). US National Archives and Records Administration. https://www.archives.gov/milestone-documents/dred-scott-v-sandford
8. Robson, M. (Director). (1945). *Isle of the dead*. IMDb. https://www.imdb.com/title/tt0037820/characters/nm0237655
9. *Theodore Parker and the "moral universe."* (2010, September 2). NPR. https://www.npr.org/2010/09/02/129609461/theodore-parker-and-the-moral-universe
10. University of Pittsburgh, Department of Classics. (n.d.). *Ancient philosophy and science*. Retrieved February 12, 2025, from https://www.classics.pitt.edu/research/ancient-philosophy-and-science
11. Ibid.
12. Savolainen, R. (2009). Everyday life information seeking. In *Encyclopedia of library and information sciences* (3rd ed.). CRC Press.
13. LaFrance, A. (2015, April 29). Long live secrets on the internet. *Atlantic*. https://www.theatlantic.com/technology/archive/2015/04/long-live-secrets-on-the-internet/391910/3-120043920
14. Mangan, L. (2009, November 12). A brief history of agony aunts. *Guardian*. https://www.theguardian.com/lifeandstyle/2009/nov/13/agony-aunts
15. Falk, D. (2024, July 1). AI chatbots seem as ethical as a *New York Times* advice columnist. *Scientific American*. https://www.scientificamerican.com/article/ai-chatbots-seem-as-ethical-as-a-new-york-times-advice-columnist/
16. Terwiesch, C., Meincke, L., & Nave, G. (2023). *The AI ethicist: Fact or fiction? The Wharton School Research Paper*. https://doi.org/10.2139/ssrn.4609825
17. Lee, S. (1962, August 1). *Amazing Fantasy* (Vol. 1, iss. 15). Marvel Comics.
18. Lovett, J. (2021, May 17). *"X-Men Red" is the best X-Men story of the past decade*. ComicBook. https://comicbook.com/marvel/news/x-men-red-best-series-tom-taylor-marvel/
19. Connolly, K. (2010, August 12). *Teaching philosophy with Spider-Man*. BBC News. https://www.bbc.com/news/world-us-canada-10900068
20. Dailey, J., & Davidson, R. (2012). *The law of superheroes.* Gotham.
21. Waggener, G. (Director). (1941). *The Wolf Man*. IMDb. https://www.imdb.com/title/tt0034398/quotes/?item=qt0306388
22. Markkula Center for Applied Ethics, Santa Clara University. (2022). *The character of AI: A technology ethics conference.* https://www.scu.edu/ethics/events/the-character-of-ai–a-technology-ethics-conference/

23. Markkula Center for Applied Ethics, Santa Clara University. (n.d.). *Space ethics*. YouTube. Retrieved February 13, 2025, from https://www.youtube.com/watch?v=zrjodaqIN-Y
24. Markkula Center for Applied Ethics, Santa Clara University. (2022). *Virtual reality, real virtues, and augmented norms and laws*. YouTube. https://www.youtube.com/watch?v=jQzSGrW2TkQ
25. Markkula Center for Applied Ethics, Santa Clara University. (2009, May). *A framework for thinking ethically.* https://web.archive.org/web/20110219225544/http://www.scu.edu/ethics/practicing/decision/framework.html
26. Ibid.
27. Associated Press. (2017, June 8). Chimpanzees do not have same legal rights as humans, US appeals court rules. *Guardian*. https://www.theguardian.com/world/2017/jun/09/chimpanzees-do-not-have-same-legal-rights-as-humans-us-appeals-court-rules
28. Shanahan, Ed. (2022, June 14). Happy the elephant isn't legally a person, top New York court rules. *New York Times*. https://www.nytimes.com/2022/06/14/nyregion/happy-elephant-animal-rights.html
29. Nonhuman Rights Project. (n.d.). *About NHRP.* Retrieved February 12, 2025, from https://www.nonhumanrights.org/about-us
30. Levine, M. (2021, March 31). Federal bill to limit CEO pay marks one step toward fairness and equity. *Nonprofit Quarterly*. https://nonprofitquarterly.org/federal-bill-to-limit-ceo-pay-marks-one-step-toward-fairness-and-equity/
31. Markkula Center for Applied Ethics, Santa Clara University. (2009, May). *A framework for thinking ethically.* https://web.archive.org/web/20110219225544/http://www.scu.edu/ethics/practicing/decision/framework.html
32. Ibid.
33. Oxford Reference. (n.d.). *Virtue*. Retrieved February 12, 2025, from https://www.oxfordreference.com/display/10.1093/oi/authority.20110803120019992
34. Hursthouse, R., & Pettigrove, G. (2022, October 11). Virtue ethics. *Stanford Encyclopedia of Philosophy*. https://plato.stanford.edu/entries/ethics-virtue/
35. Kwan, J. (2023, May 5). *Care ethics.* Markkula Center for Applied Ethics, Santa Clara University. https://www.scu.edu/ethics/ethics-resources/ethical-decision-making/care-ethics/
36. Ibid.
37. Markkula Center for Applied Ethics, Santa Clara University. (2021, November 8). *A framework for ethical decision making.* https://www.scu.edu/ethics/ethics-resources/a-framework-for-ethical-decision-making/
38. Ibid.
39. The Ethics Institute at Kent Place School. (n.d.). *Ethics everywhere.* Retrieved February 13, 2025, from https://www.ethicsatkentplace.org/
40. *The Ethics Institute at Kent Place School.* (n.d.). Retrieved February 13, 2025, from https://www.youtube.com/watch?v=LDvYzxnwMGA&t=2s
41. Flanagan, L. (2019, May 24). *What students gain from learning ethics in school*. KQED. https://www.kqed.org/mindshift/53701/what-students-gain-from-learning-ethics-in-school
42. Ibid.
43. Ibid.
44. *The Ethics Institute ethical decision-making model.* (2022). Retrieved from ibid.
45. The Ethics Institute at Kent Place School. (n.d.). *Training and curriculum.* Retrieved February 13, 2025, from https://www.kentplace.org/ethics-institute/training-and-curriculum
46. *Perplexity*. (n.d.). Retrieved February 13, 2025, from https://www.perplexity.ai/search/give-me-an-example-of-an-ethic-1caYy2ZbS4qWcjafxOscIA

Chapter 2

Ethics Codes and Guardrails

Introduction: Ethics Guardrails

Going-to-the-Sun Road in Montana's Glacier National Park is one of North America's most stunningly beautiful drives. Built between 1921 and 1933, it is considered "a great feat of engineering" due to the numerous challenges engineers and laborers had to confront and solve during its construction.[1] This scenic 51-mile Rocky Mountain road winds and wends its way higher and higher to the 6,466 foot (2,025 m) alpine Logan Pass. Lucky observers of this "Big Sky Country" majesty are treated to spectacular panoramic views of the surrounding aquamarine glacial lakes, cedar forests, and verdant valleys far below. For drivers focused on staying within the byway's designated lines, it's reassuring to see the several-foot high stone-block guardrails that have been cut from native materials and intentionally stacked by human hands along some sections of the road (see Figure 2.1). These defensive shields stand between the breathtakingly steep drop-offs and the road's edge. Those sitting in the passenger seats—passively and trustingly taking in these high-country tableaus—may be even more relieved than their drivers to see these protective stone walls. Indeed, I remember the iconic drive well, as a teenage boy during a memorable family trip to Glacier in the mid-1970s.

Similar to the rock-hewn guardrails along Going-to-the-Sun Road, ethics codes and frameworks can serve as reassuring guardrails. Though not guaranteeing safety from all calamities, guardrails of both the road and ethics kinds represent forethought and preparedness, the ultimate aim of which is to promote safety and well-being. Road guardrails are also sometimes used as visual cues to denote areas of elevated concern and risk for motorists. In a similar way, ethical principles can telegraph issues of potential concern and danger. On the flip side, ethical principles can be used to signify values and issues that are of high priority.

It is also important to acknowledge the limitations of guardrails; namely, that the existence of a road guardrail doesn't guarantee it will be strong enough to ensure absolute safety or deliverance from all mishaps. What are some possible "Achilles' heels" of guardrails? A guardrail can fall into disrepair by way of neglect. Guardrails can also suffer from a lack of appropriate inspection, maintenance, and repair. Sometimes, too, they need updating and replacement. In the same vein, ethics guardrails can't guarantee that bad things will never happen. Ethics guardrails, in the form of principles and precepts, need to be periodically inspected, evaluated, and even changed, as needed. Later in this chapter, we look at some examples of ethics codes that have been examined, revised, and updated.

Figure 2.1 Guardrail along the Going-to-the-Sun Road, West Glacier, Flathead County, Montana. (Tango Images/Alamy Stock Photo)

Though they are not a panacea from all adversities, ethics codes and principles can be of significant value. Ethics principles may provide *some* level of protection, depending on their enforcement mechanisms. Moreover, ethics codes may be better than having no manner of protection at all, though opinions vary on their utility and efficacy. What ethics codes can do, in most instances, is to serve as an acknowledgment that some degree of thinking and communication has been conducted in devising those principles. Ethics codes also can remind us that just because we *can* do something doesn't mean we *should* do something. Potter Stewart, former justice of the US Supreme Court, famously expressed this distinction when he wrote, "Ethics is knowing the difference between what you have a right to do and what is right to do."[2]

Like anything created by human design, the value, durability, and integrity of ethics principles are often most shaped by and dependent on the people who had a hand in bringing them to life: the thinkers and engineers who envision and construct ethical frameworks, the policymakers who manifest their implementation, and the administrators who oversee, assess, and enforce their use. In this chapter, we examine some of the common features of ethics codes, the ways they vary, and examples of codes and principles. We also examine the roles that additional stakeholders can play in influencing ethics codes.

"A Rose by Any Other Name"

One of the first things to note about ethics codes is that they can have differing names. Many *are* simply referred to as "ethics codes." However, sometimes the terms "code of conduct" or "code of conduct and responsibility" may be used; in substance, though, these kinds of codes often function and look like ethics codes. The ethics rules that guide and govern the actions and decision-making of attorneys provide a useful example. Attorneys must adhere to ethics rules for the specific jurisdictions (e.g., state, territory, or Native American nation) where they are licensed to engage in the practice of law. The ethics rules for lawyers in a particular state are generally modeled after the American Bar Association (ABA) Model Rules of Professional Conduct, adopted in 1983.[3] As an example regarding the variability of names used for collections of ethics rules, prior to the 1983 adoption of the ABA Model Rules, the rules schema in use was known as the 1969 Model Code of Professional Responsibility.[4] Before that 1969 model, a framework called the 1908 Canons of Professional Ethics was used.[5]

Licensed and Nonlicensed Professions

Both licensed and nonlicensed professions have elements of ethics that are incorporated within their philosophies and practices. Some professions require their members to be licensed in order to be practicing members. These groups tend to have ethics codes or codes of conduct that are purposefully designed to have more compliance mechanisms and enforcement power than professions that do not require licenses. Doctors, nurses, lawyers, and certified public accountants (CPAs), for example, are all vocations that are licensed.

In contrast to licensed professions, information professionals—librarians, archivists, museum staff—are not licensed. Some states, however, do have certification processes and requirements. Such certifications largely are meant to verify that some variation of the master of library and information science (MLIS) degree has been earned. Verifying that applicants have received their degrees, via a certification application process, can be an attractive element for an employer when hiring a job applicant. This process can enable an employer to reasonably rely on the expectation that a job applicant or hired staff member has acquired various types of foundational or specialized knowledge in the course of earning their degree. Certification also warrants that an individual has completed an accredited MLIS degree. For example, I received certification from the states of Ohio and Pennsylvania while working as a library administrator in those states.

Another distinction between licensed and unlicensed professions is that the former require their members to complete continuing education courses on an annual basis, as was mentioned in the preface. Unlicensed professions, like information professionals, frequently are encouraged to engage in professional development, but it is rarely if ever a requirement. In the United States, states select and designate approved educators and educational entities to provide learning opportunities that enable licensees to take, complete, and report the completion of continuing education courses. This practice is not immune from controversy. In September 2023, a Connecticut lawyer had his Maine license suspended for a year for engaging in unethical behavior related to continuing legal education (CLE) courses. FindLaw explained that the lawyer claimed to have watched two live CLE course webcasts simultaneously: "Through a complaint, the bar asked [the lawyer] about it, and he responded that he watched the programs simultaneously, one on his computer and the other on his iPad. He said he didn't realize that attending multiple CLEs at the same time violated the professional rules of conduct."[6]

Continuing education courses include a specified number of credits that must derive from ones that have ethics components. We know from prior discussion in the preface, for example, that jurisdictions that license lawyers require a certain number of CLE credits to be designated as ethics credits. This underscores the importance that licensing associations (in the case of lawyers, state bar associations) place on lawyers staying current on ethics rules and maintaining and improving their competence. Similarly, doctors (who must complete annual continuing medical education [CME] credits) and other licensed professions require that annual continuing education courses and credits include a set number that address ethics.

Undergirding the philosophy of continuing education and ethical competency for all of these professions is a recognition of the crucial nature of lifelong learning and ongoing professional development. When we consider how technologies change societies and the ways that all professions must adapt, in order to effectively operate in the face of these new technological realities, the importance of professional development becomes even more understandable and persuasive. Indeed, ethics codes and codes of conduct frequently include mention of some aspect of professional development. This is true for licensed and unlicensed professions. Principle 8 of the American Library Association's Code of Ethics, as an example, codifies several aspects of lifelong learning: "We strive for excellence in the profession by maintaining and enhancing our own knowledge and skills, by encouraging the professional development of co-workers, and by fostering the aspirations of potential members of the profession."[7]

Another significant distinction between the ethics codes of licensed and unlicensed professions involves enforcement power. For example, ethics codes covering doctors and lawyers can generally be described as having "teeth." As *Collins Dictionary* states, having teeth refers to "something, such as an organization or a law, [that] has the necessary authority or power to make people obey it."[8] As this definition explains, the "teeth" aspect encompasses not only a law, policy regulation, or ethic principle that sets forth enforcement power, but also the administrative body itself that oversee those rules. Compliance and enforcement powers mean that oversight bodies have the ability (though not necessarily executed) to take meaningful disciplinary actions against rules-breaking members of their professions, when warranted. Such disciplinary measures may include actions such as

- censures
- sanctions
- suspensions
- removal of licenses (for lawyers, this is called disbarment; for doctors, this may be referred to as "license surrendered," "license revoked," or "license terminated")

When evaluating how substantive or nonsubstantive ethics rules and codes are, it's necessary to determine whether the entities that oversee the enforcement of ethics rules actually engage in meaningful oversight and the meting out of disciplinary actions. One can, for instance, have an ethics code that on its face may look or talk "tough" and meaningfully. However, in practice that code may essentially be worthless if it isn't acted on. Additionally, an ethics code may be replete with lofty rhetoric and yet essentially amount to nothing more than performative "ethics code washing." The bottom line: ethics codes and rules are only as useful and meaningful as the people who *choose* to live by them or to uphold them.

US Supreme Court Ethics Case Study

This case study begins with a question:

> True or False: The US Supreme Court has an enforceable, binding ethics code.
> Answer: False

Indeed, a curious and distinctive feature of the US Supreme Court is that it is the *only* judicial body in the US federal court system that does not have a binding ethics code. Proponents of an ethics code, such as Fix the Court, a nonprofit organization that advocates for all of the federal courts to be "more open and more accountable to the American people," argue that the lack of an enforceable ethics code for US Supreme Court justices undermines public trust.[9] Trust is seen as an indispensable element of both the rule of law and democracy. Authoritarian and totalitarian systems depend on citizens not trusting in the rule of law and the legitimacy of courts. In response to long-standing calls for a binding US Supreme Court ethics code, justices like Chief Justice John Roberts and Justice Samuel Alito have long countered that they can police their own ethics-related conduct.

In the early 2020s, a growing number of high-profile ethics controversies have roiled the high court. News media outlets like ProPublica, the *New York Times*, and the *Washington Post* have played critical roles in exposing questionable ethics-related conduct by some of the justices. *Forbes* summarized the ethics controversies in a July 2024 article, "Supreme Court Ethics Controversies: All the Scandals That Led Biden to Endorse Code of Conduct."[10]

Various members of the US Congress have called for Supreme Court ethics reforms over the years. Such calls have been made more urgently and insistently following ongoing revelations involving gifts to Justice Clarence Thomas by billionaire Harlan Crow. Secret recordings of Justice Samuel Alito and his wife by a watchdog reporter, without their knowledge, as well as photos of flags controversially flown at their homes, raised concerns about conflicts of interest and calls for recusal from cases in front of the court. Other elected officials have rebutted such clamoring for reforms, stating that changes are unnecessary. These persons argue that calls for reformation of the court's ethics stances are solely a reflection of discontent among progressives with the court's shift to a 6–3 conservative supermajority. That supermajority was cemented following the confirmation of Amy Coney Barrett in October 2020, leading to the subsequent overturning on July 24, 2022, of the nearly 50-year abortion-approving decision rendered by the court in the landmark 1973 *Roe v. Wade* ruling.

Bowing to increased scrutiny, as Reuters noted via its headline "Under Fire, US Supreme Court Unveils Ethics Code for Justices,"[11] the court announced on November 13, 2023, that it had crafted its own ethics code. However, court experts were largely unmoved by the effort, which was seen by many as simply "window dressing" to tamp down public criticism of the court's unwillingness to institute a code of conduct with teeth: the court's self-crafted ethics code was nonbinding on the justices. Ongoing ethics concerns about the justices include a lack of transparency on financial issues. This includes acceptance of gifts and travel from third parties, some of whom have had legal matters brought before the court.

Additionally, a lack of clarity regarding when justices should recuse themselves from weighing in on cases where they may have conflicts of interest continues to be a significant source of contention. Some assert that issues such as conflicts of interest and the acceptance of gifts and travel can create an air of impropriety about the conduct of justices, which in turn may diminish the public's faith in the fairness and impartiality of the court.

In July 2024, shortly after announcing that he would withdraw his reelection bid, President Joe Biden announced a three-part plan "to reform the Supreme Court and ensure no president is above the law."[12] Two of the proposals were term limits and a binding code of conduct for justices. In the words of the announcement,

> President Biden believes that Congress should pass binding, enforceable conduct and ethics rules that require Justices to disclose gifts, refrain from public political activity, and recuse themselves from cases in which they or their spouses have financial or other conflicts of interest. Supreme Court Justices should not be exempt from the enforceable code of conduct that applies to every other federal judge.[13]

Decisions that are made by judges, particularly at the Supreme Court level, can have implications for all members of societies, including information professionals and their institutions and users. Consider, for example, the ways that judicial rulings can impact agency and departmental decisions that may have consequences for broadband internet access, children's privacy, and the collection and use of personal data by technology companies. Determinations that are made by courts involving censorship of books, the responsibilities of social media companies for moderating content and mitigating online harassment, and the "scraping" of copyrighted content by Big Tech from the internet for use as training data for artificial intelligence (AI) tools may also have life-changing outcomes for individuals, groups, and geographic regions.

Judges have tremendous powers. They also often possess broad discretion to decide how, when, where, why, and on whom to exercise such powers, though they can be overruled by higher courts. Hence, the role of ethics as a potential check and guardrail on the substantial powers of judges is of great significance to societies. Later in this chapter, we define concepts like transparency, accountability, and conflicts of interest and discuss why those ethics principles and activities are so consequential to all societies and nations.

Prevalence of Ethics Codes

Many organizations and professions operating in the United States today have some kind of code or framework that incorporates ethical principles. Often, though, we may not realize that an ethics code or code of conduct is in place for a particular profession or organization. We may only learn that an ethics code exists—or doesn't exist—when something controversial or newsworthy occurs.

The former was the case in January 2021, shortly after the inauguration of President Joe Biden. During a briefing by then White House press secretary Jen Psaki, media outlets uncovered and revealed that the sign language interpreter for that briefing was, according to a headline *Business Insider*, a "Trump supporter who translates far-right videos."[14] Questions regarding the character of the interpreter ensued. One such question was whether the signer, who allegedly supported Biden's predecessor, former president Donald Trump, could be trusted to accurately convey the words of the press secretary speaking to the media on behalf of her boss. One of the ethics-related facts that came out in the course of this controversy was that that there is a Code of Professional Conduct (CPC) for signing members of the Registry of Interpreters for the Deaf (RID).[15]

The RID CPC home page displays the tagline "Upholding high standards of professionalism and ethical conduct of interpreters."[16] RID and the National Association of the Deaf (NAD) coauthored the code. The code was adopted in 2005. It's important to highlight that codes should be seen as "living documents" that, like strategic plans and mission statements, are revisited and periodically revised, as deemed necessary. To that end, the CPC includes the statement that "This Code of Professional Conduct is a working document that is expected to change over time."[17] The first few pages of the CPC include brief sections on

- Scope
- Philosophy
- Voting protocol
- Adoption of the Code of Professional Conduct
- Function of the guiding principles

The most impactful aspect of the CPC is its seven so-called tenets. The tenets include elements such as showing respect for clients as well as colleagues, interns, and those studying to become interpreters, maintaining confidentiality standards, exercising ethical business practices, and a commitment to professional development. The CPC is an example of a quite detailed code, as each tenet has its own explanatory section, and the code also includes statements regarding its applicability as well as definitions of terms such as *conflict of interest* and *consumers*. A noteworthy facet of the CPC is that tenet 1 on confidentiality is identified as its guiding principle.

Evaluating Ethics Codes

Because ethics codes vary in structure, content, usefulness, and enforcement, it's crucial to examine them with discerning eyes, in order to assess their value and efficacy, or lack thereof. Consider questions like the following when critically analyzing the effectiveness and strengths of a particular code:

- When was the ethics code created? Can you determine *when* it was created, or is the date not included with the code?
- When was the code updated or revised, if at all? Is an explanation provided about why a revision occurred?
- Is the code up to date or forward looking in terms of its content, or does it appear dated and "behind the times"?
- Does the code include an explanatory statement that describes aspects such as its scope, limits, the process that was used to create it, and enforceability mechanisms?
- Are important terms defined or clearly described to promote maximal comprehensibility?
- Does the ethics code or code of conduct align with the mission, vision, and values of the profession or organization?
- Who is the code intended for (i.e., who is the audience)? What stakeholders have interests in the code?
- What ethical theories/approaches, if any, are indicated from the principles or rules that are set out?
- What values does the code encompass from a holistic standpoint? What values does the code leave out and not address?
- Does the code communicate a particular ethos or philosophy? Does the code inspire or identify values the organization or professions aspires to?
- What might you change about this code, if anything?

Examples of Ethics Codes and Codes of Conduct

Question: What do the groups listed below all have in common?

- American Correctional Association
- American Institute of Certified Public Accountants
- American Psychological Association
- Association of Zoos and Museums
- Firefighters
- Music Teachers National Association
- National Public Radio

Answer: Each of these groups, despite their differing missions and functions, has an ethics code or code of conduct.

Virtually every organization or association operating today has an ethics code or code of conduct of some kind. Examining ethics codes can reveal ways that codes and rules vary. It can also shed light on common features. Exploration of codes, too, often can provide insights into the norms, mores, and values of organizations. Analysis of supporting materials related to ethics codes, such as meeting minutes and notes, articles, and interviews, can illuminate how decisions were made by those who had a hand in creating ethics codes and codes of conduct, much like legislative histories, as noted previously.

Center for the Study of Ethics in the Professions

To gain an expansive sense of the state of ethics codes, one of the first sites to consider examining is the Center for the Study of Ethics in the Professions. Founded in 1976, the center is based at the Illinois Institute of Technology, also known as Illinois Tech, and serves as a curator and portal to a broad swath of ethics codes:

> The Ethics Codes Collection (ECC) is the largest database of codes of ethics and guidelines in the world. It contains over 2,500 individual codes from around 1,500 different organizations, and collects both current and historical versions of these documents. The ECC seeks to provide practitioners, students, scholars and the public access to codes of ethics and guidelines and in this way inform ethical decision making in professional, entrepreneurial, scientific, and technological fields.[18]

The mission of the center is "to be one of the nation's leading centers on practical and professional ethics."[19] The Ethics Center Library provides access to "ethics codes from all over the world and a large collection of ethics education materials."[20]

American Medical Association Code of Ethics

Doctors are arguably the profession in which ethical standards are the most existentially paramount. This stems from the trust that patients need to feel toward these medical experts, with whom they entrust their well-being and very lives. The American Medical Association (AMA) was founded in 1847 and adopted an ethics code in that founding year. As the AMA's Code of Medical Ethics home page describes:

> The American Medical Association was founded in part to establish the world's first national code of medical ethics. The *Code* is widely recognized as the most comprehensive ethics guide for physicians. Opinions in the *Code* address issues and challenges confronting the medical profession and represent AMA policy.
>
> Promoting adherence to the professional standards promulgated in the *Code* is essential to preserving patient trust and public confidence in the medical profession.[21]

The AMA website identifies the global significance and dynamic nature of its code:

> In May 1847, the nascent American Medical Association took what has been described as a revolutionary action by promulgating standards of ethical conduct for physicians, distinct from personal virtue or morality. Much has changed since its creation and to guide physicians in the face of continuing breakthroughs in science and changing conditions of practice, the *Code* must be a living document.[22]

The Code of Ethics of the AMA refers to a number of items that together encompass the code. In addition to the code's ethics rules that doctors must adhere to, these items include the opinions of AMA's Council on Ethical and Judicial Affairs. The council for 2023–2024 comprised nine individuals representing different kinds of medical and ethics expertise. The code explains that the council's opinions "help physicians think through what is at stake as they engage patients and work with colleagues to reach ethically sound decisions."[23] Because medical ethics questions can often be nuanced and gray in nature, without clear-cut answers, opinions like these that provide assistance with ethical decision-making can be invaluable to health-care providers facing morally complex choices and outcomes.

Another beneficial role of the code is to offer guidance on novel medical developments and questions. As an example, strides are occurring with research involving the use of animal organs in humans. This area is called xenotransplantation. The Code of Medical Ethics Opinion 6.3.1 on Xenotransplantation compiles thinking and guidance in this area. The opinion sheds beneficial light on the various interests at play as health-care providers balance the costs and benefits of this emerging medical treatment.

Medical Library Association

Investigating the AMA and its ethics code offers helpful context for examining another medical association and its code. Information professionals working in health-care settings or working with medical information can turn to the Medical Library Association (MLA) for information and guidance. Like the AMA, MLA has its origins in the nineteenth century. MLA debuted in 1898 as the Association of Medical Librarians with an octet of four physicians and four librarians. The MLA history web page explains that the group founded the association "in the office of the *Philadelphia Medical Journal* at the invitation of George M. Gould, M.D., editor of the journal."[24]

Today, MLA has grown from its humble nineteenth-century beginnings in the City of Brotherly Love to a "global, nonprofit educational organization, with a membership of more than 2,500 institutions and professionals in the health information field."[25] Information quality is of central importance to MLA, as reflected in its vision statement: "The Medical Library Association (MLA) believes that quality information is essential for improved health. MLA aspires to be the association of the most visible, valued, and trusted health information experts."[26] MLA's core values specify seven statements. Unsurprisingly, one of these values includes ethics. The last bulleted statement is "Irreproachable ethical standards (Code of Ethics for Health Sciences Librarians)."[27]

Given that MLA has memorialized ethics in the form of one of its values statements, it is no surprise that MLA has developed its own ethics code. MLA's code is known as the Code of Ethics for Health Sciences Librarianship. As of 2024, MLA's code indicates that it was updated in June 2010. Knowledge is the bedrock for MLA and health sciences librarians, spelled out in the code's very first sentence: "The health sciences librarian believes that knowledge is the *sine qua non* of informed decisions in health care, education, and research, and the health sciences librarian serves society, clients, and the institution by working to ensure that informed decisions can be made."[28] This code also communicates boilerplate language that is often seen in ethics codes, namely, that "The principles of this code are expressed in broad statements to guide ethical decision making. These statements provide a framework; they cannot and do not dictate conduct to cover particular situations."[29]

The reason why ethics code statements are intended and designed to be general, guiding precepts, rather than prescriptively determinative is the following: *Ethics dilemmas and decisions will vary from case to case.* It follows certainly then, that ethical decision-making must also be conducted on a case-by-case basis, as ethics-related decisions will hinge on the specific facts of each particular ethical scenario. As law school students well know, the outcome of an ethics hypothetical posed by a law school professor in class or on an exam will often turn on the changing of merely one fact or detail. Indeed, the devil frequently *is* in the details of ethics scenarios and legal fact patterns, channeling a well-worn adage.

American Library Association Code of Ethics

The American Library Association (ALA), founded in 1876 and headquartered in Chicago, is the world's oldest and largest library group. ALA's mission is "To provide leadership for the development, promotion, and improvement of library and information services and the profession of librarianship in order to enhance learning and to ensure access to information for all."[30] One of the ways ALA provides leadership is in facilitating access to information on areas like ethics, law, and policy.

Like the MLA, ALA has a code of ethics. Similar to the MLA code, ALA's code of ethics explains that the purpose of its code is to offer "broad statements" that can be used "to guide ethical decision making."[31] Codes like those of MLA and ALA also caution people to be cognizant of the limitations of their ethics codes. This is evident where the ALA code points out that its ethics principles "cannot and do not dictate conduct to cover particular situations."[32]

ALA's code was first enacted in 1939, at the start of World War II. Since its enactment, the code has been amended four times: 1981, 1995, 2008, and 2021. In addition to providing ethics guidance, this code is an important publicly accessible communication mechanism for two audiences: (1) the information professions and (2) the general public. The code's principles address many bedrock tenets of free societies, democracy, and informed citizenries. These include access to resources and information (principle 1); intellectual freedom and resistance of censorship (principle 2); privacy and confidentiality (principle 3); balancing of intellectual property rights and use of intellectual property (principle 4); respect, fairness, and welfare of workers (principle 5); conflicts of interest (principle 6); distinctions between what information professionals personally believe as individuals and the responsibilities they must fairly provide through the institutions where they work (principle 7); and professional development and lifelong learning for oneself and others (principle 8). In 2021, a ninth principle was added to the code:

> 9. We affirm the inherent dignity and rights of every person. We work to recognize and dismantle systemic and individual biases; to confront inequity and oppression; to enhance diversity and inclusion; and to advance racial and social justice in our libraries, communities, profession, and associations through awareness, advocacy, education, collaboration, services, and allocation of resources and spaces.[33]

Throughout this book, ALA ethics code statements that relate to chapter topics like privacy, access, and intellectual freedom will be identified and discussed, given the central place that ALA holds with librarianship and information services.

Society of American Archivists

Like ALA, the Society of American Archivists (SAA), founded in 1936 and based in Chicago, has a long history and serves an important role for information professionals and communities. SAA describes itself as "North America's oldest and largest national professional association dedicated to the needs and interests of archives and archivists."[34] One of the distinctive features of SAA is that it specifically indicates the cooperative roles of its ethics code and core values. As the SAA website states, "The *Core Values of Archivists* and the *Code of Ethics for Archivists* are intended to be used together to guide individuals who perform archival labor or who work in archival environments. These aspirational values and ethical principles help shape SAA's expectations for professional actions and engagement."[35] The current code of ethics was approved by the SAA Council in 2005. It was revised in 2012 and 2020.

Another distinguishing aspect of SAA is that it provides a number of case studies in archival ethics. SAA notes that

> These case studies are drawn from real life. They address one or more of the areas covered by the Code of Ethics for Archivists: professional judgment in carrying out basic archival tasks, protecting records authenticity, access to and use of records, professional relationships with donors or users, privacy issues, ensuring security against theft, and questions of trust in archivists' conduct.[36]

Examples of topics covered include the following:

- Freedom of Information Act (FOIA) requests
- Intellectual property
- Online exhibits
- Culturally sensitive American Indian material
- Preservation

IFLA Code of Ethics for Librarians and Other Information Workers

The International Federation of Library Associations and Institutions (IFLA) describes itself as "the global voice of the library and information profession."[37] Founded in 1927 in the United Kingdom, IFLA today is headquartered in The Hague, Netherlands, and has more than 1,500 members drawn from upward of 150 nations. IFLA leads on ethics in two significant capacities: (1) providing its own ethics code and (2) collecting and making library-related ethics codes freely accessible from a number of countries.

IFLA's Code of Ethics for Librarians and Other Information Workers was adopted in 2012. The preamble of the code explains its purpose and principal audiences: "This Code of Ethics and Professional Conduct is offered as a series of ethical propositions for the guidance of individual librarians as well as other information workers, and for the consideration of Library and Information Associations when creating or revising their own codes."[38] A singular aspect of IFLA's code is that it has a full version[39] and a short version.[40] As mentioned, IFLA, via its web page "National Codes of Ethics for Librarians by Countries," offers free access to ethics codes for librarians and information professionals.[41] The collected codes represent more than sixty nations and regions, alphabetically from Armenia to Venezuela.

IEEE Code of Ethics

The Institute of Electrical and Electronics Engineers (IEEE) bills itself as "the world's largest technical professional society."[42] Because the organization's membership has outgrown its original narrower name, the organization solely uses (other than for legal purposes) the acronym IEEE today. IEEE explains that it is composed of

- engineers
- scientists
- computer scientists
- software developers
- information technology professionals
- physicists
- medical doctors
- electrical and electronics engineering core professionals
- and others

IEEE emerged in 1963 from the merger of two organizations, the American Institute of Electrical Engineers (AIEE) and the Institute of Radio Engineers (IRE). As of 2020, IEEE had nearly 400,000 members residing in 160 nations.[43]

IEEE's code of ethics was adopted by its board of directors in 2020.[44] Its first sentence is written in the form of a "We, The People"–style pledge, with its members pledging to uphold the standards and precepts that are identified. The code addresses issues and values including but not limited to

- public safety
- integrity
- conflicts of interest
- disclosure/transparency
- technical competency
- attribution

- fairness
- respect

Many of these issues and values are investigated in more detail later in this book when we survey so-called character ethics.

Association for Computing Machinery (ACM) Code of Ethics and Professional Conduct

For contrast and comparison purposes, a readily apparent threshold difference between IEEE's code and Association for Computing Machinery's code is length: whereas the former's code fits within a one-page PDF, the latter's code comprises more than 3,500 words. The first two paragraphs of the designated preamble of the code are an amalgam of aspirational rhetoric and pragmatic description:

> Computing professionals' actions change the world. To act responsibly, they should reflect upon the wider impacts of their work, consistently supporting the public good. The ACM Code of Ethics and Professional Conduct ("the Code") expresses the conscience of the profession.
>
> The Code is designed to inspire and guide the ethical conduct of all computing professionals, including current and aspiring practitioners, instructors, students, influencers, and anyone who uses computing technology in an impactful way. Additionally, the Code serves as a basis for remediation when violations occur. The Code includes principles formulated as statements of responsibility, based on the understanding that the public good is always the primary consideration. Each principle is supplemented by guidelines, which provide explanations to assist computing professionals in understanding and applying the principle.[45]

In light of global concerns about the impacts of emerging technologies on individuals and groups, the ACM code's General Ethical Principle 1.2: Avoid Harm is a particularly conspicuous feature of this code. This "avoid harm" section defines harm and sets out some examples of harm while underscoring that "this list is not exhaustive." This section is noteworthy, too, in the explicit charges it issues, such as

> To minimize the possibility of indirectly or unintentionally harming others, computing professionals should follow generally accepted best practices unless there is a compelling ethical reason to do otherwise.... A computing professional has an additional obligation to report any signs of system risks that might result in harm. If leaders do not act to curtail or mitigate such risks, it may be necessary to "blow the whistle" to reduce potential harm.[46]

Ethics Training

To illustrate the necessity and rationale for ethics awareness and education in a wide range of professions, including information professionals, let's look at how lawyers approach ethics training and see what lessons can be gleaned and leveraged. As discussed earlier in this chapter, the Model Rules of Professional Conduct guide and govern the conduct of lawyers.

Because lawyers are expected to be competent in the law and representation of their clients, it is fitting that rule 1.1 is the competence rule addressing this relationship.

Information and knowledge are integral to lawyers attaining and maintaining competence. Ethics is a key area in which lawyers are required to stay current on an annual schedule. Hence, regular training and education in law and ethics are crucial for even the most seasoned attorney. A 2020 *ABA Journal* article, "How to Avoid 10 Common Ethics Pitfalls," describes the roles and legal and ethical responsibilities that lawyers must uphold: "Lawyers are stewards of their clients' most sensitive and personal information. They serve as officers of the court and are in positions of public trust. But these high standards can lead to steep falls, and a lawyer who doesn't carefully mind ethics obligations can quickly run afoul of the rules of professional responsibility."[47]

The article goes on to note that almost all states mandate "ethics training as part of continuing legal education requirements." Requisite ongoing education is an important mechanism to ensure that lawyers are held to high standards and fulfill competency obligations owed to their clients as well as to society in general. Still, education and training are neither antidotes to nor panaceas for legal and ethical lapses and malfeasances, as the article observes: "a quick scan of disciplinary records reveals lawyers behaving badly on a spectrum of issues—from improper advertising to mishandling private information and everything in between." One of the best practices for avoiding "ethics traps" identified by legal ethics experts consulted for the article is "to not only beware, but be aware."[48]

Training and education are the most impactful ways to steer clear of ethics mishaps and keep one's eyes open for potential and actual ethics minefields. Although information professionals are neither licensed nor required to stay current on ethics issues, it is ethically incumbent on them to proactively *be* competent and current on ethics issues and rules. Librarians, archivists, and other information professionals serve in positions of public trust. They often have access to sensitive patron/customer/user information and data. As we'll examine in a forthcoming chapter on privacy, statutes require libraries to comply with patron privacy protections. In addition to these laws though, librarians have an ethical duty to respect and affirmatively protect the privacy and confidentiality interests of their users.

Summary

Ethics codes can function as useful guardrails to guide people and organizations as well as to flag potential trouble spots and areas to be avoided. Ethics codes and principles should be periodically reviewed and updated to stay current with societal and technological changes. Enforcement mechanisms are important in order for ethics codes and codes of conduct to have teeth and function as more than ineffective rhetoric. Some professions that are licensed, like doctors and lawyers, require ongoing ethics training. Unlicensed professions, such as the information, data, and computing fields, can benefit from ensuring that these professionals, too, are grounded in ethics principles.

Questions for Reflection, Exploration, Action, Discussion (READ)

1. Choose one ethics code from the information/data/computing-related codes listed to examine and get familiar with. Then, in at least a few sentences, specifically identify at least one principle/precept from your selected code and talk about why you believe it is a beneficial or noteworthy principle.
   - American Library Association (ALA): https://www.ala.org/tools/ethics
   - Medical Library Association (MLA): https://www.mlanet.org/page/code-of-ethics
   - Society of American Archivists (SAA): https://www2.archivists.org/statements/saa-core-values-statement-and-code-of-ethics
   - American Alliance of Museums (AAM): http://www.aam-us.org/resources/ethics-standards-and-best-practices/code-of-ethics
   - Association for Computing Machinery (ACM): https://www.acm.org/code-of-ethics

2 Having read this chapter and examined some of the ethics codes, share your thoughts about one strength and one weakness of ethics codes. Regarding the weakness that you identify, suggest a way that that weakness can be addressed or mitigated.

3 Go to the website for IFLA's National Codes of Ethics for Librarians by countries (https://www.ifla.org/national-codes-of-ethics-for-librarians-by-countries) and choose an ethics code from one country—other than the United States—to explore and familiarize yourself with. In at least a few sentences, share several of your impressions/thoughts about the overall code you investigated. Some questions to consider, as starting points:
   - Is there anything in the code that could be helpful or transferrable to the US codes identified in this chapter?
   - Is there anything missing or unspoken in your selected code that you think should have been included?
   - Are there key stakeholder interests that are addressed or not addressed in your selected code?
   - Are there any strengths or weaknesses you see in your selected code?

4 Ethics education and training are important for all members of information centers.

a. Identify several ways that libraries, archives, and museums can incorporate ethics awareness raising and training at all levels of information centers, including staff, administrators, board members, and volunteers.

b. Talk about some ways to assess the effectiveness of ethics training in information centers.

Notes

1. Federal Highway Administration. (n.d.). *Going-to-the-Sun Road.* Contributions and crossroads: Our national road system's impact on the U.S. economy and way of life. Retrieved February 13, 2025, from https://www.fhwa.dot.gov/candc/factsheets/goingtothesunroad.pdf
2. *Jacobellis v. Ohio*, 378 US 184 (1964).

3. American Bar Association. (1983). *Model rules of professional conduct.* https://www.americanbar.org/groups/professional_responsibility/publications/model_rules_of_professional_conduct/
4. American Bar Association. (1969 [1981]). *Model code of professional responsibility*. https://www.americanbar.org/content/dam/aba/administrative/professional_responsibility/model-code-of-prof-responsibility1969.pdf
5. American Bar Association. (1908). *Final report of the committee on code of professional ethics*. https://www.americanbar.org/content/dam/aba/administrative/professional_responsibility/1908_code.pdf
6. Ellison, S. (2023, September 28). *Lawyer suspended for lying about taking two CLE courses simultaneously*. FindLaw. https://www.findlaw.com/legalblogs/practice-of-law/lawyer-suspended-for-lying-about-taking-two-cle-courses-simultaneously/
7. American Library Association. (2021, June 29). *ALA code of ethics*. https://www.ala.org/tools/ethics
8. Collins Dictionary. (2020, February 13). *Something has teeth*. https://www.collinsdictionary.com/us/dictionary/english-word/something-has-teeth
9. Fix the Court. (n.d.). *About us*. Retrieved February 13, 2025, from https://fixthecourt.com/about-us/
10. Durkee, A. (2024, July 29). Supreme Court ethics controversies: All the scandals that led Biden to endorse code of conduct. *Forbes*. https://www.forbes.com/sites/alisondurkee/2024/07/29/supreme-court-ethics-controversies-all-the-scandals-that-led-biden-to-endorse-code-of-conduct/
11. Chung, A., & Kruzel, J. (2023, November 14). *Under fire, US Supreme Court unveils ethics code for justices*. Reuters. https://www.reuters.com/legal/us-supreme-court-announces-formal-ethics-code-justices-2023-11-13/
12. *Fact sheet: President Biden announces bold plan to reform the Supreme Court and ensure no president is above the law.* (2024, July 29). American Presidency Project. https://www.presidency.ucsb.edu/documents/fact-sheet-president-biden-announces-bold-plan-reform-the-supreme-court-and-ensure-no
13. Ibid.
14. Ankel, S. (2021, January 30). Biden's first White House sign language interpreter is a Trump supporter who translates far-right videos. *Business Insider*. https://www.businessinsider.com/bidens-sign-language-interpreter-runs-pro-trump-translation-group-2021-1
15. Registry of Interpreters for the Deaf & National Association of the Deaf. (2005). *NAD-RID code of professional conduct.* https://rid.org/programs/ethics/code-of-professional-conduct/
16. Registry of Interpreters for the Deaf. (n.d.). *Code of professional conduct*. Retrieved February 13, 2025, from https://rid.org/programs/ethics/code-of-professional-conduct/
17. Registry of Interpreters for the Deaf & National Association of the Deaf. (2005). *NAD-RID code of professional conduct*. https://rid.org/programs/ethics/code-of-professional-conduct/
18. LibGuides. (n.d.). *Ethics Code Collection.* Religious studies: Resources. University of Mount Union. https://umu.libguides.com/religious-studies/resources
19. Center for the Study of Ethics in the Professions. (n.d.). *Mission*. Illinois Tech. Retrieved February 13, 2025, from https://www.iit.edu/center-ethics
20. Center for the Study of Ethics in the Professions. (n.d.). *About*. Illinois Tech. Retrieved February 13, 2025, from https://www.iit.edu/center-ethics/about
21. American Medical Association. (n.d.). *Code of medical ethics*. Retrieved February 13, 2025, from https://code-medical-ethics.ama-assn.org/
22. Ibid.
23. American Medical Association. (n.d.). *About*. Retrieved February 13, 2025, from https://code-medical-ethics.ama-assn.org/about

24. Medical Library Association. (n.d.). *MLA history.* Retrieved February 13, 2025, from https://www.mlanet.org/about-mla/history-of-mla/
25. Medical Library Association. (n.d.). *About MLA.* Retrieved February 13, 2025, from https://www.mlanet.org/about-mla/
26. Medical Library Association. (n.d.). *Vision, values, mission, and planning.* Retrieved February 13, 2025, from https://www.mlanet.org/about-mla/vision-values-mission-and-planning/
27. Ibid.
28. Medical Library Association. (2010). *Code of ethics.* https://www.mlanet.org/wp-content/uploads/2025/02/MLA-Code-of-Ethics.pdf
29. Ibid.
30. American Library Association. (n.d.). *About ALA.* Retrieved February 13, 2025, from https://www.ala.org/aboutala
31. American Library Association. (2021). *ALA code of ethics.* https://www.ala.org/tools/ethics
32. Ibid.
33. Ibid.
34. Society of American Archivists. (n.d.). *Who we are.* Retrieved February 13, 2025, from https://www2.archivists.org/aboutsaa
35. Society of American Archivists. (2020). *SAA core values statement and code of ethics.* https://www2.archivists.org/statements/saa-core-values-statement-and-code-of-ethics#code_of_ethics
36. Society of American Archivists. (n.d.). *Case studies in archival ethics.* Retrieved February 13, 2025, from https://www2.archivists.org/groups/committee-on-ethics-and-professional-conduct/case-studies-in-archival-ethics
37. International Federation of Library Associations and Institutions. (n.d.). *Our history.* Retrieved February 13, 2025, from https://www.ifla.org/history/
38. International Federation of Library Associations and Institutions. (2012). *IFLA code of ethics for librarians and other information workers. Long version.* https://www.ifla.org/wp-content/uploads/2019/05/assets/faife/publications/IFLA%20Code%20of%20Ethics%20-%20Long_0.pdf
39. Ibid.
40. International Federation of Library Associations and Institutions. (2012). *IFLA code of ethics for librarians and other information workers. Short version.* https://www.ifla.org/wp-content/uploads/2019/05/assets/faife/news/IFLA%20Code%20of%20Ethics%20-%20Short.pdf
41. International Federation of Library Associations and Institutions. (n.d.). *National codes of ethics for librarians by countries.* Retrieved February 13, 2025, from https://www.ifla.org/national-codes-of-ethics-for-librarians-by-countries/
42. IEEE. (n.d.). *History of IEEE.* https://www.ieee.org/about/ieee-history.html
43. Ibid.
44. IEEE. (2020). *IEEE code of ethics.* https://www.ieee.org/content/dam/ieee-org/ieee/web/org/about/corporate/ieee-code-of-ethics.pdf
45. Association for Computing Machinery. (2018). *ACM code of ethics and professional conduct.* https://www.acm.org/code-of-ethics
46. Ibid.
47. Hudson, D. L. (2020, June 1). How to avoid 10 common ethics pitfalls. *ABA Journal*. https://www.abajournal.com/magazine/article/how-to-avoid-10-common-ethics-pitfalls
48. Ibid.

Chapter 3

Professional Ethics and Character Issues

Introduction: Professional Ethics and Ethics

Embedded within most ethics codes are principles, moral values, norms, and standards that provide a kind of skeletal framework to guide and govern the culture and conduct of people and organizations. These principles and standards are sometimes referred to as "professional ethics," a term of art that can be ambiguous and fluid. Moreover, *professional* itself can be an amorphous term. *Cambridge Dictionary* defines professional via a few connotations. One sense of the word is "Relating to work that needs special training or education." Another meaning is "Having the qualities that you connect with trained and skilled people, such as effectiveness, skill, organization, and seriousness of manner."[1]

Professions are at the heart of professional ethics. The Australian Council of Professions provides a helpful encapsulation of the key elements of a profession:

> A Profession is a disciplined group of individuals who adhere to ethical standards and who hold themselves out as, and are accepted by the public as[,] possessing special knowledge and skills in a widely recognised body of learning derived from research, education and training at a high level, and who are prepared to apply this knowledge and exercise these skills in the interest of others.
>
> It is inherent in the definition of a Profession that a code of ethics governs the activities of each Profession. Such codes require behaviour and practice beyond the personal moral obligations of an individual. They define and demand high standards of behaviour in respect to the services provided to the public and in dealing with professional colleagues. Often these codes are enforced by the Profession and are acknowledged and accepted by the community.[2]

Examples of professions include lawyers, doctors, accountants, educators, clergy, scientists, and many more, including the many kinds of roles that are filled by people in the information, data, and computing professions.

We've previously looked at some definitions of ethics, like the Markkula Center for Applied Ethics' offering that identifies ethics as "well-founded standards of right and wrong that prescribe what humans ought to do."[3] Crafting a hybridized working definition then, professional ethics can be described as "standards of right and wrong that guide and govern persons working in employment as well as nonemployment settings on how they should and should not conduct themselves."

Looking at some examples of definitions of professional ethics, the Pennsylvania Department of Education (PDE) defines professional ethics as "the standards of behavior, values, and principles that inform and guide professional decision-making."[4] Notably, the PDE explains that its professional ethics are "detailed in the Pennsylvania Model Codes of Ethics for Educators (MCEE)." The PDE definition specifically mentions "decision-making," but professional ethics arguably goes well beyond the limits of decision-related matters.

The 2006 academic paper "Professionals Ethics: Does It Matter Which Hat We Wear?" makes the point that ethics and professional ethics are not the same but that both are necessary and important. The article's Carnegie Mellon University author, John Hooker, explains: "They are not the same because professional ethics is not ethics. It is the identification of expectations that professions have created. To teach professional ethics is to familiarize professionals with codes and standard practices. To teach ethics to professionals is to train them in ethical reasoning."[5]

It can be helpful to note the distinctions between professional ethics and ethics as a shorthand visual:

Professional ethics = orientation with codes and standard practices
Ethics = engagement with ethical analysis, reasoning, and decision-making

Hooker asserts that both instruction in professional ethics and instruction in ethical reasoning are a necessity. The former is vital, per Hooker, because in order to "serve its purpose" as a profession the members of that profession need "to be of one mind on how they are to conduct themselves."[6] The verbiage "of one mind" is overly idealistic and also conveys an inaccurate sense of monolithic thinking. No profession's members are ever in lockstep agreement on every single principle. For example, I have been in discussions over the years with fellow academics who take issue with some of ALA's ethics code principles. At best, an ethics code or set of standards represents a statement agreed on by the majority of the voting membership of a profession and may be revised at some future time.

One of the strongest rationales for members of professions to have at least a basic understanding of professional ethics and ethics more generally is that inevitable conflicts will arise. Professionals perform responsibilities and have duties that can come into conflict with one another. Additionally, professionals may experience conflicts with internal and external persons. Faced with such conflicts, professionals can turn to ethics codes and principles (if they, hopefully, exist) for guidance and insights. Yet codes and principles are merely statements that consequently have the limitations of statements, rather than multistep models for analysis and determinations. Professional ethics can imbue profession members with a general, overarching awareness of the ethos of a profession. It is the discipline of ethics, though, that enables people to "go deeper" and access ethical approaches and tools that can facilitate introspection, the weighing of options, decision-making, and metacognitive assessment of the individual's reasoning process and any decisions that were made. Hence, ethics is a valuable "next step" beyond ethics codes. The former includes the congruent components of ethical reasoning and decision-making that are incorporated in ethics models like the Markkula Center's Framework for Ethical Decision Making, explored in the previous chapter.

Character

The term *character* is another concept that has an important relationship with ethics. Ethics, we know, is the standards and principles of conduct that guide us in what we ought to do or ought not to do. Character is defined by *Britannica Dictionary* as "The way someone thinks, feels, and behaves."[7] Oxford Languages describes character as "The mental and moral qualities distinctive to an individual."[8]

Character has an inextricable connection with ethics. Our own mental and moral qualities necessarily interplay with the standards and principles that can guide us in deciding what to do when we are faced with ethical dilemmas. For example, how honest we are or are not will impact what we think we should do. Similarly, how forgiving we are or are not will influence our ethical choices. Character has incalculable effects on not only us but also our communities, institutions, and nations. Three thousand years ago, Greek philosopher Heraclitus famously adduced that "Character is destiny." Illustrating the staying power of this pronouncement, Senator John McCain employed it for the title of his 2005 book *Character Is Destiny: Inspiring Stories Every Young Person Should Know and Every Adult Should Remember*. The role of character in human endeavors persists. Indeed, in the July 24, 2024, Oval Office speech elucidating his decision to withdraw from the 2024 presidential election, President Joe Biden talked about the role of character in public life, posing the question, "Do we still believe in honesty, decency, and respect?"[9] In his January 9, 2025, eulogy of former president Jimmy Carter at Washington National Cathedral, Biden observed that the attribute that most summed up the 39th president was "character, character, character."[10]

In this chapter, we consider some examples of character-related ethics issues. Character-related traits are often expressed in ethics codes and codes of conduct. Such statements can be aspirational and *prescriptive*, in terms of indicating what behaviors are desirable or required. In the converse, ethics codes and principles can also be aspirational and *proscriptive*, identifying behaviors that are undesirable and which should be refrained from being associated with or doing.

Integrity

Integrity is a word often associated with ethics. *Cambridge Dictionary* defines integrity as "the quality of being honest and having strong moral principles that you refuse to change."[11] Merriam-Webster describes integrity as "firm adherence to a code of especially moral or artistic values: incorruptibility."[12] Incorruptibility is a trait linked, too, with other character-related ethical principles like transparency and accountability, which are discussed when we examine the issue of avoiding conflicts of interest later in this chapter.

Misrepresentation

One of the ways that integrity most often manifests is in the way people represent themselves to the world. Virtually everyone at some time or another has applied for a job. Submitting a resume or curriculum vitae (CV) is usually a requirement for job applicants. Hence, resumes and CVs are among the most ubiquitous ways that individuals represent themselves to the world. Problematically, studies indicate that majorities of persons are not truthful about all aspects of these documents. For example, *Forbes* reported in November 2023 that a study revealed that "70% of workers lie on resumes."[13] CNBC in December 2022 described survey results indicating that "55% of Americans have lied on their resume at least once."[14] The article went on to highlight the eight most common resume lies and the percentages of persons surveyed who admitted to those fabrications:

1. Previous work experience: 55.4 percent
2. Skills: 43.1 percent
3. College degree (or equivalent): 41 percent
4. Personal details, such as age, location, or name: 39.5 percent
5. High school details: 39.2 percent
6. Salary information: 33.6 percent
7. Job-specific software or equipment skills: 33.5 percent
8. Employer references: 21 percent

Another study, this one conducted by the reference-checking firm Checkster and reported by CNBC in February 2020, revealed an eye-popping level of actual or potential resume dishonesty: "78% of candidates who applied for or received a job offer in the last six months admit they did or would consider misrepresenting themselves on their application. Put another way, just one in six job seekers say they didn't stretch the truth in any way during their recent hiring experience."[15]

The reasons why people aren't honest about components of their resumes aren't always clear, however. Nuanced ethical dilemmas can influence the rationales people use in deciding to obscure, omit, exaggerate, or outright lie about information on their resumes. For instance, empirical research indicates that some employers may discriminate against job applicants whose first or last names tend to be associated with a particular race or ethnicity. In July 2024, news outlets, such as CNN, shared the story of a Black man named Dwight Jackson who sued the Shinola Hotel in Detroit, Michigan, "alleging he was denied a job when he applied as 'Dwight Jackson,' but later [was] offered an interview when he changed his name to 'John Jebrowski.'"[16] Furthermore, the 2022 CNBC study discussed earlier found that "names with Indian, Pakistani or Chinese origin were 28% less likely to get an interview than an anglo-sounding name, even if every qualification matched up," based on a 2017 study conducted by academic researchers in Toronto, Canada.[17]

Stress is also a factor cited as a reason why some people stretch the truth or lie on their job application materials. The story about the 2020 Checkster survey noted that "applying for a new job can be stressful, and according to one new survey, that stress is leading a majority of job seekers to lie on their resumes in order to stand out."[18] The 2020 CNBC article observes that job seekers may lie on their resumes because they hope or figure that employers will not check the details. Hiring managers were also surveyed by Checkster. Perhaps somewhat surprisingly, many of these managers said that resume truth-stretching wasn't a deal breaker: "Hiring managers seem to be lenient with candidates who stretch the truth. Just one in three would never hire a candidate who lied…. The majority of people willing to hire someone who lied say they'd do so if the candidate had a 'good explanation.'"[19]

The preceding examples of job application and hiring practices raise difficult questions, such as the following:

- Just because a hiring manager may not check a resume, does that justify not being truthful?
- What are the ethical implications of lying in the pursuit of a job?
- What are the ethical implications of lying to secure a job that deprives another applicant of that job who may have been entirely truthful in their application materials?
- What are the ethical implications for workplaces that do not exercise due diligence in checking the accuracy of job application information or treat resume falsifications or inaccuracies as trivial and excusable?
- What are the ethical duties organizations owe to the public when they screen job application materials?
- What are the ethical implications for societies that may normalize falsehoods in job application materials? Do such actions give a foothold to corruption?

Misrepresentation of educational credentials, work experience, and military background is another all-too-common and troubling practice. Politicians representing diverse backgrounds and experiences have run afoul of disseminating falsehoods about themselves. Arguably, the most notorious example of this in the modern era is George Santos. Santos won the Republican House seat in New York's Third Congressional District in November 2022. He served from January 2023 until his expulsion from the US House of Representatives in December 2023, following indictment by the US Department of Justice in October 2023 for a panoply of charges, including conspiracy, wire fraud, false statements, falsifying records, and identity theft, among others.[20] Several weeks after winning his seat in the US Congress in November 2022, investigative reporting shed light on a multitude of fallacious details about his life, including his fabrication of a Jewish identity, though a small paper had reported details, largely unnoticed, of his fabulism before the election. Many of those who had voted for Santos felt betrayed and were outraged by the lies he had constructed, peddled, and used for political and economic advantage.

Unfortunately, as the *New York Times* highlighted in a December 2022 story titled "George Santos Is in a Class of His Own. But Other Politicians Have Embellished Their Resumes, Too,"[21] Santos *was* far from alone in the pantheon of elected officials who had engaged in resume and biography truth stretching or outright deception. Some examples include

- Richard Blumenthal (D-Senate, CT)
- Madison Cawthorn (R-House, NC)
- Wes Cooley (R-State Senate, OR)
- Andy Kim (D-House, NJ)

- Mark Kirk (R-House, IL)
- Marco Rubio (R-Senate, FL)
- Herschel Walker (R-Candidate for US Senate, GA)

In the graduate ethics course I teach, students grapple with a real-world case study stemming from the educational degree falsehoods of Marilee Jones. Jones was hired by the admissions office of the Massachusetts Institute of Technology (MIT) in 1979 while in her mid-20s. Her charge, as reported by the *New York Times* in 2007, was "to recruit young women, who at the time made up only 17 percent of the institute's undergraduates, compared with nearly half today."[22] She was successful in that endeavor, became the interim dean of admissions in 1997, and was promoted to the position of dean of admissions in 1998. After serving in that capacity for nearly a decade, Jones's world fell apart when misrepresentations of her educational degrees came to light:

> Ms. Jones… resigned in 2007 after admitting that she fabricated her credentials nearly three decades earlier. She said in a press release that she had lacked "courage" to correct her résumé, which listed degrees from three institutions; she had not graduated from any of them. A roar of publicity followed. Then, the nation's most outspoken admissions dean vanished without a word.[23]

Eventually, some years after resigning, Jones resurfaced for several interviews about the broader implications of people's lies about themselves. One of these, an NPR article and interview, "Learning the Hard Truth About Lying," discussed NBC news anchor Brian Williams's controversial description of his prior wartime reporting. The article not only examined lying and other ethical lapses but also looked at what it means to try to redeem oneself and how arduous that can be: "Coming back from that kind of mistake can be harrowing and life-changing. Marilee Jones knows intimately the damage that one lie can inflict on a life."[24]

In 2019, when the United States' largest college admissions scandal in history—the Varsity Blues Scandal, also officially known by law enforcement as Operation Varsity Blues—erupted into public view, Jones was again approached for her perspectives on wrongdoing and higher education. This time though, the media inquiries were less about her own scandal and were instead focused on hearing Jones's informed experiences on why students and their parents might go to such extraordinarily immoral lengths to unfairly gain entrance to many of the world's most prestigious universities, including Stanford, Yale, and Georgetown. Jones's opinions were a source of interest as well because she had started her own business as an advisor for students applying to colleges and universities several years after her MIT departure.

The Varsity Blues Scandal, unlike Jones's ignominious actions, though, constituted not only ethical breaches but also federal law violations. Indeed, these violations led to the arrests of more than fifty parents by the US Department of Justice, ensnaring celebrities like Felicity Huffman (best known for the popular TV show *Desperate Housewives*) and Lori Loughlin (of *Full House* sitcom fame), as well as CEOs, financiers, lawyers, and other societal elites.[25] The depths to which these affluent parents were willing to stoop to get their children into America's top higher education institutions were nothing short of shocking. A March 2019 *New York Times* article on the "U.S. College Entry Fraud" case laid bare some of the most jaw-droppingly egregious legal and ethical transgressions:

> A teenage girl who did not play soccer magically became a star soccer recruit at Yale. Cost to her parents: $1.2 million.
>
> A high school boy eager to enroll at the University of Southern California was falsely deemed to have a learning disability so he could take his standardized test with a complicit proctor who would make sure he got the right score. Cost to his parents: at least $50,000.
>
> A student with no experience rowing won a spot on the U.S.C. crew team after a photograph of another person in a boat was submitted as evidence of her prowess. Her parents wired $200,000 into a special account.[26]

Ultimately, some charges were dropped against a handful of the parents charged. However, the majority who were charged with various kinds of fraud eventually pled guilty or were convicted by federal juries. Some were fired from their jobs by their employers. Some went to jail and paid fines, like Felicity Huffman, who went to jail for eleven days and was required to pay a $30,000 fine. Lori Loughlin was jailed for two months, paid a $150,000 fine, and was ordered to perform 100 hours of community service.

Despite these sentences and other penalties that were meted out to the parental perpetrators, the damage done to trust in college admissions, as well as to society, is far harder to assess and more consequential. As Dan Ariely, a Duke University professor of psychology and behavioral economics, observed in the 2015 NPR piece about the ripple effects of lying, "Honesty is a public good and... society works when people trust each other. But if you live in a society with no trust, everybody suffers.... We really have to understand that erosion in trust is an incredible erosion in society as a whole."[27] The erosive and corrosive societal effects of dishonesty on public trust that Ariely is identifying are connected to additional character ethics issues like transparency and accountability, as well as the negative impacts of conflicts of interest, which are discussed later in this chapter.

I use Marilee Jones's story as a springboard for deeper discussion and exploration of the impacts of dishonesty and misrepresentation on societies. Her example is also a vehicle for examining the roles of contrition, restitution, compassion, grace, forgiveness, redemption, and reconciliation in societies. To those ends, students exploring the Jones case study read a variety of articles to build their own foundation of familiarity with the facts and opinions. After summarizing the relevant facts, they then consider many-sided questions that often defy easy answers, including the following:

- Did Marilee Jones do anything wrong?
- Who benefited from Jones's course of action?
- Who was hurt by Jones's course of action?
- In fabricating her educational credentials in 1979, did Jones have an unfair advantage over other applicants for her initial job (assistant to director of admissions at MIT)?

- If you worked at another university/college, would you consider hiring Jones for your institution? Why or why not?
- Would you hire Jones for another type of nonprofit or for-profit entity? Why or why not?
- Today, would you pay for Jones's private consulting/speaking services? Why or why not?
- What ethical rationale(s) did you use in answering these questions?
- What broader lessons can we learn from this case study of Marilee Jones?

Given that information professionals dedicate themselves to providing accurate, high-quality information, what are the implications for them, as well as their institutions and users, when faced with misrepresentations? What roles do information centers and professionals play in promoting honesty as a public good? We examine these questions and more in a later chapter on accuracy of information, misinformation, and disinformation.

Attribution and Plagiarism

Research is as indispensable to elite academic institutions as news is to newspapers and money is to banks. Attribution is an important professional ethics issue that is integral to research integrity. We'll also delve into an activity that is antithetical to sound research: the research taboo known as plagiarism.

Attribution is one of the fundamental facets of the research process and cycle. It is defined as "the ascribing of a work or an idea to a particular author or artist."[28] Other ways of explaining attribution include "giving credit for something" or "noting where something comes from," also referred to as "provenance." Research by its very nature references and often critically analyzes and includes quoted material from the research of others. As an ethical practice, researchers are expected to give credit for the work of others that is included in their work. The practice of citing sources and references that derive from the work of others is called "citation practice." A LibGuide from SUNY Empire Online Library offers a clear explanation of what a citation is and why citations are vital to the research process and research end products:

> A citation is a reference to the source of information used in your research. Any time you directly quote, paraphrase or summarize the essential elements of someone else's idea in your work, an *in-text citation* should follow. An in-text citation is a brief notation within the text of your paper or presentation which refers the reader to a fuller notation, or *end-of-paper citation*, that provides all necessary details about that source of information. (emphasis original)[29]

Engaging in proper citation practice is essential in order to ensure that attribution and credit are given to the originator(s) of published and unpublished works, as well as their ideas and intellectual property (IP). Chapters 8 and 9 extensively explain and examine the complex areas of IP and IP-related topics like open access (OA) and indigenous knowledge (IK).

Academic Integrity

Question: What is one of the most important purposes and objectives that attribution and citation practice promote and support?

The answer is "academic integrity." Academic institutions provide definitions and information on academic integrity to educate people on engaging in high-quality research practices and avoiding the pitfalls of plagiarism. The North Carolina State University (NCSU) Office of Student Conduct provides a helpful overview of academic integrity: "Academic integrity is the cornerstone of education. The free exchange of ideas depends on the participants' trust that others' work is their own and that it was done honestly. Intellectual progress in all disciplines demands the truthfulness of all participants. Plagiarism and cheating are attacks on the very foundation of academic life, and cannot be tolerated within universities."[30]

Academic integrity is not just for students, though; rather, people at all stages of their lives and careers may intentionally or unintentionally violate academic integrity standards. Hence, colleges and universities are expected to have academic integrity and plagiarism policies to explain what conduct is and is not permissible, as well as the penalties for not upholding academic integrity. Honesty is at the heart of academic integrity, as the University of Newcastle Australia describes: "Academic integrity, honesty, and a respect for knowledge, truth and ethical practices are fundamental to the business of the University. These principles are at the core of all academic endeavour in teaching, learning and research. Dishonest practices contravene academic values, compromise the integrity of research and devalue the quality of learning."[31]

In addition to promoting academic integrity, adhering to attribution and citation practice advances the following aims:

- offering credit for intellectual property (i.e., creations of the mind, original thought)
- supporting economic systems that compensate people for their creations
- encouraging scientific and societal progress
- contributing to scholarship and knowledge creation
- organizing information and aiding in the retrieval of knowledge (citations are the foundation for organizing and retrieving knowledge; citations also provide traceable linkages to works of scholarly communication and their authors)
- uplifting the ethical values of honesty, integrity, and truthfulness
- fostering trust, which is the cornerstone of civil societies, science, education, and healthy, functioning democracies and informed citizenries

Perceptions of Academic Integrity: Harvard University Case Study

Higher education institutions have an incentive for all of their key stakeholder groups—students and their families, faculty, and staff—to (1) believe in the importance of academic integrity and (2) trust that academic integrity is being promoted and enforced at colleges and universities.

Academic integrity has a direct link to the reputation of an organization. One reason researchers study how demographic groups feel about academic integrity is to better understand the interplay between academic integrity, organizational culture, and reputation. In 2011, Harvard University conducted a survey of students "to open up a conversation about issues of plagiarism and cheating" and find out how students were thinking about academic-integrity-related matters.[32]

In terms of research design, the 2011 Harvard study anonymously surveyed students, teaching fellows, and faculty. The Center for Academic Integrity at Clemson University, the third party conducting the survey, analyzing the data and reporting back to Harvard, noted that "understanding student perception of academic integrity is crucial to reducing plagiarism and cheating.... 'We want to develop in students the sense that a degree has higher value when the reputation of the institution is intact,' said Fishman [the center's director]. 'This is all about changing the culture.'"[33]

Even in 2011, at the time of this story, higher education and administrators were concerned about a possible correlation between internet technology and higher rates of plagiarism and cheating. Following this 2011 survey, it is noteworthy that a major cheating scandal burst into public view in 2012. The scandal involved 125 Harvard University students investigated for cheating on a take-home final exam. Jay M. Harris, the dean of undergraduate education, described the case as "unprecedented in its scope and magnitude."[34] The story was ranked the most important news story to Harvard in 2012 by the *Harvard Crimson*.[35]

In the aftermath of this high-profile scandal, investigations and discipline of the offending parties occurred. Harvard embarked on efforts to address its culture, including the adoption of an honor code in 2014, which had not existed before this time.[36] Interestingly, one of the stated aims of the 2011 assessment of opinions on academic integrity was to use the survey to evaluate whether an honor code was necessary.[37] This case study can serve as a launching point for ongoing discussions about whether and why students think it is important to promote and instill a culture of integrity and what institutions can do to nurture and support these kinds of cultures.

Defining Plagiarism

Having defined and explored the positive research practices of attribution and academic integrity, let's now define the undesirable research and academic practice known as plagiarism. As a bit of interesting background and etymology, *Merriam-Webster* notes that the term "plagiarize (and plagiarism) comes from the Latin *plagiarius* 'kidnapper.' This word derived from the Latin *plaga* ('a net used by hunters to catch game'), extended its meaning in Latin to include a person who stole the words, rather than the children, of another."[38]

Merriam-Webster's transitive verb definition of plagiarism is "to steal and pass off (the words of ideas of another) as one's own: use (another's production) without crediting the source." The dictionary's intransitive verb meaning is "to commit literary theft: present as new and original an idea or product derived from an existing source." *Merriam-Webster* cites 1660 as the year of the first known use of the transitive sense of *plagiarize*, six years before the Great Fire of London.[39]

Notions of creativity and originality are crucial elements of research and academic integrity. Defining and identifying the lines between creativity and derivativeness, or distinguishing originality from that which is unoriginal, is easier said than done sometimes. Philosopher Ralph Waldo Emerson (1803–1882) opined of creativity that "the creation of a thousand forests is in one acorn." Regarding the question, "What is originality?," William R. Inge (1860–1954), dean of St. Paul's Cathedral in London, cheekily answered, "Undetected plagiarism." Nobel Prize–winning theoretical physicist Albert Einstein (1879–1955) submitted that "the secret to creativity is knowing how to hide your sources."

Amid the wit and humor of these quotes, these statements serve to shine a light on the gray nature of creativity and originality. Because creativity and originality are sometimes gray, this in turn can sometimes (though not always) make what constitutes plagiarism less than crystal clear. To that end, even expert opinions vary on what plagiarism is and is not, and whether our conceptions of plagiarism need to adapt and evolve in the digital and AI ages. Indeed, linguist John McWhorter's January 2024 *New York Times* column declared "We Need a New Word for 'Plagiarism,'" in the wake of Harvard University President Claudine Gay's resignation, in part over plagiarism instances that came to light after scrutiny of her scholarly work.[40]

A sampling of collected articles over more than a decade and a half conveys the variety of opinions on plagiarism from just a reading of the headlines alone:

- "Academic Integrity and Student Plagiarism: A Question of Education, Not Ethics"[41]
- "When Is It Plagiarism?"[42]
- "Plagiarism Lines Blur for Students in Digital Age"[43]
- "Plagiarism Is Not a Big Moral Deal"[44]
- "The Highest Form of Flattery? In Praise of Plagiarism"[45]
- "Malcolm Gladwell: Plagiarism Is Just 'Bad Manners'"[46]
- "Plagiarism Is a Big Deal: This New Case at the University of Kansas Wasn't the First"[47]
- "Plagiarism Is the Next 'Fake News'"[48]

Levels of Plagiarism

The ways institutions and even nations view and respond to plagiarism vary along a continuum of norms and mores, just as individual opinions on plagiarism vary, per the examples cited above. One approach to plagiarism is to distinguish the type of plagiarism that has been committed. For example, sometimes plagiarism may constitute "unintentional plagiarism" or "inadvertent plagiarism." Unintentional plagiarism "can be described as careless paraphrasing and citing of source material such that improper or misleading credit is given."[49] Contrasted with unintentional plagiarism is "deliberate plagiarism." Intentional plagiarism has been described as "the wholesale copying of another's paper with the intention of representing it as one's own."[50] In addition, the definition of deliberate or intentional plagiarism includes the theft of another person's ideas.

Just as criminal law makes distinctions based on the *mens rea* (i.e., intent) of an alleged perpetrator, so too is intentionality seen as material in analyzing the severity of plagiarism that has been committed. Someone who intends to willfully and wantonly engage in plagiarism is generally viewed more harshly than someone, for example, who accidentally omitted quotation marks when citing another person's quoted material or incorrectly cited another's work. While discussing gradations of plagiaristic conduct, it is necessary to also explain the differences between "plagiarism" and "infringement," two terms of art that are all too often confused or conflated. PlagiarismChecker.com offers a clear, easy-to-understand distinction between the two terms:

> Plagiarism is using someone else's work without giving proper credit—a failure to cite adequately.
>
> Copyright infringement is using someone else's creative work, which can include a song, a video, a movie clip, a piece of visual art, a photograph, and other creative works, without authorization or compensation, if compensation is appropriate.[51]

Another critical distinction between these terms is that plagiarism is an *ethics* issue, while infringement is a *legal* issue. Infringement, too, is most often associated with the legal area known as intellectual property (IP), as is explained more in later chapters.

Who Plagiarizes?

Often when people think of plagiarism, students are the first group of potential and actual plagiarizers to come to mind. However, plagiarism is an issue involving diverse types of people in all kinds of endeavors. In the past decade, numerous examples have been reported involving academic faculty and administrators whose research articles, theses, dissertations, and even data sets have been revealed as plagiarized. Writers and journalists are also not immune from the commission of plagiarism. Examples include the celebrated historians Doris Kearns Goodwin, Stephen Ambrose, and David McCullough, *New York Times* columnist Maureen Dowd, and Kaavya Wisnanathan, author of books for young adults. Newspapers like the *Hartford Courant*, comedians such as Carlos Mencia, and website entrepreneurs like Cooks Source have also been subjects of plagiarism allegations.

Digitization of previously analog-only theses and dissertations has made it easier and quicker to find plagiarized material. As a result, politicians have especially found themselves "on the hot seat" when plagiarized material has been identified in works that they may have written decades ago. Germany has almost made it a cottage industry to expose plagiarism by its leaders, with two different federal government ministers, one in education and one in defense, having been found to have plagiarized their dissertations. As a 2013 *New York Times* article, "German Fascination With Degrees Claims Latest Victim: Education Minister," explained, "Germans place a greater premium on doctorates than Americans do as marks of distinction and erudition. According to the Web site Research in Germany, about 25,000 Germans earn doctorates each year, the most in Europe and about twice the per capita rate of the United States."[52]

The story of former US senator John Walsh (D-Montana) presents one of the most striking examples of the high costs of plagiarism. In 2014, Walsh was running to remain in his US Senate seat in an August election. However, in July of that year, the *New York Times* broke news of "How Senator John Walsh Plagiarized a Final Paper." The details had career, and ultimately even degree, implications for the senator:

> Senator John Walsh of Montana took most of a 2007 final paper required for his master's degree from the United States Army War College from other sources without proper attribution. Mr. Walsh copies an entire page nearly word-for-word from a Harvard paper, and each of his six conclusions is copied from a document from the Carnegie Endowment for International Peace without attribution.[53]

Walsh withdrew from the election within weeks of the investigative reporting. Adding salt to the wound of his ethical lapse, the US Army War College rescinded his master's degree and removed his name from the bronze plaque that lists the school's graduates.

Due Process and Allegations

When considering the dimensions of character ethics issues like allegations of misrepresentation and plagiarism, it is vital to ensure that "due process" is an integral part of any investigation. Legal Information Institute (LII) states that "'due process' generally refers to the 'conduct of legal proceedings according to established rules and principles for the protection and enforcement of private rights, including notice and the right to a fair hearing before a tribunal with the power to decide the case.'"[54]

Although, as established earlier, plagiarism is an ethical matter and not a legal matter, it is essential that those accused of plagiarism are afforded the ability to have access to the allegations against them and an opportunity to respond, such as through a hearing. For this reason, schools and higher education institutions generally have policies and procedures in place to adjudicate matters involving plagiarism allegations. The University of Pittsburgh School of Computing and Information's Academic Integrity Process sets forth the rationale, instructions, and requirements for students and faculty as well as the hearing procedures.[55] Policy and procedures statements like this also specify other key aspects, such as timelines for action, sanctions/penalties, the makeup of hearing committees, and appeal rights. Having such policies and procedures in place helps to ensure that fundamental fairness and the protection of the rights and interests of any persons accused are guiding principles.

Just as we now know through the efforts of nonprofit initiatives like the Innocence Project that innocent people in criminal justice systems are sometimes unfairly accused, convicted, and even incarcerated, so too is it critical and a matter of basic fairness to have ethical guidelines in place that ensure the rights of all parties involved in accusations of plagiarism are balanced and protected. As a real-life case in point of how essential this is on a human level, the 2013 *Chronicle of Higher Education* story "UC-Berkely Exonerates Anthropologist Who Was Accused of Stealing Ideas" drives this point home:

> Terrence Deacon did not commit plagiarism. He did not steal anyone's ideas. Instead, he was the victim of a "relentless e-mail and Internet campaign" that unfairly damaged his reputation.
>
> Those are the key findings of a report released late Wednesday by the University of California at Berkeley, which investigated allegations that Mr. Deacon, a professor of anthropology at the university, borrowed and then failed to cite much of the material in his 2012 book, *Incomplete Nature: How Mind Emerged From Matter*.[56]

Examples like this remind us that flesh-and-blood human beings are inextricably at the heart of these kinds of matters. Implementing protections that safeguard the liberty interests, basic rights, and human dignity of all parties is of paramount importance.

Self-Plagiarism

Self-plagiarism is probably not as well known as other types of plagiarism, but like plagiarism it also raises ethical issues. Self-plagiarism in the context of students refers to the reuse of one's previously published or submitted work. As indicated on the website Scribbr, "Self-plagiarism misleads your readers by presenting previous work as completely new and original."[57] The website provides the following examples of self-plagiarism and instructs that any text, ideas, or data used previously should be cited to inform the reader of the reuse:

- resubmitting an entire paper
- copying or paraphrasing passages from your previous work
- recycling previously collected data
- separately publishing multiple articles about the same research[58]

Researchers, writers, and others can also commit self-plagiarism. However, a March 2024 article, "When It Is and Isn't OK to Recycle Text in Scientific Papers," from the journal *Nature Human Behavior* and reported in *Nature* explains that opinions vary on self-plagiarism: "It is widely accepted that plagiarizing the work of others without attribution is unethical in scientific publishing. However, there is less consensus when it comes to the related concept of text recycling, sometimes called self-plagiarism. Text recycling refers to the reuse (copying) of one's own previous writing in the production of a new piece of work."[59]

The article identifies some of the differing perspectives on whether the recycling of one's own text constitutes a problem, some of the ethical concerns, and asserts that "text recycling can be appropriate when it is done ethically, transparently and legally (so that it does not infringe copyright)."[60] One of the most helpful aspects of the article may be its description of a National Science Foundation–funded initiative called the Text Recycling Research Project (TRRP). The purpose of the TRRP is to investigate text recycling in STEM research. As described in more detail on the TRRP web page, "The Text Recycling Research Project is the first large-scale investigation of researchers' reuse of materials from their own prior work in new documents. Our aim is to better understand text recycling, to help build consensus among stakeholders, and to promote ethical and appropriate practice."[61]

Stakeholders

One of the steps in the Markkula Center's Framework for Ethical Decision Making is identifying the stakeholders who have an interest or stake in the matters at issue. Who then are the chief stakeholders in matters involving plagiarism and infringement? The following is a nonexhaustive list of the principal stakeholders:

- K–12 and higher education students; their families
- educators, schools, higher education systems, accrediting entities, tutors/cram school operators
- tech companies, especially inventors/owners of AI and other emerging technologies
- investors
- creators: authors, composers, artists, and the like
- content industries/content holders: publishers, media, and entertainment companies
- vendors/distributors/intermediaries/for-profit companies, such as Turnitin antiplagiarism software
- politicians
- school board and higher education board trustees
- military members
- governments and nations, developed and developing
- societies: majority cultures and minority cultures, indigenous knowledge societies (e.g., Native Americans, Australian Aborigines, Canadian First Nations, Amazonian tribes)
- general public

Plagiarism's Impacts on Reputations and Careers

Regardless of one's personal views on whether plagiarism is "a big deal" in the words of the 2022 Kansas City Star article cited earlier, is "not a big moral deal" according to academic Stanley Fish in 2010, or, per best-selling author Malcolm Gladwell in 2019, is "just 'bad manners,'" the reality is that plagiarism *does* matter to most of the majority cultures of the working world. To remain gainfully employed with positive reputational standing, serve as a public servant or elected official, or publish one's research findings, fiction, or impassioned opinions, giving attribution to the work of others is the norm and expectation. Not doing so invites unwelcome consequences.

Moreover, plagiarism carries the potential threat of muddying the good names and derailing the careers—and interrupting the revenue streams—of *anybody*, whether they are famous (i.e., public figures) or unknown to almost everyone except those in their immediate circle (i.e., private figures). As two final cases in point, consider the following cautionary tales.

The first example involves a public figure: Retired Army Lt. Gen. Robert Caslen, who was hired as the University of South Carolina's president in 2019, despite opposition from many faculty, alumni, and donors. Before taking on this leadership role in South Carolina, Caslen had been the superintendent of the US Military Academy. Plagiarism would ultimately contribute to his meteoric downfall at the public university after he gave a speech nearly identical to a 2014 commencement address by Admiral William McRaven, the Navy SEAL responsible for overseeing the mission that resulted in Osama bin Laden's death. As NPR described in a May 2021 piece, Caslen, who failed to credit McRaven's well-known speech, "delivered a speech so bungled—with the wrong school name and closing remarks lifted nearly word-for-word from another famous commencement address—that it prompted widespread criticism from social media users to state legislators in South Carolina."[62] Less than one week after the speech, he resigned from his positon. In a 2021 newspaper interview some months after his resignation, Caslen "called taking the job the biggest regret of his life."[63]

The second story is of Karen Baldwin, largely a private figure except to the community she served as school superintendent in the small town of Ridgefield, Connecticut. In 2018, a local newspaper reported that her "nearly three-year tenure came to an abrupt end this week following accusations that she plagiarized portions of letters to parents and other documents."[64] The plagiarism accusations had been presented to the eight-member school board by a high school senior who "produced four documents Baldwin had written with lines or paragraphs taken from other sources without attribution." One of these documents was a letter Baldwin had addressed to the community after a recent shooting in Parkland, Florida. "Large portions of the letter were identical to an earlier letter sent by West Hartford Superintendent Tom Moore, which Baldwin has said was shared with the Connecticut Association of School Superintendents."

Though parental reactions were mixed on how the accusations were addressed, "most said Baldwin's leaving the district is the right outcome."[65] One Board of Education member implied that Baldwin's actions reflected on her communication skills. Additionally, a former board member voiced concerns that if Baldwin were engaging in this kind of behavior, "How deep does this go? How often was this happening?"[66]

These examples and prior ones in this chapter implicate a variety of ethical principles, values, and competencies, including honesty, integrity, intellectual humility, and due diligence; they also demonstrate the need for transparency and accountability, which we probe in the next section of this chapter. One wonders, too, whether these kinds of lapses in judgment and behavior regarding attribution can be viewed as a need for more education on why attribution is not only important but essential for persons in all kinds of professional and personal roles.

For anyone wondering how relevant plagiarism is with the growing adoption and proliferation of AI in our societies, reporting and research offer some worrying data points. As one example, Axios explained in February 2024 the results of a "new report from plagiarism detector Copyleaks [which] found that 60% of OpenAI's GPT-3.5 outputs contained some form of plagiarism."[67] This high degree of plagiarism has major impacts for integrity, honesty, trust, and scientific validity in myriad settings. The article goes on to note that these examples of plagiarism have implications for creators of copyrighted works, which we dissect in chapter 11 on AI.

In addition, all of these examples and issues present opportunities for information professionals to redouble their efforts at reminding users why attribution is essential. This can be done through the kinds of information literacy, programming, and outreach/community engagement activities that varying kinds of librarians (i.e., public, academic, school, specialized/special), archivists, and other information professionals have been proactively performing for decades. Illustrating once again the interconnectedness of the ethics, information, and technology issues examined throughout this book, legal and ethical issues pertaining to attribution return in the IP-related chapters 8 and 9 as well as in chapter 11 regarding AI.

Research Integrity and Research Misconduct

Human Rights and Research Misconduct

Perhaps the greatest tragedy of the human condition is the capacity for atrocities that have been perpetrated by and against human beings throughout history. The twentieth century's mass killings of Jews, Roma, gays, and untold others, as well as the world's learning of the horrific medical experiments sanctioned by Adolf Hitler's Nazi regime, galvanized post–World War II nations like the United States to institute regulatory frameworks and enforcement mechanisms for research so that such barbarism would never again occur. In the ensuing decades, ethical dilemmas have arisen over what to do with the kinds of knowledge that was acquired through Nazi experimentation. One example is scientific knowledge on the effects of hypothermia as a result of Nazi research on purposeful submersion of nonconsenting concentration camp inmates in frigid water. Wrenching questions arise:

- What should be done with such data?
- Should it be used at all, or in part?
- Should it be annotated with explanation of its provenance (i.e., where it came from)?
- Should it not be used?
- What should be done with the data in terms of where it resides?
- Should it be openly accessible to the public or only accessible to certain classes of persons, like approved researchers?
- Should recompense be made to survivors and their descendants, if they can be ascertained?

Such ethical decisions present some of the most difficult moral questions faced by science and societies.

The controversial *Pernkopf Anatomy* medical atlas is a tangible example of these kinds of morally tainted research-related ethical dilemmas. I first learned of this medical resource in 1996 when it was described by a library science professor in a course taken at the University of Pittsburgh. Considered by many to be among the world's foremost compilations of extraordinarily detailed color depictions of the human body's mysteriously engineered organs and complex, interdependent systems, the *Pernkopf Anatomy* is mired in controversy due to its association with Germany's Nazis. Concerns about this medical resource arose in the later years of the twentieth century. A November 1996 *New York Times* article succinctly encapsulates the controversy with the headline "Doctors Question Use of Nazi's Medical Atlas."[68]

These are the known origins of the book in question: It was created at the direction of Dr. Eduard Pernkopf, dean of the medical faculty at the University of Vienna at the time that Austria was annexed by the Nazis in 1938. Pernkopf was a vocal proponent of Nazi policies of eugenics and forced sterilization. The true identities of the cadavers on which the *Pernkopf Anatomy*'s drawings are based is the crux of the controversy; namely, "whether the cadavers were the victims of Nazi concentration camps, military campaigns, and war crimes."[69]

A clear-cut answer in the form of a "smoking gun" document that could provide definitive answers has not been discovered. Yet, from what is known about the politics and personal philosophy of Pernkopf himself, as well as pertinent evidence from historical records, deeply troubling, unanswered questions about the provenance of the cadavers remain. The evidence and historical records are subject to interpretation. To some they are persuasive. To others they are merely circumstantial. In the mid-1990s, several prominent medical practitioners in the United States and Canada raised concerns about the provenance of the cadavers. These individuals—Dr. Howard A. Israel with Columbia University, Dr. William E. Seidelman at the University of Toronto, and Dr. Gerald Weissmann with New York University Medical Center—called on the University of Vienna to investigate the origins of the cadavers.

In addition to questions about the origins of the Pernkopf cadavers, though, perhaps even tougher ethical dilemmas must be faced:

- What to do with the *Pernkopf Anatomy* atlas now that it exists?
- How should its controversial origins and continuing existence and use in present-day medical schools and libraries be handled?

Sadly, the *Pernkopf Anatomy* is not the only instance of medical research abuse and ethical lapses to raise profound moral questions and provoke pain and distrust of science and medicine, particularly in historically disadvantaged communities. Medical ethics improprieties like those revealed from the Tuskegee syphilis study, the Henrietta Lacks case uncovered by investigative journalist Rebecca Skloot, and the more recent case of Guatemalans who were intentionally infected with sexually transmitted diseases by US researchers starkly deviate from the Hippocratic oath "Do no harm."

Human rights abuses like those discussed represent some of the most egregious examples of research misconduct. In the modern era, human research protection regulations specify legally codified requirements that promote the welfare of research subjects and adherence to principles like informed consent. These regulations also identify potential consequences and penalties for research misconduct violations. Additionally, institutional review boards (IRBs), situated within higher education institutions and research organizations, are charged with the responsibility of reviewing research proposals to ensure that the welfare of human and animal subjects meets legal and ethical requirements.

Other Types of Research Misconduct

Plagiarism, examined earlier in this chapter, is one kind of research misconduct. Other types include issues like fabrication and falsification of data, which is an increasing concern in the digital and AI ages. The Office of Research Integrity (ORS) within the US Department of Health and Human Services (HHS) defines research misconduct and sets out what misconduct is and is not:

> Research misconduct means fabrication, falsification, or plagiarism in proposing, performing, or reviewing research, or in reporting research results.
>
> **(a)** *Fabrication* is making up data or results and recording or reporting them.
>
> **(b)** *Falsification* is manipulating research materials, equipment, or processes, or changing or omitting data or results such that the research is not accurately represented in the research record.
>
> **(c)** *Plagiarism* is the appropriation of another person's ideas, processes, results, or words without giving appropriate credit.
>
> **(d)** Research misconduct does not include honest error or differences of opinion.[70]

Nonprofit organizations like Retraction Watch work to expose instances of research misconduct. The entity's tagline highlights its broader purpose: "Tracking retractions as a window into the scientific process."[71] Retraction Watch is situated within the Center for Scientific Integrity, a nonprofit grant-supported watchdog organization. The center's mission is "to promote transparency and integrity in science and scientific publishing, and to disseminate best practices and increase efficiency in science."[72] One of the most significant goals and contributions is to provide a searchable "database of retractions, expressions of concern and related publishing events, generated by the work of Retraction Watch. The database will be freely available to scientists, scholars and anyone else interested in analyzing the information."[73]

Throughout 2024, as in prior years, Retraction Watch raised awareness of research and publication examples that generated "red flag" concerns, as well as identified instances of research misconduct. For example, an October 15, 2024, article advised that "Springer Nature Journal Has Retracted Over 200 Papers Since September."[74] The piece explained that "the retractions come after the publisher identified problems with the articles including compromised peer review, inappropriate or irrelevant references, and nonsensical phrases, suggesting blind use of AI or machine-translation software."

Highlighting retractions of fraudulent works is an important service that groups like Retraction Watch provide. However, an ongoing concern is people who may mistakenly rely on those retracted works as "good science" and the harm that that reliance can cause. As an example, Ars Technica reported in a January 7, 2025, news story that "Science Paper Piracy Site Sci-Hub Shares Lots of Retracted Papers."[75] The article notes that once a paper is uploaded, the site does not appear to "have the ability to ensure it's kept up to date." Of more concern: "Based on a scan of its content done by researchers from India, about 85 percent of the invalid papers they checked had no indication that the paper had been retracted."

In addition to civil society watchdog organizations like Retraction Watch that strive to promote scientific integrity and illuminate persons and groups that are engaging in research misconduct, other stakeholders can assist in cultivating research integrity. Higher education institutions have research offices that can help researchers in formulating research plans and protocols, though the researchers themselves are of course responsible for conducting the research. Universities and colleges also regularly offer workshops and webinars that address a variety of "research life cycle" steps and topics, as well as warning against suspect research practices such as plagiarism and image falsification. Information professionals, particularly academic librarians within medical and science venues, can take on actives roles in aiding researchers and providing opportunities for augmenting "research integrity literacy."

Transparency, Accountability, Conflicts of Interest

Two of the chief topics discussed in this section—transparency and accountability—were briefly introduced in the previous sections on professional ethics and character. We've seen, for instance, calls for transparency vis-à-vis examples of individuals who have misrepresented themselves. Additionally, we've observed ways that individuals have (or have not) been held accountable for misrepresenting themselves or committing plagiarism. Though these two terms are quite frequently heard, it's helpful to define them to make sure we're all on the same page with their meanings.

Transparency

Taken at its most basic meaning, "transparency," per *Cambridge English Dictionary*, is "the quality of being easy to see through."[76] Transparency at a deeper level, though, symbolizes openness and accessibility. It is the ability to have an unimpeded view of something. Picture yourself looking into a car window of clear glass, rather than one that is frosted or completely opaque. *Cambridge English Dictionary* offers two additional connotations of transparency that encompass the open, viewable, accessible conceptions of transparency: "The quality of being done in an open way without secrets"; "A situation in which business and financial activities are done in an open way without secrets, so that people can trust they are fair and honest." Notice that ethical principles and values—fairness and honesty, for example—are embedded within these definitions.

Taking a moment to discern how organizations interpret these conceptions of transparency, the Arthur W. Page Center at Pennsylvania State University (PSU), as one example, describes transparency as "open, honest, and accessible communication."[77] The Page Center identifies three purposes that emanate from transparency:

1 To provide information to publics

2 To increase participation

3 To hold organizations accountable

All three of these purposes, as noted by the Page Center, have impacts on trust. Trust is a requisite element in providing information. Trust must be present in order for people to feel that they can participate in an array of civic endeavors. Trust is inseparably connected with the notion of holding people, organizations, and even nations accountable.

Transparency also has an important relationship with established and emerging technologies. One concern often raised about technologies is that because companies own the technologies embedded within their products and services, they have a proprietary interest in not disclosing the "secret sauce" of how their revenue-generating sources are designed and work. Hence, the public is expected to trust that only "good things" are carried out by and done with these tech products. For many, such trust is neither warranted nor sufficient, particularly in light of well-documented instances of bias perpetrated by technologies, which we explore later in this book. As a counter to tendencies toward opaqueness in tech design, the 2020 Rome Call for AI Ethics makes transparency one of its six principles: "AI systems must be understandable to all."[78] We return to issues of transparency and AI tech and other emerging technologies in chapters 11 and 12.

Accountability

The concept of accountability describes the rights and responsibilities that exist between people and the institutions that affect their lives, including governments, civil society and market actors. In practice, accountability can take a number of different forms, depending on the institution in question. In general, relationships of accountability have two important components:

- answerability (the right to get a response and the obligation to provide one)
- enforceability (the means to ensure some kind of action can be taken and the existence of potential redress and remedies)

Two questions emerge in thinking about accountability:

- What are some reasons accountability is important to democratic institutions and societies?
- What are some pitfalls in not making people and organizations accountable for their actions?

Returning to the example mentioned in the prior discussion of transparency, the 2020 Rome Call for AI Ethics prioritizes accountability, along with transparency, as one of its six principles: "There must always be someone who takes responsibility for what a machine does."[79] Microsoft is a signatory to the 2020 Rome Call for AI Ethics, as well as IBM. Transparency and accountability, too, are among the six principles that Microsoft says "we believe should guide AI development and use": "Transparency: AI systems should be understandable"; "Accountability: People should be accountable for AI systems."[80]

Freedom of Information Act

A brief look at the US Freedom of Information Act (FOIA) statute, a groundbreaking driver of transparency, enables us to begin to answer the two questions raised regarding accountability. FOIA today enables any citizen, including members of the press, to submit requests for information to federal agencies and departments. The history of FOIA is one that shows a steady progression from a rather novel idea in the 1950s about promoting government transparency.

This idea then led to an early version of the statute that was later strengthened at several points and became the law that exists today.

The Electronic Frontier Foundation (EFF) explicates that President Lyndon Johnson signed FOIA into law in 1966. The 1966 FOIA bill that Johnson signed provided "exemptions for national security" and included "room for interpretation."[81] These aspects, in effect, rendered FOIA without "the teeth necessary to force government agencies to comply."[82] Following the revelations of the Watergate Hotel break-in scandal and resultant resignation of President Richard Nixon in August 1974, the US Congress took concrete steps to fortify FOIA. These steps included new requirements, such as time frames for federal parties to comply with requests for information and sanctions for impermissibly withholding information from release to requesting parties. The US government describes FOIA as providing "the public right to request access to records from any federal agency. It is often described as the law that keeps citizens in the know about their government."[83]

Because democratic governments are generally founded on the principle of checks and balances, it is key to understand that there are exemptions to requests:

> Not all records are required to be released under the FOIA. Congress established nine exemptions from disclosure for certain categories of information to protect against certain harms, such as an invasion of personal privacy, or harm to law enforcement investigations. The FOIA authorizes agencies to withhold information when they reasonably foresee that disclosure would harm an interest protected by one of these nine exemptions.[84]

With more than 50 years of existence, FOIA remains a valuable legal mechanism for citizens to seek information from the US government. Moreover, because freedom of the press is a critical constitutional right, FOIA is a vital means for journalists to attain information and in turn share that information with the public. This in turn fosters informed citizenries, which are essential for robust democracies.

Founding Father James Madison was an ardent proponent of openness in government. Hence, Freedom of Information Day was started in 1986 and is annually commemorated as close to the March 16 birthday of Madison as possible. This event is also aligned with National Sunshine Week events, which focus on recognizing the importance of openness and transparency in government. The Sunshine Week organization's website explains that Sunshine Sunday started in 2002 in Florida through the efforts of the Florida Society of Newspaper Editors. This then led to the launch of National Sunshine Week in 2005 by the American Society of Newspaper Editors.[85] As its name suggests, this annual weeklong series of events comes out of a belief that sunshine, rather than darkness, enables governments to be accountable to their citizens.

Transparency is connected to the most fundamental building blocks of democracies. Science and science-centered professions, like those related to information and data, also share links with many of these building blocks, such as

- public trust
- integrity
- truth

- accuracy of information
- ability to verify or replicate data, information, and findings
- access to information, data, and technologies

In answer to the earlier question—what are some reasons accountability is important to democratic institutions and societies?—all of the listed building blocks are applicable, too. Accountability promotes trust: trust in institutions, trust in governments, and trust in the rule of law. Accountability is also essential in fostering integrity, truth, and accuracy of information. Additionally, verification of findings and conclusions promotes the principle that people and institutions may be held responsible for their actions and the work they produce. Access to information, data, and technologies can also be tracked through accountability metrics and mechanisms.

Corruption

Returning to another previous question that was raised—what are some pitfalls in not making people and organizations accountable for their actions?—corruption is perhaps *the* greatest danger when accountability is lacking. The nonprofit organization Transparency International (TI) has made it its mission to combat corruption and to instill principles of transparency and accountability; indeed, TI sees itself as not simply an organization but rather "a global movement," per its description:

> Transparency International is a global movement working in over 100 countries to end the injustice of corruption.
>
> We focus on issues with the greatest impact on people's lives and hold the powerful to account for the common good. Through our advocacy, campaigning and research, we work to expose the systems and networks that enable corruption to thrive, demanding greater transparency and integrity in all areas of public life.[86]

TI sets a high bar for its mission—"to stop corruption and promote transparency, accountability and integrity at all levels"—and vision—"a world in which government, politics, business, civil society and the daily lives of people are free from corruption."[87]

LII defines corruption as

> a dishonest, fraudulent, or even criminal act of an individual or organization, using entrusted authority or power to make a personal gain or other unethical or illegal benefits. Corruption happens not only in political fields but also in social and economic fields, such as business, education, media, and so on. A lack of transparency and effective regulations are often the main causes of corruption.[88]

LII further adds that some ways to confront corruption are "instituting rules that prevent gray areas" and establishing "transparent processes."

TI has been at the vanguard in fighting corruption through multipronged approaches. To that end, TI reported in 2022 that more than 8,000 companies throughout the world had "committed to stop corruption as a result of [their] corporate sustainability initiative."[89] TI also indicated that more than a quarter of a million people "helped to report corruption worldwide through [their] Advocacy and Legal Advice Centres (ALACs)."[90] One of the most noteworthy activities TI engages in is to compile data that is used to produce its Corruption Perceptions Index (CPI). TI explains that its CPI "ranks 180 countries and territories around the globe by their perceived levels of public sector corruption, scoring on a scale of 0 (highly corrupt) to 100 (very clean)."[91] Another useful TI resource is a glossary of terms, complete with textual explanations and pictographs illustrating the meaning of each glossary term. The ability to find anticorruption-related information on individual nations makes TI a convenient source.

The United Nations Convention Against Corruption (UNCAC) represents a coordinated effort at the global nation-state level to proactively confront corruption. UNCAC, in its own words, "is the only legally binding universal anti-corruption instrument."[92] It was adopted by the UN General Assembly in 2003 and entered into force in 2005. As of October 2023, 140 nations were signatories to the treaty and 190 countries were parties. One of the ways that UNCAC is most influential is in its relation to the UN's 2030 Agenda and Sustainable Development Goals. In 2022, TI pointed out that anticorruption efforts are vitally important in working to achieve progress on Sustainable Development Goal 16: Peace, Justice and Strong Institutions.[93]

Conflicts of Interest

Conflicts of interest are a significant ethics-related issue with direct links to transparency and accountability. TI defines a conflict of interest as a "situation where an individual or the entity for which they work, whether a government, business, media outlet or civil society organization, is confronted with choosing between the duties and demands of their position and their own private interests."[94]

Because conflicts of interest can potentially present both ethically and legally problematic situations for people and entities, organizations generally have policies in place to address conflicts of interest. Not surprisingly, ALA provides a number of resources and information on conflicts of interest. In the context of libraries, ALA's "Conflicts of Interest Q&A" explains that a conflict of interest "involves circumstances where an individual's professional actions or decisions at the library could be influenced by considerations of personal gain, usually of a financial nature, as a result of interests outside their library responsibilities."[95]

ALA definitively answers the question as to whether a library should have a conflict-of-interest policy: "All libraries should have a board- or administration-approved conflict of interest policy that is shared and signed by all members of the library's administration, staff, and volunteers. It is the responsibility of all members to be aware of and adhere to the policy. The policy must follow local, state, and national laws and regulations."[96] Helpfully, ALA also suggests the elements that a conflict of interest policy should include. In addition, the Q&A site has succinct, clear responses to questions like

- Why are we required to disclose our external activities? Why doesn't the library or its governing body simply trust its employees?

- To whom does the conflict-of-interest disclosure process apply?
- What information should new library employees receive regarding conflicts of interest?
- Are employees bound by any state/federal guidelines regarding conflicts of interest?
- What should a library do when a conflict of interest is discovered?

Whether they are potential or actual, conflicts of interest can have downsides for individuals and organizations. Some of the possible negative consequences of conflicts of interest can include but are not limited to

- damage to reputation
- diminished trust
- job suspension or termination
- cuts in financial allocations or grant monies
- appearance of impropriety
- loss of business or usage
- lower workplace morale
- negative publicity

In order to avoid or mitigate the possible negative effects of conflicts of interest, it is highly advisable to have policies that are regularly reviewed and complied with by all organizational persons and entities. Conferring with human resources or legal counsel is also good practice to avoid mistakes and lapses.

Summary

Professional ethics and character ethics provide the skeletal structure of standards and norms for professions. Examples of professional and character ethics traits include honesty, integrity, empathy, respect, and compassion. Misrepresentation, plagiarism, corruption, and conflicts of interest are behaviors that can create significant legal and ethical challenges and downsides for individuals, societies, and nations. Research misconduct is an increasing concern, as technologies like AI enable fraudulent papers to be published and proliferate online via social media and digital piracy and sites. Information professionals can take on leadership roles in promoting ethical conduct and raising awareness of the ways that ethics principles can contribute to informed citizenries and functioning democracies. Principles such as transparency and accountability are vital in fostering trust and reliability in healthy human relationships and institutions, as well as in the design and implementation of technologies that contribute to the public good.

Questions for Reflection, Exploration, Action, Discussion (READ)

1. The *Pernkopf Anatomy* medical atlas raises extraordinarily sensitive ethical concerns about what to do with research that is acquired in questionable ways.

a. Identify some of the arguments for using and not using the knowledge contained within the atlas.
b. Explanatory descriptions are sometimes included with a controversial information resource or cultural artifact to provide editorial insights and historical context. Identify pro and con arguments that you see in including an explanatory description with the *Pernkopf Anatomy* atlas.

2. Examining cultures of integrity.

a. What are some of the elements of a "culture of integrity"?
b. What are some steps that information professionals can take to foster cultures of integrity in their own institutions and in broader sectors of societies?

3. Professional and character ethics are integral aspects of professions, whether they are licensed or nonlicensed.

a. What are the responsibilities, if any, of professions for instilling professional and character ethics in their members?
b. Imagine that you are a member on an information organization committee that is working on revising its ethics code to better address professional and character ethics. What do you think are the five to eight most important professional- and character-ethics-related principles to emphasize in your organization's ethics code?

Notes

1. Cambridge English Dictionary. (n.d.). *Professional*. Retrieved February 14, 2025, from https://dictionary.cambridge.org/us/dictionary/english/professional
2. Australian Council of Professions. (2003). *What is a profession?* https://professions.org.au/what-is-a-professional/
3. Velasquex, M., Andre, C., Shanks, T., S. J., & Meyer, M. J. (2010, January 1). *What is ethics?* Markkula Center for Applied Ethics at Santa Clara University. https://www.scu.edu/ethics/ethics-resources/ethical-decision-making/what-is-ethics
4. Pennsylvania Department of Education. (2023, October). *Professional ethics (PE) program framework guidelines.* https://www.pa.gov/content/dam/copapwp-pagov/en/education/documents/educators/certifications/professional%20ethics%20program%20framework%20guidelines.pdf
5. Hooker, J. (2006). *Professional ethics: Does it matter which hat we wear?* Carnegie Mellon University. https://johnhooker.tepper.cmu.edu/hats.pdf
6. Ibid.
7. Britannica Dictionary. (n.d.). *Character*. Retrieved February 14, 2025, from https://www.britannica.com/dictionary/character
8. Oxford Languages. (n.d.). *Character*. Retrieved February 14, 2025, via Google search. (Google Dictionary uses Oxford Languages definitions.)

9. Hutzler, A. (2024, July 25). *Read the transcript of Biden's Oval Office address on decision to leave 2024 race*. ABC News. https://abcnews.go.com/Politics/read-bidens-prime-time-oval-office-address-decision/story?id=112246959
10. Associated Press. (2025, January 9). *Biden delivers eulogy at Jimmy Carter's funeral in Washington, D.C.* YouTube. https://www.youtube.com/watch?v=5PUhkzh6tG8
11. Cambridge English Dictionary. (n.d.). *Integrity.* Retrieved February 14, 2025, from https://dictionary.cambridge.org/us/dictionary/english/integrity
12. Merriam-Webster. (n.d.). *Integrity*. Retrieved February 14, 2025, from https://www.merriam-webster.com/dictionary/integrity
13. Robinson, B. (2023, November 5). 70% of workers lie in resumes, new study shows. *Forbes*. https://www.forbes.com/sites/bryanrobinson/2023/11/05/70-of-workers-lie-on-resumes-new-study-shows/
14. Schiff, M., & Martin, E. (2022, December 20). *Over 55% of people admit to lying on their resume at least once—here are the 8 most common lies*. CNBC. https://www.cnbc.com/2022/12/20/most-common-resume-lies.html
15. Liu, J. (2020, February 20). *78% of job seekers lie during the hiring process—here's what happened to 4 of them*. CNBC. https://www.cnbc.com/2020/02/19/how-many-job-seekers-lie-on-their-job-application.html
16. Mclean, N. (2024, July 10). *A Black man got a job interview after he changed the name on his resume. Now, he's suing for discrimination*. CNN. https://www.cnn.com/2024/07/10/us/dwight-jackson-employment-discrimination-lawsuit-reaj/index.html
17. Schiff, M., & Martin, E. (2022, December 20). *Over 55% of people admit to lying on their resume at least once—here are the 8 most common lies*. CNBC. https://www.cnbc.com/2022/12/20/most-common-resume-lies.html
18. Liu, J. (2020, February 20). *78% of job seekers lie during the hiring process—here's what happened to 4 of them*. CNBC. https://www.cnbc.com/2020/02/19/how-many-job-seekers-lie-on-their-job-application.html
19. Ibid.
20. US Attorney's Office, Eastern District of New York. (2023, October 10). *Congressman George Santos charged with conspiracy, wire fraud, false statements, falsification of records, aggravated identity theft, and credit card fraud.* US Department of Justice. https://www.justice.gov/usao-edny/pr/congressman-george-santos-charged-conspiracy-wire-fraud-false-statements-0
21. Vigdor, N. (2022, December 28). George Santos is in a class of his own. But other politicians have embellished their resumes, too. *New York Times*. https://www.nytimes.com/2022/12/28/us/politics/george-santos-resume-lies-politicians.html
22. Lewin, T. (2007, April 27). Dean at M.I.T. resigns, ending a 28-year lie. *New York Times*. https://www.nytimes.com/2007/04/27/us/27mit.html
23. Hoover, E. (2009, December 6). Truth and admissions: Former MIT dean seeks to reclaim her name. *Chronicle of Higher Education*. http://chronicle.com/article/TruthAdmissions-Former/49366/
24. Martin, R. (2015, March 8). *Learning the hard truth about lying*. NPR. http://www.npr.org/2015/03/08/391610430/learning-the-hard-truth-about-lying
25. Chappell, B., & Kennedy, M. (2019, March 12). *U.S. charges dozens of parents, coaches, in massive college admissions scandal*. NPR. https://www.npr.org/2019/03/12/702539140/u-s-accuses-actresses-others-of-fraud-in-wide-college-admissions-scandal

26. Medina, J., Benner, K., & Taylor, K. (2019, March 12). Actresses, business leaders and other wealthy parents charged in U.S. college entry fraud. *New York Times*. https://www.nytimes.com/2019/03/12/us/college-admissions-cheating-scandal.html
27. Martin, R. (2015, March 8). *Learning the hard truth about lying*. NPR. http://www.npr.org/2015/03/08/391610430/learning-the-hard-truth-about-lying
28. Middle Georgia State University. (n.d.). *Attribution.* Retrieved February 14, 2025, from https://www.mga.edu/plagiarism/terminology.php
29. SUNY Empire State University. (n.d.). *What is a citation? Research skills tutorial.* Retrieved February 14, 2025, from https://subjectguides.sunyempire.edu/researchskillstutorial/citationparts
30. North Carolina State University, Office of Student Conduct. (n.d.). *Academic integrity: Overview.* Retrieved February 14, 2025, from https://studentconduct.dasa.ncsu.edu/academic-integrity-overview/
31. University of Newcastle Australia. (n.d.). *Academic integrity and ethical academic conduct policy.* Retrieved February 14, 2025, from https://policies.newcastle.edu.au/document/view-current.php?id=35
32. Robbins, R. D. (2011, February 1). Survey assesses student academic integrity. *Harvard Crimson*. https://www.thecrimson.com/article/2011/2/1/academic-cheating-students-integrity/
33. Ibid.
34. Pérez-Peña, R., & Bidgood, J. (2012, August 30). Harvard says 125 students may have cheated on exam. *New York Times*. https://www.nytimes.com/2012/08/31/education/harvard-says-125-students-may-have-cheated-on-exam.html
35. Crimson News Staff. (2012, December 31). Top 12 news stories of 2012. *Harvard Crimson.* https://www.thecrimson.com/article/2012/12/31/top-news-stories-2012/
36. Harvard College's honor code. (2014, May 7). *Harvard Magazine.* https://www.harvardmagazine.com/2014/05/harvard-college-adopts-honor-code
37. Robbins, R. D. (2011, February 1). Survey assesses student academic integrity. *Harvard Crimson*. https://www.thecrimson.com/article/2011/2/1/academic-cheating-students-integrity/
38. Merriam-Webster. (n.d.). *Plagiarize.* Retrieved February 14, 2025, from https://www.merriam-webster.com/dictionary/plagiarize
39. Ibid.
40. McWhorter, J. (2024, January 23). We need a new word for "plagiarism." *New York Times*. https://www.nytimes.com/2024/01/23/opinion/plagiarism-claudine-gay-ackman-oxman.html
41. Blum, S. (2009, February 20). Academic integrity and student plagiarism: A question of education, not ethics. *Chronicle of Higher Education*. https://www.chronicle.com/article/academic-integrity-and-student-plagiarism-a-question-of-education-not-ethics/
42. Shepard, A. (2010, May 24). *When is it plagiarism?* NPR. https://www.npr.org/sections/publiceditor/2010/05/24/127017266/when-is-it-plagiarism
43. Gabriel, T. (2010, August 1). Plagiarism lines blur for students in digital age. *New York Times*. https://www.nytimes.com/2010/08/02/education/02cheat.html
44. Fish, S. (2010, August 9). Plagiarism is not a big moral deal. *New York Times*. https://archive.nytimes.com/opinionator.blogs.nytimes.com/2010/08/09/plagiarism-is-not-a-big-moral-deal/
45. Gavron, J. (2018, February 24). The highest form of flattery? In praise of plagiarism. *Guardian*. https://www.theguardian.com/books/2018/feb/24/straightjacket-originality-homage-plagiarism
46. Stern, M. (2019, March 11). Malcolm Gladwell: Plagiarism is just "bad manners." *Daily Beast*. https://www.thedailybeast.com/malcolm-gladwell-plagiarism-is-bad-manners

47. Miller, T. (2022, February 10). Plagiarism is a big deal: This new case at the University of Kansas wasn't the first. *Kansas City Star*. https://www.kansascity.com/opinion/readers-opinion/guest-commentary/article258215363.html
48. Warzel, C. (2024, January 11). Plagiarism is the next "fake news." *Atlantic*. https://www.theatlantic.com/technology/archive/2024/01/claudine-gay-plagiarism-ai-culture-war/677092/
49. John Waltman, quoted in Lathrop, A., & Foss, K. (2000). *Student cheating and plagiarism in the internet era: A wake-up call*. Libraries Unlimited, p. 163.
50. Ibid.
51. PlagiarismChecker.com. (2005). *Plagiarism and copyright infringement: Is copying illegal?* https://subr.libguides.com/ld.php?content_id=1112525
52. Kulish, N., & Cottrell, C. (2013, February 9). German fascination with degrees claims latest victim: Education minister. *New York Times*. http://www.nytimes.com/2013/02/10/world/europe/german-education-chief-quits-in-scandal-reflecting-fascination-with-titles.html
53. Martin, J. (2014, July 23). How Senator John Walsh plagiarized a final paper. *New York Times*. https://www.nytimes.com/interactive/2014/07/23/us/politics/john-walsh-final-paper-plagiarism.html
54. Legal Information Institute. (n.d.). *Amdt14.S1.7.1.1, overview of personal jurisdiction and due process*. Cornell Law School. Retrieved February 14, 2025, from https://www.law.cornell.edu/constitution-conan/amendment-14/overview-of-personal-jurisdiction-and-due-process
55. University of Pittsurgh, School of Computing and Information. (2018). *Academic integrity policy*. https://www.sci.pitt.edu/student-resources/policies/academic-integrity-policy
56. Bartlett, T. (2013, January 31). UC-Berkeley exonerates anthropologist who was accused of stealing ideas. *Chronicle of Higher Education*. https://www.chronicle.com/article/uc-berkeley-exonerates-anthropologist-who-was-accused-of-stealing-ideas/
57. George, T., & Caulfield, J. (2022, February 7). *What is self-plagiarism? Definition & how to avoid it*. Scribbr. https://www.scribbr.com/plagiarism/self-plagiarism/
58. Ibid.
59. When it is and isn't OK to recycle text in scientific papers. (2024). *Nature Human Behaviour, 8*(3), 401–401. https://doi.org/10.1038/s41562-024-01862-3
60. Ibid.
61. Text Recycling Research Project. (n.d.). *Home*. Retrieved February 14, 2025, from https://textrecycling.org
62. Sullivan, B. (2021). *University of South Carolina president resigns after plagiarizing part of speech*. NPR. https://www.npr.org/2021/05/13/996523535/university-of-south-carolina-president-resigns-after-plagiarizing-commencement-s
63. Associated Press. (2021, October 4). *Bob Caslen says he regrets taking job as University of South Carolina president*. WFAE 90.7. https://www.wfae.org/south-carolina/2021-10-04/bob-caslen-says-he-regrets-taking-job-as-university-of-south-carolina-president
64. Quinn, A. (2018, March 12). Ridgefield superintendent forced out after plagiarism claims. *News-Times*. https://www.newstimes.com/news/article/Ridgefield-school-board-votes-for-12748282.php#photo-15211864
65. Ibid.
66. Ibid.
67. Morrone, M. (2024, February 22). *New report: 60% of OpenAI model's responses contain plagiarism*. Axios. https://www.axios.com/2024/02/22/copyleaks-openai-chatgpt-plagiarism

68. Wade, N. (1996, November 26). Doctors question use of Nazi's medical atlas. *New York Times*. https://www.nytimes.com/1996/11/26/science/doctors-question-use-of-nazi-s-medical-atlas.html
69. Ibid.
70. Department of Health & Human Services, Office of Research Integrity. (n.d.). *Definition of research misconduct*. Retrieved February 18, 2025, from https://ori.hhs.gov/definition-research-misconduct
71. Retraction Watch. (n.d.). *Home page*. Retrieved February 18, 2025, from https://retractionwatch.com
72. Retraction Watch. (n.d.). *The Center for Scientific Integrity*. Retrieved February 18, 2025, from https://retractionwatch.com/the-center-for-scientific-integrity/
73. Ibid.
74. Retraction Watch. (n.d.). *Springer Nature journal has retracted over 200 papers since September*. https://retractionwatch.com/2024/10/15/springer-nature-journal-has-retracted-over-200-papers-since-september/
75. Timmer, J. (2025, January 7). *Science paper piracy site Sci-Hub shares lots of retracted papers*. Ars Technica. https://arstechnica.com/science/2025/01/science-paper-piracy-site-sci-hub-shares-lots-of-retracted-papers/
76. Cambridge Dictionary. (n.d.). *Transparency*. Retrieved February 18, 2025, from https://dictionary.cambridge.org/us/dictionary/english/transparency#
77. Arthur W. Page Center/Public Relations Ethics. (n.d.). *Transparency*. Retrieved February 18, 2025, from https://archive.pagecentertraining.psu.edu/public-relations-ethics/transparency/transparency/transparency/
78. *The call*. (n.d.). RenAIssance Foundation. Retrieved February 18, 2025, from https://www.romecall.org/the-call/
79. Ibid.
80. Microsoft. (n.d.). *Empowering responsible AI practices*. Retrieved February 18, 2025, from https://www.microsoft.com/en-us/ai/responsible-ai
81. Electronic Frontier Foundation (EFF). (n.d.). *History of FOIA*. Retrieved February 18, 2025, from https://www.eff.org/issues/transparency/history-of-foia
82. Ibid.
83. US Department of Justice. (n.d.). *What is FOIA?* Retrieved February 18, 2025, from https://www.foia.gov/about.html
84. US Department of Justice. (n.d.). *What are FOIA Exemptions?* Retrieved February 18, 2025, from https://www.foia.gov/faq.html#exemptions
85. Sunshine Week. (n.d.). *About Sunshine Week*. Retrieved February 18, 2025, from https://sunshineweek.org/about
86. Transparency International. (n.d.). *About*. Retrieved February 18, 2025, from https://www.transparency.org/en/about
87. Ibid.
88. Legal Information Institute. (n.d.). *Corruption*. Cornell Law School. Retrieved February 18, 2025, from https://www.law.cornell.edu/wex/corruption
89. Transparency International. (n.d.). *What we have done so far*. Retrieved February 18, 2025, from https://www.transparency.org/en
90. Ibid.
91. Transparency International. (2023). *Corruption perceptions index*. https://www.transparency.org/en/cpi/2023

92. United Nations, Office on Drugs and Crime, Corruption and Economic Crime Branch. (n.d.). *Learn about UNCAC*. Retrieved February 18, 2025, from https://www.unodc.org/corruption/en/uncac/learn-about-uncac.html
93. Transparency International. (n.d.). *Our impact*. Retrieved February 18, 2025, from https://www.transparency.org/en/our-impact
94. Transparency International. (n.d.). *Conflict of interests*. Retrieved February 18, 2025, from https://www.transparency.org/en/corruptionary/conflict-of-interests
95. American Library Association. (n.d.). *Conflicts of interest Q&A*. https://www.ala.org/tools/ethics/conflictsofinterestqa
96. Ibid.

Part 2

Core Values and Opposing Forces

Chapter 4

Access and Divides

Introduction: Doorways and Documenting

History and cultural lore tell us that ancient humans shared information orally and through the form of stories. The ability to record the written word did not yet exist. Access to knowledge and information in oral cultures was and is dependent on a "sender" and a "receiver." Knowledge transmission in oral cultures often occurs through storytelling, in which information is orally communicated to another person in the form of a story. With no tangible information record or data backup capability, the continuation of the information chain is dependent on the humans who possess the information and are able to convey it to another person. Remove a sender or receiver from that process and the knowledge may disappear or be materially altered if too few people remain to carry the stories on. The conveyance of knowledge and information is then dependent on continually cultivating receivers who will become eventual senders of knowledge to new receivers.

Anyone familiar with the concept of the "telephone game," in which one person passes on a piece of information to someone, who then tells it to another person, and so on down the line, knows that the last person to receive the assembly-line-conveyed message often possesses information that is unrecognizable to the original teller. As oral information is conveyed, its meaning can change in the course of successive communication. Hence, oral knowledge, by nature of the impermanent form in which it exists, may be more transitory and ephemeral than recorded knowledge. (It is important to note as a qualifying point that recorded information in some formats like books has the potential to be edited, changed, or written over.) However, numerous examples exist of exclusively oral knowledge societies, such as Native Americans and Aboriginal Australians, that have conveyed information to one another orally for generations. In chapter 9, we delve more into these types of knowledge areas that are referred to as traditional knowledge (TK), indigenous knowledge (IK), and traditional cultural expressions (TCE).

Recording and Preserving Information

Since ancient times, human-dervived technologies have been central to the ability of peoples to record information in tangible mediums of expression. Surviving histories and artifacts from around the world evince the technological innovations that societies have devised and made manifest to document, provide access to knowledge, and preserve information. Over thousands of years, human beings have demonstrated uncanny ingenuity in conceptualizing, shaping, and deploying native raw materials to record and harness knowledge. A few examples include

- stone tablets[1] and papyrus scrolls[2] of Middle Eastern and Mediterranean civilizations
- woven *Quipu* records of the Incans in South America[3]
- paper—invented in China and eventually adopted in much of the world to record bonafide history as well as works of fiction, like Japan's eleventh-century *The Tale of Genji* (*Genji Monogatari*), considered to be the world's first full novel[4]
- fabrics like France's *Bayeux Tapestry* that recount actual historical events[5]
- illuminated manuscripts of Europe, like Ireland's *Book of Kells*[6]

Prehistoric Knowledge and Information

The oldest known vestiges of human-generated knowledge and information survive in mediums of rock scattered around the globe. Some of the best-known examples are the cave paintings of Lascaux (southwestern France)[7] and Altamira (northern Spain),[8] estimated to have been created by prehistoric humans 15,000–20,000 years ago. Some scholars have suggested that cave paintings like these may have served as a kind of "tribal encyclopedia," transmitting vital knowledge to current and future tribe members about animal species and migration routes. Their exact purpose, though, remains a mystery. Aboriginal Australian rock carvings, some depicting now extinct megafauna, are thought to be even older than the cave paintings of southern Europe.[9] Other civilizations for whom we do not have extant information artifacts may too have recorded tribal and societal knowledge through methods and mediums that proved less durable and lasting. Climatic factors, like heat and humidity, may have doomed such information objects to crumbling, withering, and returning to dust, causing them to remain unknown and inaccessible to us today.

Houses for Knowledge

The transition of societies from oral to written cultures catalyzed the need for tools and technologies to affix, communicate, and preserve knowledge for present and future conveyances of knowledge. One of the chief impetuses for this was the realization that knowledge came to be seen as a resource of great value that can confer opportunity and advantage. Recorded knowledge might be of value in military campaigns, negotiations between rival lands, or more mundane purposes like tax collection. Once collected and recorded, knowledge needed physical places to be housed. The housing and safeguarding of information was necessary for several reasons:

- Controlling access to knowledge and making it only available to authorized persons
- Protecting knowledge from human threats and environmental conditions that could result in its loss, degradation, or destruction
- Ensuring its continued survival

From such needs, humans first conceived the notion of the *library*. At its core, a library is a *collection of organized recorded knowledge with a system for access*. Access to knowledge and information *is* the most foundational of the core principles for information professionals and users. Without access, there is no "written" or "tangible" freedom of expression. Without tangible information objects, privacy is only relevant with regard to oral transmission of knowledge.

Access in Many Forms

Information professionals recognize a number of different connotations of "access." Each of these connotations raises its own ethical issues, some overlapping and some distinct. One conceptualization is the ability to *attain* access to knowledge. This is the kind of access identified earlier in this chapter: access to recorded knowledge and information in some kind of "fixed medium of expression" (which is a required element of most of the world's copyright statutes). The inspirational words of Nobel Prize–winning author Toni Morrison, displayed for all visitors to see on a wall at an entrance to New York Public Library, encapsulate the importance of access: "Access to knowledge is the superb, the supreme act of truly great civilizations. The New York Public Library is, in this regard, both symbol and act of what the best civilization has to offer."[10] Though focused on print books, access to knowledge and information is also the broader theme of Ray Bradbury's *Fahrenheit 451* (1953). In that book's dystopian setting, books are burned by firefighters and access to knowledge is strictly controlled through government surveillance and censorship.

Access can also denote access to the *physical space* where the information is housed, preserved, and made accessible. "Brick and mortar" structures like the New York Public Library, extolled in the Morrison quote, are examples of these physical spaces. *Kreimer v. Bureau of Police for Town of Morristown* (1991) is a landmark "library law" case that at its heart is a dispute over access.[11] *Kreimer*, in short, addressed whether denying access to a public library by an unhoused person was justifiable under certain sets of facts. In this example, access related to what circumstances might be used to determine whether a person would or would not be permitted to access and inhabit, even temporarily, a physical setting.

Another aspect of access stems from whether a person has the *physical capability* to enter a physical or virtual space. A commanding driver of the legislative intent behind the groundbreaking Americans With Disabilities Act (ADA), signed into law in 1990 by President George H. W. Bush, was to promote and in some cases mandate access for persons with physical challenges.[12] Another important aim of the ADA was to provide legal penalties for discrimination against persons with physical and mental disabilities. In 2015, the 25th anniversary of the ADA's signing and implementation, President Barack Obama strengthened the ADA with additional protections and mandates, particularly with regard to mental health. In the digital age, access has also come to include access to virtual spaces. In the past decade, for example, various organizations for the visually impaired have brought lawsuits against for-profit companies and academic institutions for alleged deficiencies regarding access to web portals and e-textbooks.

Access can also denote *equal opportunities*: to be present, to be represented within an organization or a space, and to share one's voice and "have a place at the table." In the wake of the US antisegregation movement of the 1950s and 1960s as well as other civil rights and social justice movements, more and more organizations today identify diversity and inclusion as integral components of their missions and visions, core values, organizational cultures, and strategic planning. While "diversity" denotes recognition of the varying visible and invisible characteristics that make up a person, "inclusion" is predicated on extending the recognition of differentness a step further. Inclusion's aim is to "include" more voices in the conversation, to provide a "place at the table" for persons who, for a variety of reasons such as historic discrimination, have faced and continue to face barriers to access and full equality. As Candi Castleberry-Singleton, vice president for global diversity, equity, and inclusion at Amazon, has explained, in workplace culture, inclusion means "ensuring every employee is treated with dignity and respect and has the opportunity for professional growth and development."[13]

Over the past decade, technology industries have been roiled by external and internal data analytics reports revealing disproportionately low numbers of women, African Americans, and Hispanics within these organizations, relative to their demographic numbers within US society. The numbers of these persons in management and top leadership positions is even lower. Opinions vary as to the causes of the low numbers, but some of the reasons stem from historic discrimination to overt or unconscious bias. "Bro culture" is also cited as a factor. This term represents the notion of an insular, predominantly White male majority within an organization. Some bro cultures may affirmatively and actively create and maintain barriers to access and opportunities of the kind that members of the bro cadre enjoy. Other bro cultures may take a passive role vis-à-vis barriers to access by non-bro members of the organization, while still reaping the rewards of their privileged status. Technology companies and data and information professionals continue to grapple with how to provide more *access opportunities* for existing and prospective employees, customers, and users.

With the advent and adoption of digital technologies over the past few decades, access also encompasses the ability to obtain knowledge and information *digitally and virtually*, through such means as the internet, mobile applications, and social media. Access encompasses not just the passive acquisition of information through digital means but also the *ability to proactively participate and contribute* to analog and digital spaces and communities. More recently, AI technologies and tools are serving as new conduits for the creation of content and participation possibilities. For example, in a US Copyright Office (USCO) virtual public forum seeking comments about AI I attended in the early 2020s, two budding female screenwriters revealed ways that AI tools were enabling them to "level the playing field." They explained that entry barriers to entertainment-creation opportunties for novice screenwriters were formidable but that the use of AI tools to brainstorm and develop ideas enabled them to pitch their scripts and compete with more established screenwriters, who are often aided with the upper hand that insiders and the well-connected readily possess and wield. In recent years, there has been more awareness of the inequities of generational wealth, which disadvantaged significant numbers of historically marginalized persons from being able to share in wealth-generating opportunities like home buying. Hence, AI tools and devices may facilitate more opportunities for historically underrepresented people to get a "foot in the door" and leverage their unique qualities in the creation of content.

Access can also be a *safety* issue with life-influencing ethics dimensions. On September 4, 2024, another US school shooting resulted in fatalities and injured students and teachers. This one transpired at Apalachee High School in Winder, Georgia, when a 14-year-old student killed two teachers and two students and seriously injured nine people. *Education Week* maintains a School Shooting Tracker that provides data relating to school shooting incidents. As of December 19, 2024, there were, in that year,

- thirty-nine school shootings with injuries or deaths
- seventy-seven people killed or injured
- eighteen people killed: eight students or other students; ten school employees or other adults
- fifty-nine people injured[14]

Later in this book's ethics of social media chapter, we look at the issues of how smartphones are contributing to cyberbullying and mental health issues for children and adults. As a result of research-based concerns about the harmful effects of smartphones on the well-being of children, some schools have taken the administrative step of banning or restricting access to smartphones in schools. However, the issue of whether to ban or permit children to have access to their smartphones during school attendance hours is not as clear-cut as it may appear on its face. The September 7, 2024, NBC News article "Lifeline or Distraction? Georgia Shooting Reignites Debate Over Cellphones in Schools" captures the major pros and cons of this debate, as well as some concerns that have not garnered as much consideration and press—namely, that phone use can distract students from focusing on directives as a shooting is taking place. The article quotes the president of a consulting firm that focuses on school security and emergency preparedness training: "If you have 20 kids in a classroom and they're texting, calling parents, livestreaming—they're not paying full attention to the directions of adults and not being fully situationally aware of things they may need to quickly do to save their lives.... You have seconds to follow directions and move locations."[15] Moreover, the article identifies more novel concerns, such as the "unwanted attention" that the sounds from smartphones—pinging, buzzing, ringing—can create as students and school employees are endeavoring to hide or escape. Many people making smartphone calls during an incident may also overwhelm 911 networks and hinder the ability of first responders to communicate and render assistance. Yet the NBC News article highlights the abject fear that some students experienced who were unable to access their phones and call their parents as the shooting was occurring: "I was shaking, scared,... and I couldn't communicate with my mom for a half an hour.... I didn't know what was going to happen because you could hear the gunfires right down the hall. I don't know if it was, like, the last chance I had to talk with her."[16]

Access to smartphones during a crisis can also be an *access to information* issue. For example, smartphone access can enable people to learn important facts and instructions within or in proximity to an active shooter situation. Following the April 2007 massacre at Virginia Polytechnic Institute and State University (i.e., Virginia Tech) in Blacksburg, perpetrated by an undergraduate student attending the school, the response time of law enforcement during the attack was heavily scrutinized and debated. One of the contentious issues that eventually resulted in lawsuits filed against Virginia Tech by the families of student victims was the allegation that lives could have been saved if more timely warnings had been issued regarding the gunman, who was shooting people at various locations on the campus. The Jeanne Clery Act is a federal law with relevance to this issue: "The Clery Act requires colleges and universities to report campus crime data, support victims of violence, and publicly outline the policies and procedures they have put into place to improve campus safety."[17]

[WARNING: The following brief description of the genesis of the Jeanne Clery Act includes information that may be upsetting to victims of violence and other kinds of violative actions.] The Jeanne Clery Act was born from tragedy: "Jeanne Clery was 19 years old when she was raped and murdered in her college dormitory. Her parents, Connie and Howard Clery, could not have known the danger she was in; standards for campus crime reporting simply did not exist in 1986."[18] Through intensive lobbying efforts, Jeanne Clery's parents worked with allies to secure the passage and signing of the Jeanne Clery Act by President George H. W. Bush. They also established the Clery Center as an ongoing 501(c)(3) non-profit organization: "Our mission: Working together with college and university communities to create safer campuses. When institutions of higher education seek guidance on campus safety regulation, they turn to Clery Center."[19]

The Clery Act has a specific section on Timely Warnings and Emergency Notifications:

> When a crime covered by the Clery Act occurs, campus officials are required to evaluate if there is a serious or ongoing threat to the campus community to determine if a timely warning needs to be issued to all staff and students.
>
> In the event of an immediate, significant danger to the health or safety campus community (e.g. weather, disease outbreak), campus officials may issue an emergency notification. This notification can include the entire campus, or be limited to a specific area deemed to be at risk.[20]

Educators, policy makers, administrators, parents, students, and other stakeholders have many interests to weigh in considering the issue of smartphone access in schools. As a coda to this discussion, stemming from increases in active shooter incidents in the past decade, I have instructed graduate students in my synchronous courses that they may leave their cellphones on vibrate during class sessions in order to ensure that information about potentially exigent circumstances can be accessed and acted on quickly. Each instructor must weigh various factors and decide their own stance, in relation, also, to any applicable policies of their parent organization.

Analog and Digital Access

The arrival of the World Wide Web in the 1990s made "access" a two-pronged challenge for libraries and information professionals: namely, how to provide access to both analog and digital content. Not only was this a challenge in terms of how to enable people to physically and virtually "get their hands" on print and digital formats, but it was also terrifically daunting in terms of the costs. Ethical and policy considerations were identified regarding how, when, and where to provide access. Moreover, knowledge institutions were confronted with deeply challenging financial decisions about whether they could even afford to provide access to, for example, the internet, as computers were not afforadable for many libraries. This included even large and venerable public library systems such as Andrew Carnegie's eponymous Carnegie Library of Pittsburgh, which took advantage of Gates (of Microsoft fame) Grants to acquire many of the computers it made available to its patrons/customers. Some of the difficult decisions that libraries faced included dilemmas like

- Whether to retrospectivelty convert already existing analog content to digital formats
- Whether to discontinue analog content or continue collecting both analog and digital formats where both options were possible
- How to manage so-called born-digital content that had never existed in analog format

These kinds of thorny considerations and decisions continue to be faced by information professionals today. Such issues also raise a host of ethical implications:

- What to collect
- How much to collect
- What to retain (accession)
- What to let go of (deaccession)
- Which stakeholders' needs (or the loudest, most powerful, and persuasive voices) will prevail in making these kinds of decisions

In such scenarios, institutions must do utilitarian analyses of the pros and cons of each option, weighing the impacts on stakeholders and determining whose interests are paramount. Ethics issues, hence, will inevitably arise and are inherent.

Access as Ethical Principle and Core Value

Access has long been a core ethical principle of the American Library Association. ALA's Code of Ethics addresses "access" explicitly via statement #1: "We provide the highest level of service to all library users through appropriate and usefully organized resources; equitable service policies; equitable access; and accurate, unbiased, and courteous responses to all requests."[21] Throughout the last century and into the digital age, access and its democratizing influences have been a key focus of the service that librarians provide. As Vickery Bowles, city librarian at Toronto Public Library, explains, "Access to information and pathways to learning were the great equalisers of the 20th century.... In the 21st century, we're increasingly dependent on access to online services, and understanding of and comfort with that technology."[22] Karen Muller, former librarian and knowledge management specialist for ALA, also notes librarianship's long role in addressing "the information needs of everyone" and the profession's commitment to providing "tools for librarians seeking to eliminate barriers to information access."[23] The

International Federation of Library Associations and Institutions (IFLA) also recognizes access as a core value. Indeed, access is identified explicitly in the first three of IFLA's four core values statements:

1 the endorsement of the principles of freedom of access to information, ideas and works of imagination and freedom of expression embodied in Article 19 of the Universal Declaration of Human Rights
2 the belief that people, communities and organizations need universal and equitable access to information, ideas and works of imagination for their social, educational, cultural, democratic and economic well-being
3 the conviction that delivery of high quality library and information services helps guarantee that access[24]

At the global level, the United Nations has promoted internet access and collectively made statements, including a nonbinding resolution passed in 2016, condemning the actions of nations that have restricted or banned internet access. As stated in an article appearing in the *Verge*, "The resolution builds on the UN's previous statements on digital rights, reaffirming the organization's stance that 'the same rights people have offline must also be protected online,' in particular the freedom of expression covered under article 19 of the Universal Declaration of Human Rights."[25]

In July 2021, the nonprofit freedom-of-expression advocacy organization Article 19 reported on

> the adoption of a resolution on "the promotion, protection and enjoyment of human rights on the Internet" at the UN Human Rights Council. The resolution—led by a core group of Brazil, Nigeria, Sweden, Tunisia and the United States, and co-sponsored by 70 countries from all regions—was adopted by a vote with strong support at the Council on 13 July 2021.
>
> This iteration of the resolution has the overarching thematic theme of bridging digital divides.[26]

The Digital Divide and Broadband Access

With the growth of internet usage in the 1990s, some observers noted the beginnings of a technology gap. This gap—between "haves" able to access the internet, computers, and technology and "have-nots" who lacked such access—came to be called the "digital divide." Following concerted pushes by civic leaders, schools, libraries, technology companies like Microsoft, and philanthropic organizations, such as the Bill and Melinda Gates Foundation, to bridge the digital divide throughout the first decade of the twenty-first century, a sense emerged for some that the digital divide was over. Or that at least "enough Americans" were now online—whether via their own computers and mobile devices or through more computer access in public libraries and discounted or free WiFi programs in some cities. A 2011 now nonexistent web page sums up well this cooling of interest in the digital divide: "The 'digital divide' site has been retired from pbs.org."

Pew Research Center has been tracking internet usage in the United States since 2000. Almost 50 percent of adults were online at that time. The percentage has climbed dramatically to approximately 90 percent now.[27] More Americans than ever before have access to the internet, closing the gap on that digital divide. As a June 2018 article in CNET describes, "More than 13 million users—including low-income families, seniors, African Americans and Hispanics—have signed on to the internet over the past two years," per the National Telecommunications and Information Administration division of the US Department of Commerce.[28] "The digital divide is showing signs of closing," it asserts.

It's important to note, though, that there isn't just one kind of digital divide. The *speed* with which Americans are able to access the internet has created a new and seemingly intractable digital divide: namely, one of broadband access. As Pew Research reports: "The proportion of American adults with high-speed broadband service at home increased rapidly between 2000 and 2010. In recent years, however, broadband adoption growth has been much more sporadic. Today, roughly two-thirds of American adults have broadband internet service at home."[29]

Rural Americans are disproportionately affected by the broadband digital divide. In the same vein as urban neighborhood "food deserts" that lack access to grocery stores and fresh fruits and vegetables, large swaths of the country lacking broadband access are effectively internet deserts. At the same time though, Americans have become increasingly dependent on internet access for a broad array of activities and services:

> As broadband internet becomes more and more important in the U.S.—the way Americans do everything from apply for jobs to chatting with their relatives to watching TV—one gap has become more glaring: the difference between those who have broadband and those who don't. An estimated 24 million people, about 8 percent of Americans, still have no home access to high-speed internet service, defined by the Federal Communications Commission as a download speed of 25 megabits per second. (That's what the FCC [Federal Communications Commission] says allows telecommuting or streaming high-definition video.) The overwhelming majority of those people live in rural areas.[30]

The data is especially stark for Native American communities. As *Politico* observed in a special 2018 report on the digital and technology divides, American Indian communities have the lowest rates of high-speed internet access in the entire country: "About 35 percent of Americans living in tribal lands lack broadband access, according to the most recent report by the FCC. In Idaho, the FCC estimates that 83 percent of the tribal population lacks broadband, making the Nez Perce tribe among the least-connected groups in the country."[31]

The paucity of broadband access exists on a continuum of inadequate to nonexistent service, as *Politico* identifies in a 2018 article that contains two maps illustrating the country's digital divide, "Some areas lack access to an internet connection at all. Others have service [too] creaky to let a student take an online course, or an adult working from home to join a videoconference."[32] There are profound economic and medical implications for people and business within the boundaries of these broadband deserts. Rural areas of Pennsylvania, one of the United States' most populous states, present an illustrative and disconcerting example of the inequities that digital have-nots experience on a daily basis, as covered by the *Pittsburgh Post-Gazette*: "Even as businesses in Pittsburgh compete to commercialize artificial intelligence, just a three-hour drive away people struggle with dial-up connections—if there are internet connections at all. . . .

'We're basically being cut off from the 21st century,' Huntingdon County's planning director, Mark Colussy, said."[33] The *Post-Gazette* article elaborated that "troubling correlations have been drawn between poor health and the areas of the country where broadband is lacking.... Based on 2015 data, the least-connected counties in the U.S. generally have the highest rates of chronic disease." The article quotes former FCC commissioner Mignon Clyburn: "In sum, the data show that the rural/urban broadband health gap appears to be growing."

Ameliorating these internet access disparities will require a broad-based approach bringing together for-profit telecommunication companies, government agencies like the FCC, local governments and community leaders, watchdog organizations, and nonprofit institutions, such as libraries and schools. Federal programs like Lifeline, "which helps to make broadband internet and both landline and mobile phone services available to low-income Americans," including many Native Americans who benefit significantly from the subsidies, are also being cut back under FCC Chairman Ajit Pai.[34] Broadband-deficient communities are increasingly finding that they need to assume lead roles in addressing the problem. As a 2017 special report on the digital divide in *Motherboard* magazine concludes, "Big Telecom has little interest in expanding to small towns and farmlands, so rural America is building its own solutions."[35]

Net Neutrality

Another potential barrier to access of information and technology is net neutrality. As IFLA explains, net neutrality "refers to the principle that all data or traffic on the Internet should be treated equally."[36] In June 2018, new rules for net neutrality went into effect. These rules, championed by FCC Chairman Pai, were supported by a majority of the US Congress and the Trump administration. One of the main concerns about the new rules is that they enabled telecommunications providers to charge different rates for access to the internet. Another concern is that telecommunications companies can engage in so-called throttling, controlling the speed at which consumers are able to access the internet and charging different rates for internet speeds, based on who is the internet user. IFLA noted that this can allow companies to involuntarily put some people "in the slow lane" and to "prioritize" some information and voices over that of others, raising free speech and censorship concerns.[37] Additionally, IFLA warned that "this poses a challenge not just to the broader values of libraries—notably free and equal access to information and free expression—as well as a potentially direct threat to the ability of libraries to interact with users on the best connection possible."[38]

The Biden administration, which took control of the executive branch in late January 2021, endeavored to return net neutrality to the Obama administration's "Open Internet" rules of 2015. However, in late summer 2024, that effort suffered a judicial setback: "The Sixth Circuit U.S. Court of Appeals [in Cincinnati, Ohio], which had temporarily delayed the [net neutrality] rules, said on Thursday it would temporarily block net neutrality rules and scheduled oral arguments for late October or early November on the issue, dealing a serious blow to President Joe Biden's effort to reinstate the rules."[39] On January 2, 2025, the Sixth Circuit invalidated the FCC's Open

Internet Order. In the words of Electronic Frontier Foundation (EFF), "the court removed net neutrality protections for all Americans and took away the FCC's ability to meaningfully regulate internet service providers."[40]

Initiatives for Access to Information and Technology

With the rise of the internet over the past two-and-a-half decades and the increasing penetration of digital technology into the lives of persons around the world, technology leaders and their organizations have spearheaded efforts to provide technology to those lacking access. As discussed earlier, this was evinced through efforts by philanthropic institutions like the Bill and Melinda Gates Foundation, whose grants enabled public libraries to acquire desktop computers to combat digital divide 1.0. Two initiatives, One Laptop Per Child (OLPC) and Facebook's Free Basics, represented different attempts to "bring technology" to those deemed to need and want it. Both initiatives present the promises and pitfalls of access to information and technology and raise thought-provoking ethical considerations.

One Laptop Per Child

In 2005, MIT Media Lab founder Nicholas Negroponte launched OLPC. Its mission? "To empower the world's poorest children through education."[41] The tool for achieving that mission, as envisioned by Negroponte, was to create a durable, environmentally forward $100 children's laptop. In the words of OLPC,

> We aim to provide each child with a rugged, low-cost, low-power, connected laptop. To this end, we have designed hardware, content and software for collaborative, joyful, and self-empowered learning. With access to this type of tool, children are engaged in their own education, and learn, share, and create together. They become connected to each other, to the world and to a brighter future.[42]

As Negroponte saw it, OLPC represented an opportunity to collaborate with like-minded partners, using open-source software, to sell its laptops to developing countries for a fraction of what the big technology companies were charging for their electronic devices.

Over the next decade, however, Negroponte found that his vision was harder to achieve and that not everyone saw the same vision. Concerns were raised about whether the laptop could be produced for $100, whether developing countries should even spend that amount on technology over other needs, and about the quality of the device. On a broader level, critics questioned the propriety of prioritizing access to information technology by providing laptops to children who often lacked access to clean water, health care, affordable housing, and education in countries such as Brazil and Cambodia. Though still in existence at present, OLPC's mission and vision remain unrealized and controversial. As an April 2018 article observed in reviewing OLPC's history and legacy, "Thirteen years ago, OLPC told the world that every child should get a laptop. It never stopped to prove that they needed one."[43]

Facebook's Free Basics

Like Nicholas Negroponte, Facebook CEO Mark Zuckerberg has nurtured a vision for providing technology and electronic content to digital have-nots on a global scale. Where the former's vision was to hook kids up with digital devices, Zuckerberg's aim has been to bring the actual *internet hookup* to devices in developing nations through a mobile application called Free Basics. Free Basics is a "Facebook-developed mobile app that gives users access to a small selection of data-light websites and services. The websites are stripped of photos and videos and can be browsed without paying for mobile data."[44]

As with One Laptop Per Child, opinions vary on the pros and cons of Free Basics. Facebook and proponents of Free Basics view the service as a way to provide internet access and content at discounted prices to new users in developing markets. Others see Free Basics as a Facebook-curated form of "digital colonialism," promoting Western content, limiting linguistic diversity, and violating the open internet ethos of net neutrality. In 2016, Mark Zuckerberg made an all-out push for adoption of Free Basics in India. Despite Zuckerberg himself journeying to India and engaging in intensive high-profile lobbying efforts, ultimately Free Basics was roundly rejected in India. As the BBC reported in the wake of that rejection, "In centuries gone by, colonialism was about exploitation of resources. In the modern world, it's digital—moving in, setting up companies and building insurmountable user bases before any other company can."[45]

Access for Underserved Communities

Persons with physical and visual disabilities are faced with a number of digital and technology divides related to their respective disabilities. Visually impaired individuals and the organizations that advocate for their interests have especially been at the forefront in challenging barriers to access. As digitally formatted books have proliferated in recent years, blind persons have successfully challenged book publishers for access to e-readers. Additionally, visually impaired persons have brought successful legal actions against various universities to adopt digital textbooks that are accessible for the blind.

Websites also present access barriers for persons with disabilities. For-profit companies are increasingly finding themselves on the receiving end of legal complaints and negative publicity for not providing accessibility to their websites and online buying portals. ADA web accessibility compliance lawsuits have included some of the nation's largest companies and top brands, such as Bank of America, Target, Safeway, Charles Schwab, Foot Locker, Toys "R" Us, Brooks Brothers, and the NBA.[46]

Web Content Accessibility Guidelines

The US Access Board is a little-known independent government agency with an important public interest mission of "advancing full access and inclusion for all."[47] The Access Board describes itself as promoting "equality for people with disabilities through leadership in accessible design and the development of accessibility guidelines and standards for the built environment, transportation, communication, medical diagnostic equipment, and information technology."[48] With regard to information technology, it sets out standards for "electronic and information technology procured by the federal government" under Section 508 of the Rehabilitation Act and guidelines for "telecommunication products and services" covered by Section 255 of the Communications Act.[49] In January 2018, the Access Board upped the ante for digital accessibility compliance by "requiring all federal agencies to have websites that are compliant with Web Content Accessibility Guidelines."[50] The Web Content Accessibility Guidelines (WCAG) were created by the World Wide Web Consortium (W3C)

> with a goal of providing a single shared standard for web content accessibility that meets the needs of individuals, organizations, and governments internationally.
>
> The WCAG documents explain how to make web content more accessible to people with disabilities. Web "content" generally refers to the information in a web page or web application, including:

- natural information such as text, images, and sounds
- code or markup that defines structure, presentation, etc.[51]

Marrakesh Treaty

Like the W3C's Web Content Accessibility Guidelines, the Marrakesh Treaty represents another effort to break down access barriers for disabled persons. Following years of multination negotiations within the United Nations' (UN) World Intellectual Property Organization (WIPO), the Marrakesh Treaty for facilitating access to published works was adopted by participating nations. Negotiated in Marrakesh, Morocco (hence, its shortened designation), the treaty is formally known as the Marrakesh Treaty to Facilitate Access to Published Works for Persons Who Are Blind, Visually Impaired, or Otherwise Print Disabled. As Jonathan Band explains in a 2013 *American Libraries* article, "How the Marrakesh Treaty Opens Vistas for Print-Disabled Readers," the treaty is intended "to achieve the objective spelled out in its title by obligating the countries (known as contracting parties) that sign it to make exceptions in their copyright laws for the creation and distribution of accessible-format copies both domestically and across borders.... [The] agreement represents a significant development in international copyright law."[52]

Following adoption by participating member states in 2013, countries needed to take efforts to ratify the treaty in their respective nations. In June 2018, the US Senate voted to ratify the treaty, and in October 2018, President Donald Trump signed the Marrakesh Treaty Implementation Act.[53] At the time of this writing in early 2025, 98 nations have ratified the Marrakesh Treaty; the number of individual nations is higher because the European Union joined as a bloc. The Eastern European nation of Georgia is the most recent, with the treaty entering into force on February 26, 2025.[54] Nations continue to take steps to operationalize by, in many cases, amending their respective copyright laws to permit the exceptions that are necessary under the treaty obligations. On August 28, 2024, the Caribbean nation of Jamaica officially entered into the

Marrakesh Treaty. A governmental minister observed that "the Treaty represents a significant advancement in promoting inclusivity and access to knowledge for one of the most vulnerable groups in our society."[55]

In a February 22, 2021, Library of Congress *Copyright Creativity at Work* blog post, "The Marrakesh Treaty in Action: Exciting Progress in Access to Published Works for the Blind and Print-Disabled Communities," Register of Copyrights and Director of the US Copyright Office Shira Perlmutter remarked that "the Marrakesh Treaty has proven to be a success in advancing global copyright policy and achieving meaningful results for those in need. Exciting progress has been made in the United States and abroad in fulfilling the Treaty's mandate."[56] Perlmutter further identifies the need and significance of the treaty, as well as the joy that was present when consensus for the agreement was reached:

> This treaty was prompted by the widespread recognition of what has been called the global "book famine"—the dearth of books published in formats accessible to the hundreds of millions of people around the world who are blind or visually impaired. I had the honor to be part of the U.S. interagency team that worked with the affected communities, stakeholders, and other countries to develop the text and negotiate the treaty. I will never forget the cheers that reverberated in the hallway when final agreement was reached at the diplomatic conference in Marrakesh.

Naming, Agency, and Self-Determination

Nominative determinism, according to *Collins Dictionary*, is "a jocular theory which suggests that people end up doing jobs that suit their surname."[57] It's also highly questionable as a scientific theory, as briefly summarized in the September 2024 *New York Times* op-ed essay "Is Your Name Your Destiny?" by author Jesse Singal:

> I've dug into the evidence for nominative determinism, or the theory that a person's name influences his choice of occupation, interests or spouse, and I think there are good reasons to be skeptical of it. But the continued interest in the idea—across centuries and, arguably, against the evidence—is in itself revealing, highlighting humans' deep-seated desire for order in a chaotic universe and the role science plays in satisfying that need.[58]

As Singal observes, human beings have a long-standing inclination to order their environments through classification and naming conventions. Further, despite the dubious hypothetical assertions of nominative determinism vis-à-vis our vocational choices, the ways that we choose and implement naming designations and categorial representations for people and things can have significant desirable and undesirable impacts. To this point, increasing numbers of internet-age information professionals—librarians, archivists, and persons working in museums and other cultural heritage institutions—have been acculturated through their respective degree programs and workplace cultures to be better attuned to the ways that subject matter classifications and terms can deleteriously portray, influence, and offend people and societies. Indeed, a growing number of library, archives, and information studies degree programs incorporate courses with learning outcomes that are focused on providing students with greater awareness of the powers and privileges held by the people in societies who organize and classify information. Such powers can render information findable, hidden, or nonexistent.

A significant history of scholarship exists that expounds on pejorative library catalog subject classifications and headings, such as examples created and used by the Library of Congress. Founded in 1800, the Library of Congress (LOC) is the world's largest library, "with millions of books, films and video, audio recordings, photographs, newspapers, maps and manuscripts in its collections. The Library is the main research arm of the U.S. Congress and the home of the U.S. Copyright Office."[59] LOC also functions as the de facto national library for the United States and offers a globally accessible catalog to its colossal collection. As part of its information organizing and retrieval processes, LOC, like other libraries, created subject matter classifications. Some of these subject matter headings referred to racial, gendered, sexual, and immigrant minority communities in ways that members as well as nonmembers of those communities found to be insulting and insensitive. It's important to note that in most cases these designations were instituted without consulting with members of these groups as well as not seeking their consent. Despite civil rights advances in the 1960s, subject headings for these groups remained mired in the past through much of the twentieth century. These classifications functioned as artifacts of entrenched systems of oppression, ignorance, bias, and inequity. Moreover, omissions of peoples and historical events also represent another dimension of bias in subject headings and information retrieval systems. Controversial subject headings were and remain a manifestation of long-standing inequities between majority and minority cultures.

As a result of hard-fought and hard-won civil rights advances over the last roughly 75 years, minority cultures (often with the collaborative work of allied individuals and organizations) have sought and gained more power to exercise greater self-determination. One such self-determinative area involves having a voice and role in determining the names and classifications that are attached to groups themselves. (Another example of this is seen in *Matal v. Tam*, the Slants trademark case, which is discussed in chapter 8 on intellectual property.) Greater degrees of agency and autonomy beget more desire and ability by disenfranchised people to dually demand and take on more active roles in defining themselves. Advocates—like long-crusading subject heading reform crusader Sanford "Sandy" Berman, head cataloger at Minnesota's Hennepin County Library (HCL), until his involuntary retirement in 1999—successfully strove and continue to work to modernize subject headings in library catalogs. These designations are seen as outdated at best and, in the extreme, offensive and injurious to historically marginalized individuals and communities. They are also, arguably, harmful to societies at large. In a 2022 piece, "Sandy Berman: The Making of a Radical Librarian," for the Minnesota Interview Project, Berman gives firsthand testimony that poignantly accentuates the significance and calling for this type of remedial work:

> As Head Cataloger at Hennepin County Library I discovered there were historical events our card catalogue denied the existence of, like the German genocides in Southwest Africa from 1904 to 1908. We created new headings: *Native American Holocaust (1492 to 1900)*, *Romany Holocaust* and *Gay Holocaust* because these also happened during the Nazi era. We created

new categories: *Affordable Housing*, *Mass Incarceration*, *Middle Passage* and *White Privilege*. There were no headings for those subjects![60]

As with so many areas in life, including law, ethics, policy, *and* the information professions, words can and *do* matter.

Access Lessons Learned from the COVID-19 Pandemic

Inarguably, no event in the twenty-first century better demonstrated the capabilities that internet access could offer than the COVID-19 pandemic. In March 2020, initiatives to mitigate the spread of the COVID-19 virus by widely shutting down physical buildings and restricting in-person attendance at venues that included K–12 schools, colleges and universities, for-profit and nonprofit workplaces, libraries and cultural heritage organizations, government agencies, and more illuminated the striking advantages for those with access to online technologies. The impacts of these changes on education were phenomenally unfamiliar and jarring: seemingly overnight, students, K–12 teachers, and higher education instructors were meeting through the technology of online platforms like Zoom. Graduations transpired online. Parent and teacher conferences took place digitally, rather than in person. Moroever, virtually everyone who could do so turned to online platforms for work and personal use. Underlying all of this hastily constructed infrastructure, though, was the expectation that users—children and adults—would be able to secure technology, successfully use technology, and maintain access to technology.

It is important to underscore, however, that not everyone was able to "shelter in place" and turn to their smartphones and other electronic devices for connectivity with the world outside of their apartments and houses. So-called essential workers and first responders continued to provide services at places like grocery stores, restaurants (those that remained open or resorted to takeout-only service), hospitals, assisted living facilities, and nursing homes. Indeed, billboards and broadcast ads bore witness to the sacrifices that health-care providers and other essential workers were courageously taking on, reminding the public that "not all heroes wear capes" and that everyone should wear personal protective equipment (PPE) like masks and gloves and "stay six feet apart" to minimize exposure and transmission. Mask mandates were instituted and generated controversy, and even lawsuits, among some people and places that resisted such efforts at stemming the virus's spread. COVID-19 vaccines would not become available in the United States until early 2021. Misinformation and disinformation were a constant problem, fueled by the ready power and ease of the internet in amplifying and disseminating errors, falsehoods, and conspiracy theories. Accuracy of and timely access to high-quality information about the virus became a flashpoint regarding daily and weekly briefings from the White House Coronavirus Task Force, created in January 2020 by the Trump administration.

Throughout the pandemic, the presence of ethics and ethical dilemmas in decisions being contemplated and made during this unprecedented public health crisis was inescapable:

- Who could work from home?
- Who was an "essential worker"?

- Who were the populations most vulnerable to the virus?
- Which businesses would or would not be granted government assistance because of adverse impacts from the pandemic?
- How do we fairly allocate PPE and address PPE shortages?
- Whether, when, and where do we require people to wear masks?

Perhaps no issue more starkly demonstrated the gravity of these kinds of dilemmas than the life-and-death decisions some medical providers and frontline caregivers faced regarding the allocation of the highly in-demand and at times hard-to-find ventilators that were needed to care for the most ill of those who were hospitalized with the virus.

Though deaths from the COVID-19 virus have dimiminished at the time of this writing, mutations of the virus continue to cause significant numbers of illness and deaths. Regrettably, access to reliable data on COVID infections in the United States was made voluntary in early 2024, making it impossible to know the full extent of infection rates beyond anecdotal observations coupled with data that can be gleaned from virus rates detected within wastewater samples of some municipalities. Per the Centers for Disease Control and Prevention (CDC): "Effective May 1, 2024, hospitals are no longer required to report COVID-19 hospital admissions, hospital capacity, or hospital occupancy data to HHS through CDC's National Healthcare Safety Network (NHSN). CDC encourages ongoing, voluntary reporting of hospitalization data."[61] In early September 2024, US federal agency officials reversed the earlier decision: "Hospitals in the United States will soon be required to report hospital admissions related to COVID-19, restoring a mandate which was lifted earlier this year, according to federal officials. The new rules are due to take effect on November 1."[62] Roughly one in three hospitals had elected to voluntarily participate in submitting COVID-related data.

> Insights and first-hand observations gained from pandemic and post-pandemic field work have buttressed arguments in support of access to information and technology connectivity. Anne-Marie Gray, executive director and CEO of USA for UNHCR, a national partner of the Office of the United Nations High Commissioner for Refugees, in a December 2020 *UN Chronicle* piece "The Case for Connectivity, the New Human Right" persuasively shared some of the lessons learned at that time: "If this pandemic has taught us anything, it is that connectivity is absolutely necessary—for communities in the United States and around the world. And it has tragically laid bare two vastly different realities for those who are connected and those who are not."[63]

AI Divides

The past few years have seen rapid spread of AI technologies in societies. AI chatbots are now ubiquitous on many smartphones and computers. As with digital divides and access to other prior emerging technologies, though, ethical concerns and thorny questions will likely arise as to who will have access to technologies like AI chatbots and who will not. What roles can information professionals and centers play in mitigating disparities in access to AI tools? Later in this book we examine AI in greater detail.

Summary

Humans have been organizing and preserving information for thousands of years, from the earliest known cave paintings and rock carvings to present-day digital cloud storage systems. Information organizing activities foster access to information. Access to information is central to the basic human rights of individuals. Healthy democracies also depend on informed citizenries that have access to high-quality information. Information organizations, such as the American Library Association (ALA), recognize access to information as an essential ethical principle and core value.

Throughout history, individuals and groups have not enjoyed equal access to information. Efforts like the Marrakesh Treaty seek to remedy historical inequities and promote access to information for groups like visually impaired persons. The proliferation of technologies in the modern age has similarly seen imbalances for those who have and do not have access to technology. Access to computers and broadband internet access continue to present digital divides for many parts of the developed and developing world. As the adoption of AI technologies like chatbots and generative AI advances in societies around the world, concerns are being raised about new divides that are arising for those who will and will not have access to these revolutionary tools and products. AI divides present ethical issues for individuals and communities. Access to information and technology also raises many thorny issues relating to safety and well-being, particularly regarding the issue of access to smartphones in K–12 and higher education settings, due to the increase in school shooting incidents and concerns about the impacts of smartphones on mental health.

Questions for Reflection, Exploration, Action, Discussion (READ)

1. Thinking about information access.

a. Discuss some of the ways that access to information and technology impacts democratic principles and ideals.
b. What are some ways that information professionals can provide greater access to information?
c. Identify some reasons that access to information can be both a boon and a bane for democracies.

2. Examining subject headings.

a. Identify some of the ethics issues stemming from Library of Congress subject headings that were given to historically disadvantaged communities.
b. What are some lessons that can be learned from prior examples of pejorative subject headings that were given to marginalized peoples?

3. Persons with ability challenges (e.g., physical, mental) have historically experienced inequities regarding access to print books. Describe some of the strongest ethics-centered arguments for breaking down barriers that ability-challenged individuals face in having equal access to books and other print materials.

4. Huge disparities continue to exist for millions of people in developing nations (i.e., Global South countries) who are unable to access technology.

a. Discuss the ethics pros and cons of Big Tech companies being involved in efforts to bring technologies to Global South communities.

b. Are there ethics-centered actions that information centers and professionals can take to address technology divides in the Global South?

5. What is the strongest ethics rationale for the recognition of access to information as a basic human right?

6. Imagine you are the principal in a US middle school who has been asked to brief your local school board and community, at an upcoming publicly accessible school board meeting, on whether smartphones should be banned or restricted at your school. Identify the pros and cons of the issues, including the ethics arguments for and against the ban or restrictions. Then, state your recommended course(s) of action and decision(s). Clearly articulate the rationale(s) for your determination.

7. Investigate and learn more about issues involving AI divides (i.e., those who do and do not have access to AI and AI tools).

a. Identify some ethical concerns regarding those who have access to, for example, AI chatbots and those who don't.

b. What are some ways to provide AI chatbot access to those who don't have access?

c. What initiatives, if any, have information professionals already developed to address AI chatbot divides, and what ideas do you have?

Notes

1. *Oldest known stone tablet inscribed with the Ten Commandments sells for over $5M.* (2024, December 18). Associated Press. https://apnews.com/article/ten-commandments-sothebys-auction-f68da9c26bb38bf62c6d98982a8561e3
2. Kamrin, J. (2015, March). *Papyrus in ancient Egypt*. Met. https://www.metmuseum.org/toah/hd/papy/hd_papy.htm
3. Smithsonian. (2011, November 8). *Quipu*. https://www.si.edu/newsdesk/snapshot/quipu
4. *Summary of* The Tale of Genji. (n.d.). taleofgenji.org. Retrieved February 18, 2025, from http://www.taleofgenji.org/summary.html
5. Bayeux Museum. (n.d.). *The Bayeux Tapestry*. Retrieved February 18, 2025, from https://www.bayeuxmuseum.com/en/the-bayeux-tapestry/
6. Trinity College Dublin. (n.d.). *The Book of Kells experience.* Retrieved February 18, 2025, from https://www.visittrinity.ie/book-of-kells-experience/
7. Ministère de la Culture. (n.d.). *Lascaux*. Retrieved February 18, 2025, from https://archeologie.culture.gouv.fr/lascaux/en
8. UNESCO World Heritage Centre. (n.d.). *Cave of Altamira and Paleolithic cave art of Northern Spain.* Retrieved February 18, 2025, from https://whc.unesco.org/en/list/310/
9. National Museum Australia. (n.d.). *First rock art.* Retrieved February 18, 2025, from https://www.nma.gov.au/defining-moments/resources/first-rock-art
10. Drayton, L. (2022, September 14). *Truth is trouble: Toni Morrison's advocacy against censorship.* New York Public Library. https://www.nypl.org/blog/2022/09/14/truth-trouble-toni-morrisons-advocacy-against-censorship
11. *Kreimer v. Bureau of Police for Town of Morristown*, 765 F. Supp. 181 (D.N.J. 1991). (1991, May 22). Justia Law. https://law.justia.com/cases/federal/district-courts/FSupp/765/181/2351202/

12. National Archives. (n.d.). *25th anniversary of the Americans With Disabilities Act.* Retrieved February 18, 2025, from https://www.archives.gov/calendar/ada25
13. Gannon, J. (2008, October 12). Experience guides diversity officer's efforts to go beyond traditional programs. *Pittsburgh Post-Gazette*. https://www.post-gazette.com/business/businessnews/2008/10/12/Experience-guides-diversity-officer-s-efforts-to-go-beyond-traditional-programs/stories/200810120204
14. Education Week. (2024, December 19). *School shootings this year: How many and where.* https://www.edweek.org/leadership/school-shootings-this-year-how-many-and-where/2024/01
15. Chuck, E. (2024, September 7). *Lifeline or distraction? Georgia shooting reignites debate over cellphones in schools*. NBC News. https://www.nbcnews.com/news/us-news/lifeline-distraction-shooting-reignites-debate-phones-schools-rcna169920
16. Ibid.
17. Clery Center. (n.d.). *The Jeanne Clery Act*. Retrieved February 18, 2025, from https://www.clerycenter.org/the-clery-act
18. Clery Center (n.d.). *About*. Retrieved February 18, 2025, from https://www.clerycenter.org/about
19. Ibid.
20. Clery Center. (n.d.). *The Jeanne Clery Act.* Retrieved February 18, 2025, from https://www.clerycenter.org/the-clery-act
21. American Library Association. (2021). *ALA code of ethics.* https://www.ala.org/tools/ethics
22. Besner, L. (2018, June 15). Risotto, robotics and virtual reality: How Canada created the world's best libraries. *Guardian*. https://www.theguardian.com/cities/2018/jun/15/risotto-robotics-and-virtual-reality-how-canada-created-the-worlds-best-libraries
23. Muller, K. (2018, June 1). Information access for all. *American Libraries*. https://americanlibrariesmagazine.org/2018/06/01/information-access-for-all/
24. International Federation of Library Associations and Institutions. (2008). *Response from IFLA to the European Commission's green paper "Copyright in the knowledge industry."* https://www.ifla.org/wp-content/uploads/2019/05/assets/clm/position_papers/IFLA_Response_Green-paper-copyright.pdf
25. Vincent, J. (2016, July 4). UN condemns internet access disruption as a human rights violation. *Verge*. https://www.theverge.com/2016/7/4/12092740/un-resolution-condemns-disrupting-internet-access
26. Article 19. (2021, July 15). *UN: Human Rights Council adopts resolution on human rights on the internet.* https://www.article19.org/resources/un-human-rights-council-adopts-resolution-on-human-rights-on-the-internet/
27. Pew Research Center. (2024, November 13). *Internet/broadband fact sheet.* http://www.pewinternet.org/fact-sheet/internet-broadband/
28. Benz, K. (2018, June 6). *The digital divide narrows*. CNET. https://www.cnet.com/tech/mobile/the-digital-divide-narrows/
29. Pew Research Center. (2024, November 13). *Internet/broadband fact sheet.* http://www.pewinternet.org/fact-sheet/internet-broadband/
30. McGill, M. H. (2018, February 7). The least connected people in America. *Politico*. https://www.politico.com/agenda/story/2018/02/07/rural-indian-reservations-broadband-access-000628
31. Ibid.
32. Hendel, J., & Doherty, T. (2018, February 7). Graphic: America's digital divide, in 2 maps. *Politico*. https://www.politico.com/agenda/story/2018/02/07/digital-divide-in-america-graphic-000639
33. Mamula, K. B., & Wardarski, J. (2018, June 26). Lifeline offline: Unreliable internet, cell service are hurting rural Pennsylvania's health. *Pittsburgh Post-Gazette*. https://newsinteractive.post-gazette.com/telemedicine/

34. Dellinger, A. J. (2018, July 7). FCC refuses to back down from plan to strip phone and internet subsidies for American Indians. *Gizmodo*. https://gizmodo.com/fcc-refuses-to-back-down-from-plan-to-strip-phone-and-i-1827418859
35. Rogers, K. (2017, August 289). Rural America is building its own internet because no one else will. *Vice*. https://motherboard.vice.com/en_us/article/paax9n/rural-america-is-building-its-own-internet-because-no-one-else-will
36. International Federation of Library Associations and Institutions. (n.d.). *Net neutrality toolkit.* Retrieved February 18, 2025, from https://www.ifla.org/publications/net-neutrality-toolkit/
37. Ibid.
38. Ibid.
39. Shepardson, D. (2024, August 1). *US court blocks Biden administration net neutrality rules*. Reuters. https://www.reuters.com/legal/us-court-blocks-biden-administration-net-neutrality-rules-2024-08-01/
40. Liu, C., & Trendacosta, K. (2025, January 7). *Sixth Circuit rules against net neutrality; EFF will continue to fight*. Electronic Frontier Foundation. https://www.eff.org/deeplinks/2025/01/sixth-circuit-rules-against-net-neutrality-eff-will-continue-fight
41. One Laptop Per Child. (n.d.). *About the project: Mission*. Retrieved February 18, 2025, from https://web.archive.org/web/20140327095407/http://one.laptop.org/about/mission
42. Ibid.
43. Robertson, A. (2018, April 16). OLPC's $100 laptop was going to change the world—then it all went wrong. *Verge*. https://www.theverge.com/2018/4/16/17233946/olpcs-100-laptop-education-where-is-it-now
44. Solon, O. (2017, July 27). "It's digital colonialism": How Facebook's free internet service has failed its users. *Guardian*. https://www.theguardian.com/technology/2017/jul/27/facebook-free-basics-developing-markets
45. Lee, D. (2016, February 11). *Facebook "colonialism" row stokes distrust in Zuckerberg*. BBC News. https://www.bbc.com/news/technology-35547416
46. Donnelly, G. J. (2018, January 18). *Starting today, all federal websites must be accessible to those with disabilities; the business community should take note*. Masslive. https://www.masslive.com/opinion/index.ssf/2018/01/starting_today_all_federal_age.html
47. US Access Board. (2014, May). *Outdoor developed areas.* https://www.access-board.gov/files/aba/guides/outdoor-guide.pdf
48. US Access Board. (n.d.). *Guidelines and standards.* Retrieved February 18, 2025, from https://www.access-board.gov/guidelines-&-standards/learn-more/
49. US Access Board. (n.d.). *Information and communication technology.* Retrieved February 18, 2025, from https://www.access-board.gov/ict.html
50. Donnelly, G. J. (2018, January 18). *Starting today, all federal websites must be accessible to those with disabilities; the business community should take note*. Masslive. https://www.masslive.com/opinion/index.ssf/2018/01/starting_today_all_federal_age.html
51. W3C Web Accessibility Initiative. (2005). *WCAG 2 overview.* https://www.w3.org/WAI/standards-guidelines/wcag/
52. Band, J. (2013, September 10). How the Marrakesh Treaty opens vistas for print-disabled readers. *American Libraries*. https://americanlibrariesmagazine.org/2013/09/10/how-the-marrakesh-treaty-opens-vistas-for-print-disabled-readers/
53. American Intellectual Propery Law Association. (2018, October 12). *President Donald J. Trump signs Marrakesh Treaty into law.* https://www.aipla.org/detail/news/2018/10/12/president-donald-j.-trump-signs-marrakesh-treaty-into-law

54. World Intellectual Property Organization. (n.d.). *WIPO-administered treaties.* Retrieved February 18, 2025, from https://www.wipo.int/wipolex/en/treaties/ShowResults?search_what=C&treaty_id=843
55. Ministry of Industry, Investment and Commerce. (2024, September 4). *Jamaica accedes to the Marrakesh Treaty.* https://www.miic.gov.jm/content/jamaica-accedes-marrakesh-treaty
56. Perlmutter, S. (2021, February 22). The *Marrakesh Treaty in action: Exciting progress in access to published works for the blind and print-disabled communities*. Library of Congress Blogs. https://blogs.loc.gov/copyright/2021/02/the-marrakesh-treaty-in-action-exciting-progress-in-access-to-published-works-for-the-blind-and-print-disabled-communities/
57. Collins Dictionary. (n.d.). *Nominative determinism.* Retrieved February 18, 2025, from https://www.collinsdictionary.com/us/submission/18319/nominative+determinism
58. Singal, J. (2024, September 8). Is your name your destiny? *New York Times*. https://www.nytimes.com/2024/09/08/opinion/names-destiny-nominative-determinism.html
59. Hayden, Carla. (n.d.). *Welcome message from Carla Hayden, 14th Librarian of Congress*. Retrieved February 18, 2025, from https://www.loc.gov/about/
60. Berman, S. (2022, March 8). *Sandy Berman: The making of a radical librarian*. Minneapolis Interview Project: A Hennepin History Museum Archive. https://turtleroad.org/2022/03/08/sandy-berman/
61. Centers for Disease Control and Prevention. (2024, May 10). *Recent updates to COVID data tracker.* https://covid.cdc.gov/covid-data-tracker/#datatracker-home
62. BNO News. (2024, September 6). *U.S. to resume mandatory reporting of COVID hospitalizations.* https://bnonews.com/index.php/2024/09/u-s-to-resume-mandatory-reporting-of-covid-hospitalizations/
63. Grey, A.-M. (2020, December 10). *The case for connectivity, the new human right*. United Nations. https://www.un.org/en/un-chronicle/case-connectivity-new-human-right

Chapter 5

Privacy and Surveillance

Introduction: Privacy in Authoritarian and Democratic Regimes

To understand how necessary privacy is to healthy human existence—and what can happen to people when they're forced to live without it—we don't need to imagine dystopian science fiction futures. Instead, we can simply look to the recent past: namely, the post–World War II nation of East Germany (more formally referred to as the German Democratic Republic [GDR]) that existed from 1949–1990, until its reunification with West Germany. *The Lives of Others* (2006), the 2007 Academy Award winner for Best Foreign Language Film, gives a chilling glimpse into the degradations and damage inflicted by state-sponsored surveillance. Through the eyes of a handful of fictionalized characters, this 1984-set film depicts the historically accurate surveillance role of East Germany's Ministry of State Security (called the Stasi), which functioned as a secret police force. Employing widespread surveillance, intimidation, incarceration, interrogation, and abusive practices like bullying and beatings, the principal aim of the Stasi was to detect and stamp out dissent and political unrest, by whatever means deemed necessary. To carry out these objectives, the East German government mobilized legions of enforcers, informants, and administrative underlings to collect and leverage its most valuable commodity against its own people: information.

The amount of information collected, organized, and made retrievable by the Stasi in the pre-internet era is nothing short of staggering. As the *Guardian* recounted in a 2011 article on Germany's efforts to acquire Stasi-related documents held in the UK, "The Stasi archives already encompass 69 miles (111km) of files, including 39m index cards, 1.4m photos and 34,000 video and audio recordings."[1] To further put the sheer amount of that data collection into a scale conveying the numbers of humans impacted, ProPublica reported in 2014 that the Stasi "amassed dossiers on about one quarter of the population of the country during the Communist regime."[2] Reinforcing that stunning statistic, nearly one out of every four East Germans was the target of an intelligence-gathering effort orchestrated by their own government. Statista indicates that from the period of 1950 to 1989, "the population of East Germany declined from 18.4 million to just 16.4 million."[3] Perhaps not surprisingly, considering the allure of freedom in the West and the oppressiveness and austerity of Cold War–era life in East Germany, Statista notes that the population of West Germany rose "from 51 million in 1950 to 62.7 million in 1989."

One of the most strategic and insidious acts the Stasi perpetrated was their mapping of social networks. Most of us today would count ourselves among the billions of people on the planet who have interconnected tendrils with a variety of social media networks or platforms: Facebook, YouTube, TikTok, Instagram, Snapchat, WhatsApp, among many options. Now imagine if one or more of those networks were weaponized: namely, anyone with whom one has a link within that network could be a potential target, reeled in along with the initially targeted party and subjected to a stop-and-frisk-style search. This is the kind of pre-internet surveilling that the Stasi conducted, but without the speed, ease, and information search and storage capabilities made possible via digital age technologies. Julia Angwin, author of the ProPublica piece cited earlier, was able to obtain a "hand drawn social network graph" as well as other files from Berlin's Stasi Archive.[4] Angwin provides an example: "The graphic shows forty-six connections, linking a target to various people..., places..., and meetings," and indicates that she was surprised by the "crude" techniques employed, consisting mainly of mail and telephone surveillance as well as informants.

Germany's *Spiegel International*, in a 2009 article, "Stasi Files Reveal East Germany's 'Dirty Reality,'" provides beneficial context and summary for Germany's ongoing reckoning with the Stasi's past actions and legacy:

> At the height of its powers in the late 80s, the Stasi, considered to be one of the effective intelligence agencies in the world, [*sic*] employing an estimated 91,000 people to keep a close eye on the population in the former communist state. And hundreds of thousands more—perhaps even as many as two million, according to some researchers—also spied on friends, family and colleagues. The information gathered on up to 5.6 million individuals was filed at the Stasi's central catalogue and used to stifle political dissent. It also fostered a repressive climate of fear in East German communities because what the Stasi knew about you could, and most likely would, affect your educational, professional and even recreational opportunities.[5]

The Stasi's brandishing of information that could or would influence critical parts of people's lives serves as a kind of analog precursor to *the Black Mirror* TV series episode "Nosedive" (2016), starring Bryce Dallas Howard. In the fictionalized smartphone-based era of "Nosedive," people are able to rate each other via their phones, essentially constituting an informant-driven social ranking system that is dependent on the crowd-sourcing of real-time data. These rankings then result in life-changing implications for whom people can date, where they can live, and what jobs they can do.

Several of the main characters in the film *The Lives of Others*—knowingly and unknowingly manipulated and spied on for years, with devastating consequences—serve as a social network stand-in for the more than five million East Germans subjected to abridgements of their personal privacy and dignity from 1945 to 1990. This 45-year repressive period stemmed from Nazi Germany's World War II surrender in 1945 and the subsequent division of Germany into two distinct nations: West Germany, allied with the United States and its allies, and East Germany, under the hegemony of the Soviet Union and aligned with the Eastern Bloc nations. Following years of unrest, protest, political struggles, and eventual changes, the Berlin Wall was torn down on November 9, 1989. Less than a year later, East Germany was reunified with West Germany on October 3, 1990.

Figure 5.1 Surveillance sign at the University of Maryland. (© 2021 Kip Currier)

In the ensuing decades since rejoining the West, supervised access to GDR archival records has enabled the world to peel back some of East Germany's curtains and uncover the information-gathering practices and surveillance-craft secrets of the Stasi and its administrative overlords. The archives reveal aspects of the scrupulously detailed, intimate lives of the East German citizens the Stasi watched and reported on. Germany provides limited access to these kinds of records through the Stasi Records Archive, as explained in its FAQ.[6] The FAQ provides information on who may access the records in certain circumstances, such as records that may exist for oneself or one's family members. As individuals have been able to review their records, they, as well as the world, have gained insights into the brutal and lasting damage inflicted by the Stasi on their own countrymen. One victim of these practices, Frederic L. Pryor, who was able to review the records the Stasi had compiled on him, articulated in a 1994 piece ("On Reading My Stasi Files") the profound implications for himself and his fellow citizens, particularly the residual psychological impacts: "It seems clear to me that the constant need for watchfulness and the paranoia was the most debilitating aspect of everyday life in the GDR, morally more disturbing than the material difficulties arising from the constant economic shortages or the difficulties of traveling to the West."[7]

The Stasi files present an instructive case study for how to create an archival framework that strives to balance complex, competing privacy interests of multiple stakeholders. At the same time, such a framework must also offer guidance on how to manage thorny archival preservation and access issues, such as the treatment of sensitive documents and difficult histories. Rachel E. Beattie's 2009 paper "The Poisoned Madeleine: Stasi Files as Evidence and History" examines these often gray interests and issues and the lessons that can be applied by other countries and societies:

> The Stasi records of mass surveillance and punishment of citizens serve as both a record of the East German people's oppression and a vital tool for coming to terms with the communist dictatorship in East Germany. Debates about opening versus sealing these files allowed Germans to analyze the disconnect between rights of privacy and rights of information. The success of the laws created to both open and restrict the use of the Stasi files can be attributed to the innovative way that these laws were able to accommodate those who needed the information as well as those concerned about their privacy....
>
> Ultimately, the German resolution to the problem of the Stasi files serves as an example for other governments struggling with the disposal of extremely sensitive documents.[8]

As with the privacy interests and surveillance issues presented via the Stasi case study, this chapter highlights the inherent tensions that the competing natures of privacy and surveillance present for democratic and authoritarian regimes. Privacy and surveillance also have major implications for societal interests and actions, such as

- freedom of movement
- free speech and expression
- free and independent presses
- rights of dissent
- freedom of religion
- education
- intellectual freedom
- the historical record
- cultural heritage institutions

- public safety
- national security
- data collection and use
- access to information and technologies

This chapter underscores that democratic nations consider privacy to be a fundamental human right. Examples referenced, too, shed necessary light on the imperiled state of privacy in much of the developed and developing world today and the pervasive and invasive role that surveillance plays in contributing to privacy's endangerment and virtual extinction in some regions and countries.

Cybertracking is inextricably linked with privacy and surveillance in much of the internet-dependent world. To that end, this chapter identifies issues involving data collection and use and the need for laws, ethics, and policies regarding privacy interests and more data autonomy for citizenries. Information centers have a long history in championing privacy and speaking out about surveillance concerns. The chapter also considers ways that information professionals can promote privacy and greater user access to and agency over their own data.

Definitions and History of Privacy

Privacy is a word heard frequently in the digital age. This is not surprising, given the effects that electronic technologies exert on our individual and collective privacy interests. Like the term *access* discussed in a previous chapter, it has a number of contextual meanings and interpretations. Examples below illustrate these varying connotations:

- The quality of being secluded from the presence or view of others; the condition of being concealed or hidden.[9]
- The condition of being left alone, out of public view and in control of information that is known about you.[10]
- The right of an individual to be secure from unauthorized disclosure of information about oneself that is contained in documents.[11]

Privacy and *confidentiality*, as terms and concepts, have overlapping features but are sometimes confused and conflated. It's important to distinguish between them, as the University of Pittsburgh Institutional Review Board (IRB) does:

> "Privacy" refers to the *individuals*' right to control access to themselves. Not only does this include access to the individual's personal information, but it also includes access to their body (e.g., any of their biological specimens). On the other hand, "confidentiality" refers to how private *information* provided by individuals will be *protected by the researcher* from release. Describing just how the confidentiality of research information will be maintained is an important element of the consent process.[12]

Cross-Cultural Distinctions

The concept of privacy varies, too, from culture to culture. Robust scholarship exists on cross-cultural differences and privacy. For example, a 2022 chapter in a Springer Open Access publication examines cross-cultural privacy differences. The author's abstract explains, in part, that the chapter "covers major cross-cultural differences that have been reported in privacy research" and also "briefly reviews the concept of culture, discusses the cross-cultural differences in privacy management, and recommends design implications on privacy design in the international context."[13] An interesting 2022 piece in the *Conversation* looked at how people's perspective on privacy has evolved over time:

> While privacy rights or values may be modern notions, examples of privacy norms and privacy-seeking behaviors abound across cultures throughout human history and across geography.
>
> As privacy researchers who study information systems and behavioral research and public policy, we believe that accounting for the potential evolutionary roots of privacy concerns can help explain why people struggle with privacy today. It may also help inform the development of technologies and policies that can better align the digital world with the human sense of privacy.[14]

As we see with these two scholarly communication examples, issues of technological design and privacy are important areas of inquiry. This is not at all unexpected when we consider the range of opinions on, for example, how tech companies should address privacy concerns for adult and minor users of their technologies and products.

Development of Privacy in the United States

Arguably, the most influential writing on privacy in the United States was an 1890 *Harvard Law Review* article titled "The Right to Privacy."[15] It was written by Samuel D. Warren II and Louis Brandeis. Brandeis would go on to be appointed as a US Supreme Court justice by President Woodrow Wilson and eventually confirmed in 1916. At the time of the 1890 essay, privacy was largely articulated by the authors and other thinkers as the right "to be let alone." These concerns arose from the intrusive practices of nineteenth-century tabloid journalists, presaging the later surveillance and stalking practices of modern paparazzi photographers. In their piece, Warren and Brandeis noted the role of technology, which they call "mechanical devices," in facilitating intrusions on the right to be left alone. Their words presciently speak to current and continuing concerns about the roles and impacts of technologies on peoples' privacy: "Instantaneous photographs and newspaper enterprise have invaded the sacred precincts of private and domestic life; and numerous mechanical devices threaten to make good the prediction that 'what is whispered in the closet shall be proclaimed from the house-tops.'"

Static and Evolving Attitudes on Privacy

Research underscores that people have differing feelings and attitudes on privacy, depending on their individual perspectives and circumstances. Feelings range across a spectrum of emotions that can include more than one feeling at the same time, such as apathy, uncertainty, confusion, disempowerment, anxiety, fear, and anger. Backing up this assertion, in 2019 a Pew Research Center (PRC) study, "Americans and Privacy: Concerned, Confused and Feeling Lack of Control Over Their Personal Information," found that "roughly half of Americans (48%) say they feel as if they have no control over who can access the search terms they use, and 41% say the same about the websites they visit."[16] The same study found that overwhelming majorities of people are able to weigh and determine the risks and benefits of data that is collected about them through their online activities:

> 81% of Americans think the potential risks of data collection by companies about them outweigh the benefits, and 66% say the same about government data collection about them. Relatedly, 72% of adults say they personally benefit very little or none from company data collection about them, and 76% say this about the benefits they might get from government data collection.[17]

One might reasonably expect, given such large numbers of Americans expressing legitimate, well-founded concerns about their privacy and the data that tech companies collect about them, that we might see strong legislative and regulatory frameworks to address these issues in the digital age. As we discuss later, however, that is far from the case in the United States. In contrast, the European Union (EU) has enacted stringent privacy and data collection regulations in the recent past.

To get a better handle on how people view privacy, and how feelings about privacy may or may not have changed over the last decade or two, it is revealing to look at both research data on privacy and reporting on the topic of privacy. For example, does media reporting on issues like privacy influence and shape perceptions, or is media reporting a true and accurate reflection of how users feel, irrespective of that reporting?

Characterizations of privacy can vary widely. To that end, it's important to be mindful of the positionality of those who are writing about privacy (i.e., what are their educational and vocational backgrounds, what are their areas of interest and expertise, what are their life experiences?). Additionally, where people work, the organizations they are associated with, and the causes they are passionate about can provide insights into how issues like privacy and technology are framed and messaged. To illustrate variations in reporting and data on privacy, consider the following two articles.

A March 2010 CNET piece used the provocative, attention-grabbing title "Why No One Cares About Privacy Anymore." The piece states that "norms are changing, with confidentiality giving way to openness. Participating in YouTube, Loopt, FriendFeed, Flickr, and other elements of modern digital society means giving up some privacy, yet millions of people are willing to make that trade-off every day."[18] Questions arise from this assertion:

- Were people in the 2010 time period fully cognizant of ways that their privacy was impacted when they used social media and Internet sites?
- Did they understand the so-called trade-off of their privacy when using these sites?
- Were they given any kinds of meaningful choices about these trade-offs?
- Did people understand privacy settings at that time?
- Were people given a series of options vis-à-vis privacy settings, or was it a binary, "yes or no" option provided by technology companies?

In contrast to the CNET article, a February 2011 *USA Today* article, "Most Google, Facebook Users Fret Over Privacy," indicated that "nearly seven out of 10 Facebook members surveyed—and 52% of Google users—say they are either 'somewhat' or 'very concerned' about their privacy while using the world's most popular social network and dominant search engine."[19] Even then, in 2011, significant numbers of people were registering concerns about their own privacy. The *USA Today* article's polling data is a sharp rebuttal to the CNET article's assertion that no one cares about privacy anymore. Moreover, the *USA Today* article further explains that, per "technologists and privacy experts," most people do not have a "clear grasp" of the risks of being online, they don't understand who is able to acquire their data, and they don't understand what that data is being used for. Those conclusions indicate a need and desire for greater awareness and education on privacy-related matters, which is far different from indifference and not caring about privacy. This triggers questions such as the following:

- Is there an interest or benefit in disseminating messaging that people don't care about privacy?
- Who stands to reap financial incentives or other competitive advantages if a lack of interest in privacy is normalized?
- What are the disincentives, if any, to privacy education and awareness?
- Who are the audiences for articles about privacy-related issues?

Methodologies of studies are crucial to understand:

- What research method or methods was/were used?
- When was the data collected?
- What was the sampling size?
- What are the margins of error?

How questions are phrased can also shape and yield markedly different results. Consider, for example, the distinctions between the following two questions on privacy:

- Are you concerned about the privacy of you and your family?
- Are you concerned about technology companies like Google and Facebook collecting and using data about you and your family when you visit internet sites?

The ways these questions are constructed are very different: The first one is much more abstract and general, while the second is significantly more specific, personal, and relatable. The second question does a better job of "unpacking" the concept of privacy and how it touches people's everyday lives. Particularly in the absence of knowing how questions are asked and additional mechanics of research studies, maintaining a critical, questioning stance is always a good rule of thumb.

Over the course of several decades, Pew Research Center (PRC) has extensively studied attitudes about privacy from many angles and continues to do so. The following is a selective sampling of explicative data points from several PRC privacy-related studies conducted in the United States. Many PRC studies like these also implicate surveillance, data collection and use by technology companies, and online safety of minors and adults:

- September 2013 PRC study, "Anonymity, Privacy, and Security Online": "Most internet users would like to be anonymous online at least occasionally, but many think it is not possible to be completely anonymous online. New findings in a national survey show: 86% of internet users have taken steps online to remove or mask their digital footprints."[20]
- April 2020 PRC study on "privacy concerns": "About half (52%) of U.S. adults said they decided recently not to use a product or service because they were worried about how much personal information would be collected about them."[21]
- May 2020 PRC study on "data collection": "57% of online dating users are concerned about data collection by sites and apps."[22]
- October 2023 PRC study on "key findings about Americans and data privacy":

> More than half of Americans (56%) say they always, almost always or often click "agree" without reading privacy policies. . . .
>
> Americans have little faith that social media executives will protect user privacy. Some 77% of Americans have little or no trust in leaders of social media companies to publicly admit mistakes and take responsibility for data misuse. . . .
>
> There is bipartisan support for more regulation to protect personal information. Some 78% of Democrats and 68% of Republicans think there should be more government regulation of what companies can do with customers' personal information. . . .
>
> About nine-in-ten Americans (89%) are concerned about social media sites knowing personal information about children. Most Americans are also concerned about advertisers using data about children's online activities to target ads to them (85 percent) and online games tracking children on the internet (84%).[23]

Codes of Ethics: Information Organizations and Privacy

Now that we've identified some ways of defining privacy and looking at perceptions of privacy cross-culturally, though predominantly from a US perspective, we turn to examining ethics stances on privacy-related issues. The focus here is on ethics codes from information organizations.

Libraries, archives, museums, and other cultural heritage institutions have a well-documented and long-standing history of championing privacy. These institutions have implemented and safeguarded privacy-oriented policies and ethics principles addressing their staffs as well as users of their organizations. Moreover, they serve an integral role in advocating for privacy protections for societies at large. Laws also play a role in incentivizing these organizations to avoid risk by advancing privacy. For example, laws govern what libraries must do to protect patrons' records.

American Library Association

Privacy has long been a core ethical principle of the American Library Association (ALA) and is codified in the ALA Code of Ethics. Of the code's nine principles, principle 3 is the one that specifically addresses privacy: "We protect each library user's right to privacy and confidentiality with respect to information sought or received and resources consulted, borrowed, acquired or transmitted."[24] ALA articulates the importance and wide-reaching influences of its core values:

> The American Library Association's Core Values articulate the profession's principles and highest aspirations. They guide and unite library workers, including staff, volunteers, trustees, advocates, and others who contribute their talents, expertise, and dedication to furthering the library mission.
>
> The Core Values uplift and support other foundational documents, the Library Bill of Rights, the Code of Ethics, and Libraries: An American Value. These foundational documents represent the common ground upon which we can build policies, make decisions, set goals, and prioritize actions to bring equitable library access to all users. The Core Values describe the rights we seek to protect and the ethical guidelines we look to for direction.[25]

ALA recognizes privacy as one of the core values of the profession. In February 2024, as mentioned in the previous chapter on access, ALA added five values statements to its existing Core Values.[26] One of the five new values statements addresses intellectual freedom and privacy. This new statement also specifically mentions that the right to privacy includes freedom from surveillance.[27]

As pointed out earlier in this book, core values of the information professions—like access, privacy, and intellectual freedom—are interdependent. Like carnival bumper cars, these interests often bump up against one another. Restricting or expanding one often results in restrictions or expansions of other values.

Special Library Association

Special libraries (more precisely referred to as specialized libraries) are best conceptualized as encompassing any library or information center that is *not* a public library, academic library, or school library. The Special Library Association (SLA) brings together information professionals from diverse organizations that include business, government, academia, medicine, law, music, art, and other "specialized" settings. SLA was founded in 1909 by maverick library leader John Cotton Dana, who, among many contributions, pioneered library services for children and broader public access to library materials. SLA's "Competencies for Information Professionals" document, approved by its board of directors in 2016, lays out six core competencies. Competency 6, "Information Ethics," includes privacy and confidentiality, among other issues like intellectual property and intellectual freedom.[28]

Society of American Archivists

The Society of American Archivists (SAA), founded in 1936, bills itself as "North America's oldest and largest national professional association dedicated to the needs and interests of archives and archivists."[29] Like SLA, SAA is made up of archivists working in an array of settings that includes museums, historical societies and other cultural heritage organizations, libraries, businesses, government, and higher education. SAA's Code of Ethics for Archivists was most recently revised in 2012. The code sets out seven distinct areas, one of which addresses a number of privacy dimensions, such as concern for the privacy interests of indigenous communities in relation to culturally sensitive archived material.[30]

International Federation of Libraries Associations and Institutions

The International Federation of Libraries Associations and Institutions (IFLA) has more than 1,400 members in more than 140 nations. Founded in 1927 in Edinburgh, Scotland, IFLA's headquarters are in the Royal Library in the Netherlands. IFLA's Code of Ethics for Librarians and Other Information Workers was approved in 2012 and last updated in 2016. Translated versions of the code are available in approximately 20 languages, from Amharic to Ukrainian. The code is available in a long version and a short version. Its format is a preamble with seven sections, one of which is "Privacy, Secrecy, and Transparency":

> Librarians and other information workers respect personal privacy, and the protection of personal data, necessarily shared between individuals and institutions.
>
> The relationship between the library and the user is one of confidentiality and librarians and other information workers will take appropriate measures to ensure that user data is not shared beyond the original transaction.
>
> Librarians and other information workers support and participate in transparency so that the workings of government, administration and business are opened to the scrutiny of the general public. They also recognise that it is in the public interest that misconduct, corruption and crime be exposed by what constitute breaches of confidentiality by so-called "whistleblowers."[31]

IFLA's website also provides links to the collected national codes of ethics for librarians from more than sixty countries. These codes are available in English as well as languages for each respective country.[32] Interestingly, the code serving as the example provided for the United

States is the ALA Code of Ethics. Exploration of these codes reveals areas of commonality and areas of emphasis, divergence, and gaps. Varied statements that address privacy are included in these globe-spanning codes.

Global Case Studies: Surveillance Accountability

The Stasi files are but one example of how individual people, communities, and countries are grappling with and finding ways to learn from state-sanctioned experiences that fundamentally violate the privacy rights of people through surveillance and human-rights-offending practices. Real-life case studies like these also shed important light on the amassing and weaponizing of data on peoples who are surveilled. Other examples of these types of efforts include the following:

- Archival projects about the roughly 1960s to late-1990s period of strife in Northern Ireland, referred to as "the Troubles"; this era also involved surveillance and networks of informants. The UK's National Archives collection on the Troubles in Northern Ireland provides extensive access to a wide range of records.[33]
- Wits University's (Johannesburg, South Africa) Apartheid Archives Project. Begun in 2008, this project includes an international team of researchers from South Africa, Australia, the UK, and the United States. The project "aims to examine the nature of the experiences of racism of particularly 'ordinary' South Africans under the old apartheid order and their continuing effects on individual and group functioning in contemporary South Africa... [and] collects documents, analyses and provides access to personal or narrative accounts of the impact of apartheid on the lived realities of their authors."[34] Like the examples of East Germany's Stasi Files and Northern Ireland's Troubles, South Africa's apartheid era framework was built and reinforced via state-sponsored surveillance, racial profiling, and data collection.

Countries and cities take a variety of stands on the continuum of maximal privacy at one end and maximal surveillance at the other. The following is a sampling of nations and cities that have garnered attention for their controversial stances on surveillance.

Many democratic nations within the international community have expressed disapproval of China's state-approved actions in its northwestern region known as Xinjiang. Surveillance plays a vital part in keeping tabs on the several million Turkic and Muslim Uyghurs of that so-called autonomous region. In May 2022, more than 10 gigabytes of hacked materials (referred to as the Xinjiang Police Files) were "obtained by a third party from confidential internal police networks" and made available, shedding light on the Chinese government's actions in Xinjiang and its internment camps there.[35]

Though little is publicly known about the specifics of North Korea's storied surveillance apparatus, it remains the most isolative and secretive nation in the world, aptly earning its sobriquet as the Hermit Kingdom. One of the ways its rulers maintain a de facto hermetically sealed barrier of secrecy around its physical and digital borders is through societally interwoven networks of people and technologies that monitor virtually all aspects of North Korean lives. Humanitarian groups like Amnesty International have referred to North Korea as "The Surveillance State."[36]

As for cities, a 2020 *U.S. News & World Report* article, "The Most Surveilled Cities in the World," focusing on closed-circuit television cameras (CCTCs), observed that nine out of the top 10 surveilled cities were in China, with its capital city, Beijing, at the top of the list. China contained over half (54 percent) of CCTCs used globally. Beijing, with a population of 20 million, had 1.15 million cameras installed.[37] Among Western cities, London in the UK and Chicago in the United States are two of the most heavily surveilled cities, as has been well documented. The same 2020 *U.S. News & World Report* piece indicated that London was the only non-Chinese city to make the top 10. "The capital of the United Kingdom has 627,727 cameras for a population of 9.3 million—equal to one camera for every 14 residents."[38] The American Civil Liberties Union (ACLU), in a 2021 article headlined "Ten Years After First Warning, Chicago's Massive Surveillance Camera System Continues to Pose an Unregulated Threat to Privacy," raised alarms that Chicago's network of public and private cameras had skyrocketed from approximately 10,000 in 2011 to "by some estimates, more than 30,000."[39]

Depending on one's viewpoint, city-supported surveilling is decried or lauded. Israel employs an extensive network of checkpoints and advanced surveillance through such strategies as facial recognition software in the West Bank and Gaza, as has been widely reported. *Wired*, in

Figure 5.2 Flame of Knowledge on rotunda of Library of Congress and nearby surveillance cameras, 2006. (© Takomabibelot via Flickr, Creative Commons License CC BY 2.0)

a November 2023 article following the October 7, 2023, Hamas attacks on Israel and Israel's subsequent ground invasion into Gaza on October 27, referred to the West Bank as "Israel's surveillance laboratory."[40]

Regulatory Efforts

The world's most comprehensive privacy protection framework is the European Union's General Data Protection Regulation (GDPR). In the words of the EU, it's also "the toughest privacy and security law in the world."[41] The GDPR has been in effect since May 2018. Just as Warren and Brandeis were influenced by the combination of newspaper reporting practices and photographic technology in writing their seminal privacy essay, so too were the European Union nations influenced by their World War II and Cold War experiences with surveillance and violations of personal privacy to eventually elevate privacy to a fundamental human right in the digital age. The EU's Data Protection web page explains why the EU views privacy and data protection as "vital components for a sustainable democracy": "In the EU, human dignity is recognised as an absolute fundamental right. In this notion of dignity, privacy or the right to a private life, to be autonomous, in control of information about yourself, to be let alone, plays a pivotal role. Privacy is not only an individual right but also a social value."[42]

The GDPR can be viewed then as an offensive and defensive manifestation of that ethos, designed as both shield and sword to protect individual and societal privacy and data autonomy interests and rights. The GDPR is a legal schema that has "teeth," as discussed earlier regarding ethics codes with compliance and enforcement capabilities. The legal framework enables the assessment of significant fines against violators. Since its enactment, a number of Big Tech companies have already faced accountability for alleged GDPR infractions. Lawsuits for damages can also be brought by individuals covered under the GDPR who believe and can prove that their privacy, data, and security interests have been violated. On that point, as much as the GDPR was conceived and implemented for privacy and security reasons, it is also a mechanism for promoting individual data autonomy—namely, the ability of people to have a say in when third parties can access their personal data and what those entities can do with that data. Personal data is defined quite broadly:

> Personal data is any information that relates to an individual who can be directly or indirectly identified. Names and email addresses are obviously personal data. Location information, ethnicity, gender, biometric data, religious beliefs, web cookies, and political opinions can also be personal data. Pseudonymous data can also fall under the definition if it's relatively easy to ID someone from it.[43]

In the United States, a number of states have enacted or are moving to enact legislation that has commonalities with the GDPR. At their root, these laws, like the GDPR, seek to give individuals more power to protect their privacy and proactively direct what happens to their personal data. The most significant of these US efforts is the California Consumer Privacy Act (CCPA) of 2018. The CCPA includes "new privacy rights for California consumers," such as

- The right to know about the personal information a business collects about them and how it is used and shared
- The right to delete personal information collected from them (with some exceptions)
- The right to opt out of the sale or sharing of their personal information
- The right to nondiscrimination for exercising their CCPA rights[44]

It's important to note that privacy and data rights under the CCPA have expanded even more in recent years. As a result of California voters passing Proposition 24, also known as the California Privacy Rights Act (CPRA), in late 2020, additional privacy protections went into effect in January 2023. The CPRA can best be understood as an amendment to the CCPA. The CPRA added the following protection rights to the CCPA:

- The right to correct inaccurate personal information that a business has about an individual
- The right to limit the use and disclosure of sensitive personal information collected about an individual[45]

How effective legislative efforts like the CCPA will prove to be for California citizens, as well as how privacy and data protection initiatives will continue to evolve in other states, will remain of great interest to a variety of nonprofit and for-profit stakeholders. Privacy and data collection and use issues also implicate a host of ethical considerations, as we have explored in this chapter and will continue to examine in subsequent chapters, such as sections involving emerging technologies. Given the long and distinguished history that many information organizations have in publicly recognizing and safeguarding the privacy and data interests of their users, as well as in taking leadership roles on privacy education and advocacy for societies at large, information professionals will certainly want and need to keep track of and remain informed about ongoing developments in this legal, ethical, and policy space.

Privacy and Emerging Technologies in the AI Age

Today's digital data is the equivalent of gold in the nineteenth century, or oil-fueled rushes and booms: a valuable commodity that companies are prospecting for, mining, and monetizing. Emerging technologies, and their products, like AI chatbots, depend on data to exist and grow. Self-driving vehicles, examined later in this book, collect immense amounts of data, too, in order to operate. They also by their very nature involve tremendous privacy and security risks, such as data breaches via malicious hackers. These risks derive from the many kinds of data that autonomous vehicles collect, such as via lidar detection systems, external and internal microphones, and other data collection capabilities. For liability and insurance reasons, autonomous vehicles may, at some point, mandate surveillance and monitoring of people in these cars. Potential developments like these raise challenging questions regarding, as examples, how long and where this collected data will be kept and for what purposes it may be used.

Later in this book we investigate AI in detail and look at the Biden administration's groundbreaking 2023 AI executive order. This order contains eight principles, two of which pertain to privacy and security. The US government is working on best practices stemming from this order. The aim of this best practices framework is to serve as a model that balances innovation and economic goals with civil liberties, like privacy, and public safety protections.

Summary

Privacy is vital to healthy human existence and is viewed by many democracies as a fundamental right. Warren and Brandeis's influential 1890 *Harvard Law Review* essay "The Right to Privacy" is seen as the first US print publication calling for a "right to be let alone." Their article was in response to the proliferation of new technologies like the camera, presaging modern-era concerns about infringement of privacy rights and data collection excesses enabled by tech company practices and tools. Valuable insights can be learned from the harmful ways that surveillance and information gathering practices have been used, such as by the Stasi secret police in East Germany. In the digital age, surveillance and data collection activities used in developed and developing nations continue to raise civil liberty and human rights concerns. Information professionals and organizations, like the American Library Association and the Society of American Archivists, have long championed and advocated for the privacy rights of their users and all citizens. Privacy is a core value and ethical principle recognized in the ethics codes of leading information organizations. Recent regulatory efforts such as the European Union's General Data Protection Regulation and the California Consumer Privacy Act serve as models for privacy protection and data autonomy efforts. Emerging technologies like AI chatbots and autonomous vehicles are raising new concerns about privacy and data collection activities and uses by corporate interests, governments, and other stakeholders.

Questions for Reflection, Exploration, Action, Discussion (READ)

1. Privacy is viewed by many democracies as a fundamental human right.

a. Do you view privacy as a fundamental human right in the digital and AI ages? Why or why not?
b. Distill your viewpoint into a 60-second explanation you can articulate to a library, archive, or museum board member wanting to know your succinct position on privacy.

2. Privacy is an ethics principle specified in the ethics codes of numerous information-center organizations.

a. Investigate and discuss some key developments that are challenging privacy practices in libraries today.
b. Identify one tangible strategy that information professionals can use to strengthen privacy practices in libraries in the digital age.

3. Review and refamiliarize yourself with the Pew Research Center (PRC) data, discussed earlier in this chapter and also freely accessible via the PRC website, on privacy attitudes of the public over the past couple of decades.

a. Briefly summarize, in a written format, three takeaways you can make from your analysis of the data.

b. From your takeaways on public perceptions about privacy, talk about some concrete steps information professionals can take to raise awareness of privacy issues?

4. Apps collect large amounts of personal data about users. Libraries, archives, museums, and other information centers are increasingly creating and providing access to a wide range of apps that users can engage with for services, programming, gaming, community events, and more.

a. What is your position on the ethics of information centers making apps available to their users that can collect and potentially infringe on their privacy and data autonomy interests?

b. What are some actions that information professionals can take to minimize infringement of their users' privacy rights if they use apps that are made accessible through information centers?

Notes

1. Pidd, H. (2011, December 28). Stasi files row as Britain refuses to return documents to Germany. *Guardian*. https://www.theguardian.com/world/2011/dec/28/stasi-files-row-britain-germany
2. Angwin, J. (2014, February 11). *You know who else collected metadata? The Stasi*. ProPublica. https://www.propublica.org/article/how-the-stasi-spied-on-social-networks
3. O'Neill, A. (2024, August 9). *Population of East and West Germany 1950–2016*. Statista. https://www.statista.com/statistics/1054199/population-of-east-and-west-germany/
4. Angwin, J. (2014, February 11). *You know who else collected metadata? The Stasi*. ProPublica. https://www.propublica.org/article/how-the-stasi-spied-on-social-networks
5. Schaer, C. (2009, July 10). Stasi files reveal East Germany's "dirty reality." *Der Spiegel*. https://www.spiegel.de/international/germany/the-world-from-berlin-stasi-files-reveal-east-germany-s-dirty-reality-a-635486.html
6. Federal Archives. (n.d.). *FAQs*. Retrieved February 18, 2025, from https://www.bundesarchiv.de/en/faqs/
7. Pryor, F. (1994). On reading my Stasi files. *National Interest*. https://nationalinterest.org/legacy/on-reading-my-stasi-files-1028/
8. Beattie, R. E. (2009). The poisoned madeleine: Stasi files as evidence and history. *Faculty of Information Quarterly, 1*(4).
9. National Cancer Institute. (n.d.). *Privacy*. Retrieved February 18, 2025, from https://evsexplore.semantics.cancer.gov/evsexplore/concept/ncit/C17385
10. Oak Ridge National Laboratory. (n.d.). *Privacy*. Retrieved February 18, 2025, from https://web.archive.org/web/20050205005628/http://www.ornl.gov/sci/techresources/Human_Genome/publicat/genechoice/glossary.html
11. Oregon State University. (2000, February). *Privacy*. Archives & records management handbook. https://scarc.library.oregonstate.edu/rm-handbook.pdf
12. University of Pittsburgh, Institutional Review Board. (n.d.). *Privacy versus confidentiality*. Retrieved February 18, 2025, from https://web.archive.org/web/20100610200710/http://www.irb.pitt.edu/Guidance/PrivacyvConfidentialityREVISED070705.pdf
13. Li, Y. (2022). Cross-cultural privacy differences. In B. P. Knijnenburg, X. Page, P. Wisniewski, H. R. Lipford, N. Proferes & J. Romano (Eds.), *Modern socio-technical perspectives on privacy*. Springer. https://doi.org/10.1007/978-3-030-82786-1_12

14. Acquisti, A., & Brandimarte, L. (2022, February 11). Your sense of privacy evolved over millennia—that puts you at risk today but could improve technology tomorrow. *Conversation*. https://theconversation.com/your-sense-of-privacy-evolved-over-millennia-that-puts-you-at-risk-today-but-could-improve-technology-tomorrow-175474
15. Warren, S. D., & Brandeis, L. D. (1890). The right to privacy. *Harvard Law Review,* 4(5), 193–220. https://doi.org/10.2307/1321160
16. Auxier, B., Rainie, L., Anderson, M., Perrin, A. Kumar, M., & Turner, E. (2019, November 15). *Americans and privacy: Concerned, confused and feeling lack of control over their personal information*. Pew Research Center. https://www.pewresearch.org/internet/2019/11/15/americans-and-privacy-concerned-confused-and-feeling-lack-of-control-over-their-personal-information
17. Ibid.
18. McCullagh, D. (2020, March 12). *Why no one cares about privacy anymore*. CNET. https://www.cnet.com/tech/tech-industry/why-no-one-cares-about-privacy-anymore/
19. Acohido, B. (2011, February 9). Most Google, Facebook users fret over privacy. *USA Today.*
20. Rainie, L., Kiesler, S., Kang, R., & Madden, M. (2013, September 5). *Anonymity, privacy, and security online*. Pew Research Center. http://www.pewinternet.org/2013/09/05/anonymity-privacy-and-security-online/
21. Perrin, A. (2020, April 14). *Half of Americans have decided not to use a product or service because of privacy concerns*. Pew Research Center. https://www.pewresearch.org/fact-tank/2020/04/14/half-of-americans-have-decided-not-to-use-a-product-or-service-because-of-privacy-concerns/
22. Turner, E., & Anderson, M. (2020, May 29). *Roughly six-in-ten online daters in the U.S. are concerned about data collection*. Pew Research Center. https://www.pewresearch.org/fact-tank/2020/05/29/roughly-six-in-ten-online-daters-in-the-u-s-are-concerned-about-data-collection/
23. Faverio, M. (2023, October 18). *Key findings about Americans and data privacy*. Pew Research Center. https://www.pewresearch.org/short-reads/2023/10/18/key-findings-about-americans-and-data-privacy/
24. American Library Association. (2021, June 29). *ALA code of ethics.* http://www.ala.org/tools/ethics
25. American Library Association. (2024, January). *Core values of librarianship.* https://www.ala.org/advocacy/advocacy/intfreedom/corevalues
26. American Library Association. (2024, February 5). *American Library Association updates core values.* https://www.ala.org/news/press-releases/2024/02/american-library-association-updates-core-values
27. American Library Association. (2024, January). *Core values of librarianship.* https://www.ala.org/advocacy/advocacy/intfreedom/corevalues
28. Special Libraries Association. (n.d.). *Competencies for information professionals.* Retrieved February 19, 2025, from https://sla.org/page/competencies
29. Society of American Archivists. (n.d.). *Who we are.* Retrieved February 19, 2025, from https://www2.archivists.org/aboutsaa
30. Society of American Archivists. (2020, August). *SAA core values statement and code of ethics.* https://www2.archivists.org/statements/saa-core-values-statement-and-code-of-ethics
31. International Federation of Library Associations and Institutions. (2012). *IFLA code of ethics for librarians and other information workers.* https://repository.ifla.org/server/api/core/bitstreams/bb629889-3883-4c7b-8a40-37c02dc89146/content
32. International Federation of Library Associations and Institutions. (n.d.). *Professional codes of ethics for librarians.* Retrieved February 19, 2025, from https://www.ifla.org/g/faife/professional-codes-of-ethics-for-librarians/
33. National Archives. (n.d.). *The Troubles in Northern Ireland.* Retrieved February 19, 2025, from https://www.nationalarchives.gov.uk/help-with-your-research/research-guides/the-troubles-in-northern-ireland/

34. Wits University Research Archives. (n.d.). *Apartheid archives project.* Retrieved February 19, 2025, from http://historicalpapers-atom.wits.ac.za/apartheid-archives-project
35. Xinjiang Police Files. (n.d.). *FAQ: What are the Xinjiang police files?* Victims of Communism Memorial Foundation. Retrieved February 19, 2025, from https://www.xinjiangpolicefiles.org/press/
36. Amnesty International. (2018, January 12). *North Korea, the surveillance state.* https://www.amnesty.org.uk/north-korea-surveillance-state-prison-camp-internet-phone-technology
37. Keegan, M. (2020, August 14). The most surveilled cities in the world. *US News & World Report*. https://www.usnews.com/news/cities/articles/2020-08-14/the-top-10-most-surveilled-cities-in-the-world
38. Ibid.
39. *Ten years after first warning, Chicago's massive surveillance camera system continues to pose an unregulated threat to privacy.* (2021, February 11). ACLU of Illinois. https://www.aclu-il.org/en/news/ten-years-after-first-warning-chicagos-massive-surveillance-camera-system-continues-pose
40. Bennett, T. (2023, November 7). Intensified Israeli surveillance has put the West Bank on lockdown. *Wired*. https://www.wired.com/story/israel-surveillance-state-out-of-control-west-bank/
41. Wolford, B. (n.d.). *What is GDPR, the EU's new data protection law?* GDPR.eu. Retrieved February 19, 2025, from https://gdpr.eu/what-is-gdpr/
42. European Data Protection Supervisor. (n.d.). *Data protection.* Retrieved February 19, 2025, from https://www.edps.europa.eu/data-protection/data-protection_en
43. Wolford, B. (n.d.). *What is GDPR, the EU's new data protection law?* GDPR.eu. Retrieved February 19, 2025, from https://gdpr.eu/what-is-gdpr/
44. Bonta, R. (2024, March 13). *California Consumer Privacy Act (CCPA)*. https://oag.ca.gov/privacy/ccpa
45. Ibid.

Chapter 6

Intellectual Freedom, Free Speech, and Censorship

Introduction: Free Societies and Something to Offend Everyone

It's difficult for most people living in the world today to truly appreciate how dangerous the act of reading can be: like the threat of losing a finger or a hand for simply glancing at a page of a book. That's exactly what happens in an episode of the fictional TV series *The Handmaid's Tale*, based on the eponymous 1985 work by Canadian author Margaret Atwood. In Atwood's reimagined near-future dystopian world, an ultrareligious group has successfully overthrown and driven the US government into exile in Alaska, renamed the country the Republic of Gilead, and instituted a theocratic social order in which (among other draconian edicts) fertile women are chosen to carry children for the societal leaders (all males) whose wives are infertile. In a season two scene, one of the top leaders, Commander Lawrence, catches the handmaid known as Ofglen (whose real name is Emily) peeking at the cover of Art Spiegelman's 1986 Pulitzer Prize–winning graphic novel *Maus*, which is sitting on his coffee table. He asks her if she knows "the penalty for reading." Ofglen/Emily replies, "A finger for the first offense." Smugly, the Commander retorts, "It used to be a hand."[1] Interestingly, both *Maus* and *The Handmaid's Tale* are regularly challenged by people and groups expressing concerns about the content of those works and their inclusion in school and public library collections.

Fears of losing an appendage due to literacy-related activities aren't merely fanciful speculation by fiction writers, though. In most of the nineteenth-century slaveholding states of America, for example, reading came with severe risks and penalties for slaves. It also brought jeopardy and consequences for anyone involved with efforts to teach slaves to read. As Cornell University's "Abolitionism in America" explains: "In most southern states, anyone caught teaching a slave to read would be fined, imprisoned, or whipped. The slaves themselves often suffered severe punishment for the crime of literacy, from savage beatings to the amputation of fingers and toes."[2]

Why was the ability to read forbidden by the persecutors of slaves and their teachers? The answer, in part, includes two age-old tools of oppression: fear and control. Slaveholders and their enablers feared the sense of empowerment and information-rich connections with kindred others that slaves could experience through reading. They also sought to control the narratives that were disseminated about slavery:

- to forestall the means for slaves to reveal their own experiences of the firsthand horrors of slavery
- to cut off access to truthful information about slavery
- to block efforts to expose the falsehoods about slavery that slaveholders perpetuated by controlling primary information sources.

As the Cornell University reference explicates, reading was about intellectual freedom as well as physical freedom: "For many slaves, the ability to read and write meant freedom—if not actual, physical freedom, then intellectual freedom—to maintain relationships amongst family members separated by the slave trade. A few wrote slave narratives, which, when published, powerfully exposed the evils of slavery."[3]

Another of the many things that books can do is enable people to travel to places in the real world (and fictional worlds, too) that they might never be able to see in the present or future—and certainly not in the past, unless H. G. Wells' mythical time machine should someday be invented. In Ray Bradbury's *Fahrenheit 451* nightmarish alternate-world setting, books and reading are forbidden. Firefighters, like the book's protagonist, named Montag, deploy fire to burn the world's books and extinguish the light of accumulated knowledge. Professor Faber, one of the self-anointed saviors of books who works to convince Montag to join his renegade brotherhood of books, perspicaciously observes one of the democratizing benefits of books and literacy: "Most of us can't rush around, talking to everyone, know all the cities of the world, we haven't time, money or that many friends. The things you're looking for, Montag, are in the world, but the only way the average chap will ever see ninety-nine per cent of them is in a book."[4]

Another interpretation of this idea is that books and literacy can be a means for promoting access to information, discussed earlier in this book. No one has enough hours in a lifetime to read everything that's ever been written. Even if human lifespans didn't come with expiration dates, most people probably still would not want to read everything that's ever been written. Moreover, they would still want to have the right to choose what they read and to think for themselves about what they read. Intellectual freedom (IF) emanates from the natural law belief that each person has the inherent right—*the human right*—to make these kinds of decisions over their own lives.

The freedom to think and read what we want stands among the most hallowed principles of US democracy. As both a legal principle and as the living embodiment of a philosophical precept of the Founding Fathers, this freedom has been enshrined in the US Constitution's Bill of Rights since its ratification in 1791. As writer Oliver Wendell Holmes Sr. (father of the famed US Supreme Court justice bearing his name) expressed in 1860, "The very aim and end of our institutions is just this: that we may think what we like and say what we think."[5]

So too is IF the single most important value associated with libraries. Libraries have long supported the freedom to think for oneself and be able to access information that represents diverse, even diametrically opposed, viewpoints and voices. The late Robert Croneberger, director of the Carnegie Library of Pittsburgh from 1986 to 1998, was an ardent defender of IF and an advocate against analog and digital censorship. As just one example, he testified in court as an expert witness in a case challenging the 1996 Communications Decency Act, which was

Figure 6.1 Banned books. (© Antonio Zugaldia, Creative Commons License CC BY 2.0)

declared unconstitutional. Croneberger also frequently espoused the fervent belief, shared by other IF proponents, that a library is not doing its job if it doesn't have at least one item that offends each and every person. Underlying this deeply held commitment to IF is the belief that, in all but a few exceptions discussed in this chapter, the aims of IF should prevail over someone's offense from a book.

Definition, Foundations, and Guiding Principles of Intellectual Freedom and Free Speech

Given the importance of IF, how do we define it? What are the elements that make up IF? The American Library Association (ALA) defines IF as "the right of every individual to both seek and receive information from all points of view without restriction. It provides for free access to all expressions of ideas through which any and all sides of a question, cause, or movement may be explored."[6]

In the United States, the legal and philosophical underpinnings for intellectual freedom and free speech emanate from the US Constitution's First Amendment: "Congress shall make no law respecting an establishment of religion, or restricting the free exercise thereof; or abridging the freedom of speech, or of the press; or the right of the people peaceably to assemble, and to petition the Government for a redress of grievances."[7] Despite the widespread misunderstanding that freedom of speech endows one with the right to say anything one wants at all times, the US Supreme Court has long held that the First Amendment does not protect all speech. Several categories of speech that are not protected include "obscenity, child pornography, defamation, and 'fighting words,' or speech that incites immediate and imminent lawless action. The government is also allowed to enforce secrecy of some information when it is considered essential to national security, like troop movements in time of war, classified information about defense, etc."[8]

The UN has also codified bedrock statements supporting the principles of intellectual freedom and free speech. Articles 18 and 19 of the 30-article United Nations Universal Declaration of Human Rights (1948) specifically address intellectual freedom and freedom of expression. Article 19 also speaks to access to information:

> Article 18.
>
> Everyone has the right to freedom of thought, conscience and religion; this right includes freedom to change his religion or belief, and freedom, either alone or in community with others and in public or private, to manifest his religion or belief in teaching, practice, worship and observance.
>
> Article 19.
>
> Everyone has the right to freedom of opinion and expression; this right includes freedom to hold opinions without interference and to seek, receive and impart information and ideas through any media and regardless of frontiers.[9]

Though the UN stance on these issues is admirable, the reality today is that billions of people around the world do not enjoy these rights. The mere act of selling, buying, and reading books that have been prohibited by governmental authorities can land one in prison. Journalists are increasingly under siege. News sites, print and digital, are censored, extinguished, and outlawed, impeding access to information and the ability of citizens to be informed. In the United States, corporate media consolidation is adversely impacting coverage of local news stories and diminishing access to diverse views and voices, particularly for underserved communities in small towns and rural areas. We delve into these issues later in this chapter.

ALA Positions on Intellectual Freedom

Issues involving intellectual freedom are explicitly and implicitly addressed in all six of the principles articulated in the ALA Library Bill of Rights, which can be accessed via the ALA website.[10] ALA's Code of Ethics also, unsurprisingly, designates intellectual freedom as a foundational precept on which its other principles are anchored. ALA explains that "the Code of Ethics is the document that translates the values of intellectual freedom that define the profession of librarianship into broad principles that may be used by individual members of that profession as well as by others employed in a library as a framework for dealing with situations involving ethical conflicts."[11] The Code of Ethics sets out eight principles. Principle 2 specifically and emphatically covers intellectual freedom, as well as censorship: "We uphold the principles of intellectual freedom and resist all efforts to censor library resources."[12]

Censorship

Censorship is the inverse of intellectual freedom. ALA provides an illustrative, multifaceted definition: "Censorship is the suppression of ideas and information that some individuals, groups, or government officials find objectionable or dangerous."[13] It is no more complicated than someone saying, "Don't let anyone read this book, or buy that magazine, or view that film, because I object to it!" Censors try to use the power of the state to impose their view of what is truthful and appropriate, or offensive and objectionable, on everyone else. Censors pressure public institutions, like libraries, to suppress and remove from public access information they judge inappropriate or dangerous, so that no one else has the chance to read or view the material and make up their own minds about it. The censor wants to prejudge materials for everyone.

When most people think of censorship, they probably picture it in their mind as the banning of a book or other resource. Book bans are indeed a major kind of censorship. However, censorship exists on a continuum and can manifest in many subtle forms, while still impeding access to an information item. ALA's "Intellectual Freedom and Censorship Q&A" provides both recognizable and less familiar examples of censorship, broken down by academic library, public library, school library, and business. This public library example is one that might not appear to be censorship but nevertheless is: "Requiring a user to ask for access to regularly circulating materials (e.g., reshelving, behind a desk)."[14]

Requirements like these can be exacted against books and magazines written by all kinds of authors and that include all types of content. However, they have disproportionately high impacts on books that include LGBTQ+ and BIPOC content or which are written by LGBTQ+ and BIPOC authors.

Reasons why people, institutions, and governments censor vary. Irish playwright George Bernard Shaw (1856–1950) suggests one key reason: maintaining the status quo. In his 1902 preface to *Mrs. Warren's Profession* (1894), Shaw explains: "All censorships exist to prevent anyone from challenging current conceptions and existing institutions. All progress is initiated by challenging current conceptions, and executed by supplanting existing institutions. Consequently, the first condition of progress is the removal of censorships. There is the whole case against censorship in a nutshell."[15]

Robust scholarship exists examining humankind's history of censorship in societies throughout the world. Whether it is in 1950s America, when novels like D. H. Lawrence's *Lady Chatterley's Lover* (1928) were banned, or present-day censorship of books and films in places like China, Russia, the United Arab Emirates (UAE), and the United States, censorship at its core is about maintaining control over information and ideas that are deemed by the majority culture to be unacceptable and impermissible. Censorship, too, often involves fear or disapproval of "the other." The late Dr. Sara Fine, a psychologist by training and a professor in the library and information science department at the University of Pittsburgh School of Information Sciences, posited that censorship is a mixture of both fear and caring. The censor, in Fine's view, fears information outside of her or his comfort level and experience, while also caring about the impact that that information may or will have on people. Rather than taking a strident, knee-jerk defensive response to a censor's challenge, Fine suggests something that is perhaps more radical and unexpected—*listening*:

> Listening works. I don't mean that listening will change minds; listening opens up dialogues. It is important to hear a person's needs, to listen to the music rather than to respond to the words. When someone says, "I can't believe you have this smut in your library where my kids can find it," if only we could ignore the words and listen to the music: "You must care a lot about your kids."... Confrontation has been replaced by dialogue.[16]

Despite the reasons underlying censorship, however, no matter how ill- or well-intentioned, censorship is by its nature a subjective act. *I* is the operative word, as in "*I* know what is objectionable and offensive" or "*I* know what is best." The censor places the importance of their perception of materials over what others may think about the same item. Rather than deciding not to borrow an objectionable book or closing a book that isn't what one expected or wanted, the censor often appoints themselves to decide which books will and will not be included and made accessible in a collection. These kinds of subjectively motivated actions are antithetical to the core principles of (most) public libraries, library associations, and information professionals.

ALA Office for Intellectual Freedom

ALA's Office for Intellectual Freedom (OIF), a unit within ALA, is arguably the most visible organization advocating for and collecting and communicating data related to IF in the United States and abroad. OIF's website has a succinct statement summarizing its founding, mission, and goals:

> Established December 1, 1967, the Office for Intellectual Freedom is charged with implementing ALA policies concerning the concept of intellectual freedom as embodied in the Library Bill of Rights, the Association's basic policy on free access to libraries and library materials. The goal of the office is to educate librarians and the general public about the nature and importance of intellectual freedom in libraries.[17]

Judith Krug was OIF's first leader. A tireless advocate for IF, freedom to read, and unfettered access to information and ideas, Krug passed away in 2009. A 2009 memorial video uploaded to YouTube features clips from some of the numerous speaking appearances made by Krug throughout her career. One sentence included in that tribute video, stated by Krug herself, crystallizes her belief in the crucial role of libraries in facilitating access to information and IF: "We have to serve the information needs of everybody."[18] Those who have succeeded Krug in leadership of OIF have continued to champion IF. OIF's work is focused on four chief areas: support, training, publications, and public awareness. These are discussed further in the following sections.

Support

Regarding the support area, OIF offers "confidential support to anyone addressing a challenge to materials, programming, policy, or services."[19] The office also has an online "challenge report form."[20] This form enables anyone "to report censorship and request assistance with the challenge."[21]

Training

In the training area, OIF has "subject matter experts" who "are available to speak at workshops and professional development sessions."[22] The topic areas include "collection development, professional ethics, and law and policy concerning intellectual freedom, censorship, privacy, internet filtering, the First Amendment and more."[23] OIF also creates and delivers webinars. These webinars are tailored for information professionals as well as members of the public. Many recordings of the webinars can be accessed via the OIF YouTube channel.[24]

Publications

The *Journal of Intellectual Freedom and Privacy* (*JIFP*)[25] is a digital-only quarterly open-access publication of OIF. It publishes a wide range of IF-related writings, including essays, peer-reviewed articles, legal briefs, court rulings, book-banning incidents, editorials, book reviews, and committee reports. Another significant publication of OIF is the periodically updated *Intellectual Freedom Manual*.[26]

Public Awareness

Banned Books Week is a high-profile annual public awareness campaign, typically held in late September or early October. This initiative arose from an idea of OIF's Judith Krug and has been a continuous IF staple since its launch in 1982: "By focusing on efforts to remove or restrict access to books, Banned Books Week draws national attention to the harms of censorship."[27] Banned Books Week is a multistakeholder effort of "librarians, educators, authors, publishers, booksellers, and readers of all types" to raise greater awareness of "the value of free and open access to information" and the freedom to speak, read, and think.[28]

Book bannings are not a new phenomenon. Throughout human history, books have been contested, banned, and even burned. As Harvard's Gutman Library notes, many works that are now considered "classics" and essential parts of the of literary canon have been viewed as controversial, heretical, blasphemous, or obscene.[29] Examples include Martin Luther's translation of the *Bible*, burned in Germany in 1624 by papal decree; an original version of William Shakespeare's play *King Richard II*, opposed by Queen Elizabeth I in 1597; and Roman Emperor Caligula's AD 35 objections to the reading of Homer's *Odyssey* for promoting "Greek ideas of freedom."[30]

In conjunction with Banned Books Week, each fall OIF releases its yearly list of the top 10 most challenged books, representing challenges over the course of the previous year. ALA distinguishes between a challenged book and a banned book: "A challenge is an attempt to remove or restrict materials, based upon the objections of a person or group. A banning is the removal of those materials."[31] From where does the data on challenged books come? ALA explains that "ALA compiles data on book challenges from reports filed by library professionals in the field and from news stories published throughout the United States."[32] The data that OIF collects and communicates derives from incidents that are reported. Many challenged and banned books are unreported and unidentifiable. Hence, the data is an approximation of challenged books but is necessarily only a portion of the full picture of censorship challenges in the United States.

What rationales are offered for why books are challenged or banned? ALA states that book challenges frequently arise from "a desire to protect children from 'inappropriate' sexual content or 'offensive' language."[33] ALA lists the top reasons for challenging materials as

1 The material was considered to be "sexually explicit"
2 The material contained "offensive language"
3 The material was "unsuited to any age group"[34]

Book Challenges, Removals, and Bans: A Florida Case Study

In the 2020s, an unprecedented number of censorship efforts have roiled the United States. Steeped in the rhetoric and rationale of parental rights, political figures and advocacy groups have achieved significant successes in challenging books and having them removed from classrooms and school and public libraries. March 2024 data released in a report and press release by ALA underscored this remarkable rise: "The number of titles targeted for censorship surged 65 percent in 2023 compared to 2022, reaching the highest levels ever documented by the American Library Association."[35] More recent data from ALA and PEN America on book challenges and bans are discussed later in this chapter, following some additional contextual foundation laying. We also examine various IF, free speech, and censorship-related issues, stakeholders, and developments in the next sections to facilitate deeper analysis and understanding.

No state has been more at the center of the so-called culture wars in the 2020s than Florida. Books—a technology dating back more than ten centuries—have become a fiercely contested asset in those wars. Libraries, schools, and universities—repositories and living laboratories for books, ideas, and knowledge—have, lamentably, all too often become the de facto "theaters of conflict" for these skirmishes. Access to information, intellectual freedom, and the freedoms to speak, think, and even "be" have been assaulted and battered by these conflicts. A diverse array of stakeholders, too, have experienced collateral damage from these culture wars. Many of these "interested parties" have had little if any say in these clashes over values and essential freedoms, chief among them being minors and young children who lack the ability to vote because of their ages. The overarching aim of looking at Florida as a case study is to uncover ethics and legal issues, principles, and determinations that may or may not be applicable to and present in other locales beyond that state.

Ron DeSantis

No recent politician has more influenced intellectual freedom, freedom of expression, and censorship than Florida Republican Governor Ron DeSantis. As PEN America stated in a November 2023 report, "The Florida Effect: How the Sunshine State Is Driving the Conservative Agenda on Free Expression," "Under Governor Ron DeSantis, Florida has faced the country's most intensive rollback of Americans' right to freely express themselves. Recent legislative and policy changes impact individuals' ability to receive and share information online, provide an open and pluralistic education system for their children, participate in peaceful protest, and engage in artistic expression."[36] DeSantis's impacts have been multifold:

- calling for book removals and bans and making concerns about book and pedagogicial content a high-profile issue for the American public, not just in Florida but nationally
- generating momentum for verified book removals and bans in significant numbers
- securing passage of legislation in Florida that makes it easier and legally incumbent on schools and libraries to remove content that is deemed objectionable
- fomenting the passage of similar book-banning laws in other states

First elected in 2018 and then reelected in 2022, DeSantis unsuccessfully ran as a Republican Party presidential candidate in the 2024 election cycle before dropping out of the race in January 2024. Due to Florida term limits, he is set to leave office in January 2027.

Moms for Liberty

Before delving further into the specifics of the DeSantis administration's impacts on IF, free speech, and censorship and the ways that Florida's leading role on these issues has served as a kind of "look what we've done" role model for some states, it's necessary to weave into this narrative the advocacy group known as Moms for Liberty, which sprang from the Sunshine State and has had close links with DeSantis and the Republican Party. Moms for Liberty emerged in January 2021, during the COVID-19 pandemic. Founded by three Florida school board members, Tina Descovich, Tiffany Justice, and Bridget Ziegler, the group initially coalesced to protest two governmental mandates: requiring people (1) to be inoculated with COVID vaccines, which were just becoming available to consumers in early 2021, and (2) to wear masks when in public and at gatherings, schools, businesses, and workplaces to stem the spread of the virus. Vaccine skeptics, like Moms for Liberty, disputed the safety of COVID-19 vaccines as well as the necessity for people to wear masks. These public health requirements were viewed as infringing on their individual rights to abstain from government-sanctioned compliance.

From that original organizing principle, Moms for Liberty expanded its mission scope to welcome "all that have a desire to stand up for parental rights at all levels of government."[37] Its membership rapidly grew. The nonprofit, nonpartisan Brookings Institution published an October 2023 article, "Moms for Liberty: Where Are They, and Are They Winning?," that provided information on the size and range of the group, which it abbreviates as M4L:

> Moms for Liberty is a national organization that oversees dozens of semi-autonomous county chapters. To understand the types of communities where these chapters operate, we collected a list of local M4L chapters from the national M4L Facebook page. We used the number of members in these local chapters' Facebook groups to obtain county-level estimates of the organization's membership. Overall, this amounts to about 103,000 members across 278 chapters in 45 states. These numbers are similar to those reported by M4L in May 2023 (115,000 members across 285 chapters in 45 states).[38]

The group's founders, in speaking to CNN for a December 2021 story, indicated that

> the duo's initial goal was to establish 67 chapters, one in each Florida county. Two weeks after the organization launched, however, a mom from New York reached out, expressing interest in starting a chapter there, Descovich said.
>
> "I said, 'Yes, let's do it," Justice said. "From that moment on, it's been like a wildfire all across the country."
>
> The duo credits "word of mouth" for their growth, which has also spurred groups of parents who oppose their stance. The Moms for Liberty use social media and their website to spread their message.[39]

The communicated core focus of M4L is "parental rights." Indeed, it achieved fame early on for its well-documented T-shirt and sign slogan "We do NOT CO-PARENT with the GOVERNMENT."[40] Its mission statement in February 2025 said, "Moms for Liberty is dedicated to fighting for the survival of America by unifying, educating and empowering parents to defend their parental rights at all levels of government."[41] Under the umbrella of parental rights, M4L has centered its attention on influencing (1) school curricula and (2) books in school libraries. Specifically, the organization works against school curricula that include any elements related to LGBTQ+ communities, critical race theory (explained later), and racial, gender, and ethnic identities and histories. Similarly, M4L strives to remove or outright ban books in school libraries that include references to the issues noted. The group has also advocated against books that address such issues being included and accessible within public library collections. In late 2021, M4L referred to Ron DeSantis as "the Parents' Governor."[42] In December 2023, cofounder Ziegler left M4L "following accusations of sexual assault against her husband, Christian Ziegler, by a woman with whom he and Bridget had a three-way sexual relationship."[43]

The Southern Poverty Law Center (SPLC) is a nonprofit group, founded in 1971, that works to advance human rights and "racial justice in the South and beyond" and "to dismantle white supremacy."[44] It designated M4L as an "extremist" organization in its 2022 "Year in Hate and Extremism Report."[45] In a June 2023 NPR story about M4L having been named "extremist" by the SPLC, Rachel Carroll Rivas, SPLC's deputy director for research, reporting, and analysis, opined that M4L seeks "to undermine public education holistically and to divide communities."[46] That same article observed that although M4L has made book challenges and bans "a central campaign" of its membership, polling research found "that most Americans, including nearly half of Republicans, oppose banning certain books in schools." In a June 2, 2023, piece, NPR explicated the polling methodology and findings on the book ban views of parents and K–12 teachers:

> One poll, of the general public, included 1,316 respondents with an oversampling of K-12 parents (452); the other surveyed 510 K-12 teachers. We sorted through the results and smooshed them thematically into a handful of the most interesting takeaways....
>
> When it comes to state lawmakers removing certain books from schools, such bans have the support of just 5% of Democrats, 16% of Independents and 35% of Republicans. Fifty-two percent of Republicans oppose such efforts.[47]

Reflecting M4L's continuing intersection with conservative values and GOP officials and candidates, on August 30, 2024, Donald Trump spoke at the group's annual convention in Washington, DC.[48] In late September 2024, M4L's cofounder Tiffany Justice shared that she "is interested in serving as former President Donald Trump's education secretary—or in another key position—should he win a second term. 'I would be honored to serve,' Justice said in an interview with *Education Week*."[49]

One of the writers whose published work has been attacked by M4L and other censors is critically acclaimed author and illustrator Art Spiegelman, who spoke to CBS News in an April 2023 story on the fight over banning books. His words expressed a mix of indignation and irony, nuanced acknowledgment, clear-eyed analysis, and frank advice:

> "Stop it!" said cartoonist Art Spiegelman. "I mean, talk about Orwellian, you know? Calling this organization Moms for Liberty, when it's actually for suppression, is about as basic as you could find in '1984,' which I think is listed as a young adult novel still, and probably has been banned in lots of places."
>
> Spiegelman has been speaking out ever since the McMinn County, Tennessee school board voted unanimously last year to ban "Maus," his Pulitzer Prize–winning graphic novel about the Holocaust, citing violence, profanity, and... nudity.[50]

Spiegelman acknowledged that a parent may decide they don't want their child reading a particular book in class. However, he strongly objects to a parent being able to stop other children from reading that book or being able to borrow it from the library. That, he asserted, is "not liberty," but rather is "suppression and authoritarianism."[51] He also urged interested people to become more active in the work of school boards, where decisions on these kinds of issues are often made.

New Laws in Florida

In March 2022, DeSantis championed the successful passage and signing of legislation that restricted public school teachers from including instruction or mentioning of LGBTQ+ content. This statute was named the Parental Rights in Education Act. However, opponents of the law nicknamed it the "Don't Say Gay" law because of its restrictions, in part, on the use of trans-friendly pronouns in schools. The *Hill* reported via a March 2022 article that under the new law

> classroom instruction related to sexual orientation or gender identity will be restricted in the state's primary schools [Kindergarten through 3rd grade]. Educators of all grade levels are prohibited from engaging in instruction on those topics in a manner that is not "age appropriate or developmentally appropriate" for children.
>
> While international media outlets, LGBTQ+ advocacy groups, Hollywood actors and even the White House have condemned the law as hateful and discriminatory, supporters have repeatedly argued that the legislation aims to strengthen the rights of parents and protect young children from being "indoctrinated."[52]

A November 2022 article by Edward Swidriski, assistant counsel for the American Association of University Professors (AAUP), published on the American Bar Associations' website, is titled "Florida's 'Don't Say Gay' Law Raises Serious Legal Questions."[53] In it, Swidriski points out the law's "lack of clarity," its "extraordinary enforcement mechanisms," its impacts of "stigmatizing and silencing LGBTQ+ teachers and students," and adversely influencing "basic principles of academic freedom." Moreover, he observes that "the law arguably runs afoul of the First Amendment's stringent prohibition on viewpoint discrimination and imposes an unconstitutional chilling effect on disfavored speech."

Emboldened by passage of the 2022 Parental Rights in Education Act ("Don't Say Gay" law), DeSantis and fellow Florida legislators turned their attentions to combatting the teaching of critical race theory (CRT) and "wokeism" in schools. In 2022, DeSantis signed HB7, the Individual Freedom Act. This statute is better known as the Stop WOKE Act, also designated the Stop Wrongs to Our Kids and Employees Act. As CNN explained, "The act was one of several bills the Republican governor signed in 2022 as part of his war on 'woke ideology.'"[54] It was intended to prevent teachings or mandatory workplace activities that suggest a person is privileged or oppressed based on their race, color, sex, or national origin. A December 2021 "Stop WOKE Hand-Out," created and disseminated by the governor's office, bullets the aims of the statute:

For schools:

- Codifies the Florida Department of Education's prohibition on teaching critical race theory in K–12 schools.
- Prohibits school districts, colleges and universities from hiring woke CRT consultants.

For corporations:

- Protects employees against a hostile work environment due to critical race theory training.

Consequences:

- Provides employees, parents and students a private right of action.
- Strengthens enforcement authority of the Florida Department of Education.[55]

It is important to note that the Stop WOKE Act would have applicability to a variety of venues and stakeholders, as indicated. For example, it seeks to provide *restrictions*, or *guidance*, depending on one's perspective, on the issue of mandatory trainings in workplaces. The DeSantis administration, for instance, has argued that diversity, equity, and inclusion (DEI) trainings, which may include critical race theory, discussed in more detail in the next section, infringe on the First Amendment rights of persons to *not* be required to participate in activities that violate their constitutional rights in order to be gainfully employed at an establishment that mandates such trainings. The Stop WOKE Act also would apply to Florida public higher education institutions. Examples of universities that would be covered under these sections of the statute include

- University of Florida (Gainesville)
- Florida State University (Tallahassee)

- University of Central Florida (Orlando)
- University of South Florida (Tampa)
- Florida International University (Miami)

DeSantis has spoken extensively about concerns that students are being indoctrinated to liberal and left-leaning ideologies by progressive faculty members at the majority of US universities and colleges. Faculty members have voiced concerns about the whittling away of academic freedom and curtailments on the open pursuit of knowledge that has been the bedrock of higher education for centuries.

"Wokeness," Critical Race Theory, and Social Justice

The concept "being woke" conveys the sense of becoming attuned to and enlightened about issues like overpolicing of minority neighborhoods and the controversial deaths of persons of color, particularly Black Americans, at the hands of law enforcement. "Wokeness" also, though, encompasses other social justice issues and movements, like LGBTQ+ rights, particularly of trans people, greater awareness of sexual assaults of women via the #MeToo movement, and racial justice. *Merriam-Webster Dictionary* defines woke as "aware of and actively attentive to important societal facts and issues (especially issues of racial and social justice)."[56]

"Wokeness" or "wokeism" gained more traction in the public consciousness following the murder of 46-year-old African American George Floyd on May 25, 2020, by Minneapolis, Minnesota, police officer Derek Chauvin and three other officers. Video and audio footage of the incident, including Floyd's desperate, repeated uttering of the phrase "I can't breathe" and the final words "Mama… I'm through" as a plaintive call to his mother who had passed on two years earlier, galvanized protests and riots against police brutality in many US cities for approximately one year. The murder of George Floyd was a pivotal event within the Black Lives Matter movement's protests and calls for police reforms. Yet Floyd's tragic death also needs to be viewed systemically within the larger context of controversial deaths of African Americans at the hands of law enforcement officials in a wide-ranging number of other states, such as Florida and Kentucky.

To more fully understand the contexts that gave rise to elevated rates of book challenges since about 2020, one must take note of other factors that were vital ingredients for these contested areas. One of those undeniable factors was critical race theory (CRT). *Britannica* summarizes CRT:

> Intellectual and social movement and loosely organized framework of legal analysis based on the premise that race is not a natural, biologically grounded feature of physically distinct subgroups of human beings but a socially constructed (culturally invented) category that is used to oppress and exploit people of colour. Critical race theorists hold that racism is inherent in the law and legal institutions of the United States insofar as they function to create and maintain social, economic, and political inequalities between whites and nonwhites, especially African Americans. Critical race theorists are generally dedicated to applying their understanding of the institutional or structural nature of racism to the concrete (if distant) goal of eliminating all race-based and other unjust hierarchies.[57]

The beginnings of CRT can be traced to the 1960s and 1970s, building on prior research examining systemic injustices in legal institutions that "serve the interests of the wealthy and powerful at the expense of the poor and marginalized."[58]

George Floyd's murder in 2020, as well as the pre- and post-Floyd murders of other Black Americans by police officers, coincided with several highly influential scholarly works that intersect with CRT. One of these, Dr. Ibram X. Kendi's 2019 book *How to Be an Antiracist*, is a hybrid of nonfiction memoir and pragmatic how-to guide for actively acknowledging, taking on, and dismantling the facilitators of racism. A *Christian Science Monitor* reviewer described the book: "Deftly weaving his personal experiences with the history of racism in America and current racial inequalities, he offers assessments of the ways that racism in America is shaped and perpetuated by power structures, ethnicity, culture, behavior, class, color, gender, sexuality, and ethnicity, and the steps he thinks we need to take to address them."[59]

Interestingly, in terms of timing and impact, the first date of publication for Kendi's *Antiracist* tome was August 13, 2019, one day before the launch of what would become a similarly influential scholarly contribution. On August 14, "The 1619 Project: A New Origin Story," debuted in the pages of the *New York Times*. In the descriptive words of the newspaper itself, "The 1619 Project is an ongoing initiative from *The New York Times Magazine* that began in August 2019, the 400th anniversary of the beginning of American slavery. It aims to reframe the country's history by placing the consequences of slavery and the contributions of black Americans at the very center of our national narrative."[60] The 1619 Project is a collaboration of Nikole Hannah-Jones, the *New York Times*, and the *New York Times Magazine*. Following their respective publications in the late summer of 2019, works like those of Hannah-Jones and Kendi gradually built audiences of readers and thinkers.

Following the murder of George Floyd nine months later, in May of 2020, and subsequent protests in American cities by citizen activists calling for police reforms, CRT, in conjunction with similarly critical frameworks suggested by Kendi, Hannah Jones, and others, like the intersectional theory work of Dr. Kimberlé Crenshaw, became highly controversial topics in conservative media outlets and circles. DeSantis was one among numerous politicians for whom vociferous denouncement of CRT and related works became a rallying cri de coeur. Indeed, CRT was used by DeSantis as a kind of "original sin" premise for lobbying for new laws such as 2022's Parental Rights in Education and Stop WOKE Act.

Additional Legislative Efforts

In early 2023, the Florida legislature approved legislation, designated HB 1069, that placed restrictions on books in school libraries. This law also made it easier for any person to challenge the inclusion of any book they deemed objectionable. DeSantis signed HB 1069 into law in May 2023, and it went into effect in July 2023. This law resulted in the removal of thousands of books from school libraries. In August 2024, six publishers, five best-selling authors, two parents, and two students brought suit against the State of Florida.

Earlier in this chapter, we looked at some ways that the Stop WOKE Act (2022) has attempted to address free speech in K–12 schools and corporations. Controversially, DeSantis has also made it part of his administration's mission to take measures to rein in free speech at public universities in Florida. As one part of the Stop WOKE Act, DeSantis has continued, as of September 2024, efforts at shaping how faculty can talk about race and gender in classrooms, with the force of law behind his ambitions. Moreover, DeSantis has sought to control whether faculty members can even criticize him or his administration. Yet, since November 2022, "federal judges have repeatedly blocked the law from impacting universities."[61] A coalition of students, faculty, and legal advocacy groups have joined together to have the law's curbs on what can be taught in the classroom declared unconstitutional by the courts. However, as *Inside Higher Ed* reported in a June 2024 article, "Florida Argues It Could Stop Professors From Criticizing Governor," Florida hired prominent conservative lawyer, Charles J. "Chuck" Cooper, to advocate for its position that a state can require professors not to endorse views that are different than those of the state. During arguments before the 11th Circuit Court of Appeals in Atlanta,

> One of the three judges on the panel, a Donald Trump appointee, later asked a question that showed how far Cooper's argument could extend. The judge posed a hypothetical about how much the state could limit classroom teaching if the judges were to accept Cooper's arguments: "Could a legislature prohibit professors from saying anything negative about a current gubernatorial administration?"
>
> Cooper replied: "I think, your honor, yes, because in the classroom the professor's speech is the government's speech and the government can restrict professors on a content-wide basis and restrict them from offering viewpoints."[62]

A number of organizations have challenged Florida's stance. The American Association of University Professors' (AAUP) general counsel Risa Lieberwitz asserted that Florida's intended actions under the Stop WOKE Act have no merit under the First Amendment's protections of free speech and "would eliminate academic freedom completely in the classroom."[63]

In July 2024, Chief US District Judge Mark Walker blocked the parts of the Stop WOKE Act relating to trainings in the workplace. This section at issue "was intended to prevent teachings or mandatory workplace activities that suggest a person is privileged or oppressed based on their race, color, sex or national origin."[64] The judge's order granted a "permanent injunction" on the statute's workplace training restrictions, holding that this section "violates free speech rights under the First and Fourteenth Amendments to the US Constitution."[65]

During his failed bid to become the Republican nominee for president in the 2024 election period, DeSantis continued to make his fight against "wokeness" a signature issue of his platform. With a second gubernatorial election in November 2022, DeSantis's January 2023 inauguration speech avowed that "we will never surrender to the woke mob. Florida is where woke goes to die."[66] As of late 2024, the judicial verdicts on the DeSantis administration's so-called war on woke have been anything but decisive wins, with several key losses notched.

In addition to Florida, numerous other US states have considered or signed bills into law that have impacted intellectual freedom in public and school libraries. Some of these laws have even threatened or codified civil and criminal penalties for librarians and others who provide minors with materials that are deemed to be inappropriate or pornographic. At the time of this writing, a number of these cases have not been decided.

In Florida, there is an ongoing case bringing together a coalition of publishers, authors, parents, and students suing a school district for the alleged unconstitutional removal of school library books in Escambia County (county seat Pensacola) in the state's Panhandle. However, in September 2024, a settlement was announced between a similar coalition of author, parent, and student plaintiffs who had sued the school district and school district officials in Nassau County, near Jacksonville, in northeastern Florida. As a *Politico* article noted, "A northeast Florida school district this week agreed to restore 36 books that were challenged and previously pulled from campus libraries in a settlement of a federal lawsuit fighting how local officials carried out the state's policies for shielding students from obscene content."[67]

The Selendy Gay PPLC law firm represented, among others, Peter Parnell and Justin Richardson, the authors of the consistently challenged 2005 children's picture book *And Tango Makes Three*. A September 12, 2024, announcement by Selendy Gay explicates the settlement details:

> The settlement requires the restoration of students' public school library access to *Tango* and 35 other books challenged by Citizens Defending Freedom (CDF), a national special interest group leading the charge on book banning throughout Florida. These 35 books include significant works by Toni Morrison, Jonathan Safran Foer, Alice Sebold, Jodi Picoult and Erika Sanchez, which address racism in America, as well as the life experiences of immigrants, first-generation Americans, trans Americans and other underrepresented communities and individuals.
>
> The settlement prevents the School Board from unlawfully removing public school library books during private, closed doors meetings—a violation of Florida's public disclosure statutes, which are intended to ensure transparent government decision-making. Additionally, the settlement undoes the School Board's use of unlawful pretextual bases for removing public school books that have viewpoints with which the Board and its representatives disagrees.
>
> Significantly, the Board acknowledges in the settlement that *Tango* has pedagogical value, is appropriate for students of all ages, and contains no "obscene" material—facts that another Florida School Board—in Escambia County—is still contesting.[68]

Relevant to some of the major ethics topics examined in this book, Selendy Gay attorney Lauren Zimmerman emphasized the information access issues implicated by censorship actions like those taken by Nassau County's school board: "Students will once again have access to books from well-known and highly lauded authors representing a broad range of viewpoints and ideas."[69] Of pertinence to intellectual, free speech, and freedom of expression interests, Katie Blankenship, senior director of PEN America Florida, observed that the settlement "is an historic moment in the fight to protect free speech and expression. This settlement sends a clear message to those who seek to ban and limit access to literature, art, and the freedom to read."[70]

Insights From the Florida Case Study: What Comes Next?

Though access to diverse books was preserved via a settlement decision in Nassau County, Florida, the outcomes of lawsuits challenging similar restrictions of books in public libraries are yet to be determined. Additionally, it is unclear how newly enacted IF and censorship laws in states such as Idaho, Utah, Texas, and more will be applied, contested, upheld, modified, or struck down. Hence, issues involving IF, censorship, and access to books and other collection materials in some cities and states can reasonably be expected to remain highly fluid, unpredictable, and fraught with legal, ethical, and policy implications. This Florida case study has intentionally focused on that state to present a "snapshot in time" of one US geographical region that has, at the very least, elevated IF and censorship issues to a higher profile at the national level. The legal and ethical principles, contentious issues, and stakeholders are, in many ways, applicable to other US states that find themselves at the forefront of battles over fundamental precepts like the freedoms to read, think, and speak that are at the very foundation of democracy. With the departure of Joe Biden from the 2024 presidential race on July 21, 2024, Democratic Party candidate Vice President Kamala Harris ran for president, and she and her vice presidential pick, Governor Tim Walz of Minnesota, regularly included mentions of conservative agenda-pushed book bans within their campaign stump speeches.

Later in this chapter we examine some additional ways to fight back against censorship. Some of these strategies include testimony by IF scholars and advocates before the US Congress and governing bodies like school boards and the making and airing of documentary films to raise awareness of IF, as well as groundbreaking collaborations between libraries and publishers to make access to banned books possible via digital means.

Self-Censorship/Internal Censorship

Self-censorship, also called internal censorship, is another aspect of censorship. Self-censorship involves a person unconsciously or consciously censoring themselves. This contrasts with censorship, or a request/demand for censorship that derives from an external source, such as a customer, supervisor, employer, or governing body. Though self-censorship may be unconsciously done, it still has impacts on collections, individuals, and communities, just as when it is purposefully done. Self-censorship decreases access to information. Some reasons or rationales for internal censorship may include the following:

- Desire to avoid potential controversy or negative publicity about a work
- Belief that certain items are dangerous or fall outside of community standards and acceptable norms and mores
- Trepidation about the consequences of possessing or acquiring a work, such as loss of revenue from funding and granting bodies
- Acquiescence to the "chilling effects" of potentially negative impacts associated with an information object
- Fears regarding cyberbullying/online trolling, doxing, swatting, and physical safety of oneself and family
- Self-protection or self-defense
- Concern(s) regarding damage to one's reputation; disapproval from one's family, friends, community, peers, and superiors; loss of gainful employment due to termination or suspension; or experiencing an undesirable downward change in status

- Greater emphasis on one's own needs than the needs of others
- Lack of information, awareness, or empathy for others

Self-censorship is often described as "insidious" because it is more hidden and subtle than censorship. Censorship often manifests through a verbal or written complaint from an individual or group, which creates a physical record and paper trail. In recent years in the United States, people and groups objecting to materials have been more prone than in previous years to holding press conferences about items they want removed from collections, speaking at organizational and community board meetings, conversing with the media via articles and interviews, writing op-eds, and posting content to social media sites that explains their concerns and objections.

To the contrary, the self-censor and internal censor may never speak of or acknowledge in any way the censorship in which they are engaging. The effects of internal censorship on intellectual freedom, nevertheless, create results that are akin to overt censorship, namely, less access to information, ideas, and other voices. Longtime *New York Times* reporter Tom Wicker observed that "self-censorship silences as effectively as a government decree."[71] It often leaves no paper trail or evidence of self-censorship. The unseen thoughts and beliefs within the mind of the internal censor drive the censoring behavior. A lack of certain materials within a collection, or a pattern and practice of "disappearance" of materials that were previously collected, may enable one to infer or prove internal censorship. However, other explanations and variables, such as budgetary factors, outdated formats, changing tastes, collection development priorities, community needs, as well as deaccessioning and weeding, can make it difficult to render clear-cut conclusions about self-censorship.

One hallmark of internal censorship is that even though an organization may have policies in place for users to submit objections to certain items within collections, such policies may in fact not be followed. Rather, internal censors may decide ad hoc, "in the moment," whether to keep or remove an item from a collection without adhering to any kind of formal, transparent, substantive review process and communication of decisions. I worked in a public library where this practice occurred. Items were removed from the collection (in library parlance, *deaccessioned*) with no explanation.

Sometimes explanations for censorship may be provided that offer a window into the mind of the censor or would-be censor. Indeed, I experienced firsthand the flagging and subsequent temporary sequestering of a new book by humorist/detective novelist and 2006 (unsuccessful) Texas independent gubernatorial candidate Kinky Friedman at the hands of a public library staff person working in a collection development capacity. What was the explanation given for why the new book was flagged when it arrived at the library for cataloging and addition to the library's fiction collection? Two things: the staff person's unfamiliarity with the author and that individual's admitted discomfort with the writer's unusual first name. On the contrary, the MLIS-holding librarian who had selected the book for inclusion was familiar with the author and had consulted mainstream publications that provide reviews of books, as well as having looked at peer institutions in the library's shared network to see who had purchased the book.

Internal censorship can occur in all types of information centers. In addition to public libraries, Debra Lau Whelan, writing for *School Library Journal*, calls out the "dirty little secret" of self-censorship in K–12 libraries:

> Self-censorship. It's a dirty secret that no one in the profession wants to talk about or admit practicing. Yet everyone knows some librarians bypass good books—those with literary merit or that fill a need in their collections. The reasons range from a book's sexual content and gay themes to its language and violence—and it happens in more public and K–12 libraries than you think.
>
> "It's probably fairly widespread, but we don't have any way of really knowing, because people who self-censor are not likely to broadcast it," says Pat Scales, president of the Association of Library Services to Children and author of *Protecting Intellectual Freedom in Your School Library* (ALA Editions, 2009). And since most people think librarians are the best champions of books, adds Scales, their jobs give them the perfect cover.[72]

It is often said in management circles that it's not enough to just name a problem, but rather important to offer some possible solutions. What, then, are some ways to begin to (better) tackle the challenges and consequences associated with self-censorship?

- Talking about self-censorship
- Ensuring it is an integral part of information studies programs' core curricula and electives
- Including self-censorship in staff orientation and organizational handbooks to ensure it is addressed during on-boarding, as well as addressing self-censorship during annual reviews, as applicable, to ensure that clear expectations are set and maintained throughout all stages of an employee's tenure with the institution
- Raising awareness of ethics resources and advice hotlines that are available within information-related professional associations at local, state/provincial, and national levels
- Providing online, in-person, and hybrid trainings, webinars, and forums for information professionals at all levels to build trust and to learn and be able to talk about these issues in safe environments
- Promoting greater awareness within the general public as to why well-developed, diverse collections are important and how these collections benefit individuals and contribute to healthier, more inclusive, better-educated communities
- Implementing and adhering to clear, balanced, transparent collection development policies and procedures that users can file for objecting to materials
- Listening to the perspectives and feelings of others, in the vein of Sara Fine's observations about the importance of listening

Piercing the secretiveness that pervades self-censoring behaviors is a way to also diminish the shame and isolation around these challenges by bringing them into the open. Because self-censorship is not easily discernable by others, it is incumbent on information professionals to hold themselves to the highest ethical principles in the course of building and maintaining the collections that internal and external users entrust them to select, provide access to, and steward on their behalf.

Collection Development and Selection

One of the seminal readings in the literary canon of library and information science is Lester Asheim's 1953 article "Not Censorship but Selection."[73] The crux of Asheim's article is that librarians *select* materials, in contrast to censors who *censor* materials. ALA's website further clarifies the distinction between selection and censorship in answering the question, "Don't librarian's censor everything they choose not to buy for the library?"

> No library can make everything available, and selection decisions must be made. Selection is an inclusive process, where the library affirmatively seeks out materials which will serve its mission of providing a broad diversity of points of view and subject matter. By contrast, censorship is an exclusive process, by which individuals or institutions seek to deny access to or otherwise suppress ideas and information because they find those ideas offensive and do not want others to have access to them. There are many objective reasons unrelated to the ideas expressed in materials that a library might decide not to add those materials to its collection: redundancy, lack of community interest, expense, space, etc. Unless the decision is based on a disapproval of the ideas expressed and desire to keep those ideas away from public access, a decision not to select materials for a library collection is not censorship.[74]

The selection/censorship distinction can be conceptually challenging, and ALA's "Intellectual Freedom and Censorship Q&A" teases out some further helpful clarification on how censorship and selection differ:

> Selection is following a collection development policy. Library workers use their expertise to select items for the community they serve. There are various constraints and factors that library workers consider when they are purchasing materials for their communities' collections. Censorship occurs when a library worker strays from the collection development policy to not select certain materials based on personal beliefs or fear of controversy.[75]

Defining Collection Development

If collections are the foundations of libraries and archives, it follows that development of those collections is one of the most important activities that librarians, archivists, and other information professionals perform and oversee. The Online Dictionary of Library and Information Science (ODLIS), a free digital resource created by Joan M. Reitz and published in paperback and hardcover print formats by ABC-Clio's Libraries Unlimited, defines collection development as

> the process of planning and building a useful and balanced collection of library materials over a period of years, based on an ongoing assessment of the information needs of the library's clientele, analysis of usage statistics, and demographic projections, normally constrained by budgetary limitations. Collection development includes the formulation of selection criteria, planning for resource sharing, and replacement of lost and damaged items, as well as routine selection and deselection decisions.[76]

The selection of materials and development of collections have consequences for many values and ethical considerations that are central to the missions and visions of information centers. Front and center is IF. Access to information is also directly influenced by selection and collections. Diversity, inclusion, and representation of viewpoints, communities, and lived experiences are undoubtedly impacted by information center decisions about what and whose works to choose and collect. Indeed, I believe that what an information center chooses *not* to include in its collections often reveals more about an organization's culture and values than what it does provide access to:

- Whose voices are not being heard?
- What works are being excluded?
- Whose information needs are being unmet?
- Why are these voices, works, and needs being unmet?

Data and Sources for Collection Development

Information professionals can turn to a variety of tools, resources, and strategies to ensure they are proactively creating well-balanced collections, per the ODLIS definition provided in the previous section. One resource is US Census Bureau data. The US Constitution requires a nationwide census every 10 years, called the decennial census. The decennial census is necessary for apportioning members of the US House of Representatives. Districts in states that lose population may commensurately lose representatives to the House, and states that experience population growth may similarly gain representation. What isn't as well known, though, is that the US Census Bureau collects and reports various types of data continuously, not just through the requisite decennial census. Hence, the US Census is a useful source for examining all kinds of demographic data, which can provide insights about a community on, say, the numbers of people with a college degree, breakdowns of homeowners and renters, and racial, gender, and age statistics.

The US Census Bureau is useful, but it probably does not provide a complete picture of the community or region. Less visible or invisible persons and communities may exist, which traditional reporting mechanisms may not observe. Ethnographic studies, grant-funded research, reports from social service agencies and nonprofits, focus groups, task forces, mainstream and independent media publications, and local for-profit businesses can provide snapshots and insights into individuals and communities who may be unserved in terms of collections, programming, and broader strategic planning.

Some trade magazines and tools that information professionals use for collection development include, but are not limited to

- *Booklist*
- *Kirkus Reviews*
- *Library Journal Book Review*

- *New York Times Book Review*
- *Publishers Weekly*
- *School Library Journal*

One of the tangible things that information professionals, as well as media sources reporting on information centers, can do to better inform the public about the care and expertise that is taken in selecting materials and building collections is to emphasize the knowledge that most MLIS-program students attain on aspects of selection and collection development through their courses. Additionally, as a general practice, information professionals are reading thoughtful, nuanced, trustworthy reviews from authoritative publications like those listed. Selectors are also looking at library catalogs to see who else owns works and determining where gaps in coverage exist.

Sanitization and Bowdlerization

History records that some humans throughout the ages have felt the need to appoint themselves as sanitizers of creative works. Even Shakespeare's celebrated Renaissance-era works have not been immune from the language cleanup police. A 2017 *Smithsonian Magazine* piece highlights one of the most famous examples, so famous, in fact, that the family name became a verb—*to bowdlerize*—to describe this kind of book sanitizing behavior: "Thomas Bowdler is best-remembered for being the credited author of *The Family Shakespeare*, a book first published in 1807 'in which nothing is added to the original Text: but those words and expressions are omitted which cannot with propriety be read allowed [*sic*] in a Family.'"[77]

In the modern digital age, Bowdler's motivations are carried on by others. In 2011, Alan Gribben, an Auburn University English professor, published a new edition of Mark Twain's *Huckleberry Finn*. Because the book's copyright period has expired, it is in the public domain and can be changed in any way a creator desires and serve as a building block for new works. One of the changes Gribben made was to replace the N-word with the word *slave*. For context, a 2015 MTV News story points out that the "N-word shows up 219 times in Mark Twain's classic 1885 tale of childhood adventure 'The Adventures of Huckleberry Finn,' which tracks Huck's flight from his drunken, abusive dad and journey down the Mississippi River with an escaped slave name Jim."[78] In a 2011 *New York Times* article, Gribben explained that he thought "his edition would be welcomed by schoolteachers and university instructors who wanted to spare the reader from a racial slur that never seems to lose its vitriol."[79]

In that same article, *New York Times* book reviewer Michiko Kakutani provides historical context for Gribben's bowdlerized work. She also presciently frames the free speech and culture wars of both the right and the left that were beginning to roil college campuses and that have substantially increased in the past few years, issues like who can be invited to campuses to speak, whether to rename campus buildings named for proslavery historical figures, and the removal of controversial statues:

> Efforts to sanitize classic literature have a long, undistinguished history. Everything from Chaucer's "Canterbury Tales" to Roald Dahl's *Charlie and the Chocolate Factory* have been challenged or have suffered at the hands of uptight editors. There have even been purified versions of the Bible (all that sex and violence!). Sometimes the urge to expurgate (if not outright ban) comes from the right, evangelicals and conservatives, worried about blasphemy, profane language and sexual innuendo.[80]

In addition to sanitization of books, such efforts also encompass other creative works as well as the venues where those works are made available. The high-school musical and play, a staple of the young adult educational experience for thousands of students for decades, has increasingly become embroiled in the censorship wars. This is particularly true as modern musicals and dramas, like *Rent*, *The Laramie Project*, *August: Osage County*, and *The 25th Annual Putnam County Spelling Bee*, have become more inclusive and representative of the range of human experiences. A 2023 New York Public Library article, "Censorship on Stage: Plays and Musicals Challenged at High Schools," summarizes a number of illustrative examples. One example involving censorship of language is *Legally Blonde: The Musical*:

> In 2021 a planned production of the musical at Maryville High School in Missouri was postponed by the school board less than 24 hours before the opening curtain in order to request changes in language and choreography from the play's license holder, Music Theatre International. While some changes were granted, the request to cut the song "Gay or European" was rejected.[81]

Sanitization of works raises thorny ethics issues, such as

- What are the ethical interests of the creators (living and deceased) whose works may be sanitized? What are the interests of their estates?
- What are the ethical distinctions between sanitizing works that are still covered by copyright and those that are in the public domain?
- What are the ethics interests of persons who access works? Do people who consume and engage with works have the right to not be offended by works or to be protected from the impacts of those works?
- What are the archival roles of the historical record and memory vis-à-vis creative works?
- Who decides what is "age appropriate"?
- What are the middle-ground positions that exist between sanitizing a work and using the original version of a work? Should changes or redactions be explained, such as via annotations? Should explanatory notes be included, say, as footnotes, as is done with some publications of Shakespearean works to explain archaic language and historical references?
- What are the ethical duties, if any, of information professionals in weighing whether to provide access to sanitized works?

Creator Interests

Opinions on sanitization of works span the gamut of viewpoints. Even what we call this practice is not without disagreement: is it sanitizing, or bowdlerizing, or retouching, or something else?

The editorial board of the *Washington Post* did not hide *its* stance on this contentious issue with a June 2023 op-ed titled "21st-Century Editors Should Keep Their Hands off 20th-Century Books." The board referenced recent-past examples of sanitization of works by the estates or copyright-holding entities of acclaimed and deceased authors such as Agatha Christie, Roald Dahl, and Ian Fleming. In the board's words, this kind of "retouching" is an affront to multiple interests: "It's a threat to free expression, to historical honesty and, indeed, to readers themselves for contemporary editors to comb through works of fiction written at different moments and rewrite them for today's mind-set."[82]

The editorial board also raises important questions about the nature of authenticity and the rights of authors to express their voices, warts and all. Additionally, the board observes that these works, whether they are controversial or not, serve as "cultural artifacts" that represent snapshots of different moments in time. These artifacts provide insights into norms, mores, and beliefs of the authors themselves and the societies and time periods that they inhabited. To alter or delete these references is akin to exercising a self-anointed "I know better than you," Command + Z role over the world's historical record and memory.

The following example offers insight into how some creators may feel about their works being altered or sanitized. Amid today's personal brand-focused social media "influencers" world, singer-songwriter Janis Ian's 1975 coming-of-age song "At Seventeen" is as timely and mournfully resonant as ever. Ian's song "Society's Child," written and recorded in 1965 when she was only 14, depicts the all-too-human costs of a taboo teenage interracial romance and remains achingly evocative. As a 2016 *New York Times* article observes, "Has there ever been a major pop hit more brutally self-lacerating than Janis Ian's confessional song 'At Seventeen,' in which she used the words 'ugly duckling girls like me' to describe the misery of not being a high-school golden girl? The wounds of adolescent rejection and cruelty may eventually heal, but they leave scars."[83]

After reading the June 2023 *Washington Post* op-ed on "keeping hands off of works," I perused some of the comments in the piece and serendipitously chanced on the comment below that had been posted and signed by Janis Ian herself (with the username Janis17):

> As a songwriter and author, I dread what might be changed in my work after my death. Perhaps someone would try to make "At Seventeen" less painful? Perhaps "Society's Child" should have a different ending? Objectionable words should be used to educate. Changing them is worse than banning the book outright. If a book is banned, you can still find copies. If words are changed, sooner or later all editions will be changed, and the original—along with the author's intent, framed within the context of their times—irretrievably lost.[84]

Creator Responses to Sanitization Efforts

What recourse do authors have in contesting proposed or actual changes made to their works? The frequent answer that lawyers give when asked a legal question is "That depends," which signifies the variety of variables and "fact pattern" subtleties that may influence an answer or recourse. Copyright law affords one avenue for authors. For example, copyright law, which is discussed in detail in a later chapter, gives authors the rights to control the content of their works (including translations, prequels, and sequels, all of which fall under what in copyright parlance are called *derivative works*). However, authors sometimes "sign over" their copyright rights when they sign contracts with publishers, particularly at the start of their writing careers before they gain negotiating clout and notoriety. So authors may be at the mercy (or munificence) of their publishers to determine whether their works will be sanitized or not. Additionally, works of authors who are deceased may be controlled by all kinds of parties, including friends or family members, or "estates" that are often made up of corporate interests.

Moral rights is a subset within copyright law that may offer a possible course of action.[85] US copyright law only recognizes some narrow moral rights, particularly with regard to specific kinds of "visual art," as codified in the Visual Artists Rights Act (VARA) of 1990. Moral rights give authors two key rights: (1) the right of attribution and (2) the right of integrity. "Attribution" denotes the right of the creator to have their name acknowledged in conjunction with their creative work. "Integrity" signifies the right of the creator to protect the integrity of their work and preclude others from altering, distorting, or destroying it. Aside from the VARA exception cited, though, US copyright law largely does not incorporate rights of attribution and integrity. In part, this is because of the fair use defense/doctrine of US copyright law, which in some instances legally enables people to make changes to copyrighted works that are not infringing creator's or copyright holder's copyright rights. As an example of the ways that courts and countries vary regarding their stances on moral rights, interesting case law exists on issues involving whether colorization of black-and-white films violates the right of integrity.

Additionally, readers and "the public" may influence decisions on sanitization and retouching. In 2023, for example, the estate of the late children's book author Roald Dahl set off a firestorm when it announced that the publisher Penguin Random House was changing some words deemed offensive within beloved books like Dahl's 1964 *Charlie and the Chocolate Factory*. As a 2023 BBC article "Roald Dahl: The Fierce Debate Over Rewriting Children's Classics" described it, "The [*Telegraph*] newspaper found that *Charlie and the Chocolate Factory*, *The Witches*, *Matilda* and more than a dozen other titles had words relating to weight, height, mental health, gender and skin colour removed. Some of the changes seemed baffling."[86] Following outcries about the proposed changes, the publisher ultimately agreed to provide two versions of each Dahl book at issue: an amended version and the original, unexpurgated version. As a February 2023 *USA Today* headline announced, "Penguin Will Sell 'Classic' Versions of Roald Dahl Books After Rewrite Censorship Backlash."[87]

Controversies like the Dahl one can be expected to continue to periodically arise as the interested stakeholders—authors, their estates, publishers, and the public—weather evolving norms and sensibilities, amid a sea of ethics issues. In the words of the June 2023 op-ed by the *Washington Post*'s editorial board, though, "Literature is often meant to be provocative."[88]

Strategies for Championing Intellectual Freedom and Confronting Censorship

Faced with unprecedented numbers of book challenges and bans in the past few years, intellectual freedom advocates can feel overwhelmed at times. What, if any, actions can be taken? As Mahatma Gandhi observed, "Be the change that you wish to see in the world." The following are some tangible examples of individuals, communities, and organizations that are taking action to advance the freedom to read and pushing back against censorship:

Toni Morrison Courage Award

In June 2024, the *New Republic*, the writers group PEN America and other partners conferred the first annual Toni Morrison Award, named for the late Pulitzer Prize–winning author. The award recognizes individuals and organizations that have demonstrated a commitment to intellectual freedom and the freedom to read. Morrison championed access to information, as noted via her New York Public Library–displayed quote about access that was spotlighted in chapter 4.

One of the five award recipients was Florida's Broward County Library Director Allison Grubbs. Broward County created so-called book sanctuaries in all thirty-six branches: "designated areas in their libraries where endangered stories are protected and made accessible for exploration or check out."[89] Other award recipients included a Texas history teacher, two Texas booksellers who challenged a Texas law requiring all booksellers to rate books that are sold to schools, and Newberry Award–winner Jacqueline Woodson, whose 2014 book *Brown Girl Dreaming* has been challenged in Florida and Texas.

Congressional and State Legislature Testimony

Testifying before federal and state legislatures can serve at least three useful purposes:

1. Raising awareness in a public forum of issues, like IF and book censorship
2. Potentially securing support from elected legislators
3. Contributing to the public record by providing sworn testimony

On September 12, 2023, the US Senate Committee on the Judiciary heard from experts on the topic "Book Bans: Examining How Censorship Limits Liberty and Literature." One of those who testified was Dr. Emily J. M. Knox, associate professor at the University of Illinois at Urbana-Champaign School of Information Sciences and chair of the board of the National Coalition Against Censorship (NCAC). Knox's research expertise includes intellectual freedom, censorship, access to information, and information ethics. Her testimony included brief descriptions of book banning and book challenges in the United States, IF and freedom of expression, autonomy, censorship practices, and the role of the library as "the physical embodiment of the right to freedom of expression in a community."[90] Additionally, Knox included in her appendices two chapters from her 2023 book *Foundations of Intellectual Freedom*.

TV Segments and Book Clubs

MSNBC journalist Ali Velshi has been including a Banned Books Club segment as part of his Saturday morning program *Velshi* since the culture wars over books in schools and libraries heated up in 2023. In addition to talking each week with the author of a challenged or banned book, spotlighting an intellectual freedom advocate, or discussing a challenged classic like George Orwell's *1984*, the initiative also invites people to read these books: "'Velshi Banned Book Club' is a multi-platform series rooted in literary and cultural analysis. Read (and watch, and listen) along with Ali as he discusses books that range from the courageous and illuminating to the misunderstood and powerful. Never boring, but always banned, this post will keep updating as we add new books to the list."[91]

On the September 7, 2024, segment, Velshi spoke with young adult author Mindy McGinnis a former school librarian, about her 2019 book *Heroine*. *Heroine* grippingly exposes the impacts of drug use, as a blurb by New Castle/Henry County Public Library in Indiana describes: "Heroine is a gut-wrenching exploration of addiction, friendship, and the unraveling of a promising young athlete's life.... that serves as a cautionary tale for readers."[92]

Opponents of the inclusion of *Heroine* in some libraries have argued that it glorifies drug use. McGinnis ardently rebuts such assertions, explaining that from the very first pages of the book, the negative effects of drug use are made clear. In the *Velshi* segment, McGinnis shared that one young reader of the book wrote to the author to tell her how much the book had enabled her to better understand her mother, who is incarcerated for drug-related convictions.

Documentaries

Documentary films are an important tool for raising awareness of incidents involving book challenges and bans, the people who are squaring off against censorship, and varying stakeholder perspectives. One example is *The ABCs of Book Banning*, which features 101-year-old Grace Linn. She inspired the documentary after speaking against book removals at a school board meeting in Martin County, Florida: "The 27-minute short documentary highlights young children who talk about the books that have been challenged, restricted and banned and what they think about the decisions to pull certain books from libraries."[93] The film was nominated for but did not win the Oscar for "Best Short Documentary" at the 96th Academy Awards in March 2024.[94]

Another example is the 31-minute documentary *To Be Destroyed* (2024). As described on the Tribeca Film Festival website:

> *To Be Destroyed* follows author Dave Eggers as he meets with students and teachers in the Rapid City, SD school district where his novel *The Circle* was pulled from shelves along with works by Alison Bechdel, Stephen Chbosky, Bernardine Evaristo and Imbolo Mbue. Through footage of heated school board meetings, community rallies and interviews with locals, we learn how these books were designated "to be destroyed."[95]

Panel Discussions

As with testimony given in front of legislature bodies, panel discussions can be effective in making people more aware of issues such as IF and book challenges. Panel discussions can be conducted in a number of formats and modalities. They also can provide opportunities for interesting interactions and the sharing of viewpoints among the moderator and panelists, as well as the chance to take questions from audience members.

In addition to panels that are oral in nature, articles can serve panel discussion purposes by interviewing a variety of people with expertise in the subject matter at hand. For example, Teachers College (TC) Columbia University provided the opportunity for four TC experts to talk about "What You Need to Know About the Book Bans Sweeping the U.S.," via a September 2023 article.[96] The domain area expertise of the four faculty members spanned various disciplines, as evinced by their TC affiliations, facilitating interdisciplinarity and diversity of perspectives:

- Alex Eble, assistant professor of economics and education
- Sonya Douglass, professor of education leadership
- Michael Rebell, professor of law and educational practice
- Ansley Erickson, associate professor of history and education policy

In April 2023, I participated in a recorded panel discussion for the University of Pittsburgh (Pitt) on the topic "Haven't We Been Here Before: A Panel Discussion on Banning LGBTQIA+ Books," as well as participating in a live and recorded Banned Books Week Panel at Pitt in September 2019. Additionally, I presented on "Why in the World Would We Want a Program on Banned Books? Using R³ECALL to Champion Intellectual Freedom" at the 2023 Association for Library and Information Science Education (ALISE) Conference in Milwaukee, Wisconsin. (R³ECALL is an acronym for a model I created for the conference; it stands for Reframe/Response/Research, Education, Collaboration, Advocacy, Legal Action, and Listening.)

Scholarship/Scholarly Communication

In the digital age, scholarship can manifest in a variety of traditional (e.g., peer-reviewed articles and books) and nontraditional (e.g., website articles and blog postings) formats. The digital site the *Conversation* is a venue for academics and other professionals to share research and viewpoints on a wide swath of topics. Dr. Nicole A. Cooke, Baker Endowed Chair and Professor of Library and Information Science at the University of South Carolina, researches IF and censorship. In July 2023, Cooke wrote the web article "How Book-Banning Campaigns Have Changed the Lives and Education of Librarians—They Now Need to Learn How to Plan for Safety and Legally Protect Themselves."[97] This information not only can make people more aware of the human toll of censorship efforts but also provide practical information and recommendations for addressing a novel issue (i.e., the dangers faced by librarians due to polarization over censorship efforts and ways that library science programs are responding to these threats):

> The drive to challenge, ban or censor books has not only changed the lives of librarians across the nation. It's also changing the way librarians are now educated to enter the profession. As a library school educator, I hear the anecdotes, questions and concerns from library workers who are on the front lines of the current fight and are not sure how to react or respond. What once, and still is, a curriculum that includes book selection, program planning and serving diverse

communities in the classroom, my faculty colleagues and I are now expanding to include discussions and resources on how students, once they become professional librarians, can physically, legally and financially protect themselves and their organizations.[98]

Collaborative Partnerships

Many stakeholders are concerned about and have overlapping interests in supporting intellectual freedom and pushing back against censorship initiatives. Collaborations and partnerships of publishers, authors, and booksellers, for example, are one way to champion the freedom to read and access different viewpoints and lived experiences via books in print and digital formats.

Books Unbanned is a program to provide ebook access to banned books for US teens who reside in areas where books have been banned:

> In April 2022, Brooklyn Public Library joined those fighting for the rights of teens nationwide to read what they like, discover themselves, and form their own opinions. Inspired by the American Library Association's Freedom to Read Statement, Books Unbanned is a response to an increasingly coordinated and effective effort to remove books from library shelves.

The following libraries have joined Books Unbanned as crucial partners in defending and expanding the freedom to read.

- Boston Public Library
- LA County Library
- San Diego Public Library
- Seattle Public Library[99]

In September 2024, *Library Journal* observed via "hard numbers" that the program is "still growing strong more than two years after launch":

> Since launching Books Unbanned in April 2022, Brooklyn Public Library has given 8,200 teens and young adults aged 13 to 21, from all 50 U.S. states, full access to its extensive catalog of ebooks, e-audiobooks, and online learning database collections. These cardholders, all of whom have written personal emails explaining their need for access to the library's collection, have checked out over 270,000 books.[100]

Publisher Simon & Schuster launched its Books Belong multipronged initiative to coincide with ALA's annual Banned Books program in October 2023:

> This program will highlight the merits of books that have been subject to bans and challenges, and will provide educators, parents, librarians and students with tools and resources to preserve and strengthen our right to read.
>
> Launching in October in conjunction with national Banned Books Week 2023, the initiative includes a new resource website, Books Belong, which will host reading group guides and videos, book lists, giveaways, exclusive author and expert content, and links to additional resources, including those from Unite Against Book Bans and PEN America. Supporters of

> the right to read will find resources for them to take action when faced with a challenge in their community, as well as guidance on how to easily incorporate banned and challenged books into classroom, library, and family reading time.[101]

Books Belong continued in 2024, featuring items such as a list of 486 Simon & Schuster books "that have been targeted for removal from bookshelves around the nation."[102] Examples of these banned and challenged books include:

- *Go Ask Alice* by Anonymous
- *Heaven's Prisoners* by James Lee Burke
- *Between the Lines* by Jody Picoult
- *All the Light We Cannot See* by Anthony Doerr
- *Close Range* by Annie Proulx
- *The Sun Also Rises* by Ernest Hemingway
- *Dracula* by Bram Stoker
- *The Glass Castle* by Jeannette Walls

Digital Public Library of America Banned Book Program

In 2023, the Digital Public Library of America (DPLA) debuted a Banned Book Program. This program offers "free nationwide access to books restricted in schools or libraries" accessible for digital downloading via an app.[103] The app is based on "GPS-based geo-targeting." A user types in their zip code, which then reveals a list of books that are banned in that area. The specific banned books in that area can be accessed and downloaded.

Firsthand Testimonial Books, Articles, and Reports

Personal stories about the experiences of individuals who are both advocating for the freedom to read and fighting censorship efforts can raise awareness and potentially open minds. *That Librarian: The Fight Against Book Banning in America* is a 2024 book by Louisiana middle school librarian Amanda Jones. Sometimes one day can change a life forever. That day for Jones was July 19, 2022, when she spoke at a public library board meeting in her hometown of Livingston Parish in support of diverse books. Though others spoke against censorship at that meeting, too, Jones was relentlessly cyberbullied by right-wing extremists who labeled her as "a groomer." *Merriam-Webster Dictionary* defines a groomer as "someone who grooms... a minor for exploitation and especially for nonconsensual sexual activity."[104]

Making false claims about a person can constitute defamation, which in essence functions as the legal term for disinformation. In the next chapter, we probe what disinformation is, as well as misinformation, and identify ways to confront it. In a later chapter on the ethics of social media, we also define and investigate cyberbullying and social-media-fueled harassment.

A September 23, 2024, *New York Times* essay, "Looking for a Superhero? Check the Public Library," notes that Jones has sued her online harassers for defamation and has cofounded an intellectual freedom advocacy group called Louisiana Citizens Against Censorship (LCAC).[105] LCAC's website states it is

> an all-volunteer, grassroots organization dedicated to fighting attempts to ban or restrict books based on subjective standards of inappropriate content. We believe that informed citizens are good citizens, and access to information is the cornerstone of a functioning democracy. To that end, we oppose any legislation aimed at restricting citizens' First Amendment Rights, as well as the right to read freely.[106]

The *New York Times* essay further explains that others are resisting efforts to erase books involving BIPOC and LGBTQ+ stories, characters, and perspectives, as well as classic literary tomes:

> Black parents are taking a public stance in defense of Black authors and historically accurate depictions of racism. In Florida, arguably the epicenter of the current book-banning movement, a group of students, parents and the authors of the oft-challenged picture book "And Tango Makes Three" sued to force Nassau County to return banned books to school libraries. (The county settled, restoring 36 titles, including "Tango.") Several publishing houses are suing the state of Florida for limiting access to "timeless classics," such as *I Know Why the Caged Bird Sings* by Maya Angelou.[107]

At the heart of these battles against erasure are issues of representation and equity. Jones believes that children need "to see themselves in the books they read."[108]

Earlier in this section, Books Unbanned was spotlighted for the groundbreaking work it is doing. An April 2024 "In Their Own Words" report compiled and published by Seattle Public Library (SPL) and Brooklyn Public Library, two of the participating library systems, provided both qualitative and quantitative insights into the experiences of the anonymous young people whose lives have been aided through access to Books Unbanned.[109] "In Their Own Words" analyzes 855 stories shared by young people, ages 13 to 26, who signed up for a free Books Unbanned e-card from either BPL or SPL from April 2022 to December 2023. University of Washington Information School graduate students categorized and tagged the stories to identify common themes. The stories analyzed for the report are a subset of thousands that the two libraries have received in the past two years from every state in the nation, as well as DC, Guam, and Puerto Rico.

A key theme uncovered in the analysis was how formal acts of censorship, bans, challenges, and restrictions to reading material can create a climate of fear and intimidation for young people. Cardholders also mentioned the impact of more "hidden" forms of censorship—self-censorship, soft or "quiet" censorship (preemptively restricting access as a means to avoid controversy), or the threat of future censorship—in affecting their freedom to read. Young people told stories of librarians commenting on their checkouts, of books being locked up and unavailable, and of small, limited collections not containing the kinds of books they would most like to read.

Some cardholders reported the lack of access to representative materials, especially for LGBTQIA+ (lesbian, gay, bisexual, transgender, queer, intersex, asexual) and BIPOC (Black, indigenous, people of color) cardholders, and how it exacerbated feelings of isolation and the sense that their identities are under attack. For these young people, digital access to books is essential to the ability to read freely.[110] Reading and digesting the selected quotes from anonymous young people, solely described by their age and state of residence, it can be difficult to decide which testimonials are more compelling. The statements are at once uplifting because of the positive differences these readers say Books Unbanned is having in their lives, yet they're also heartbreaking because the frank admissions illustrate the isolation and secrecy these young people are experiencing. Moreover, these "one-sentence story clips" show the stark realities of these youths' information access needs. The fact that they must turn to Books Unbanned to satisfy these information needs, rather than being able to do so via libraries and information professionals in their own geographic communities, cannot help but convey a feeling of abandonment and betrayal by all of the "adults" and political, educational, and information and cultural heritage systems in those regions. Here are some of these testimonies:

- "I identify as queer. I feel like the bans and culture surrounding LGBT+ stuff is isolating.... I really like reading but want to read about people like me finally."—Age 14, Ohio
- "The library closest to me is very underfunded, and it is very conservative. It has a plethora of Christian novels, but their novels surrounding people of color and [other] religions is very limited. As a person of color, it sucks to not be able to see myself in novels I read."—Age 17, Texas
- "There are books that I cannot take home because they would put me in danger. Reading digitally allows me to keep myself safe but still give[s] me the ability to read freely."—Age 19, Virginia
- "I am severely visually impaired and can't read physical book copies. Having a wide selection of audiobooks and accessible ebooks helps me access the same amount of literature as my sighted peers."—Age 15, Washington[111]

This sampling dramatically illustrates not only the IF needs of these young people but also implicates issues of access, including disabilities-related access, diversity, equity, inclusion, representation, privacy, surveillance, and safety and security. Notice, too, that the respondents come from politically "red" (Republican) *and* "blue" (Democrat) states: censorship does not hold to clear-cut assumptions that all persons in predominantly blue states can access the books they want and need. California persuasively shows this, via the September 2024 *Library Journal* update on the success of Books Unbanned:

> LA County Librarian and CEO Skye Patrick noted that while California is nationally perceived as a very progressive state, "there is somewhat of a political divide between the west side of the state and the east side of the state. You have two very different communities, generally speaking.... We've had plenty of challenges here in California following along the same path that you see around the country," leading to a need for Books Unbanned within the nation's most populous state.[112]

Ongoing IF and Censorship Battles

This chapter's initial examination focused on Florida, its governor Ron DeSantis, and Florida-based Moms for Liberty as pivotal incubators and evangelists for the escalating book challenges and bans that have spread to numerous other communities and states. Demonstrating the considerable ongoing challenges faced by IF and freedom-to-read advocates in responding to mushrooming censorship efforts, in June 2024 the *Guardian* reported on South Carolina's enactment of "one of the most restrictive book ban laws in the US, enabling mass censorship in school classrooms and libraries across the state. Drafted by Ellen Weaver, the superintendent of education and close ally of the far-right group Moms for Liberty, the law requires all reading material to be 'age or developmentally appropriate.'"[113] The article more broadly describes the law as "part of an alarmingly broader nationwide fight against literature exploring race, sexuality, or anything seemingly contentious or divisive."

Despite restrictive laws on books like the one enacted in South Carolina, as well as other censorious laws implemented on books in Idaho, Utah, Oklahoma, Texas, Arkansas, and other states, two recent reports offer some potential glimmers of encouragement for IF advocates, though some differing interpretations of the data exist. As part of 2024's Annual Banned Books Week, on September 23, 2024, ALA released preliminary data indicating fewer numbers of book challenges for the first eight months of 2024 compared to the same period in 2023. ALA elucidated its findings and offered some possible, though as yet inconclusive, explanations:

> The American Library Association has released preliminary data documenting attempts to censor books and materials in public, school, and academic libraries during the first eight months of 2024 in preparation for Banned Books Week (September 22–28, 2024).
>
> Between January 1 and August 31, 2024, ALA's Office for Intellectual Freedom tracked 414 attempts to censor library materials and services. In those cases, 1,128 unique titles were challenged. In the same reporting period last year, ALA tracked 695 attempts with 1,915 unique titles challenged. Though the number of reports to date has declined in 2024, the number of documented attempts to censor books continues to far exceed the numbers prior to 2020. Additionally, instances of soft censorship, where books are purchased but placed in restricted areas, not used in library displays, or otherwise hidden or kept off limits due to fear of challenges illustrate the impact of organized censorship campaigns on students' and readers' freedom to read. In some circumstances, books have been preemptively excluded from library collections, taken off the shelves before they are banned, or not purchased for library collections in the first place.
>
> "As these preliminary numbers show, we must continue to stand up for libraries and challenge censorship wherever it occurs," said American Library Association President Cindy Hohl. "We know library professionals throughout the country are committed to preserving our freedom to choose what we read and what our children read, even though many librarians face criticism and threats to their livelihood and safety. We urge everyone to join librarians in defending the freedom to read. We know people don't like being told what they are allowed to read, and we've seen communities come together to fight back and protect their libraries and schools from the censors."[114]

On the same day as the ALA report above was released, September 23, 2024, PEN America noted a "dramatic rise in book bans" that are "off the charts":

> Over 10,000 books were banned in public schools during the 2023–2024 school year, according to PEN America's preliminary findings.

> This dramatic increase is nearly triple the number from the previous school year, when PEN America recorded 3,362 bans nationwide. The final count for the 2023-2024 school year will be released later this fall along with a public *Index of School Book Bans*. PEN America will also release a detailed content analysis of titles banned during the 2023–2024 school year.[115]

A few other key findings from the PEN America report:

- More book bans were indicated for Florida (#1) and Iowa (#2) than any other US states and territories.
- The data is "certainly an undercount, as stories of book bans often go unreported."
- "Soft censorship" includes reasons like "chilling effects" on selecting books due to fears of courting controversy and "cancellations of author visits and book fairs"[116]

The PEN America report provides summaries of additional findings as well as new policies that are likely to affect the coming school year in Utah, South Carolina, and Tennessee.

Seeing *Oneself*, Seeing *the Other*

Discussions of challenged and banned books can seem like abstractions at times: we see a book listed on a challenged or banned book list that we may have read or not read, but we frequently don't know what motivated the author to write that particular book. The all-too-rare times when we *can* learn the stories behind an author's published story—the lived experiences and motivations that led the author to write their book—can sometimes be as exquisitely revealing and forceful as the book's content itself.

African American queer author George M. Johnson is the author of the 2020 book *All Boys Aren't Blue: A Memoir-Manifesto*. *All Boys* collects personal essays about experiences the author had from age five to adulthood, addressing topics like toxic masculinity and gender identity. The book has been the subject of numerous challenges and bans over the past few years. It was listed as the second most challenged book on ALA's list of most challenged books of 2023, with eighty-two documented challenges under the claimed rationales of "LGBTQIA+ content" and being "sexually explicit."[117] In a September 27, 2024, appearance with MSNBC host Joy Reid during her show *The ReidOut*, Johnson poignantly divulged who they had written *All Boys* for: themself. Johnson explained that they had written the book for their 14-year-old self because they had been unable to find any books that reflected their identity. Confronted with no representation, they exercised agency and instead chose to represent themselves in their own book, so that others might also perhaps see themselves represented in Johnson's book.

Additionally, having appeared on the first episode of MSNBC's Ali Velshi Banned Books Club broadcast on February 13, 2022, Johnson returned for the 100th episode on September 28, 2024. Fittingly, that was the last day of 2024's Annual Banned Books Week. During the interview, Velshi posed the question:

> Who loses more when you ban a book? The kids who can't see themselves reflected or everyone else who can't learn about those people in the first place because they don't live your experience?

Johnson responded:

> That is a good question. It is like we all lose at the end of the day. I do feel like those who need to be seen need to be able to read about themselves in a book and will lose out more. But I think as a society we lose when people don't understand those who are different from them.[118]

Building on this point, Johnson then shared their experience of having visited rural Ouray, Colorado, for a book event, where, to Johnson's surprise, 100 people showed up. They observed that many of the attendees may have never met a Black openly queer person before. This example hammers home the argument for "representation via book," if and when "representation via an in-person human" isn't possible. As Johnson remarked, "I can't go to every city, but my book can."[119] Books like *All Boys Aren't Blue*, thus, can perform the dual purpose of enabling marginalized people to see others like themselves in books, while also enabling people not within such groups to learn about the lives of unknown or little-known others.

As a coda, in the September 2024 conversation with Velshi, Johnson noted that even in fairly reliably "blue" Cobb County, a suburb of Atlanta, Georgia, a school district had just recently chosen to ban *All Boys*. Yet, despite the reality and specter of continuing book bans and censorship, Johnson was undeterred, emphasizing the need for soldiering on and sharing the voices of diverse peoples. As an inflection point to that stark actuality, Johnson discussed the September 2024 publication of their new book of essays, *Flamboyants: The Queer Harlem Renaissance I Wish I'd Known*, celebrating the "writers, performers, and activists from 1920s Black America whose sexualities have been obscured throughout history."[120] Both Velshi and Johnson openly ventured that *Flamboyants*, like its predecessor *All Boys Aren't Blue*, would unfortunately and likely face a similar spate of book challenges and bans.

Summary

Like access, intellectual freedom and freedom of expression are bedrock principles of democracy. Healthy, well-functioning, and responsive democracies depend on informed citizenries. The freedoms to access, read, think, and express oneself are essential to fostering and sustaining the ability of citizens to become and remain informed. Intellectual freedom and free speech are among the most cherished and vital tenets set forth in the US Constitution's First Amendment. Correspondingly, these precepts are recognized in Articles 18 and 19 of the UN Declaration of Human Rights (1948). Censorship is the opposite of intellectual freedom and access: it involves the suppression of information and ideas and restrictions on access to information and ideas. Libraries, archives, and information centers specifically recognize freedom of thought and expression in their codes of ethics and core values and are committed to opposing censorship efforts. Self-censorship, like censorship, deprives societies of the full range of human expression and creativity. The internet and social media have presented opportunities for and threats to intellectual freedom and expression. Information professionals and advocates for intellectual freedom are engaging in a wide range of initiatives to champion the rights of people to read, think, and express themselves.

Questions for Reflection, Exploration, Action, Discussion (READ)

1. Listen to the 29-minute NPR/1A podcast "The Strain of Censorship on Public Libraries," recorded on September 28, 2022: https://www.npr.org/2022/09/28/1125746234/the-strain-of-censorship-on-public-libraries. Then, in at least a few sentences, identify your most significant takeaway insight/lesson from this conversation.
2. Imagine you are on an IF committee planning a one-hour program on self-censorship and ethics for next year's annual conference. A key goal of the committee is for attendees to leave the session knowing more about self-censorship and learning about some of the ethics-related resources that can help them to be better prepared to mitigate self-censorship. Create an action plan of three to six steps you and your committee colleagues can use in planning this program and promoting it.
3. Familiarize yourself with the American Library Association's Office for Intellectual Freedom: https://www.ala.org/aboutala/offices/oif. Then, list and describe something you found to be of interest or usefulness, and explain why.
4. Examining ethics and sanitized works.

a. Identify some of the key ethical issues that are involved in consideration of whether libraries should purchase and provide access to sanitized works that remove or change words that are injurious to individuals and groups.

b. What ethical theory or theories can be used to justify or not justify the purchase of sanitized works?

c. In the following hypothetical scenario, imagine you are a library manager selecting fiction resources for your midsize public library and its four branch libraries. Your library director has asked you to purchase sanitized editions of *The Adventures of Huckleberry Finn* as well as any other sanitized editions of "classic works" that you can find.

- What additional information, if any, would you like to know?
- What do you do?

Explain your thinking process and the rationale(s) for your decision.

5. Evaluating Banned Books Week and raising awareness.

a. Discuss one strength/upshot and one weakness/shortcoming of the annual Banned Books Week program.

b. If you have any suggestions/recommendations for modifying/changing Banned Books Week or phasing out the initiative or replacing it with something else, share your thoughts.

c. Are there other ideas *you* have for raising awareness of intellectual freedom and censorship issues?

Notes

1. Reid, D. (Director). (2018, July 4). Postpartum. *The Handmaid's Tale*, season 2, episode 12. IMDb. https://www.imdb.com/title/tt7435260/
2. *"I will be heard!": Abolitionism in America.* (2002). Cornell University Library. https://rmc.library.cornell.edu/abolitionism/narratives.htm
3. Ibid.

4. Bradbury, R. (1953). *Fahrenheit 451*. Ballantine.
5. *Oliver Wendell Holmes quotes*. (n.d.). Famous Poets and Poems. Retrieved February 19, 2025, from http://famouspoetsandpoems.com/poets/oliver_wendell_holmes/quotes
6. American Library Association. (2022). *Challenge support.* https://www.ala.org/tools/challengesupport
7. *US Constitution, First Amendment.* (1791). Constitution Annotated. Congress.gov. https://constitution.congress.gov/constitution/amendment-1/
8. American Library Association. (2023). *Intellectual freedom and censorship Q&A.* http://www.ala.org/advocacy/intfreedom/censorship/faq
9. United Nations. (1948). *Universal Declaration of Human Rights.* http://www.un.org/en/universal-declaration-human-rights/
10. American Library Association. (2019). *Library Bill of Rights.* https://www.ala.org/advocacy/intfreedom/librarybill
11. American Library Association. (2021). *ALA code of ethics.* http://www.ala.org/tools/ethics
12. Ibid.
13. American Library Association. (2021). *First Amendment and censorship.* https://www.ala.org/advocacy/intfreedom/censorship
14. American Library Association. (2023). *Intellectual freedom and censorship Q&A.* http://www.ala.org/advocacy/intfreedom/censorship/faq
15. Shaw, G. B. (1894). *Mrs Warren's profession*. Project Gutenberg. https://www.gutenberg.org/files/1097/1097-h/1097-h.htm
16. Fine, S. (1996). How the mind of a censor works: The psychology of censorship. *School Library Journal, 42*(1), 23–27.
17. American Library Association. (n.d.). *Office for intellectual freedom.* Retrieved February 19, 2025, from https://www.ala.org/aboutala/offices/oif
18. American Library Association. (2009, April 21). *Judy Krug memorial.* YouTube. https://www.youtube.com/watch?v=OD6kfKAsRWQ
19. American Library Association. (2022). *Challenge support.* https://www.ala.org/tools/challengesupport
20. American Library Association. (2022). *Challenge reporting.* Retrieved February 19, 2025, from https://www.ala.org/tools/challengesupport/report
21. American Library Association. (n.d.). *Office for intellectual freedom.* Retrieved February 19, 2025, from https://www.ala.org/aboutala/offices/oif
22. Ibid.
23. Ibid.
24. American Library Association. (n.d.). *ALA Office for Intellectual Freedom.* YouTube. Retrieved February 19, 2025, from https://www.youtube.com/channel/UCqruep4YtE-2ioIiG2MsWZg
25. American Library Association. (n.d.). *Journal of Intellectual Freedom and Privacy*. Retrieved February 19, 2025, from https://www.ala.org/aboutala/offices/oif/JIFP
26. Garnar, M., & Magi, T. (Eds.). (2021). *Intellectual freedom manual* (10th ed.). American Library Association, Office for Intellectual Freedom.
27. American Library Association. (n.d.). *Banned books week*. Retrieved February 19, 2025, from https://www.ala.org/advocacy/bbooks/banned
28. Ibid.
29. Gutman Library, Harvard University. (n.d.). *Banned books*. Retrieved February 19, 2025, from https://guides.library.harvard.edu/c.php?g=1269000&p=9306840
30. Freedom to Read. (n.d.). *Bannings and burnings in history*. Retrieved February 19, 2025, from https://www.freedomtoread.ca/resources/bannings-and-burnings-in-history/

31. American Library Association. (n.d.). *About banned & challenged books*. Retrieved February 19, 2025, from https://www.ala.org/advocacy/bbooks/aboutbannedbooks
32. American Library Association. (n.d.). *Book ban data.* Retrieved February 19, 2025, from https://www.ala.org/advocacy/bbooks/book-ban-data#methodology
33. American Library Association. (n.d.). *About banned & challenged books.* Retrieved February 19, 2025, from https://www.ala.org/advocacy/bbooks/aboutbannedbooks
34. Ibid.
35. American Library Association. (2024, March 14). *American Library Association reports record number of unique book titles challenged in 2023.* https://www.ala.org/news/press-releases/2024/03/american-library-association-reports-record-number-unique-book-titles
36. Blankenship, K., & Tager, J. (2023, November 28). *The Florida effect: How the Sunshine State is driving the conservative agenda on free expression*. PEN America. https://pen.org/report/the-florida-effect/
37. Moms for Liberty. (n.d.). *Who we are*. Retrieved February 19, 2025, from https://www.momsforliberty.org/about/
38. Sinha, S., Zerbino, N., Valant, J., & Perera, R. M. (2023, October 10). *Moms for Liberty: Where are they, and are they winning?* Brookings. https://www.brookings.edu/articles/moms-for-liberty-where-are-they-and-are-they-winning/
39. Santiago, L., & Weisfeldt, S. (2021, December 17). *How two Florida moms spearheaded a parent movement aiming to impact the 2022 elections*. CNN. https://www.cnn.com/2021/12/16/politics/florida-parent-power-elections-moms-for-liberty/index.html
40. Ibid.
41. Moms for Liberty. (n.d.). *Who we are.* Retrieved February 19, 2025, from https://www.momsforliberty.org/about/
42. Santiago, L., & Weisfeldt, S. (2021, December 17). *How two Florida moms spearheaded a parent movement aiming to impact the 2022 elections*. CNN. https://www.cnn.com/2021/12/16/politics/florida-parent-power-elections-moms-for-liberty/index.html
43. Anderson, Z. (2023, December 2). Police records detail sex assault allegations against Florida Republican Party chair. *Sarasota Herald-Tribune*. https://www.heraldtribune.com/story/news/politics/2023/12/02/records-detail-assault-allegations-against-florida-republican-party-chair/71778421007/
44. Southern Poverty Law Center. (n.d.). *About*. Retrieved February 19, 2025, from https://www.splcenter.org/about/
45. Southern Poverty Law Center. (2023). *The year in hate & extremism 2022.* https://www.splcenter.org/year-hate-extremism-2022
46. Yousef, O. (2023, June 7). *Moms for Liberty among conservative groups named "extremist" by civil rights watchdog*. NPR. https://www.npr.org/2023/06/07/1180486760/splc-moms-for-liberty-extremist-group
47. Turner, C. (2023, June 2). *Poll: Americans say teachers are underpaid, about half of Republicans oppose book bans*. NPR. https://www.npr.org/2023/06/02/1177566467/poll-teachers-underpaid-republicans-book-bans
48. McCreesh, S. (2024, August 31). Conservative moms, charmed by Trump, would rather avoid his misogyny. *New York Times*. https://www.nytimes.com/2024/08/31/us/elections/conservative-moms-for-liberty-trump.html
49. Klein, A. (2024, September 25). Moms for Liberty co-founder would be "honored" to be Trump's education secretary. *Education Week*. https://www.edweek.org/policy-politics/moms-for-liberty-co-founder-would-be-honored-to-be-trumps-education-secretary/2024/09
50. Teichner, M. (2023, April 23). *War of words: The fight over banning books*. CBS News. https://www.cbsnews.com/news/war-of-words-the-fight-over-banning-books/

51. Ibid.
52. Migdon, B. (2022, March 28). *Florida's "Don't Say Gay" bill signed into law*. Hill. https://thehill.com/changing-america/respect/equality/598999-floridas-dont-say-gay-bill-signed-into-law/
53. Swidriski, E. (2022, November 22). *Florida's "Don't Say Gay" law raises serious legal questions*. American Bar Association. https://www.americanbar.org/groups/labor_law/publications/labor_employment_law_news/fall-2022/florida-do-not-say-gay-law/
54. Bishop, S., Royal, D., & Suarez, C. (2024, July 30). *Judge permanently blocks part of Florida's "Stop WOKE Act."* CNN. https://www.cnn.com/2024/07/29/us/florida-ron-desantis-stop-woke-act-part-overturned/index.html
55. Office of Governor Ron DeSantis. (2021, December). *Stop W.O.K.E. Act handout.* https://www.flgov.com/wp-content/uploads/2021/12/Stop-Woke-Handout.pdf. (Page no longer active. Handout retrieved February 19, 2025, from https://vdhtesting.vanderhouwen.com/wp-content/uploads/2023/02/Florida-Stop-Woke-Act.pdf)
56. Merriam Webster. (n.d.). *Woke*. Retrieved February 19, 2025, from https://www.merriam-webster.com/dictionary/woke
57. Britannica. (n.d.). *Critical race theory.* Retrieved February 19, 2025, from https://www.britannica.com/topic/critical-race-theory
58. Ibid.
59. Hartle, T. (2019, September 20). "How to Be an Antiracist" opens a vital dialogue on race. *Christian Science Monitor*. https://www.csmonitor.com/Books/Book-Reviews/2019/0920/How-to-Be-an-Antiracist-opens-a-vital-dialogue-on-race
60. Hannah-Jones, N. (2019, August 14). The 1619 project. *New York Times*. https://www.nytimes.com/interactive/2019/08/14/magazine/1619-america-slavery.html
61. Quinn, R. (2024, June 21). Florida argues it could stop professors from criticizing governor. *Inside Higher Education*. https://www.insidehighered.com/news/faculty-issues/academic-freedom/2024/06/21/fla-argues-it-could-stop-professors-criticizing
62. Ibid.
63. Ibid.
64. Bishop, S., Royal, D., & Suarez, C. (2024, July 29). *Judge permanently blocks part of Florida's "Stop WOKE Act."* CNN. https://www.cnn.com/2024/07/29/us/florida-ron-desantis-stop-woke-act-part-overturned/index.html
65. Ibid.
66. Dixon, M., & Finout, G. (2023, January 3). "Where woke goes to die": DeSantis, with eye toward 2024, launches second term. *Politico*. https://www.politico.com/news/2023/01/03/desantis-2024-second-term-00076160
67. Atterbury, A. (2024, September 12). Florida county restoring dozens of books to school libraries after "book ban" lawsuit. *Politico*. https://www.politico.com/news/2024/09/12/florida-book-ban-lawsuit-00178981
68. *Selendy Gay secures major settlement in book censorship battle in Florida.* (2024, September 12). Selendy Gay. https://www.selendygay.com/news/general/2024-09-12-selendy-gay-secures-major-settlement-in-book-censorship-battle-in-florida
69. Ibid.
70. Ibid.
71. Wicker, T. (1978). *On press*. Viking Press.
72. Whelan, D. (February 1, 2009). A dirty little secret: Self-censorship. *School Library Journal*. https://www.slj.com/story/a-dirty-little-secret-self-censorship

73. Asheim, L. (1953, September). Not censorship but selection. *Wilson Library Bulletin, 28*, 63–67. Retrieved from https://www.ala.org/advocacy/intfreedom/NotCensorshipButSelection
74. American Library Association. (2023). *Intellectual freedom and censorship Q&A.* https://web.archive.org/web/20090712074247/http://www.ala.org/ala/aboutala/offices/oif/basics/ifcensorshipqanda.cfm
75. Ibid.
76. Reitz, J. M. (2013). *Online dictionary for library and information science*. https://odlis.abc-clio.com/
77. Eschner, K. (2017, July 11). The Bowdlers wanted to clean up Shakespeare, not become a byword for censorship. *Smithsonian Magazine*. https://www.smithsonianmag.com/smart-news/bowdlers-wanted-clean-shakespeare-not-become-byword-censorship-180963945/
78. Kaufman, G. (2015, December 14). *Here's why banning "Huck Finn" over the N-word sends the wrong message*. MTV News. https://web.archive.org/web/20220625223647/https://www.mtv.com/news/2687298/friends-central-school-huckleberry-finn-curriculum-dropped-n-word/
79. Kakutani, M. (2011, January 6). Light out, Huck, they still want to sivilize you. *New York Times*. https://www.nytimes.com/2011/01/07/books/07huck.html
80. Ibid.
81. Steinberg, G. (2023, October 19). *Censorship on stage: Plays and musicals challenged at high schools*. New York Public Library. https://www.nypl.org/blog/2023/10/19/censorship-stage-plays-and-musicals-challenged-high-schools
82. Editorial Board. (2023, June 12). 21st-century editors should keep their hands off 20th-century books. *Washington Post*. https://www.washingtonpost.com/opinions/2023/06/12/books-editing-retouching-free-expression/
83. Holden, S. (2016, February 7). Review: Janis Ian, singer of "At Seventeen," now at 64 and in solo concert. *New York Times*. https://www.nytimes.com/2016/02/08/arts/music/review-janis-ian-singer-of-at-seventeen-now-at-64-and-in-solo-concert.html
84. Editorial Board. (2023, June 12). 21st-century editors should keep their hands off 20th-century books. *Washington Post*. https://www.washingtonpost.com/opinions/2023/06/12/books-editing-retouching-free-expression/
85. World Intellectual Property Organization. (2016). *Understanding copyright and related rights*. https://www.wipo.int/edocs/pubdocs/en/wipo_pub_909_2016.pdf
86. Armstrong, N. (2023, May 30). *Roald Dahl: The fierce debate over rewriting children's classics*. BBC. https://www.bbc.com/culture/article/20230530-roald-dahl-the-fierce-debate-over-rewriting-childrens-classics
87. Lawless, J. (2023, February 24). Penguin will sell "classic" versions of Roald Dahl books after rewrite censorship backlash. *USA Today*. https://web.archive.org/web/20230302011112/https://www.usatoday.com/story/entertainment/books/2023/02/24/roald-dahl-books-classic-versions-censorship-backlash/11337687002/
88. Editorial Board. (2023, June 12). 21st-century editors should keep their hands off 20th-century books. *Washington Post*. https://www.washingtonpost.com/opinions/2023/06/12/books-editing-retouching-free-expression/
89. Broward County Libraries. (2024, May 8). *Broward County Library honored for championing intellectual freedom*. PR Newswire. https://www.prnewswire.com/news-releases/broward-county-library-honored-for-championing-intellectual-freedom-302140343.html
90. Knox, E. J. M. (2023, September 12). *Testimony before the U.S. Senate Committee on the Judiciary*. https://www.judiciary.senate.gov/imo/media/doc/2023-09-12_-_testimony_-_knox.pdf
91. Holland, H. (2024, September 12). *Every single banned book from Ali Velshi's "Banned Book Club."* MSNBC. https://www.msnbc.com/ali-velshi/read-every-banned-book-ali-velshi-banned-book-club-rcna156215

92. Stanich, K. (2024). *Heroine by Mindy McGinnis*. https://nchcpl.org/heroine-by-mindy-mcginnis/
93. Fontenay, B. (2024, March 8). "I had the guts": 101-year-old woman inspires Oscar-nominated documentary on book bans. *Treasure Coast Newspapers*. https://www.tcpalm.com/story/opinion/columnists/blake-fontenay/2024/03/08/101-year-old-grace-linns-book-banning-talk-inspires-oscar-nomination/72809791007/
94. Montesano, G. (2024, March 11). Oscars 2024: "The ABCs of Book Banning" loses to "The Last Repair Shop." *Treasure Coast Newspapers*. https://www.tcpalm.com/story/entertainment/2024/03/11/101-year-old-grace-linn-walks-the-oscars-red-carpet-for-the-abcs-of-book-banning/72929166007/#
95. Tribeca TV. (2024). *The turning point: To Be Destroyed*. https://tribecafilm.com/films/turning-point-to-be-destroyed-2024
96. Eble, A., Douglass, S., Rebell, M., & Erickson, A. (2023, September 6). *What you need to know about the book bans sweeping the U.S*. Teachers College, Columbia University. https://www.tc.columbia.edu/articles/2023/september/what-you-need-to-know-about-the-book-bans-sweeping-the-us/
97. Cooke, N. A. (2023, July 20). How book-banning campaigns have changed the lives and education of librarians—they now need to learn how to plan for safety and legally protect themselves. *Conversation*. https://theconversation.com/how-book-banning-campaigns-have-changed-the-lives-and-education-of-librarians-they-now-need-to-learn-how-to-plan-for-safety-and-legally-protect-themselves-205743
98. Ibid.
99. Brooklyn Public Library. (n.d.). *Recognizing two years of Books Unbanned*. Retrieved February 19, 2025, from https://www.bklynlibrary.org/books-unbanned
100 Enis, M. (2024, September 23). Books Unbanned still growing strong more than two years after launch. *Library Journal*. https://www.libraryjournal.com/story/books-unbanned-still-growing-strong-more-than-two-years-after-launch
101. Simon & Schuster. (2023, September 20). *Simon & Schuster to launch Books Belong initiative in response to ongoing book bans*. https://about.simonandschuster.biz/news/books-belong/
102 Simon & Schuster. (n.d.). *Books Belong*. Retrieved February 19, 2025, from https://www.simonandschuster.com/p/bannedbooksweek
103. Empson, O. (2023, August 18). "Knowledge is power": New app helps US teens read books banned in school. *Guardian*. https://www.theguardian.com/us-news/2023/aug/18/us-teens-banned-books-schools
104 Merriam-Webster. (n.d.). *Groomer*. Retrieved February 19, 2025, from https://www.merriam-webster.com/dictionary/groomer
105. Renkl, M. (2024, September 23). Looking for a superhero? Check the public library. *New York Times*. https://www.nytimes.com/2024/09/23/opinion/book-bans-librarians.html
106. Louisiana Citizens Against Censorship. (n.d.). *Who are we?* Retrieved February 19, 2025, from https://www.la-cac.org
107. Renkl, M. (2024, September 23). Looking for a superhero? Check the public library. *New York Times*. https://www.nytimes.com/2024/09/23/opinion/book-bans-librarians.html
108. Ibid.
109. Brooklyn Public Library and Seattle Public Library. (2024, April). *In their own words*. https://booksunbanned.com/documents/Books%20Unbanned%20Teen%20Testimonials.pdf
110. Seattle Public Library. (2024, April 11). *"In their own words": Books Unbanned report details the impact of censorship on young people*. https://www.spl.org/about-us/news-releases/books-unbanned-report-details-the-impact-of-censorship-on-young-people-across-the-united-states

111. Brooklyn Public Library and Seattle Public Library. (2024, April). *In their own words.* https://booksunbanned.com/documents/Books%20Unbanned%20Teen%20Testimonials.pdf
112. Enis, M. (2024, September 23). Books Unbanned still growing strong more than two years after launch. *Library Journal.* https://www.libraryjournal.com/story/books-unbanned-still-growing-strong-more-than-two-years-after-launch
113. Empson, O. (2024, June 30). South Carolina implements one of US's most restrictive public school book bans. *Guardian*. https://www.theguardian.com/us-news/article/2024/jun/30/south-carolina-public-school-book-ban
114. American Library Association. (2024, September 23). *American Library Association reveals preliminary data on 2024 book challenges.* https://www.ala.org/news/2024/09/american-library-association-reveals-preliminary-data-2024-book-challenges
115. Meehan, K., & Baêta, S. (2024, September 23). *Memo on school book bans from PEN America*. https://pen.org/memo-on-school-book-bans-2023-2024-school-year/
116. Ibid.
117. American Library Association. (n.d.). *Top 10 most challenged books of 2023.* https://www.ala.org/bbooks/frequentlychallengedbooks/top10
118. Velshi. (2024, September 28). *"I can't go to every small city, but my book can": 100th meeting of the Velshi Banned Book Club*. MSNBC. https://www.msnbc.com/ali-velshi/watch/-i-can-t-go-to-every-small-city-but-my-book-can-100th-meeting-of-the-velshi-banned-book-club-220365893890
119. Ibid.
120. Johnson, G. M. (2024). *Flamboyants: The queer Harlem Renaissance I wish I'd known*. https://www.amazon.com/Flamboyants-Queer-Harlem-Renaissance-Known-ebook/dp/B0CM23M8KY

Chapter 7

Accuracy, Misinformation, and Disinformation

Introduction: Convenience, Information Environments, and Quality Control

"Convenience decides everything"—so said Evan Williams, a cofounder of Twitter and its CEO from 2008 to 2010.[1] Williams is also the founder of Medium and Blogger, the Google tool I have used since 2008 for my ethics, IP, and other blogs. The ease and pervasiveness of internet access for many people—whether via smartphones or computer and tablet devices—has profoundly affected the ways people seek and acquire the information they need and want. Every person with a smartphone now possesses the power of a computer in the palm of their hand. That portable computer power each of us now has at our fingertips brings with it the powers—and yes, the pitfalls—of *convenience*.

Want to find that book you remember reading in the 1990s about a desert shepherd boy who discovers his life's calling? (Answer: Paulo Coelho's *The Alchemist*, 1988.) What's the correct number of the X-Men issue that you referenced in the testimonial you gave at the 2004 or 2005 wedding of two fellow X-Men aficionados? (Answer: X-Men #30, the wedding of Jean Grey and Scott Summers in 1994.) In years past, where those of us of a particular generation might have been inclined to seek out the answers to such questions from the trusty librarians at the "reference desk" (which in many libraries would be rechristened a "help desk" to dejargonize that service whose very name could be seen as a barrier to people seeking help from experts willing to give it), we can now search for and find the answers to such queries on our phones in mere seconds.

Behavioral changes in the ways people seek information, combined with technological innovations, have correspondingly influenced the *accuracy* of the information that people seek and attain. Before the internet, humans had far fewer choices for where to find information than today. A variety of print media—books, magazines, encyclopedias, newspapers—offered sources for people to actively read and peruse. Information could also passively be acquired through the twentieth-century inventions of radio (by listening) and television (by listening/watching). Many, though not all, nonfiction and even fiction print sources were (and still are, though to lesser degrees) vetted by editorial experts and peer reviewers before being made available to readers, a layer of status inspection between the acts of information creation and information distribution. The aim of this level of scrutiny was to discover any errors (i.e., untruths or mistakes) or potentially problematic words (i.e., terms or phrasings that could be viewed as inappropriate or unacceptable in relation to accepted norms and standards).

An adjacent goal of editorial oversight and review is honing and improving: the idea of striving to make the product better, even if it is not in error or prone to provoking offense. This vetting and oversight process instilled—borrowing a contract law concept—a certain *implied warranty* of quality assurance and control: the notion that *someone* behind the scenes of these information products (most often unseen, unheard, and unknown) was inspecting them for soundness and affirming their integrity. As a related aside, the Comics Code Authority (CCA), launched in 1954, was created as an editorial oversight layer for the comic books medium. The CCA's intent was to assuage the public's fears that comic books (particularly the genre of horror-related comics) were damaging to the moral character of vulnerable young readers. *Moral panics* had been fueled by psychiatrist Frederic Wertham's 1954 comic-book-demonizing book *Seduction of the Innocent*, asserting links between comic books and juvenile delinquency, and Wertham's testimony before the US Senate on the influences of comics on America's youths.

Similarly, in the pre-internet period, radio and television programs, particularly with regard to news, employed internal staff whose chief roles were to ensure accuracy and truthfulness. Broadcast *standards and practices* (S&P), also referred to as "editorial S&P," was an accepted part of media (and print) industries. Public trust and truthfulness were viewed as intangible, highly sought-after currencies that the extant news broadcast companies of the times focused on earning, growing, and leveraging as lucrative elements of their brands. This kind of "brand assurance of news quality" was perhaps best evinced via veteran newsman Walter Cronkite's trust-inducing signoff at the end of his nightly CBS Evening News broadcasts, "... And that's the way it is," followed by the day and date of the broadcast, as when Cronkite signed off for the last time on Friday, March 6, 1981.

As with all human endeavors, periodic lapses in truth and integrity *did* inevitably occur in the pre-internet broadcast media ecosystem, however. Such failures resulted in public exposure and shame, though scanty legal repercussions. As one example of an especially unsavory *willful* category of transgressions to truth, the television industry was rocked by game show scandals in 1950s America when it was revealed that some game show questions and winners were rigged in advance (dramatized in the 1994 Oscar-nominated Robert Redford–directed film *Quiz Show*). Who was hurt most by these crooked quiz shows (precursors for internet-era reality show programming)?

> The true victims were not the contestants, [Manhattan Assistant District Attorney Joseph] Stone writes, many of whom made money, but the television viewers hoodwinked into believing that the fiction they were watching was actually real-life drama. But even television viewers seemed not to mind. A straw poll conducted at the time by the *Miami Herald* showed that the public was not outraged by the quiz hoaxes. Most Miamians wanted the shows back, regardless of all that had taken place. "Everything on TV is somewhat of a lie," one such individual explained to the *Herald*, "but it's still entertainment."[2]

As a result of the quiz show scandals, the Federal Communications Commission (FCC) licensing renewals were held over the heads of television companies like the mythical sword of Damocles to ensure compliance with laws and standards. The "what, me worry?" sentiments referenced in the *Miami Herald* survey data eerily foreshadowed two concerning developments: (1) the trend of some media outlets in the ensuing decades to commingle and serve up truth and facts with fiction and entertainment, and (2) the public's appetite for that fusion cuisine, which though appealing, has in some cases proven to be unhealthy for them and the societies within which they consume such fare.

For centuries, libraries have held a singular place in societies as trusted conduits and storehouses for information and humanity's collective knowledge. Reference desks and the librarians who staff them have been recognized year after year as places and persons to whom patrons can reliably turn for all manner of questions. Librarians, too, routinely rank among the most trusted members of societies. As just one example, Book Riot and EveryLibrary Institute conducted "a series of research studies exploring parental perceptions of the library" in 2023 and found that "public and school librarians ranked in the top five of most trustworthy professions."[3]

With more and more people since the mid-1990s able to access the web on their own through digital technologies, some have questioned the continuing need for libraries and librarians. Indeed, I clearly recall an example of this in 2011, when comedian and talk show host Bill Maher via his HBO weekly Friday night show *Real Time* asserted that "we have the Internet. We don't need a library at all."[4] Paradoxically, Maher's statement was made at a time when libraries were experiencing record numbers of users, especially job seekers and career changers, who were navigating the rocky seas of the 2008 economic downturn in the United States and much of the world following the subprime mortgage crisis. As librarian Teri Markson pointedly observed in a rebuttal column on *Huffington Post*, "For a political guy who fancies himself a progressive and seems buoyed by the Occupy movement, this was a totally arrogant and disappointing statement for him to make, particularly during lean economic times when public libraries are busier than ever."[5]

It *is* inarguable that in the past 20-plus years, search engines like Google and social media platforms like Twitter (renamed X in 2023), YouTube, Facebook, and, more recently, TikTok have made it possible for millions of people to search for and find information on their own. What Google, Twitter, Facebook, and other platforms have not done, though, is to make it as easy and reliable for self-searchers to evaluate the *quality* of the information their searches scoop up. Questions arise:

- How "good" or reliable is the information in the first ten "hits" that Google ranks?
- How accurate, truthful, and timely is the information on X, Facebook, and TikTok?
- Is the information a YouTube influencer is talking about in a video actually misinformation or disinformation?

In this chapter, we learn more about accuracy of information, misinformation, and disinformation. In the next chapter, we delve more into ethical issues of social media, which intersect with the focus of this chapter. As *American Gods* and *The Sandman* author Neil Gaiman trenchantly opined in 2017, "Google can bring you back 100,000 answers. A librarian can bring you back the right one."[6]

With the maturation of the internet over the past two decades and increasing concerns about the truthfulness of news and information, various tools and strategies for information assessment and fact-checking have been and are continually being developed and deployed. Yet the oldest, most reliable example of such fact-checking tools is the librarian. Assessing the veracity and quality of information has been one of a librarian's most important roles, and helping users to find accurate, truthful information continues to be a mainstay of librarianship. These activities are mainstays of "information literacy," a practice that has been led by and continues to be soundly associated with librarianship in the digital age.

Facts, Evidence, and Truth

John Adams, a Harvard-educated attorney who would eventually come to be known to posterity as one of the United States' Founding Fathers, famously remarked in 1770, "Facts are stubborn things; and whatever may be our wishes, our inclinations, or the dictates of our passions, they cannot alter the state of facts and evidence."[7] Though that adage about facts is fairly well known, its context is less recognized. As tensions simmered between the colonies and England throughout the 1760s, incidents between the British and the American colonists regrettably but inevitably happened. One such incident was the Boston Massacre. The National Park Service's Adams National Historical Park in Boston, Massachusetts, explains:

> On March 5, 1770, British soldiers fired into a crowd of civilians, killing five people. Crispus Attucks, Samuel Gray, and James Caldwell died that night. Seventeen-year-old Samuel Maverick died the next morning. Patrick Carr died from his wounds nearly two weeks later. This event sparked outrage in Bostonians who later called this "the Boston Massacre."
>
> Shortly after the Massacre, the soldiers (William Wemms, Hugh White, Hugh Montgomery, James Hartigan, William McCauley, Matthew Kilroy, William Warren, and John Carroll) and their captain (Thomas Preston) were arrested and jailed. Royal government officials decided to delay the trial to let tensions in the town cool. Both sides doubted the soldiers would be given a fair trial. Loyalists feared that colonists would want vengeance; colonists feared that loyalists would arrange for charges to be dismissed.[8]

One day after the massacre, a loyalist named James Forrest approached John Adams, a practicing lawyer. Forrest told Adams that the accused Capt. Preston was to be tried separately from the eight accused men under his command. Yet no attorney in all of Boston could be secured to represent Preston; that is, until Adams agreed: "John Adams firmly believed that everyone had the right to a lawyer and a fair trial, so he willingly agreed to represent the soldiers even if it meant risking his reputation."[9]

In defending his client, Adams faced formidable obstacles. Neither the prosecution nor the defense had physical evidence: instead, both "had to rely almost entirely on eye-witness testimony."[10] Contradictory versions of events were heard as the eyewitnesses gave their testimony. Lack of a consistent narrative made the eyewitnesses "seem less credible" and sowed "seeds of doubt in the jury."[11] All was not lost for the accused Preston, however:

> The testimonies that ultimately saved Preston were from multiple witnesses who swore that they had seen the captain standing in front of the soldiers when they began to fire.

> One of these witnesses was a free Black man, Newton Prince. Prince testified that he saw Captain Preston standing in front of the soldiers and "heard no orders given to fire." If Preston had indeed given the order, he surely would have been wise enough to stand behind his soldiers.
>
> On October 30, 1770, the jury delivered its verdict: Captain Thomas Preston was found not guilty.[12]

It was in the course of delivering his client's defense that Adams uttered his now famous words: "Facts are stubborn things; and whatever may be our wishes, our inclinations, or the dictates of our passions, they cannot alter the state of facts and evidence."[13] The facts and evidence Adams presented to the jury led them to conclude that Capt. Preston was telling the truth about the Boston Massacre. That truth, as the jury saw and heard it, propelled them to render their not guilty verdict. This historical example underscores the power that facts, evidence, and truth possess and the ways that these elements can constitute the difference between liberty or confinement, and even life and death.

Legal Information Institute (LII) provides definitions of the words *fact* and *evidence*:

> Fact: "A fact is an event that actually happened, or a statement presented as objective truth"[14]

> Evidence: "Evidence is an item or information proffered to make the existence of a fact more or less probable. Evidence can take the form of testimony, documents, photographs, videos, voice recordings, DNA testing, or other tangible objects."[15]

The term *truth* is, arguably, more difficult to define. Philosophers dating back to antiquity have striven to describe, conceptualize, and communicate the nature of truth. Donald Davidson's June 1996 article "The Folly of Trying to Define Truth" in the *Journal of Philosophy* identifies some of the challenges and differing views of thought vis-à-vis definitions of truth:

> For the most part, the concepts philosophers single out for attention, like truth, knowledge, belief, action, cause, the good and the right, are the most elementary concepts we have, concepts without which (I am inclined to say) we would have no concepts at all. Why then should we expect to be able to reduce these concepts definitionally to other concepts that are simpler, clearer, and more basic? We should accept the fact that what makes these concepts so important must also foreclose on the possibility of finding a foundation for them which reaches deeper into bedrock.
>
> We should apply this obvious observation to the concept of truth: we cannot hope to underpin it with something more transparent or easier to grasp. Truth is, as G. E. Moore, Bertrand Russell, and Gottlob Frege maintained, and Alfred Tarski proved, an indefinable concept. This does not mean we can say nothing revealing about it: we can, by relating it to other concepts like belief, desire, cause, and action.[16]

While not offering a definition for "truth," *Black's Law Dictionary* defines true as

> conformable to fact; correct; exact; actual; genuine; honest[.] "In one sense, that only is true which is conformable to the actual state of things. In that sense, a statement is untrue which does not express things exactly as they are. But in another and broader sense, the word 'true' is

> often used as a synonym of 'honest,' 'sincere,' 'not fraudulent.'" Moulor v. American L. Ins. Co.. Ill U. S. 345, 4 Sup. Ct. 400, 28 L. Ed. 447.[17]

In contrast to Donaldson's article, *Merriam-Webster Dictionary* dares to reveal a defined conception of truth: "The property (as of a statement) of being in accord with fact or reality."[18] For the working purposes of this book, the conventional sense of the term is used, with the acknowledged limitation that "fact" and "reality" can be subject to inherent and divergent biases and branches of human systems and interpretations.

Acknowledging the diverse interpretations of "truth," in the 2024 book *Warra Warra Wai: How Indigenous Australians Discovered Captain Cook, and What They Tell About the Coming of the Ghost People*, Aboriginal Australian Darren Rix and Caucasian Australian Craig Cormick (respectively referring to themselves as "a blackfella" and "a whitefella") recount their life-changing experience, "seeking the other side of history," by retracing the journey of eighteenth-century British explorer Captain James Cook along the East Coast of Australia.[19] Their goal:

> We knew the journey of Captain James Cook, but we wanted to ask what stories First Nations people tell about the coming of Captain Cook....
>
> We travelled the same journey that the signal fires and message sticks travelled, all along the east coast, warning different peoples that this strange vessel was approaching their land and sea—and we listened.
>
> Many communities told us: "Yes, we have stories about the British explorer, but what we really want to talk about is truth telling."[20]

Misinformation, Disinformation, and "Fake News"

Despite Adams's discerning declaration about the immutability of facts and evidence, the information ecosystem of the internet age—and now the AI age—unrelentingly tests the universality of that admonition's conclusion. Indeed, his statement's conclusion rests on premises of information quality, that is, the reliability and validity of the "facts" at hand in a given situation, that are determinative. Many terms exist that define and describe various states of *information quality*. Examples include *information*, *misinformation*, *disinformation*, *fake news*, *hoax*, and *propaganda*. Information centers can take on a vital role in aiding people in understanding the varying meanings that relate to information quality. The College of Staten Island Library, for instance, provides a research guide "adapted from a guide developed by the Research Center at the CUNY Graduate School of Journalism" for "Thinking Critically About Information Sources."[21] This guide provides definitions for some key terms associated with information quality:

- "Information is the presentation of facts about a subject."
- "Misinformation is the sharing of inaccurate and misleading information in an unintentional way."
- "Disinformation is the deliberate dissemination of false or inaccurate information in order to discredit a person or organization."

- "Fake News: the intentional falsification and fabrication of news-based information with the purpose to harm and deceive people."
- "Hoax: something intended to deceive or defraud"
- "Propaganda: information deliberately spread to influence or raise awareness to a particular political cause or value position. Propaganda may resemble advertisements since both are highly visual, but propaganda does not engage in selling products or services."[22]

The guide also offers definitions for terms relating to information quality, such as *advertising*, *bias*, *clickbait*, *confirmation bias*, *parody*, and *satire*. This plethora of ways to define and explain information quality illustrates the complexity of this topic. It also underscores the importance of understanding the nuances of these terms and the ways that they overlap and are similar and dissimilar.

Distinguishing Misinformation and Disinformation

Building on the point regarding the importance of recognizing nuances in information quality terms, it is vital, too, to exercise precision when using these terms. For example, *misinformation* and *disinformation* are routinely used incorrectly, even by some media entities. Sometimes they are conflated. Other times (many times, actually), what is actually disinformation is erroneously labeled as misinformation. The critical element for understanding how these terms are different is *intentionality*. Misinformation relates to information that is mistakenly communicated but in an *unintentional* way. Disinformation, on the other hand, involves information that is communicated with the *purposeful intent* to deceive. *Mens rea* is a Latin term that translates to "guilty mind." As LII states:

> The mens rea requirement is premised upon the idea that one must possess a guilty state of mind and be aware of his or her misconduct; however, a defendant need not know that their conduct is illegal to be guilty of a crime. Rather, the defendant must be conscious of the "facts that make his conduct fit the definition of the offense." See: Staples v. United States, 511 US 600 (1994).[23]

The law distinguishes between criminal acts that are intentional and unintentional. Greater culpability, liability, and punishment consequences are meted out for crimes that are intentional. For example, in the area of homicide, the law deems voluntary manslaughter (a crime of intentionality) to be a more serious crime and subject to more legal jeopardy than, say, involuntary manslaughter. Just as the criminal law makes distinctions between acts that are intentionally done and those that are not, so too must all of us keep the intentionality aspect in mind when we are determining whether something constitutes misinformation or disinformation.

Tactics of Disinformation

Disinformation—"false information that is intended to mislead"—is not a new phenomenon.[24] History abounds with examples of leaders and would-be leaders who have intentionally utilized untruths and conspiracy theories to further their agendas and aims. In the 1930s and 1940s, Germany's Adolf Hitler used lies about Jews, Romani, homosexuals, and other groups to demonize "the other" and advance his views of White Aryan superiority, directly resulting in the murder of millions, world war conflagrations, and devastating destruction. Throughout the mid-twentieth century, the Soviet Union's Joseph Stalin exercised disinformation and propaganda to further his vision of empire, forcibly resettling millions of peoples against their will and causing the deaths of millions through repressive policies and purges:

> The Soviet Union was—unlike past imperialist states—to be a harmonious multiethnic construction. "Friendship of the peoples" was one of its most ubiquitous slogans, and Soviet leaders claimed they had solved the "nationalities problem." However, between 1939 and his death in 1953, Stalin implemented policies of forced resettlement, or deportations, against many of the Soviet Union's ethnic minorities.
>
> He used the brutal policy against Poles, Romanians, Lithuanians, Latvians, Estonians, Volga Germans, ethnic Finns, Crimean Tatars, Crimean Greeks, Kalmyks, Balkars, Karachays, Meskhetian Turks, Koreans, Chechens, Ingush, and others.
>
> It is estimated about 6 million people were subjected to forced migration, of whom about 1.5 million died.[25]

Today, Russian leader Vladimir Putin, former head of his country's KGB-cum-FSB intelligence service, employs Stalin-esque multipronged disinformation campaigns and information warfare throughout the world against institutions of democracy and to clamp down on any dissent within Russia or its satellites. Indeed, disinformation is one of the chief tactics in the playbook of so-called strongmen authoritarians, whether past or present. One of Putin's aims has been to rehabilitate the reputation of Stalin—and recapture the expansionistic vision of Stalin's Soviet Union—which had precipitously declined after his death in 1953. In the latter part of the twentieth century, and accelerated by the dissolution of the Soviet Union in 1991, Russia and the world were able to access previously classified records and data evincing the state-sanctioned horrors of the Stalin era. However, the Putin regime has cut off access to such archival records and even shut down museum exhibitions that reveal Stalin's true actions. Unsurprisingly, knowledge about the Stalin era has diminished:

> The percentage of people who admit that they "know nothing" about the Stalin period has increased from 30 percent in 1988 to 70 percent in 2012. Sociologists explain Stalin's rehabilitation by citing the change of generations; the loss of personal experience of the Stalin period; and the generally positive portrayal of Stalin and the Stalin period in the media, films, and literature. The state has also "taken control" of the teaching of history in the schools, and the main lesson taught is that Stalin was able to create a global superpower despite some "excesses."[26]

Sadly, the United States is not immune from disinformation. The eighteenth-century American colonies wielded disinformation about Native Americans and Blacks in a deliberate effort to unite the 13 colonies against their common foe, the British. Robert Parkinson, assistant professor of history at Binghamton University, State University of New York, writes in the 2016 book *The Common Cause: Creating Race and Nation in the American Revolution* about the use of "fake news and fear-based political dialogue" by the United States' Founding Fathers. He describes how these tactics were used "to unite the 13 colonies during the American Revolution" by portraying "black slaves, Indians and (for a short time) Hessian mercenaries as 'proxies' of the British."[27] Parkinson also reveals a fascinating story showing that the Founding Fathers, including Benjamin Franklin, who "wrote a fake letter from naval great John Paul Jones," were not above using disinformation to achieve political goals.[28] Later in this chapter, we look at some additional examples of disinformation in the modern age. This chapter also heavily overlaps with the next chapter on the ethics of social media, as disinformation and social media both benefit from a symbiotic relationship with one another.

Two of the most common disinformation strategies are "false equivalence" and "whataboutism." Dictionary.com defines false equivalence as "a logical fallacy in which one assumes or asserts that two things are the same or equal when, while alike in some ways, they are not sufficiently similar to be considered equivalent."[29] False equivalence has long been used to place competing ideas, facts, or theories on equal footing when that equivalent parallel weight is not warranted. A 2019 PBS article, "False Equivalence: Why It's So Dangerous," focused on helping teachers respond to misinformation, explained the ramifications: "Not every topic warrants a 'both sides' approach. Some viewpoints are simply not backed by empirical evidence or are based on false ideas. Journalists and anyone who works with facts have to be careful not to present them as legit debates. If they do, they are creating a 'false equivalence.'"[30]

Whataboutism is a cousin of false equivalence. *Merriam-Webster* describes whataboutism as "the act or practice of responding to an accusation of wrongdoing by claiming that an offense committed by another is similar or worse."[31] In other words, if someone were to raise concerns about issue X, another party responds with a statement like, "What about Y?" Vladimir Putin is renowned for exercising whataboutism as a disinformation tactic, having honed the technique through years in counterintelligence and spycraft. Scholar Lee McIntyre elaborates, in his 2023 book *On Disinformation: How to Fight for Truth and Protect Democracy*:

> The tactic of whataboutism is another Putin favorite, where—when your opponent begins to make too much sense—you change the subject to something unrelated, such as when an American reporter challenged Putin in Geneva in 2021, pointing out that "the list of your political opponents who are dead, in prison, or jailed is long… What are you so afraid of?" Putin responded by bringing up the January 6 insurrection, offering that "many countries are going through exactly what we're going through… we sympathize with what was happening in the States, but we do not wish that to happen in Russia."[32]

Techniques such as false equivalence and whataboutism are integral components of disinformation ambitions. We see this especially regarding elections, as both a tactic to sway voters toward a certain viewpoint or to suppress voters from even casting a vote. False equivalence and whataboutism contribute to nihilistic cynicism and apathy: the sense that everything is corrupt, nothing matters, and neither anyone nor anything can be trusted or believed. If nothing can be believed or trusted and they develop a sense that "nothing I do will matter anyway," voters may decide to opt out of voting. While not voting affects elections for individual candidates, the larger issue is the impacts that not voting has on democracy in general, as democracies depend on the principles of (1) having people vote and (2) fostering the belief that their vote matters.

McIntyre explains in *On Disinformation* that another "truth killer" type of disinformation strategy is called the "firehose of lies." He asserts that Trump "borrowed" both whataboutism and the firehose of lies "from Putin's disinformation-fueled leadership":

> The firehose of lies is when someone advances a series of multiple and contradictory explanations, usually in response to criticism of a situation that makes them look bad. According to Kiril Avramov, an expert from the University of Texas on Soviet and Russian disinformation tactics, this is disorienting and intended to disrupt the use of logic to find truth. A prime example is when Putin responded to the poisoning of Sergei Skripal in Britain in 2018, by saying it was the fault of Britain, "and/or Ukraine. And/or it was an accident. And/or it was suicide. And/or it was a revenge killing by relatives. And/or Russia did not produce the nerve agent that was used."[33]

The Cybersecurity and Infrastructure Security Agency (CISA), a division of the Department of Homeland Security, has developed informative educational materials on disinformation, such as a "Tactics of Disinformation" explanatory document. The resource states what disinformation tactics are:

> Disinformation actors use a variety of tactics and techniques to execute information operations and spread disinformation narratives that pose risk to critical infrastructure. Each of these tactics are designed to make disinformation actors' messages more credible, or to manipulate their audience to a specific end. They often seek to polarize their target audience across contentious political or social divisions, making the audience more receptive to disinformation.[34]

Examples of tactics that CISA identifies are

- cultivating fake or misleading personas and websites
- creating deepfakes and synthetic media
- devising or amplifying conspiracy theories
- abusing alternative platforms
- exploiting information gaps
- manipulating unsuspecting actors
- spreading targeted content

Another tactic CISA notes in this section is "astroturfing and flooding the information environment":

> Disinformation campaigns will often post overwhelming amounts of content with the same or similar messaging from several inauthentic accounts. This practice, known as astroturfing, creates the impression of widespread grassroots support or opposition to a message, while concealing its true origin. A similar tactic, flooding, involves spamming social media posts

> and comment sections with the intention of shaping a narrative or drowning out opposing viewpoints.[35]

This tactic has some degree of relationship to a disruptive strategy epitomized by conservative firebrand Steve Bannon. Media commentator Brian Stelter, formerly with CNN Business, elucidates Bannon's tactic in a November 2021 article, "This Infamous Steve Bannon Quote Is Key to Understanding America's Crazy Politics." Stelter explains that Bannon revealed his playbook while talking with writer Michael Lewis in 2018: "'The real opposition is the media. And the way to deal with them is to flood the zone with shit.' That's the Bannon business model: Flood the zone. Stink up the joint. As Jonathan Rauch once said, citing Bannon's infamous quote, 'This is not about persuasion: This is about disorientation.'"[36]

In a February 2020 Vox article titled "Flood the Zone With Shit: How Misinformation Overwhelmed Our Democracy," Sean Illing's observations shed light on how transgressive actors like Bannon are often able to effectively manipulate modern media environments: "We live in a media ecosystem that overwhelms people with information. Some of that information is accurate, some of it is bogus, and much of it is intentionally misleading. The result is a polity that has increasingly given up on finding out the truth."[37] As Sabrina Tavernise and Aidan Gardiner put it in a *New York Times* piece, "people are numb and disoriented, struggling to discern what is real in a sea of slant, fake, and fact."[38] This is partly why an earth-shattering historical event like a president's impeachment did very little to move public opinion. Illing's thesis, though, is that some of the zone flooding that is occurring in the United States is a result of increased tribalism and a function of the "new media ecosystem," with emphasis on the right's media networks:

> There's "us" and "them," and the possibility of persuasion is off the table.
>
> It's worth noting that this polarization is asymmetric. The left overwhelmingly receives its news from organizations like the New York Times, the Washington Post, or cable news networks like MSNBC or CNN. Some of the reporting is surely biased, and probably biased in favor of liberals, but it's still (mostly) anchored to basic journalistic ethics.
>
> As a recent book by three Harvard researchers explains, this just isn't true of the right. American conservative media functions like a closed system, with Fox News at the center. Right-wing outlets are less tethered to conventional journalistic ethics and exist mostly to propagate the bullshit they produce.[39]

Illing asserts that the 24/7 news cycle "is easily commandeered by misinformation, innuendo, and outrageous content" and "the political economy of media makes it very hard to ignore or dispel bullshit stories."[40] As a result, "bullshit stories" are amplified and normalized. He states that the mainstream media's orientation is "to conquer lies by exposing them," but today "there are too many claims to debunk and too many conflicting narratives." Functioning democracies depend on some level of shared agreement on basic facts, beliefs, and values that undergird those democracies. Illing's conclusion starkly assesses that "liberal democracy cannot function without a shared understanding of reality. As long as the zone is flooded with shit, that shared understanding is impossible."

It is important to point out that CISA includes a statement that distinguishes between disinformation efforts and First Amendment rights of free speech:

> The Cybersecurity and Infrastructure Security Agency (CISA) produced this graphic to highlight tactics used by disinformation campaigns that seek to disrupt American life and the critical infrastructure that underlies it. CISA's publication of informational materials about this issue are intended for public awareness, and are not intended to restrict, diminish, or demean any person's right to hold, express, or publish any opinion or belief, including opinions or beliefs that align with those of a foreign government, are expressed by a foreign government-backed campaign, or dissent from the majority. CISA respects the First Amendment rights of all U.S. persons and publications.[41]

This statement, as we see at other places in this book, illustrates tensions that exist where issues like combating disinformation test where the line is between a fundamental right, such as freedom of expression, and censorship. Later in this chapter, as well as in the next chapter, we look at some examples involving specific creators and disseminators of misinformation and disinformation.

Ethics Codes and Codes of Conduct

In previous chapters we've observed that ethics codes and codes of conduct specifically address issues like access, privacy, intellectual freedom, free speech, and censorship. Many codes that have relationships to the information professions, like those identified in this section, include statements addressing accuracy and truth. Additional elements that intrinsically relate to accuracy and truth—integrity, trust, accountability, and transparency—are also often present in such principles.

Society of Professional Journalists

The Society of Professional Journalists, founded in 1909, is a professional association whose members wrestle with issues of truth, accuracy, misinformation, and disinformation on a daily basis. One of the ways the organization offers guidance to its members is through its code of ethics, which was revised in 2014. The code's preamble states that "members of the Society of Professional Journalists believe that public enlightenment is the forerunner of justice and the foundation of democracy. Ethical journalism strives to ensure the free exchange of information that is accurate, fair and thorough. An ethical journalist acts with integrity."[42] Four principles form the foundation of "ethical journalism" for the society:

1. Seek truth and report it.
2. Minimize harm.
3. Act independently.
4. Be accountable and transparent.

The society provides a series of thought-provoking ethics case studies for discussion and education.[43] Though focused on real-life journalistic-related events and issues, these case studies have broad relevance for other contexts outside of media, like the information professions. Sample topics include whether music magazine reporters should or should not provide links to pirated music, the pros and cons of publishing drunk drivers' mug shots, whether reporters should be able to attend fundraising events for political candidates or march in events like gay pride parades without questions of conflicts of interest arising, and the responsibilities of media outlets to ferret out plagiarism.

American Library Association

The American Library Association (ALA) Code of Ethics does not specifically use the word *truth* in its nine statements or principles, but it does refer to "accuracy" of "responses" in its first principle: "1. We provide the highest level of service to all library users through appropriate and usefully organized resources; equitable service policies; equitable access; and accurate, unbiased, and courteous responses to all requests."[44]

Medical Library Association

The Medical Library Association Code of Ethics provides "broad statements to guide ethical decision making," in the vein of most ethics codes.[45] Interestingly, the code does not specifically refer to "accuracy" or "truthfulness." The closest it comes to speaking in that area is a statement that indicates health science librarians should provide the best available information to clients. This code was updated in 2010.

American Association of Law Librarians

Noteworthy with regard to the American Association of Law Libraries's (AALL) ethical principles is that there is no mention of "accuracy" or "truthfulness." The first two sentences of the preamble to the principles delineate "expert knowledge" and "legal information" but do not address the accuracy or truthfulness of that knowledge and information: "Members of the American Association of Law Libraries (AALL or Association) possess expert knowledge of sources of law, legal research methods, and the use of information technologies. They apply their expertise to select, collect, organize, and preserve sources of legal information and guide researchers in using that information in an effective, efficient and discerning manner."[46] In 2019, the principles were revised and approved by AALL members.

Combatting Misinformation and Disinformation

Concerns about information quality, misinformation, and disinformation have catalyzed the creation of tools and processes to help information professionals and users better assess the accuracy of information and proactively respond to misinformation, disinformation, and conspiracy theories. This section explores some of these fact-checking and information evaluation tools and strategies, including watchdog organizations.

ALA

ALA provides a useful digital LibGuide that brings together various resources for "Evaluating Information."[47] The LibGuide provides helpful historical context, underscoring that misinformation (and disinformation) are not new phenomena:

> Misleading information published as news is not new to the 21st century. In the late 19th century we called it "yellow journalism," and its practitioners used sensational headlines and outright fraudulent stories to increase sales. Today, with increasing reliance on both digital news outlets and social media for news, sifting through the messages for non-biased sources requires attention, and possibly reviewing multiple sources—including seeking out a reliable original source.[48]

The LibGuide includes tips, information literacy standards, resources, and tools from varied sources, such as National Public Radio (NPR) and academic and public libraries. Archived webinars, such as two that were presented by ALA following the 2016 presidential election in early 2017, "Post-Truth: Fake News and a New Era of Information Literacy"[49] and "Using Media Literacy to Stop the Fake News Cycle,"[50] are useful sources of cutting-edge information and expertise from practitioners and researchers. The LibGuide also includes examples of ALA policy statements. "Resolution on Access to Accurate Information" (2017) addresses "fake news, personalized newsfeeds, web search algorithms and the delay of Freedom of Information Act [FOIA] requests."[51] The policy emphasizes, perhaps somewhat counterintuitively, that "access to accurate information, rather than censorship, is the best way to counter disinformation and media manipulation."[52]

International Federation of Library Associations and Institution

The International Federation of Library Associations and Institutions (IFLA) document "How to Spot Fake News" provides an eight-step process for identifying fake news via a freely accessible and downloadable infographic (see Figure 7.1).[53] IFLA explains the genesis for and rationale behind the development of its fake news–spotting strategy:

> Critical thinking is a key skill in media and information literacy, and the mission of libraries is to educate and advocate its importance.
>
> Discussions about fake news has [*sic*] led to a new focus on media literacy more broadly, and the role of libraries and other education institutions in providing this.
>
> When Oxford Dictionaries announced *post-truth* was Word of the Year 2016, we as librarians realise action is needed to educate and advocate for critical thinking—a crucial skill when navigating the information society.
>
> IFLA has made this infographic with eight simple steps (based on FactCheck.org's 2016 article "How to Spot Fake News") to discover the verifiability of a given news-piece in front of you. Download, print, translate, and share—at home, at your library, in your local community, and on social media networks. The more we crowdsource our wisdom, the wiser the world becomes.[54]

Figure 7.1 "How to Spot Fake News" infographic. (Used with permission from IFLA under Creative Commons License CC-BY 4.0. Icons available individually under FreePik licence)

The infographic is available in English and more than three dozen languages, spanning the Romanized alphabet from Arabic, Bulgarian, Chinese, and Dutch to Russian, Turkish, Ukrainian, and Welsh.

NewsGuard

NewsGuard is a browser plug-in for search engines, such as Mozilla Firefox, Google Chrome, and Microsoft Edge. It deploys a pool of journalists and media administrators from diverse outlets to assess and then provide a "trustworthiness" score on various heavily visited sites on the internet. Just as the rationale for nutrition labels is to arm consumers with more information about foods and drinks, NewsGuard strives for readers to be better equipped to make decisions about the information they choose or don't choose to consume. In their own words, "NewsGuard combines human expertise and technology to provide data, analysis and journalism that helps enterprises and consumers identify reliable information online."[55]

A defining feature of NewsGuard is that evaluation of the websites is performed by humans rather than AI algorithms, bucking the trend of farming out tasks for AI systems to perform. This human element makes the evaluation process more labor intensive and time consuming. According to NewsGuard, though, having information assessment conducted by human specialists is worthwhile in both the short term and long term. This is also in keeping with the well-established roles information experts have played in curating information. As longtime news executive Steve Brill noted during an MSNBC interview about the tool, "That's what librarians have been doing since the invention of the library."[56] In NewsGuard's own description from its website it says, "Our work is conducted by a team of journalists trained as information reliability analysts using a transparent and apolitical approach and a multi-layered review process to ensure high information quality."[57]

Snopes

Snopes is a fact-checking organization that traces its 1994 launch to the beginnings of the internet as we know it,

> investigating urban legends, hoaxes, and folklore. As demand for reliable fact checks grew, so did Snopes. Now it's the oldest and largest fact-checking site online, widely regarded by journalists, folklorists, and readers as an invaluable research companion. Snopes.com is an independent publication under the company "Snopes Media Group Inc.," which is owned by Chris Richmond and Drew Schoentrup (each owning 50%).
>
> Snopes is a member of IFCN and in compliance with the International Fact Checking Network's highest standards of combating misinformation online.[58]

The site has an FAQ. Perhaps the most important question is, "What is Snopes' fact-checking process?"

Since the material we fact-check can range from everything to analyzing whether an image has been digitally manipulated to explicating the text of a Congressional bill, we can't describe any single method that applies to all our fact-checking efforts.

In general, each entry is assigned to one of the members of our editorial staff who undertakes the preliminary research and writes the first draft of the fact check.

Our research generally begins with (whenever possible) attempting to contact the source of the claim for elaboration and supporting information. We also attempt to contact individuals and organizations who would be knowledgeable about, or have relevant expertise in, the subject at hand, as well as searching out printed information (news articles, scientific and medical journal articles, books, interview transcripts, statistical sources) with bearing on the topic.

We attempt to use non-partisan information and data sources (e.g., peer-reviewed journals, government agency statistics) as much as possible, and to alert readers that information and data from sources such as political advocacy organizations and partisan think tanks should be regarded with skepticism.

Depending upon the nature and complexity of the topic, other members of the editorial staff may contribute additional research (or their own personal expertise) and editing. The final product will pass through the hands of at least one editor. Any piece that is not deemed up to our standards by one or more editors is subject to further revision and review before being released for publication.[59]

Case Study: False Claims of Pet-Eating in Ohio

Let's look at an example of the types of claims that fact-checkers investigate and endeavor to clarify. In this example, both Snopes and NewsGuard investigated and rated claims made that Haitian immigrants in the Ohio community of Springfield were illegally taking people's pet cats, as well as park swans and ducks, and eating them. The claims quickly went viral on social media. GOP vice presidential candidate J. D. Vance (US senator for Ohio) repeated the allegation via a post made to the social media platform X (formerly Twitter). A number of conservative media personalities, as well as politicians like US Rep. Jim Jordan (R-Ohio), also "elevated" the claim. On September 9, 2024, Snopes published the results of its investigation, determining that "these charges had no basis in fact—they were unsourced internet rumors."[60]

Following this one-sentence determination, Snopes provides a detailed several-paragraphs-long explanation of the controversy, which credits the local newspaper, the *Springfield News-Sun*, with tracking the claim back to its origin as "a rumor posted to a Springfield Facebook group." The newspaper reported: "A social media post originally from a Springfield Facebook group went viral nationally in recent days. The original poster did not cite first-hand knowledge of an incident. Instead they claimed that their neighbor's daughter's friend had lost her cat and found it hanging from a branch at a Haitian neighbor's home being carved up to be eaten."[61]

That original Facebook post was shared on various social media platforms, going viral, and was amplified by Senator J. D. Vance and others, as mentioned. During a September 10, 2024, debate between Vice President Harris and former president Trump, Trump brought up the spurious Springfield, Ohio, claim in the course of responding to a question on immigration from moderator David Muir: "In Springfield, they're eating the dogs. The people that came in. They're eating the cats. They're eating—they're eating the pets of the people that live there. And this is what's happening in our country. And it's a shame."[62] The subsequent exchange between Muir and Trump is illustrative:

> DAVID MUIR: I just want to clarify here, you bring up Springfield, Ohio. And *ABC News* did reach out to the city manager there. He told us there have been no credible reports of specific claims of pets being harmed, injured or abused by individuals within the immigrant community—
>
> FORMER PRESIDENT DONALD TRUMP: Well, I've seen people on television
>
> DAVID MUIR: Let me just say here this…
>
> FORMER PRESIDENT DONALD TRUMP: The people on television say my dog was taken and used for food. So maybe he said that and maybe that's a good thing to say for a city manager.
>
> DAVID MUIR: I'm not taking this from television. I'm taking it from the city manager.
>
> FORMER PRESIDENT DONALD TRUMP: But the people on television say their dog was eaten by the people that went there.
>
> DAVID MUIR: Again, the Springfield city manager says there's no evidence of that.
>
> FORMER PRESIDENT DONALD TRUMP: We'll find out.[63]

The impacts of disinformation can exact serious tolls. Indeed, a September 12, 2024, *Newsweek* article, "Donald Trump, MAGA 'Eating Pets' Rumor Sparks Fear for Springfield Haitians," indicated that "Haitians in Springfield, Ohio, are considering leaving the city over unfounded claims some have been eating pets and livestock, adding to a growing backlash against immigrant communities."[64] Sadly, these kinds of falsehoods directed at cultural groups are not historically new but now, in the digital age, can be amplified and are able to virally reach millions at the tap of a smartphone or computer key. Canada's CBC reported in its September 12, 2024, story "Springfield, Subject of Baseless Pet-Eating Claim Spread by Trump at Debate, Receives Bomb Threats":

> Food-related scorn and insults have been hurled at various immigrant communities in U.S. history, from the Chinese on the West Coast in the late 1800s as they started coming to the United States in larger numbers, and in later decades to other Asian and Pacific Islander communities like Thai or Vietnamese. As recently as last year, a Thai restaurant in California was hit with the stereotype, which caused such an outpouring of undeserved vitriol that the owner had to close and move to another location.[65]

The night of September 12, Trump held a rally in Arizona. At that rally, despite having had Springfield city officials and Ohio Governor Mike DeWine denounce Trump's falsehoods about pet eating and wild duck eating, Trump continued to disseminate disinformation, as reported by the *Guardian* ("Trump Attacks Migrants and Claims Victory Over Harris at First Rally Since Debate"):

> He also focused heavily on immigration, a typical target for his rallies, stoking fears about waves of criminals crossing into the country and again accusing Haitian immigrants of eating animals in Springfield, Ohio.
>
> "They take in the geese... and even walk off with their pets," Trump said, repeating the unfounded and racist smear that has been challenged by the town's officials and has inflamed tensions and attacks against members of the Haitian community.[66]

NewsGuard investigated the origin of the Springfield pet-eating claims and published its results on September 12. That investigation revealed that the original source of the pet-eating claim in Ohio admitted that they had "no first-hand knowledge" of the story and alleged facts that they communicated.[67] NewsGuard created a graphic that showed how this rumor spread: from an alleged cat owner to the cat owner's acquaintance to the friend of an acquaintance to the friend's neighbor to the conservative influencers to Senator J. D. Vance to President Trump.[68] To better understand how these unfounded claims started and were subsequently amplified by private and public figures, it is informative to look at what NewsGuard's examination revealed when the fact-checking organization located and then exclusively interviewed the two people pivotal to the pet-eating allegations:

> Erika Lee, the Springfield resident who wrote the original Facebook post, and Kimberly Newton, the neighbor who had provided her with a third-hand account of the rumor, making Lee's social media post a fourth-hand account: the alleged acquaintance/cat owner; Newton's friend; Newton; and Lee, who posted it on Facebook....
>
> The interviews reveal just how flimsy and unsubstantiated the rumor was from the beginning—based entirely on third hand hearsay. Yet it quickly gained traction and, remarkably, found its way to Trump's lips on a national stage.[69]

Newton acknowledged to NewsGuard that she was "not the most credible source" because she wasn't acquainted with the lost cat's owner.[70] Despite that lack of knowledge, Newton told the rumor to her neighbor, Lee, who then posted the rumor to Facebook. The rumor eventually made its way to Vance and Trump, who amplified the message on the national stage.

Poynter Institute

Poynter Institute is a journalism school that was started by *St. Petersburg Times* (name changed to the *Tampa Bay Times*) newspaper owner Nelson Poynter in 1975.[71] Originally, the name was the Modern Media Institute, which was changed to the Poynter Institute for Media Studies (Poynter Institute, for short) in 1984.[72] It functions in two nonprofit capacities: (1) as a journalism school and (2) as a research entity. Poynter Institute also became the owner of the *Tampa Bay Times* newspaper upon Nelson Poynter's death.[73]

The institute describes itself as "a global nonprofit that strengthens democracy by improving the relevance, ethical practice and value of journalism. Through teaching, publishing, convening, fact-checking and media literacy, Poynter creates a crossroads where communities come together to use journalism to confront society's complex problems."[74] One of Poynter Institute's chief goals is to foster public trust. To advance that aim, the institute prioritizes media literacy and fact-checking. In 2015, it launched the International Fact-Checking Network (IFCN):

> The network advocates for information integrity in the global fight against misinformation and supports fact-checkers through networking, capacity building and collaboration. IFCN's network reaches more than 170 fact-checking organizations worldwide through advocacy, training and global events. Our team monitors trends in the fact-checking field to offer resources to fact-checkers, contribute to public discourse and provide support for new projects and initiatives that advance accountability in journalism. We believe truth and transparency can help people be better informed and equipped to navigate harmful misinformation.[75]

In January 2021, IFCN was nominated for a Nobel Peace Prize. Fact-checking can be dangerous work, as Poynter explained in its response to being nominated:

> While we recognize that this nomination is a long way from being shortlisted for this unparalleled award, we see it as an important validation of the work of fact-checkers worldwide. Simply put: Facts matter and fact-checking can save lives. Fact-checkers are working worldwide, often under threat or attack, to provide quality information and combat misinformation, often made deliberately, that pollutes society or inhibits freedom.[76]

IFCN strives to "promote higher standards in fact-checking and to advocate for reliable and accurate information across all mediums."[77] It developed a code of principles for fact-checkers. Poynter states that its code "is for organizations that regularly publish nonpartisan reports on the accuracy of statements by public figures, major institutions, and other widely circulated claims of interest to society."[78] To ensure the upholding of standards, Poynter says that IFCN audits "the quality of the work being done by our members every year."[79] Poynter also operates a digital media literacy initiative called MediaWise.[80] Additionally, the Craig Newmark Center for Ethics and Leadership at Poynter "is dedicated to improving the reliability of journalism and transparency between newsmakers and news consumers."[81]

Correkt

AI chatbots are discussed in detail in Chapter 11. For this chapter, though, it's important to explain that diverse kinds of stakeholders are engaging in efforts to harness AI in ways that promote accuracy of information and ferret out misinformation and disinformation. As announced in a May 2024 article, a group of University of California, Santa Barbara (UCSB), students from the computer science, engineering, and data science disciplines developed an app called Correkt whose aim is to counter "misinformation dissemination." The app does this by only being connected "to reputable data sources—newspapers, textbooks and peer-reviewed journals. The AI model currently draws its information from an expansive data bank of over 180 million well-established, trustworthy resources—a number set to grow with time."[82] The second way Correkt addresses misinformation is by having "a built-in citation function that details the precise location where every piece of information it presents to the user was retrieved. Essentially, Correkt is a hybrid between a search engine and an AI chatbot."[83] This example shows some of the ways that students are innovatively responding to the need for "more ethical usage of information" and to promote greater transparency about the sources of information. As the Correkt article concludes, "projects like 'Correkt' give hope that this complex balance between technological advancement and ethics can be navigated."[84]

Pre-bunking

Pre-bunking is a proactive strategy for combatting mis- and disinformation. It differs from *debunking*, a reactive approach, in which a rebuttal "to set the record straight" or respond to a falsehood or error is made *after* something has been said or has occurred. For example, ongoing falsehoods about the 9/11 attacks on the Pentagon and New York City's World Trade Center in 2001, so-called false flag school shootings like the tragic killings of students and teachers at Connecticut's Sandy Hook Elementary School in 2012, the origins of the COVID-19 virus, and the "Big Steal" election lies that fueled the January 6, 2021, insurrection at the US Capitol continue to require debunkers to point out the inaccurate premises, alternate facts, and unsubstantiated conclusions on which such conspiracy theories rely. Pre-bunking takes a different approach. A May 2024 *Washington Post* article, "Debunking Misinformation Failed. Welcome to 'Pre-bunking,'" explained that, "Modeled after vaccines, these campaigns—dubbed 'prebunking'—expose people to weakened doses of misinformation paired with explanations and are aimed at helping the public develop 'mental antibodies' to recognize and fend off hoaxes in a heated election year."[85]

Litigation lawyers habitually employ pre-bunking tactics when trying cases before judges and juries. For example, a prosecuting attorney may state in their opening argument that "you (i.e., the jury or the judge) are going to hear the defense say X. However, the truth that the prosecution and the evidence will show is Y." By doing this, the attorney is laying the groundwork for the listener to mentally watch out for the information that may be coming from the opposing side, in order to be prepared to mentally process and better make sense of that information in ways that are desired by the pre-bunking party.

Journalist and political pundit Anand Giridharadas observed, on a September 12, 2024, cable news program, that this pre-bunking technique was used by Vice President Kamala Harris during the September 10 debate between Harris and former president Donald Trump, when just several minutes into the debate she stated:

> What we have done and what I intend to do is build on what we know are the aspirations and the hopes of the American people. But I'm going to tell you all, in this debate tonight, you're going to hear from the same old, tired playbook, a bunch of lies, grievances and name-calling. What you're going to hear tonight is a detailed and dangerous plan called Project 2025 that the former president intends on implementing if he were elected again.[86]

By telling listeners what she thought would likely be said by her opponent later in the debate, Harris was essentially "inoculating" listeners from the effects that her adversary's words might cause, instead suggesting how listeners might interpret such statements in their own minds, in the way that she wanted them to. Using a highway metaphor, pre-bunking is akin to the cautionary road signs we see when driving, indicating possible foreseeable hazards like "deer crossing," "slippery when wet," or "winding road ahead."

Information professionals are actively involved with and engaging in pre-bunking and debunking education in a number of ways. As one commercial example, InfoBase is a subscription-based for-profit company that advertises that it has "created and curated exceptional content for educators, librarians, and learners across the globe" and provides licensed access to products such as Facts On File, Films On Demand, and the *World Almanac for Kids*.[87] Two Ohio-based academic librarians are the authors of the InfoBase white paper "Debunking & Prebunking: Strategies for Librarians to Eradicate Misinformation":

> Concerned about the effects of misinformation, librarians Evan Meszaros and Mandi Goodsett were inspired to explore how librarians can play a part in addressing this serious problem. Misinformation can impact health, financial, and voting decisions, with devastating consequences. Based on their extensive literature review, they saw a role for librarians to help stem the tide of misinformation spread, and they hoped the evidence-based conclusions of their investigation help librarians see practical ways to begin those efforts.[88]

Michelle Kraft, library director at Cleveland Clinic, authored a January 2024 lecture-cum-article for the *Journal of the Medical Library Association* on "the history of fake news, how it is spread and accepted, its impact on medical and health information, and medical librarian roles in limiting its spread and promoting correct health information."[89] She notes that though the issue of fake news has garnered much recent media notice, "misinformation and disinformation has [*sic*] been a lasting part of human history." Additionally, Kraft remarks that AI programs and the "questionable data" underlying those programs "can be fraught with bias, misinformation, and outright lies, leading to potential ethical issues regarding the data and results. Librarians should be more engaged in the understanding of the data behind the AI to help ensure factual health information is provided."

What Kraft is doing, in essence, in focusing attention on issues and concerns regarding AI training data, is to describe a kind of *pre-bunking data evaluation process* that she asserts needs to happen vis-à-vis that data. Unfortunately, in many cases, AI products, tools, and the data that are used to create them are treated by their owners as proprietary intellectual property that is not subject to outside analysis or transparency. We delve more into issues involving AI training data in this book's AI chapter. Kraft concludes by issuing a call for information professionals to take a proactive posture on misinformation and disinformation: "Librarians can't be everything at all times, but we can use our strengths to address fake news within our libraries, institutions, and the profession as a whole. Librarians must get involved; simply relying on facts to do the work for us is not a solution."[90]

"Truth Sandwich" Strategy

Developing and using strategies to respond to misinformation and disinformation is important for anyone wanting to address these antonyms of accuracy and truth. Dr. George Lakoff, professor emeritus of cognitive linguistics at the University of California, Berkeley, is credited with articulating a mis- and disinformation response therapy he coined in 2020 as a "truth sandwich." Lakoff's impetus was his observation that "media organizations are unintentionally spreading misinformation when they repeat lies or quote politicians who are asserting falsehoods."[91] In a 2020 Twitter post, Lakoff laid out the recipe for a truth sandwich, paraphrased here: First, state what "the truth" is. Next, identify what the falsehood or lie is, while not amplifying the false statement, when possible. Then, come back around to what the truthful, accurate statement is and emphasize the truth.[92]

An April 2020 PBS Standards blog story states that "accuracy includes more than simply verifying whether information is correct."[93] Situating facts within an accurate context and being vigilant about the words used to "frame the facts" are vital to avoid misleading people. At the time of the April 2020 PBS blog post, a Trump campaign advisor was resorting to "birtherism." Per *Merriam-Webster Dictionary*, birtherism is defined as "belief in or endorsement of any of various discredited claims that former U.S. president Barack Obama is not a natural-born U.S. citizen and therefore was not constitutionally eligible to be president of the United States."[94] The advisor was claiming that Senator Kamala Harris (D-California) was not eligible to be the Democratic Party's nominee for vice president. PBS explained that the truth sandwich strategy arose in the course of considering how media persons should report on President Trump, "specifically when he tweets information that is known to be untrue."[95] Yet the post also emphasizes that the truth sandwich strategy can be employed to counter a wide range of controversies, such as vaccine efficacy and climate change.

Some have claimed that the term *truth sandwich* is a misnomer because the truth steps function as the two slices of bread on the outside—with the lie placed in between the bread slides or bun halves. Sandwiches, after all, are named for what comes *between* the bread or bun, such as peanut butter and jelly, hamburger, tuna fish salad, watercress. As a technique for proactively addressing falsehoods and errors, though, rather than describing a real-life gastronomic consumable or a Food Network Guy Fieri–esque contest construction, using *truth sandwich* as nomenclature has merit and memorableness. Most would probably agree, too, that a "truth sandwich" sounds much more appetizing—and, most relevantly, positive-action-oriented—than calling the strategy a "lie sandwich" or a "misstatement burger," the latter of which should not be confused with a "nothingburger," which *Merriam-Webster Dictionary* notes is a term that "emerges in the media whenever something is downplayed as irrelevant or insignificant."[96] Far from a nothingburger, in fact, disinformation often has tangible consequences and ethical significance in the lives of real people and communities, as evinced in the earlier example involving Springfield, Ohio.

Research on Misinformation and Disinformation

To better understand fundamental aspects of internet- and social-media-enabled misinformation and disinformation, research is critically important. It's also key to developing strategies for addressing these phenomena. Some misinformation and disinformation research is conducted by tech companies and social media platforms. Other kinds of misinformation and disinformation research are done by researchers at for-profit companies that have more tangential relationships with technology than Big Tech corporations like Google/Alphabet and Facebook/Meta, academic institutions, government-affiliated agencies and labs, and civil watchdog organizations.

Facebook Whistleblower Case Study

In October 2021, an anonymous former Facebook employee came forward and filed formal "bombshell" complaints with federal officials about Facebook's research practices and findings. The former employee revealed herself, via CBS's *60 Minutes* television program, to be Frances Haugen, a data scientist with degrees in computer science and business and experience working at Google, Pinterest, and Facebook. Haugen had spirited reams of proprietary research data from Facebook on her way out the door, research that was incriminating about what Facebook had learned about the impacts it was having on users, nonusers, and the world we all inhabit:

> The complaints say Facebook's own research shows that it amplifies hate, misinformation and political unrest—but the company hides what it knows. One complaint alleges that Facebook's Instagram harms teenage girls. What makes Haugen's complaints unprecedented is the trove of private Facebook research she took when she quit in May. The documents appeared first, last month, in the Wall Street Journal. But tonight, Frances Haugen is revealing her identity to explain why she became the Facebook whistleblower.[97]

In promotional ads for the *60 Minutes* interview, Haugen unburdened herself about the ethical lapses she had witnessed firsthand that had impelled her to finally speak out: "The thing I saw at Facebook over and over again was there were conflicts of interest between what was good for the public and what was good for Facebook. And Facebook, over and over again, chose to optimize for its own interests, like making more money."[98]

Haugen's disclosures unleashed an avalanche of bad press and undesirable repercussions for Facebook and its CEO and cofounder Mark Zuckerberg. As *El Pais* reported several weeks after the revelations broke, in a piece titled "Frances Haugen, Facebook Whistleblower: 'In 10 Years We Will Wonder Why We Didn't Regulate Social Networks Sooner'":

> The ground was trembling at Facebook's headquarters in Menlo Park, California, in mid-September 2021. For several days, *The Wall Street Journal* had been publishing the Facebook Files, a journalistic investigation based on the company's internal documents which showed that executives were aware of the damage that Instagram and Facebook caused to young people. It also showed that the company was responsible for disseminating fake news and content that incited violence.
>
> A month later, in October, the world put a face to the person responsible for the leak. Engineer Frances Haugen, 39, decided to reveal her identity. In a TV interview, she explained that she had left her position at Facebook with 21,000 internal documents under her arm. The U.S. Senate called her to testify and investigations were launched into her revelations about Meta, the new name Mark Zuckerberg gave to his social media company just weeks after the leak.[99]

Following Haugen's congressional testimony, lawsuits were filed by the families of teens who had used and allegedly been harmed by Facebook. Legal pressure on Meta continued to mount: in

October 2023, "attorneys general from 41 states sued Meta for harming children with its products and failing to report those dangers."[100]

As important context, Facebook and Zuckerberg had been on the hot seat just a few years earlier with the Cambridge Analytica scandal that broke in March 2018. The essence of Cambridge Analytica could be boiled down to Facebook not having attained meaningful informed consent and having taken liberties with the privacy interests of its American users, by unethically harvesting and permitting third parties to then leverage its users' personal data. Flashing forward to 2021, Haugen's insider disclosures revealed Facebook's continuing indifference to the misinformation and disinformation harms it knew its technology was directly causing. Moreover, Haugen's use of insider data to establish a timeline of what Facebook "knew and when it knew it" was reminiscent of past "smoking gun" revelations about the tobacco and fossil fuel industries' indifference to public harms those manufacturers similarly knew were caused by their products.

Examples of Misinformation and Disinformation Research

Entire books have been and could continue to be written about mis- and disinformation research and research findings conducted over the past decade and a half since the rise of social media alone. Several examples from researchers who have spoken at the University of Pittsburgh (Pitt) highlight some of the distinct foci of scientific inquiry. The Institute for Cyber Law, Policy, and Security (Pitt Cyber, for short) has been a significant portal for raising awareness of this research, as these heath-care-related examples of events the institute has hosted show:

- "A Healthier Information Environment: The Alaska Public Health Information Response Team" (January 18, 2023). Dr. Joy Chavez Mapaye and Dr. Jennifer Meyer shared insights about collaborative efforts to stem the spread of online mis- and disinformation. As a 2021 University of Alaska Anchorage article described, "Consisting of around 30 volunteers, all of whom have professional backgrounds in health care and public health, the team responds to Facebook comments containing false or misleading data about the coronavirus."[101]
- "Battling the Infodemic: Covid-19 Mis- and Disinformation" (February 16, 2021). Pitt researchers such as Dr. Beth Hoffman and Dr. Jaime Sidani discussed "how mis- and disinformation could jeopardize control of the pandemic" as well as issues regarding mis- and disinformation about COVID-19 vaccines.[102]

Efforts Against Research on Misinformation and Disinformation

A rise in research on mis- and disinformation has precipitated pushback against the very researchers and organizations that are conducting and disseminating this research. The groups who are attempting to succeed and, in some instances, succeeding in squelching mis- and disinformation research have interests in eliminating the evidence-based voices and views of these researchers. One example, as reported by the *Washington Post* in a June 14, 2024 story, is the Stanford Internet Observatory (SIO): "The Stanford Internet Observatory, which published some of the most influential analysis of the spread of false information on social media during elections, has shed most of its staff and may shut down amid political and legal attacks that have cast a pall on efforts to study online misinformation."[103]

Stanford's Cyber Policy Center, on June 17, 2024, asserted that "Stanford has not shut down or dismantled SIO as a result of outside pressure. SIO does, however, face funding challenges as its founding grants will soon be exhausted. As a result, SIO continues to actively seek support for its research and teaching programs under new leadership."[104] However, a June 14, 2024, NPR report explained that the "Major Disinformation Research Team's Future Is Uncertain After Political Attacks."[105] The SIO was founded in 2019 with a mandate to study "some of the thorniest issues raised by the proliferation of the internet, including the way social networks such as Instagram are used for child exploitation and the spread of false and misleading information about elections and vaccines."[106] NPR noted, though, that some Republicans in Congress and the courts have accused SIO and other groups of engaging in censorship. In some cases, these Republicans efforts have been successful at chilling or curtailing research. The Election Integrity Partnership (EIP) was a collaborative initiative of SIO and the University of Washington, with the purpose of studying misinformation and falsehoods regarding the 2020 and 2022 elections. Conspiracy theories alleged that EIP "was a front for the government to suppress speech it didn't like," and EIP's web page later announced that it would "not be working on the 2024 or future elections."[107]

Because of these allegations, Stanford, University of Washington, and researchers at other institutions have been sued, have been subjected to subpoenas and document requests, and have experienced cyberbullying and cyberharassment. These actions have resulted in staggering legal costs in the millions and distractions from research. Attacks from outside groups have impacted the ability of these academic entities to secure research funding. NPR reported in its June 14 story that after it was announced that SIO would be pulling back from its work, Republican Rep. Jim Jordan of Ohio posted "Free speech wins again!" on his X social media account.[108] As the chair of the House Judiciary Committee, Jordan had been at the vanguard of efforts to undermine the work of groups like SIO, accusing them of being part of "the censorship regime."[109]

In late 2023, Harvard University was ground zero for another example of contested stakeholders, interests, and outcomes, when the *Verge* and other media outlets reported that "Disinformation Researcher Accuses Meta of Meddling in Harvard's Research":

> A prominent expert in online disinformation campaigns, Joan Donovan, says Meta pushed her former employer Harvard to shutter a major research project on media manipulation.
>
> Donovan, who led Harvard's Technology and Social Change Project (TaSC) until it ended this year, revealed her allegations earlier this week in a whistleblower complaint obtained by The Washington Post. She claims that Harvard scuttled her research after the charity of Facebook founder Mark Zuckerberg and his wife, Priscilla Chan, donated $500 million to the university in 2021. Around that time, Donovan and her colleagues at TaSC were working with the infamous Facebook Files, a cache of leaked documents that revealed the inner workings of how Facebook handled misconduct on its platform and potential risks to users.[110]

Harvard denied Donovan's assertions, explaining that the project was ended "because as a staff member rather than full faculty, she wasn't eligible to lead it. It's one of the most high-profile tussles yet over how much influence powerful tech benefactors have over academia."[111]

With the assistance of Whistleblower Aid, Donovan made public and filed with the Massachusetts Attorney General's Office and the US Department of Education a 248-page complaint, laying out her allegations and including a variety of documents as exhibits. Item 6 in the complaint places this dispute in a broader historical context of skirmishes involving governmental agencies, higher education administrators, researchers, and corporate entities/donors. "The facts in this case are similar in motive and tactics to prior industry campaigns of Big Tobacco, Oil and Gas, and Big Pharma, each of which manipulated institutions into producing 'research' products that supported their respective industries and further harmful public narratives."[112] Lobbyists and outside interest groups frequently play pivotal roles in these struggles, as well.

Whose viewpoint is "the truth," or perhaps more aptly stated "the most accurate" or "the least untrue," can be challenging for observers or even seasoned media watchers and reporters to determine. A June 2024 *Chronicle of Higher Education* article's headline asked, "Is a World-Famous Misinformation Expert Spreading Misinformation?"[113] Donovan, however, continues to speak out on mis- and disinformation, such as via her September 13, 2024, article "A New Front in the Meme Wars" in the *Atlantic*.[114]

One more example illustrates the ongoing skirmishes between those who want to study disinformation and those who would prefer that disinformation not be studied or are concerned about how free speech may be chilled by efforts taken to curb misinformation and disinformation. In April 2022, the Department of Homeland Security announced a new advisory board to study disinformation. The body was to be called the Disinformation Governance Board (DGB), and Nina Jankowicz was named as executive director. Following criticism about the mission of the DGB and concerns about its impacts on constitutional rights of free expression, plans for the DGB were "paused" and Jankowicz resigned in mid-May 2024. In April 2024, Jankowicz launched the nonprofit American Sunlight Project (APS). The *New York Times* explained that through APS, Jankowicz is leading "a new nonprofit organization intended to fight what she and others have described as a coordinated campaign by conservatives and others to undermine researchers, like her, who study the sources of disinformation."[115] The article further noted that

> the inception of the group, the American Sunlight Project, reflects how divisive the issue of identifying and combating disinformation has become as the 2024 presidential election approaches. It also represents a tacit admission that the informal networks formed at major universities and research organizations to address the explosion of disinformation online have failed to mount a substantial defense against a campaign, waged largely on the right, depicting their work as part of an effort to silence conservatives.[116]

Sanewashing

In late 2024, misinformation and disinformation continued to be major issues of concern for some voters and candidates, while largely ignored by others. A new (or at least more widely recognized) term began to be mentioned by some media outlets and reporters: *sanewashing*. Mainstream dictionaries do not yet have definitions for this ostensibly new coinage. Wiktionary provides two definitions:

1 The practice of restating someone's rhetoric to seem more palatable or acceptable.
2 Attempting to downplay a person or idea's radicality to make it more palatable to the general public.[117]

The practice of sanewashing, that is to say, to interpret or restate someone's words to make them seem more coherent or reasonable, can be viewed as a kind of misinformation or disinformation, depending on one's perspective. It can also be seen as a type of inaccuracy or effort to obscure truth. Two examples in the news elevated discussion of sanewashing amid the contentious US presidential election season. On September 5, 2024, former president Trump spoke at the Economic Club of New York. Trump was roundly criticized by some commentators, having been asked a question about what his administration would do to provide help for the high cost of childcare; he circuitously responded with a discussion of tariffs and the statement "childcare is childcare."

Another example involved the September 10, 2024, debate between Vice President Harris and former president Trump. The majority of postdebate polls indicated that by a wide margin Harris had bested Trump and that many of Trump's responses were deemed nonsensical. As described earlier in this chapter, it was at this debate, too, that Trump amplified unfounded claims of Haitian immigrants eating the pets of fellow residents in Springfield, Ohio. Some journalists and opinion writers suggested that outlets like Fox News were sanewashing Trump in order to dilute or obfuscate the content of what the former president had said in the debate. For instance, op-ed writer Dana Milbank in a September 13, 2024, *Washington Post* piece, "Fox News Cleans up Another Trump Mess," suggested that "after the debate, the network worked to keep the MAGA faithful in a state of blissful ignorance."[118] Providing examples from Fox News commentators Jesse Watters, Laura Ingraham, and Sean Hannity, among others, Milbank contrasted these responses with the comments of more liberal and progressive media voices who had acknowledged President Joe Biden's poor debate performance against Trump on June 27, 2024:

> It was a case study in how the dominant "news" organ of the right cleans up Trump's messes. When President Joe Biden had his disastrous debate, liberal outlets and commentators panned the performance and ultimately helped to force him out of the race. But when Trump had what was, objectively, a bad night, Fox News led a movement to claim it didn't happen.[119]

Voices Against Misinformation, Disinformation, and Erasure

One of the most insidious aspects of misinformation and disinformation is the people and platforms that unknowingly or knowingly amplify its falsehoods and harmful information through their actions or inaction. Hence, one of the aims of this chapter is to raise awareness and amplify the voices of those who are striving to speak out against mis- and disinformation and stem its unchecked spread. Many of these truth tellers warn of the ways, too, that disinformation serves as a gateway to authoritarianism and fascism.

Researchers and scholars, a handful of whom were highlighted earlier regarding health-related misinformation and disinformation, are at the vanguard of "disinformation literacy" efforts. Dr. Jason Stanley, the Jacob Urowsky Professor of Philosophy at Yale University, is one of these scholars sounding the alarm. His 2024 book, *Erasing History: How Fascists Rewrite the Past to Control the Future*, focuses "on the ideological, politically driven battles occurring over education nationwide."[120] As the *Los Angeles Times* interprets it,

> The goal is not simply to ignore the histories of slavery and continuing Black oppression, the genocide of Indigenous peoples, the contributions made by nonwhite immigrants, and the very existence of gay and transgender people. The larger agenda is to create a narrative in which America's greatness lies in a divine unity of purpose given to its white founders, which must be safeguarded by white people.[121]

In interviews talking about *Erasing History*, Stanley has said that we are seeing people rewrite history right before our eyes, using 1930s Nazi-style tactics of disinformation used against minorities and immigrants. Escalating efforts of ideologues to ban books in schools and libraries is another manifestation of aggressive actions to erase histories and stories of peoples whose voices and views have historically not been heard or privileged.

Two other examples of scholars who are studying and speaking about issues involving disinformation and totalitarianism are Dr. Ruth Ben-Ghiat, New York University professor of history and Italian studies, and Dr. Tim Snyder, the Richard C. Levin Professor of History and Global Affairs at Yale University. Ben-Ghiat's works include her 2020 book *Strongmen: From Mussolini to the Present*. Snyder's books include *On Tyranny: Twenty Lessons from the Twentieth Century* (2017) and *On Freedom* (2024).

The G20 and Combating Disinformation as Priority

News outlets on September 13, 2024, reported that for the first time ever the G20 nations, meeting in Maceió in northeastern Brazil, agreed to collaborate on fighting disinformation. The G20 leaders also called on social media platforms to be more transparent about their efforts to mitigate disinformation.[122]

The irony of this declaration happening while the hosting nation was locked in legal battle and wars of words with the world's richest man was the stuff of literary conceit. Just eleven days before, Brazil's Supreme Court had unanimously upheld a decision against Elon Musk's X by a lone Supreme Court justice in April 2024 ordering the "suspension of dozens of accounts for allegedly spreading disinformation."[123] One of the points of contention was Brazil's highest court's ruling that X remove roughly 100 accounts claiming that Brazil's 2022 presidential election, resulting in the unseating of Jair Bolsonaro, had been fraudulent. The court ordered X to be blocked from all access in Brazil. The *Guardian* reported on September 21, 2024, that Musk finally capitulated to the high court's directive that X appoint a legal representative in Brazil (a statutory requisite for social media companies to operate there), pay fines, and take down "more than 100 social media accounts that had been questioning whether the far-right president Jair Bolsonaro had really lost the election in 2022."[124]

Proactive Steps Against Disinformation

The key element to remember about disinformation is the word *intentionality*: people who create and distribute disinformation do so intentionally and willfully to achieve one or more aims. As described earlier, there are many purposes of disinformation. One is to confuse. Another is to overwhelm. Inciting anger is another. Disinformation can also engender feelings of apathy and disengagement. Armed with this information, what then can we do to push back against disinformation and its potential impacts?

CISA, the Homeland Security division discussed earlier in this chapter, offers a number of concrete steps people can take to affirmatively respond to and curb the dissemination of disinformation:

> Although disinformation tactics are designed to deceive and manipulate, critically evaluating content and verifying information with credible sources before deciding to share it can increase resilience against disinformation and slow its spread. Share these tips:

- Recognize the risk. Understand how disinformation actors leverage these tactics to push their agenda. Be wary of manipulative content that tries to divide.
- Question the source. Critically evaluate content and its origin to determine whether it's trustworthy. Research the author's credentials, consider the outlet's agenda, and verify the supporting facts.
- Investigate the issue. Conduct a thorough, unbiased search into contentious issues by looking at what credible sources are saying and considering other perspectives. Rely on credible sources of information, such as government sites.
- Think before you link. Slow down. Don't immediately click to share content you see online. Check the facts first. Some of the most damaging disinformation spreads rapidly via shared posts that seek to elicit an emotional reaction that overpowers critical thinking.
- Talk with your social circle. Engage in private, respectful conversations with friends and family when you see them sharing information that looks like disinformation. Be thoughtful what you post on social media.[125]

Summary

The convenience of seeking and retrieving information, largely propelled by the internet and smartphone technologies, has revolutionized information access. Technologies have diminished the gatekeeper roles of information professionals and, at the same time, have enabled users to attain information and create information objects without intermediaries. Information quality controls that promoted the accuracy and truthfulness of information have correspondingly declined. Social media platforms are a primary mechanism for misinformation and disinformation. The terms *misinformation* and *disinformation* are frequently confused. Misinformation refers to inaccurate information, through accident or unintentionality. Disinformation is false or untruthful information that is created or disseminated intentionally or deliberately. False equivalence, "whataboutism," and the "firehose of lies" are example of a plethora of misinformation and disinformation tactics. Ethics codes of information organizations vary in how they specifically address issues of information accuracy and truthfulness, but many include statements about accuracy and information quality. Snopes and NewsGuard are examples of organizations that strive to debunk erroneous information claims through fact-checking services and tools. Pre-bunking and "truth sandwiches" are proactive strategies for responding to mis- and disinformation. Researchers have provided valuable insights on misinformation and disinformation. Misinformation-and-disinformation-related research has been impacted and imperiled by technology companies who see this research as threats to their business models and revenue-generating products. Citizens have a variety of steps they can exercise to combat misinformation and disinformation.

Questions for Reflection, Exploration, Action, Discussion (READ)

1. News and information organizations often have ethics principles that speak to the quality of information.

a. Having examined some ethical principles in this chapter that relate to accuracy of information, do you think these principles are of value or not? Explain.

b. Craft a bullet or one-sentence principle of your own that you think best addresses accuracy of information.

2. Findings from empirical research have revealed that small numbers of ordinary people have been responsible for large amounts of misinformation and disinformation posted on social media sites like Twitter/X. These people are referred to as supersharers.

a. Identify some of the key ethics issues stemming from the activities of supersharers.

b. What roles and efforts, if any, can information professionals take to address the issue of supersharers posting fake news, misinformation, and disinformation?

3. Imagine you are the chief information officer (CIO) for a nonprofit, nonpartisan agency that promotes free and fair elections. One of the chief activities of your agency is providing access to accurate information. Your supervisor has directed you to give a ten-minute PowerPoint presentation to the organization's board and to address the following specific points:

  1. Give a thirty- to sixty-second explanation that defines and explains what pre-bunking is.

  2. Outline three key ethics issues involved with pre-bunking.

  3. State your opinion, based on evidence and authority, regarding whether your agency should or should not participate in pre-bunking efforts with election officials in your city and county.

4. One way to respond to misinformation and disinformation is through a strategy called a "truth sandwich." Imagine the following hypothetical scenario:

a. The Elm Street Public Library has been accused of disseminating pornography by providing access to some print and digital graphic novels in its young adult collection.

  1. When asked about these allegations by a local broadcast news outlet, consider and determine whether the library's administration should use a truth sandwich to address the allegation.

2. Construct a truth sandwich in response to the allegation that the library is "disseminating pornography by providing access to some print and digital graphic novels in its young adult collection."

b. A social media site on Twitter/X has posted false information about several librarians who work at the library to a site titled "Elm Street Public Library Groomers."

1. Should library administration use a truth sandwich to address these allegations? Why or why not?
2. What other responses or actions might library administration consider?

5. Pre-bunking is a proactive approach to addressing misinformation and disinformation.

a. Discuss the ethical rationale for information professionals to participate in efforts to pre-bunk health-care-related misinformation and disinformation.
b. Identify some specific steps that medical librarians can take to pre-bunk medical-information-related falsehoods and inaccuracies.

6. Conspiracy theories can cause harm to individuals and communities.

a. Distinguish between a "conspiracy theory" and a "theory."
b. Do you think individuals or groups should or should not be held legally responsible for (1) content they posted online that can be proven to be a conspiracy theory and (2) that can be proven to have caused harm to an individual or a community? Why or why not?
c. Do you think that information organizations do or do not have an ethical responsibility to speak out against conspiracy theories?

7. This chapter discusses various things that individuals and organizations can do to encourage accuracy of information and push back against misinformation and disinformation.

a. Drawing on ethics approaches discussed earlier in this book and ethics codes identified in this chapter, what are some arguments for tech companies to take steps to mitigate the harmful effects of disinformation on their social media platforms?
b. What are some ethics rationales that information professionals and organizations can use to advocate for elected officials to take legislative steps to address disinformation on social media platforms?
c. One of the chief concerns about addressing disinformation on social media platforms is the potential adverse impacts on freedom of expression that might occur by restricting some aspects of online discourse. Identify (1) some of the potential harms or downsides that can occur from restricting online speech and (2) some of the potential benefits or upsides that can occur from not restricting online speech.
d. Identify a concrete strategy that you can use to respond to misinformation or disinformation in your personal or professional life.

8. Whistleblowing—the sharing and exposure of information, data, and organizational actions by insiders—is an important force for transparency and accountability, especially with regard to misinformation or disinformation that may have been disseminated by an organization's leaders or staff.

a. While whistleblowing can be a force for good, it also comes with risks and downsides. Identify the ethical costs and benefits of whistleblowing.

b. Information professionals by the nature of their training and organizational ethics codes and professional codes of conduct are particularly attuned to the power and responsibilities of accurate information and truthful interpretation of data. Imagine a scenario in which an information professional becomes aware of an activity within a library, archive, or museum that raises ethical concerns and potential legal liability. Talk about the decision-making process that individual may want to consider before engaging in whistleblowing.

Notes

1. Wu, T. (2018, February 16). The tyranny of convenience. *New York Times*. https://www.nytimes.com/2018/02/16/opinion/sunday/tyranny-convenience.html
2. The aftermath of the quiz show scandal. (n.d.). *American Experience.* PBS. Retrieved February 20, 2025, from https://www.pbs.org/wgbh/americanexperience/features/quizshow-aftermath-quiz-show-scandal/
3. Jensen, K. (2023, December 1). Most parents trust, respect, and feel safe with librarians: Book censorship news, December 1, 2023. *Book Riot*. https://bookriot.com/most-parents-trust-librarians/
4. Markson, T. (2011, October 17). Shelve this, Bill Maher. *HuffPost*. https://www.huffingtonpost.com/teri-markson/bill-maher-libraries_b_1015875.html
5. Ibid.
6. Banks, M. (2017, December 19). Ten reasons libraries are still better than the internet. *American Libraries*. https://americanlibrariesmagazine.org/2017/12/19/ten-reasons-libraries-still-better-than-internet/
7. *Adams' argument for the defense.* (1770, December 3–4). Founders Online. National Archives. https://founders.archives.gov/documents/Adams/05-03-02-0001-0004-0016
8. National Park Service. (n.d.). *Boston massacre trial.* https://www.nps.gov/articles/000/boston-massacre-trial.htm
9. Ibid.
10. Ibid.
11. Ibid.
12. Ibid.
13. *Adams' argument for the defense*. (1770, December 3–4). Founders Online. National Archives. https://founders.archives.gov/documents/Adams/05-03-02-0001-0004-0016
14. Legal Information Institute, Cornell Law School. (n.d.). *Fact*. Retrieved February 20, 2025, from https://www.law.cornell.edu/wex/fact
15. Legal Information Institute, Cornell Law School. (n.d.). *Evidence*. Retrieved February 20, 2025, from https://www.law.cornell.edu/wex/evidence
16. Davidson, D. (1996, June). The folly of trying to define truth. *Journal of Philosophy,* 93 (6), 263–278.
17. Black's Law Dictionary. (1910). *True*. https://thelawdictionary.org/true/
18. Merriam-Webster. (n.d.). *Truth*. Retrieved February 20, 2025, from https://www.merriam-webster.com/dictionary/truth
19. Rix, D., & Cormick, C. (2024, September 13). We retraced Captain Cook's journey along the east coast of Australia, seeking the other side of history. It was life changing. *Guardian*. https://www.theguardian.com/books/2024/sep/14/we-retraced-captain-cooks-journey-along-the-east-coast-of-australia-seeking-the-other-side-of-history-it-was-life-changing

20. Ibid.
21. CSI Library. (n.d.). *Misinformation and disinformation: Thinking critically about information sources*. Retrieved February 20, 2025, from https://library.csi.cuny.edu/misinformation
22. Ibid.
23. Legal Information Institute, Cornell Law School. (n.d.). *Mens rea*. Retrieved February 20, 2025, from https://www.law.cornell.edu/wex/mens_rea
24. New Oxford American Dictionary. (2001). *Disinformation*. Oxford University Press.
25. *Stalin: Propaganda and reality.* (n.d.). Radio Free Europe, Radio Liberty. https://www.rferl.org/a/27774611.html
26. Ibid.
27. *Founding Fathers used fake news, racial fear-mongering to unite colonies during American Revolution, new book reveals.* (2017, January 31). BingUNews. https://www.binghamton.edu/news/story/388/founding-fathers-used-fake-news-racial-fear-mongering-to-unite-colonies-dur
28. Ibid.
29. Dictonary.com. (n.d.). *False equivalence.* Retrieved February 20, 2025, from https://www.dictionary.com/browse/false-equivalence
30. *False equivalence: Why it's so dangerous.* (2019). Above the Noise Collection. PBS Learning Media. https://wqed.pbslearningmedia.org/resource/above-the-noise-false-equivalence/above-the-noise-false-equivalence/
31. Merriam-Webster. (n.d.) *Whataboutism*. Retrieved February 20, 2025, from https://www.merriam-webster.com/dictionary/whataboutism
32. McIntyre, L. (2023). *On disinformation: How to fight for truth and protect democracy*. MIT Press, 33.
33. Ibid., 32.
34. Cybersecurity and Infrastructure Security Agency. (n.d.). *Tactics of disinformation.* Retrieved February 20, 2025, from https://www.cisa.gov/sites/default/files/publications/tactics-of-disinformation_508.pdf
35. Ibid.
36. Stelter, B. (2021, November 16). *This infamous Steve Bannon quote is key to understanding America's crazy politics*. CNN. https://www.cnn.com/2021/11/16/media/steve-bannon-reliable-sources/index.html
37. Illing, S. (2020, January 16). *"Flood the zone with shit": How misinformation overwhelmed our democracy*. Vox. https://www.vox.com/policy-and-politics/2020/1/16/20991816/impeachment-trial-trump-bannon-misinformation
38. Tavernise, S., & Gardiner, A. (2019, November 18). "No one believes anything": Voters worn out by a fog of political news. *New York Times*. https://www.nytimes.com/2019/11/18/us/polls-media-fake-news.html
39. Illing, S. (2020, February 6). *"Flood the zone with shit": How misinformation overwhelmed our democracy*. Vox. https://www.vox.com/policy-and-politics/2020/1/16/20991816/impeachment-trial-trump-bannon-misinformation
40. Ibid.
41. Cybersecurity and Infrastructure Security Agency. (n.d.). *Tactics of disinformation.* Retrieved February 20, 2025, from https://www.cisa.gov/sites/default/files/publications/tactics-of-disinformation_508.pdf
42. Society of Professional Journalists. (2014, September 6). *SPJ code of ethics.* https://www.spj.org/ethicscode.asp
43. Society of Professional Journalists. (n.d.). *Ethics case studies.* Retrieved February 20, 2025, from https://www.spj.org/ethics-case-studies/
44. American Library Association. (2021). *ALA code of ethics.* https://www.ala.org/tools/ethics
45. Medical Library Association. (2010). *Code of ethics.* https://www.mlanet.org/page/code-of-ethics

46. American Association of Law Libraries. (2019). *AALL ethical principles.* https://www.aallnet.org/advocacy/government-relations/recommended-guidelines/aall-ethical-principles/
47. American Library Association. (n.d.). *Evaluating information.* Retrieved February 20, 2025, from https://libguides.ala.org/InformationEvaluation
48. Ibid.
49. American Library Association. (2017, February 22). *Post-truth: Fake news and a new era of information literacy.* https://programminglibrarian.org/learn/post-truth-fake-news-and-new-era-information-literacy
50. American Library Association. (2017, February). *Using media literacy to stop the fake news cycle.* https://www.ala.org/yalsa/continuingeducation/icontinuous_learning
51. American Library Association. (2017, February 13). *New resolution addresses accurate information and media manipulation.* https://www.ala.org/news/2017/02/new-resolution-addresses-accurate-information-and-media-manipulation
52. Ibid.
53. International Federation of Library Associations and Institutions. (2017, March). *How to spot fake news.* https://repository.ifla.org/items/c5dbfa34-7d00-47ac-8057-0183a5056438
54. Ibid.
55. NewsGuard. (n.d.). *About.* Retrieved February 20, 2025, from https://www.newsguardtech.com/about-newsguard/
56. *Andrea Mitchell Reports.* (2018, August 24). MSNBC.
57. NewsGuard. (n.d.). *About.* Retrieved February 20, 2025, from https://www.newsguardtech.com/
58. Snopes. (n.d.). *About us.* Retrieved February 20, 2025, from https://www.snopes.com/about/
59. Snopes. (n.d.). *Frequently asked questions.* Retrieved February 20, 2025, from https://www.snopes.com/faqs/
60. Kasprak, A. (2024, September 9). *No evidence Haitian immigrants are eating ducks, geese or pets in Springfield, Ohio*. Snopes.com. https://www.snopes.com/fact-check/cats-ducks-haitians-springfield/
61. Orozco, J. (2024, September 9). Springfield police say no reports of pets stolen, after viral social media post. *Springfield News-Sun*. https://archive.ph/HmYP8
62. Hoffman, R. (2024, September 11). *Harris-Trump presidential debate transcript*. ABC News. https://abcnews.go.com/Politics/harris-trump-presidential-debate-transcript/story?id=113560542
63. Ibid.
64. Gooding, D. (2024, September 12). Donald Trump, MAGA "eating pets" rumor sparks fear for Springfield Haitians. *Newsweek*. https://www.newsweek.com/immigrants-eating-pets-cats-springfield-ohio-donald-trump-1952910
65. CBC News. (2024, September 12). *Springfield, subject of baseless pet-eating claim spread by Trump at debate, receives bomb threats*. https://www.cbc.ca/news/world/springfield-bomb-threats-1.7321289
66. Canon, G. (2024, September 12). Trump attacks migrants and claims victory over Harris at first rally since debate. *Guardian*. https://www.theguardian.com/us-news/2024/sep/12/trump-arizona-rally-kamala-harris
67. Howard, S., & Brewster, J. (2024, September 12.) *Triple hearsay: Original sources of the claim that Haitians eat pets in Ohio admit no first-hand knowledge*. NewsGuard. https://www.newsguardrealitycheck.com/p/origins-haitians-eating-pets-claim
68. Ibid.
69. Ibid.
70. Ibid.

71. Hooker, R. W. (2009, July 16). By giving away the St. Petersburg Times, Nelson Poynter may have ensured its survival. *Tampa Bay Times*. https://web.archive.org/web/20160616224512/http://www.tampabay.com/news/by-giving-away-the-st-petersburg-times-nelson-poynter-may-have-ensured-its/1018033
72. Company overview of the Poynter Instutute for Media Studies, Inc. (2018, November 16). *Bloomberg*. https://web.archive.org/web/20181117022347/https://www.bloomberg.com/research/stocks/private/snapshot.asp?privcapId=33839455
73. Hooker, R. W. (2009, July 16). By giving away the St. Petersburg Times, Nelson Poynter may have ensured its survival. *Tampa Bay Times*. https://web.archive.org/web/20160616224512/http://www.tampabay.com/news/by-giving-away-the-st-petersburg-times-nelson-poynter-may-have-ensured-its/1018033
74. Poynter Institute. (n.d.). *About Poynter*. Retrieved February 20, 2025, from https://www.poynter.org/about/
75. Poynter Institute. (n.d.). *International Fact-Checking Network*. Retrieved February 20, 2025, from https://www.poynter.org/ifcn/
76. Örsek, B. (2021, January 21). *IFCN is heartened by a Nobel Peace Prize nomination for the work of the global fact-checking community*. Poynter. https://www.poynter.org/fact-checking/2021/ifcn-is-heartened-by-a-nobel-peace-prize-nomination-for-the-work-of-the-global-fact-checking-community/
77. Ibid.
78. IFCN Code of Principles. (n.d.). *About.* Poynter. Retrieved February 20, 2025, from https://ifcncodeofprinciples.poynter.org/about
79. Örsek, B. (2021, January 21). *IFCN is heartened by a Nobel Peace Prize nomination for the work of the global fact-checking community*. Poyntor. https://www.poynter.org/fact-checking/2021/ifcn-is-heartened-by-a-nobel-peace-prize-nomination-for-the-work-of-the-global-fact-checking-community/
80. Poynter Institute. (n.d.). *MediaWise.* Retrieved February 20, 2025, from https://www.poynter.org/mediawise/
81. Poynter Institute. (n.d.). *About Poynter.* Retrieved February 20, 2025, from https://www.poynter.org/about/
82. Huang, N. (2024, May 16). Making sure it's Correkt: A group of UCSB students set out to revolutionize the ethics of AI chatbots. *Daily Nexus*. https://dailynexus.com/2024-05-16/making-sure-its-correkt-a-group-of-ucsb-students-set-out-to-revolutionize-the-ethics-of-ai-chatbots/
83. Ibid.
84. Ibid.
85. Zakrzewski, C., Menn, J., Nix, N., & Oremus, W. (2024, May 28). Debunking misinformation failed. Welcome to "pre-bunking." *Washington Post*. https://www.washingtonpost.com/technology/2024/05/26/us-election-misinformation-prebunking/
86. Hoffman, R. (2024, September 11). *Harris-Trump presidential debate transcript*. ABC News. https://abcnews.go.com/Politics/harris-trump-presidential-debate-transcript/story?id=113560542
87. Infobase. (n.d.). *Who is Infobase*? Retrieved February 20, 2025, from https://infobase.com/about-us/
88. *Debunking & prebunking: Strategies for librarians to eradicate misinformation: White paper.* (n.d.). Infobase. Retrieved February 20, 2025, from https://infobase.com/resources/debunking-prebunking-strategies-for-librarians-to-eradicate-misinformation-white-paper/
89. Kraft, M. (2024, January 16). One leg at a time: Medical librarians and fake news. *Journal of the Medical Library Association, 112*(1), 1–4. https://www.ncbi.nlm.nih.gov/pmc/articles/PMC11189140/
90. Ibid.

91. Apperson, M. (2020, April 22). *Consider using a "truth sandwich" to counter misinformation*. PBS Standards. https://www.pbs.org/standards/blogs/standards-articles/what-is-a-truth-sandwich/
92. Clark, R. P. (2020, August 18). *How to serve up a tasty "truth sandwich"?* Poynter. https://www.poynter.org/reporting-editing/2020/how-to-serve-up-a-tasty-truth-sandwich/
93. Apperson, M. (2020, April 22). *Consider using a "truth sandwich" to counter misinformation*. PBS Standards. https://www.pbs.org/standards/blogs/standards-articles/what-is-a-truth-sandwich/
94. Merriam-Webster. (n.d.). *Birtherism*. Retrieved February 20, 2025, from https://www.merriam-webster.com/dictionary/birtherism
95. Apperson, M. (2020, April 22). *Consider using a "truth sandwich" to counter misinformation*. PBS Standards. https://www.pbs.org/standards/blogs/standards-articles/what-is-a-truth-sandwich/
96. Merriam-Webster. (n.d.). *Words we're watching: "Nothingburger."* Retrieved February 20, 2025, from https://www.merriam-webster.com/wordplay/nothing-burger-nothingburger-meaning-origin-phrase
97. Pelley, S. (2021, October 4). *Whistleblower: Facebook is misleading the public on progress against hate speech, violence, misinformation*. CBS News. https://www.cbsnews.com/news/facebook-whistleblower-frances-haugen-misinformation-public-60-minutes-2021-10-03/
98. Ibid.
99. Pascual, M. G. (2023, October 31). Frances Haugen, Facebook whistleblower: "In 10 years we will wonder why we didn't regulate social networks sooner." *El País English*. https://english.elpais.com/technology/2023-10-31/frances-haugen-facebook-whistleblower-in-10-years-we-will-wonder-why-we-didnt-regulate-social-networks-sooner.html
100. Ibid.
101. Jardin, M. (2021, April 19). *UAA students and faculty collaborate to curb the spread of COVID-19 misinformation*. University of Alaska Anchorage. https://www.uaa.alaska.edu/news/archive/2021/04/students-faculty-curb-misinformation.cshtml
102. *Pitt Cyber presents: Battling the Infodemic: Covid-19 mis- and disinformation.* (2021, February 16). University of Pittsburgh. https://www.cyber.pitt.edu/events/pitt-cyber-presents-battling-infodemic-covid-19-mis-and-disinformation
103. Menn, J. (2024, June 14). Stanford's top disinformation research group collapses under pressure. *Washington Post*. https://www.washingtonpost.com/technology/2024/06/14/stanford-internet-observatory-disinformation-research-lawsuits-politics/
104. Stanford Cyber Policy Center. (2024, June 17). *Update on the Stanford Internet Observatory (SIO).* https://cyber.fsi.stanford.edu/news/update-stanford-internet-observatory-sio
105. Bond, S. (2024, June 14). *A major disinformation research team's future is uncertain after political attacks*. NPR. https://www.npr.org/2024/06/14/g-s1-4570/a-major-disinformation-research-teams-future-is-uncertain-after-political-attacks
106. Ibid.
107. Ibid.
108. Ibid.
109. Ibid.
110. Calma, J., & David, E. (2023, December 7). Disinformation researcher accuses Meta of meddling in Harvard's research. *Verge*. https://www.theverge.com/2023/12/7/23989458/harvard-disinformation-researcher-meta-meddling-joan-donovan
111. Ibid.
112. *Joan Donovan disclosure: Harvard betrayed academic freedom and the public interest to protect Meta.* (2023). Whistleblower Aid. https://whistlebloweraid.org/joan-donovan-disclosure-harvard-betrayed-academic-freedom-and-the-public-interest-to-protect-meta/

113. Lee, S. M. (2024, June 6). The distortion of Joan Donovan: Is a world-famous misinformation expert spreading misinformation? *Chronicle of Higher Education*. https://www.chronicle.com/article/the-distortions-of-joan-donovan
114. Donovan, J. (2024, September 13). A new front in the meme wars. *Atlantic*. https://www.theatlantic.com/technology/archive/2024/09/new-generation-cyberwarfare/679870/
115. Myers, S. L., & Rutenberg, J. (2024, April 22). Nina Jankowicz forms new group to defend disinformation research. *New York Times*. https://www.nytimes.com/2024/04/22/business/media/american-sunlight-project-fight-disinformation.html
116. Ibid.
117. Wiktionary. (n.d.). *Sanewashing*. Retrieved February 20, 2025, from https://en.m.wiktionary.org/wiki/sanewashing
118. Milbank, D. (2024, September 13). Fox News cleans up another Trump mess. *Washington Post*. https://www.washingtonpost.com/opinions/2024/09/13/fox-news-debate-coverage-trump-harris
119. Ibid.
120. Berry, L. (2024, September 13). How "freedom" got twisted in the American mind, fueling fascism. *Los Angeles Times*. https://www.latimes.com/opinion/story/2024-09-13/on-freedom-timothy-snyder-erasing-history-jason-stanley-book-review
121. Ibid.
122. Pessoa, G. S. (2024, September 13). *G20 nations agree to join efforts to fight disinformation and set AI guidelines*. AP News. https://apnews.com/article/brazil-g20-digital-economy-disinformation-platform-regulation-b8c7f9b5ade6954dd880864f5bad628a
123. Da Silva, J., & Buschschlüter, V. (2024, September 2). *Top Brazil court upholds ban of Musk's X*. BBC News. https://www.bbc.com/news/articles/crkmpe53l6jo
124. Tapper, J. (2024, September 21). Elon Musk backs down in his fight with Brazilian judges to restore X. *Guardian*. https://www.theguardian.com/technology/2024/sep/21/elon-musk-backs-down-in-his-fight-with-brazilian-judges-to-restore-x
125. Cybersecurity and Infrastructure Security Agency. (n.d.). *Tactics of disinformation.* Retrieved February 20, 2025, from https://www.cisa.gov/sites/default/files/publications/tactics-of-disinformation_508.pdf

Part 3

Intellectual Property Law and Equity Interests

Chapter 8

Intellectual Property

Introduction: Creators, Consumers, and Transformers

Intellectual property (IP) can be a valuable commodity. In the dystopic HBO TV series *Westworld* (2016–2022), Android "hosts," created and existing in a variety of American West frontier towns and other park-like settings, serve their human "guests" in living out hedonistic fantasies (until their eventual rebellion and escape to the outside world). Showrunner Jonathan Nolan, in a 2017 interview, pinpointed IP as *Westworld*'s sine qua non (essential ingredient): "In Westworld, the value of the park is all in its intellectual property, it's all in the code."[1] In the real world, too, IP is a major source of economic revenue and wealth generation for individuals, institutions, and nations.

IP is a legal term of art and a body of law "encompassing the products of original human thought."[2] Examples of IP include the following:

- both the lyrics and the musical composition of a song, such as "Somewhere Over the Rainbow" or John Lennon's "Imagine"
- a book, like this one
- a software program, such as Microsoft Word
- a movie, like *The Wizard of Oz*
- a musical, such as *Wicked*
- a video game, like Fortnite
- the blueprint designs of iconic buildings, such as the I. M. Pei–designed Rock and Roll Hall of Fame and Frank Lloyd Wright's Fallingwater
- characters like Popeye, Princess Leia, and Snoopy

IP presents an array of multifaceted ethical, legal, and policy issues that societies around the world are grappling with today. With few "black letter law" rules, myriad gray areas and the necessity, per judicial precedent, for case-by-case determinations, even longtime IP law practitioners and jurists view it as one of the most challenging legal areas. (Patent lawyers, working within the sui generis [i.e., of its own kind] patent law environment, are required to successfully pass a separate patent bar in addition to the general bar exam that all lawyers seeking law licenses must pass.) The advent and proliferation of new disruptive technologies, most notably, artificial intelligence (AI), threatens to render IP ever more Byzantine, as legal systems, once again, race to catch up and adapt to emerging technological developments. Indeed, an article about Harvard Law School's 2024 Artificial Intelligence and Intellectual Property Law Conference aptly inquired, "Is the Law Playing Catch-up With AI?" With a decidedly ethics-minded lens, the conference speakers and participants scrutinized and debated "what AI does, will do, should do, and shouldn't do."[3]

Because the four subareas of IP (copyrights, patents, trademarks, and trade secrets) are so poorly understood, if known at all, and are confused and incorrectly conflated by many people, it is essential to first define and explain what IP and its subareas are. Indeed, a January 2022 "Intellectual Property Awareness & Attitudes" online survey of 1,000 general population respondents (age 18 and older and weighted by age, gender, and race) conducted for the United States Intellectual Property Alliance (USIPA) revealed a striking finding, among other illuminating data points: "Respondents were asked to describe the difference between a patent, copyright, trademark and trade secret. 70% of respondents were not able to write anything."[4]

Defining Intellectual Property

IP refers to "products of the mind"—things that are created and invented through *human* creativity and ingenuity. The word *human* is italicized here for emphasis: at present, both the US Copyright Office and the United States Patent and Trademark Office (USPTO) continue to maintain the long-held statutorily mandated position that an author or inventor *must* be a human being to qualify for federal copyright, patent, and trademark registration. Whether AI can be deemed a creator remains an open question. We further explore IP and AI developments later in this chapter and in the final chapter of this book.

Under US law, and in most nations, the category of intellectual property encompasses four distinct types: copyrights, patents, trademarks, and trade secrets.

Copyrights = Original literary and artistic works fixed in a tangible means of expression (e.g., book, song, photo, movie, software, video game)

Patents = Inventions (e.g., electric light bulb, Monopoly game, baseball mitt)

Trademarks = Brands (e.g., McDonald's, Nike, Porsche, Starbucks)

Trade Secrets = Something of commercial value that is kept secret (e.g., Coca-Cola's formula or KFC's Original Recipe of 11 herbs and spices)

Value and Contestability of IP

IP is, by its very nature, rooted in economics. Copyrights, patents, trademarks, and trade secrets all possess both *potential* and *actual* commercial value. That value is significant: In the United States, a 2022 USPTO report found that "industries that intensively use intellectual property protection account for over 41% of U.S. gross domestic product."[5] The report also noted that one out of every three people (33 percent) is directly employed in IP-intensive industries. IP, thus, clearly generates economic rewards; most often in the form of money for people and organizations. It also has a policy objective of incentivizing creativity by providing creators with economic rewards, which then, it is hoped, will incentivize those creators to continue to create. The built-in capacity for IP to make money also builds in a capacity for contestability and competition. Such competitions and disputes involving IP can take many forms. One common type of dispute in the copyright area is cease-and-desist demands, also called take-down requests, in which one party asks another party to stop making use(s) of the complaining party's IP. YouTube and other social media sites must investigate and decide how to handle and resolve thousands of these requests every day.

Another common mechanism for IP disputes is legal complaints (i.e., lawsuits). Lawsuits can result in settlements or verdicts by a judge or jury, as well as injunctions, in which a party may be enjoined by a court from engaging, or not engaging, in a particular action. IP-based legal disputes invariably involve ethical questions and issues as well as policy matters. As one example, exemplifying the highly contestable nature of IP, the technology industry's smart-phone sector has been embroiled in high-stakes copyright and patent infringement disputes continuously since an Apple patent infringement lawsuit against Samsung. Other smart devices have precipitated massively expensive and consequential legal battles: in the fall of 2023, a protracted patent infringement lawsuit by tech competitor Masimo against trillion-dollar-valuated Apple imperiled the ability of consumers to purchase the latter's iWatch right before the holiday gift-buying season.

Theft of trade secrets, frequently perpetrated by individuals working on behalf of nation-states or rival corporations within fiercely competitive commerce sectors, is another increasing area of contestability and antitheft protective efforts by companies and countries. Trade secrets theft, like patent infringement, has an added degree of seriousness due to the threat of national security risks. Furthermore, theft of trade secrets arguably imperils important infrastructure and industries on which people are dependent, such as well-documented thefts of US nuclear and wind power trade secrets in the past decade. Such risks dramatically underscore the kinds of weighty legal, ethical, and policy considerations and decisions that IP can engender. IP-related decisions carry the potential to impact—for good or ill—vast numbers of people throughout the world.

Significance of IP for Information Professionals

Having established IP's multifaceted economic significance and potential for contestability, in this chapter we explore some of the key ethical issues involving copyrights, patents, and trademarks. It is also vital to explain and emphasize why these kinds of ethical issues are of importance and relevance to information professionals. Every day, information professionals across a continuum of interdisciplinary fields face complex, not easily answered IP-related questions. Often, they are required to make tough-call decisions on how to manage, pay for, protect, and provide access to IP-protected works within collections, services, and organizations. These dilemmas and decisions routinely have ethical dimensions.

As a thought exercise, picture the kinds of diverse information objects that are housed within analog and digital libraries, archives, museums, and other cultural heritage organizations: fiction and nonfiction books, journal articles, databases, computer and technology equipment, movies, music, video games, images, documents, 3D-printed items, software, maker space materials, and much more. Many of these items are subject to copyright law protection. Others may have been protected by copyright at some point but have now entered the public domain when the copyright protection period ended. If copyright-protected, these items may be accessible or usable through licensing agreements (i.e., contracts). The fair use defense to copyright infringement (sometimes also called the "fair use doctrine") may also, in certain instances, enable legally permissible uses of copyright-protected items within information venues.

Libraries of Things

Increasingly, too, libraries are creating and maintaining libraries of things (LoT). As implied by the terminology, a library of things is a collection of nontraditional items (i.e., information objects other than traditionally collected books, videos, music CDs, etc.) that can be borrowed by library users. For example, quite a number of public libraries now have collections of gardening tools[6] (e.g., kneeling pads, trowels) and cooking supplies[7] (e.g., rolling pins, whisks, knife sharpeners, juicers). Patrons are able to check these items out of the library, use them for tasks, and then return them to the library, just as they can with a physical book or video. Items within LoT collections are frequently protected by types of IP, especially patents and copyrights, and the requirements IP bestows on those items.

ALA's *American Libraries* magazine's "State of America's Libraries 2020" notes two other descriptive aspects of libraries of things: they "are community-specific and imaginative."[8] That report goes on to point out that the diverse kinds of things available for borrowing include "air mattresses, dolls, bicycles, binoculars, and accordions." Other inspired, forward-leaning items available for LoT borrowing include the following:

- smartphones to aid unhoused persons (Las Vegas–Clark County Library District, Nevada)[9]
- blood pressure monitoring devices (Greenbelt Library in Prince George's County, Maryland)[10]
- animatronic cat robot companions that can be especially comforting to persons experiencing memory and cognitive limitations, as well as loneliness (Eugene Public Library, Oregon)[11]
- accessories—neckties, handbags, briefcases—for people who are interviewing (New York Public Library's Riverside Library)[12]
- tools, including ones that are way beyond expected items like hammers and screwdrivers; some truly mind-blowing examples of tools for borrowing from the Minnesota Tool Library are a cement mixer, a log splitter, an air compressor, a miter saw, a metal detector, a battery-powered lawn mower and a 25-foot extension ladder[13]

IP Literacy

"IP literacy is not just for lawyers anymore." This assertion comes from a 2018 *Wired* magazine article, "Every College Student's Dream: An 8 am Class on Patents."[14] Author David Kline makes the case for IP education for undergraduates, highlighting one of the United States' first courses on IP offered by the University of Southern California. Interspersed with firsthand comments by students who have taken the course, Kline details ways that greater awareness of IP prepares college graduates for a host of challenges *and* opportunities they will face in their jobs and careers. Some countries, such as Japan, are starting IP education even earlier, with elementary and junior high schools introducing IP fundamentals in classrooms and textbooks. Examples of the kinds of IP basics Japanese young people might learn include "how to legally quote or sample copyrighted works."[15] Such IP awareness-raising educational efforts aim to better enable people of all ages to identify, protect, and monetize their own inventions and creations. Efforts like these are also focused on empowering people to more effectively and ethically assess risks, balance responsibilities, and exercise rights that come with uses of IP.

The bottom line when it comes to the relevance and significance of IP for information professionals and centers: whether they are aware of it or not, everyone working in an information-related position or organization is, or will at some point be, impacted by one or more of the four IP subareas, just by virtue of working within and for these establishments. Indeed, intellectual property is frequently the legal area that is seen as most impacting the day-to-day lives of information professionals. As a result, becoming more knowledgeable about IP is an increasingly essential competency, if not an imperative, for all information professionals and centers.

Patent and Trademark Resource Centers and Libraries

Later in this chapter, we delve specifically into patents and trademarks. The continuing existence and use of Patent and Trademark Resource Centers (PTRCs) within dozens of academic, public, and state libraries throughout the United States underscores the relevance of patents and trademarks to information professionals. PTRCs "are a nationwide network of public, state and academic libraries that are designated by the USPTO to disseminate patent and trademark information and to support the diverse intellectual property needs of the public."[16]

PTRC staff can assist people who want to perform patent and trademark searches to determine, for example, whether certain patents and trademarks have already been approved. These kinds of patent and trademark searches are a crucial step for would-be inventors and trademark seekers to perform so they do not infringe on patents and trademarks that have been granted to others.

The PTRC program started in 1871 "when federal statute (35 USC 12) first provided for the distribution of printed patents to libraries for use by the public."[17] The federal patent law has been updated since that time to include patents in "electronic form" as well as print. As a helpful map of PTRC locations by state illustrates, PTRCs are located in every state and Puerto Rico, except for Kentucky, Oregon, South Dakota, and Tennessee.[18]

Inclusion of IP in Ethics Codes and Core Values Statements

The ethics codes of many information professions and societies telegraph the signal priority that is placed on IP. The chief way they do this is specifically including IP issues in the principles of their codes. As one instance, the fourth principle of ALA's Code of Ethics, which has nine principles in total, specifically addresses IP. It is noteworthy that this principle communicates the need to balance stakeholder interests, as is discussed throughout this chapter: "We respect intellectual property rights and advocate balance between the interests of information users and rights holders."[19]

The code of ethics (last revised in 2020) for the Society of American Archivists (SAA) precept on access and use promotes "open and equitable access to records."[20] SAA also balances the interests of stakeholders. Though the wording does not specifically use the words *intellectual property*, aspects of IP and competing interests can be inferred from the phrasing; indeed, the wording calls to mind the issues that are explored later in this chapter's Judaica Sound Archives case study. The statement identifies use as the core rationale for maintaining archival records. Because use is key, archivists affirmatively work to enable "open and equitable access" to the archival items they oversee. Hence, they strive to keep restrictions to a minimum and augment accessibility.

The SAA Core Values Statement (as with the SAA Ethics Code, last updated in 2020) has a section on preservation that has implications for IP and access. For example, the statement on preservation includes the language "creation of physical and digital surrogates" as one activity that archivists must make determinations about, while balancing how to navigate applicable laws and best practice requirements.[21] Section 108 of the US copyright statute is the legal section that is most relevant when libraries and archives consider whether and how they are permitted to make reproductions of analog and digital information objects for further purposes.[22]

Of particular note regarding the code of ethics for museums is that it does not address IP in any of its principles. The principles are more general in nature than those of the ALA and SAA. The Museum Code includes phrases that speak to values, like "public benefit," "public interest," and "public trust."[23]

IP-Related Responsibilities for Information Professionals

Information professionals, depending on the varying natures of their specific positions, have legal and ethical responsibilities for curating, safeguarding, managing, and providing access to repositories and collections of copyright-protected and copyright-free items. They also often have responsibilities for ensuring that uses of items by internal and external users are in compliance with copyright law and licensing requirements. For example, licensing agreements for e-reserves, and compliance with the contractual terms of those agreements, are an ongoing concern for university libraries, and academic institutions in general. Digitization (e.g., whether or not to digitize analog materials) can also present thorny copyright decisions for information professionals in a wide range of organizations. On top of potentially copyright-protected examples are the wildly varying types of items within LoT collections that were spotlighted earlier: Consider that many of these items are protected by patents, unless the patent protection period has expired. Moreover, probably all of these things are protected by trademarks, which can last indefinitely, depending on whether the trademark continues to be "used in commerce" and any necessary trademark renewal requirements have been maintained.

A threshold question regarding all kinds of IP is whether they have been granted protection under state or federal law. For example, apps, APIs (application programming interfaces), various types of computer programs, and processes used within information, data, and computing organizations may fall under copyright, patent, or trade secret law. So it is critically important to determine what area of IP law applies, and then to interpret and apply the relevant law to that particular subject matter.

In addition to patented inventions and copyrighted works, information professionals need to be mindful of items within their areas of responsibility that are protected by federal and state trademark laws. For example, "scraping" someone's brand, logo, or slogan off of the internet, or from a trademarked physical item, for a commercially related purpose or activity, without permission, licensing, or attribution, is a very risky practice, in terms of potential legal liability. It is also *ethically* inadvisable to engage in practices like this.

Copyright and Ethical Issues

Having established the significance of IP to societies and nations, as well as identifying the benefits of cultivating IP literacy, we can now turn a spotlight on some of the many ethical issues related to IP. IP-related ethical issues often carry wide-ranging economic and cultural impacts. Copyrights, for example, by their very nature involve economic interests and issues (as do patents, trademarks, and trade secrets) because of their potential to generate commercial gains (i.e., money). Moreover, IP-related economic issues frequently give rise to ethics and equity considerations. To better understand the complexities and nuances of copyright-related legal and ethical issues, an overview of the origins of copyright law, as well as the exclusive rights enjoyed by copyright owners and "public interest" exceptions to those rights, is necessary before delving into specific issues and considerations.

Origins of Copyright and Fair Use

Copyright law traces its roots all the way back to 1710, in the legal manifestation of Great Britain's Statute of Anne. That law (and a broader, prescient awareness of the importance of copyrights and patents to national interests) so influenced America's Founding Fathers that they specified a patent and copyright clause in the US Constitution (1789). The very next year, in 1790, they codified and promulgated the young nation's first copyright statute. In addition to providing various *exclusive rights* to those granted a copyright, the US Copyright Act of 1790, adhering to the model of its Statute of Anne progenitor, set out a copyright protection period of 14 years. A copyright holder could also elect to file for a renewal period of 14 years, thereby creating a maximum protection period of 28 years. Once copyright protection ended, a formerly copyrighted work lost its copyright status and exclusive rights, entering the *public domain*. The public domain is a copyright-free realm, a conceptual sandbox of freely accessible ideas and building blocks that can be used for new creative expression. *Donaldson v. Becket*, a 1774 British copyright case, established that copyright is not perpetual and that copyrighted works enter the public domain at the expiration of the copyright protection term.

Longer and Longer Copyright Protection

In the ensuing years since 1790, copyright protection periods have grown steadily longer and longer. The US Copyright Act of 1976 put in place a copyright protection period of *the life of the author plus 50 years*, a marked increase from the maximum 56 years of possible protection under the US Copyright Act of 1909 (28 years of initial copyright protection, and the ability to renew for another 28 years). This law also ended the stringent requirement of copyright registration, instituting automatic copyright protection at the moment of creation of *original works fixed in a tangible medium of expression.* Though this statutory change of not requiring copyright registration had the effect of making the universe of copyrightable works exponentially larger, paving the way for the United States to eventually join the Berne Convention, the world's largest copyright treaty, with its prohibition on member states requiring copyright registration, the 1976 act created the problem of *orphan works*, discussed later in this chapter: one problem solved (i.e., unforgiving copyright registration requirements) and another created (i.e., orphan works).

Sonny Bono Act or Mickey Mouse Act?

In 1998, passage of the Sonny Bono Copyright Term Extension Act, eponymously named for the late Palm Springs, California, mayor (1988–1992), US congressman (1995–1998), and ex-husband of fellow recording artist-cum-Oscar-winner Cher, catalyzed new copyright problems and controversies. Sonny Bono had lobbied extensively for a 20-year copyright extension to the 1976 act's "life of the author plus 50 years" protection period. Bono died tragically of blunt force trauma in a January 1998 skiing accident before an extension was achieved; his fourth wife, Mary Bono, succeeded him in his House of Representatives seat and championed the continuation of her late husband's copyright extension lobbying. The act, bearing Sonny Bono's name in tribute to his legislative efforts, was passed nine months later, in October of that year.

The new copyright law was derisively shorthanded by some as the Mickey Mouse Protection Act, a sardonic jab at Disney's concerted lobbying for a twenty-year copyright protection extension. Disney's motivation in large part was a mouse named Steamboat Willie, the lead in a 1928 black-and-white eight-minute animated film with the same title as its leading mouse. Steamboat Willie the character was also the precursor to the mouse that would later be rechristened by his better-known name, Mickey Mouse, the world's most famous—and also lucratively trademarked—rodent. Without an extension to the 1976 Copyright Act, Steamboat Willie was condemned, or heroically fated, depending on one's perspective, to enter the public domain (though trademark status could continue indefinitely, with proper renewal). The enactment of the Sonny Bono Act of 1998 extended copyright protection, functioning in a sense as a "legislative stay of copyright execution" for Disney's Steamboat Willie/Mickey Mouse, as well as millions of other copyrighted works that were slated to enter the public domain. Hence, the new copyright protection period became *the life of the author plus 70 years*, still in effect at the time of this writing. This standard is also the copyright protection period for much of the developed world today, in accordance with requirements of the Berne Convention, discussed in more detail later in this chapter. As a noteworthy footnote, Disney's Steamboat Willie entered the public domain on January 1, 2024. Another version of Mickey Mouse entered the public domain a year later on January 1, 2025. On that same day, copyright-protected works from 1929 and sound recordings from 1924 entered the public domain.[24] Later versions of the world's most famous mouse remain the beneficiary of copyright and trademark protections, for now.

Copyright's Balance: *Exclusive Rights* and *Exceptions* to Exclusive Rights

In theory, the US copyright protection legal and policy framework is envisioned as a balance of rights and exceptions to those rights. At one end of the scale, copyright provides copyright holders a time-limited monopoly on their works to generate revenue, recoup investments made in the creation of those works, and incentivize creators to create new works. On the other side of the scale, copyright is intended to make it possible for the public to make use of copyrighted works when the copyright protection period ends and copyrighted works lose their privileged status and enter the copyright-free public domain—like a ballgown-and-silver-slippers-clad Cinderella returning to less rarefied status at the stroke of midnight. Indeed, this copyright-riches-to-public-domain transformation occurs for copyrighted works whose protection terms end at midnight on the first day of the new year, designated as "Public Domain Day," in years when this rule applies to works at hand. This quid pro quo–styled copyright balancing arrangement is also referred to as the "copyright bargain."

The preceding paragraph began with the words "In theory," as for many recent years in the United States, copyrighted works did not enter the public domain due to the Sonny Bono Act of 1998, which extended the copyright lifespan of works by 20 years. Critics of the 20-year extension argued that the extension diminished the replenishing of the public domain with works that can be transformed or used as building blocks to create new works. Additionally, works in the public domain are not restricted by copyright, and access to those previously copyrighted works is thus facilitated. In 2019, for the first time in two decades because of the Bono Act's 20-year extension, new works—like Cecil B. DeMille's classic film *The Ten Commandments*, Khalil Gibran's literary work *The Prophet*, and the Noel Coward musical *London Calling!*—passed into the United States' public domain on January 1.[25] Additional works have continued to enter the public domain each New Year's Day, as well-chronicled by the Center for the Study of the Public Domain, based at Duke University School of Law. In 2023, Fritz Lang's groundbreaking silent film masterpiece *Metropolis*, Virginia Woolf's classic novel *To The Lighthouse*, and Irving Berlin's musical work "Puttin' on the Ritz" joined the US public domain club.[26] Because copyright laws vary from country to country, it is vital to check the respective copyright laws of each nation, in order to ascertain the copyright or public domain status of a work.

Regarding the first part of the copyright balance, copyright holders are granted certain *exclusive rights* (sometimes colloquially called *a bundle* of copyright rights) by the US government, including

- to control who makes copies of works
- distribution of copyrighted works
- public display and performance of works
- sound recordings
- derivative works, such as sequels and translations

The US Congress, not wanting to give copyright holders absolute monopolies and veto powers on any and all uses of copyright-protected works, also codified various *exceptions* to these exclusive rights. These exceptions serve as a counterbalance to the significant copyright protections statutorily afforded to the exclusive-rights-possessing copyright owners. The most consequential of these exceptions is the *fair use defense*, codified in Section 107 of the US Copyright Act. Other exceptions include Section 108's exceptions for libraries and archives to make limited reproduction and distribution of copyrighted works for the purposes of preservation, replacement, and research, and Section 109's so-called first sale doctrine. It is the first sale doctrine that has enabled libraries (as well as the now nearly completely extinct video-lending stores like Blockbuster, but for its last-surviving location in Bend, Oregon) for years to buy physical/analog copies of books (or videos) and then loan (even for multiple times), sell, or dispose of them. No such first sale doctrine, however, exists for *digital* books, raising a host of fiscal issues—and ethical considerations—for libraries and other cultural heritage organizations that wish to provide access to such items.

Fair Use Defense: "Escape Hatch" Exception to Alleged Infringement

As indicated, fair use is an exception to a copyright holder's exclusive rights. What we today refer to as "fair use" first manifested in the 1841 Massachusetts common law case *Folsom v. Marsh*. Interestingly, the case centered on copyright infringement allegations involving the papers of America's first president, George Washington. In the holding for the case, the presiding judge, Justice Joseph Story, articulated four factors comprising what, over time, came to be called the four-part fair use test. Eventually, these four factors were statutorily codified as Section 107 in the US Copyright Act of 1976.

Previously mentioned, and worth underscoring, is that fair use is actually a *legal defense* (though often called a "doctrine") to a claim of copyright infringement; most people know similarly—from crime procedurals and detective TV shows, if nothing else—that self-defense is a legal defense to criminal or tortious allegations, such as assault or murder. Fair use has a reputation, perhaps not unearned, for being notoriously challenging to understand and apply. Greater familiarity over time and practice in applying the four factors, however, can make fair use more understandable and workable, though never wholly foreseeable. Per the US Supreme Court, the four factors are applied to each fact pattern and determinative on a case-by-case basis, with no factor designated more important than the others; some scholars have observed—cynically and candidly, or maybe more euphemistically stated, pragmatically and trenchantly—that the fourth factor, namely, market value or value of the work, often is given more weight than the other three factors. The specific language of Section 107 (located within Title 17 of the US Code) states that

> In determining whether the use made of a work in any particular case is a fair use the factors to be considered shall include—
>
> (1) the purpose and character of the use, including whether such use is of a commercial nature or is for nonprofit educational purposes;

(2) the nature of the copyrighted work;

(3) the amount and substantiality of the portion used in relation to the copyrighted work as a whole; and

(4) the effect of the use upon the potential market for or value of the copyrighted work.[27]

Because there is not a one-size-fits-all application of the four factors, the fair use test can, understandably, feel nebulous and arbitrary. Even a brief examination of fair use cases, such as those that are annotated on the website of Stanford University's Copyright and Fair Use Center,[28] can imbue a sense that case decisions are neither uniformly consistent nor predictable with reasonable certainty, even when fact patterns and circumstances are or appear to be similar.

A few observations can perhaps offer a clearer picture regarding fair use's intended purposes and the reasons for its ostensibly opaque nature. First, US congressional legislative history reveals that it was intentionally left somewhat vague and flexible, in order to avoid instilling rigid, unforgiving applications of fair use. American University's copyright and fair use scholars Pat Aufderheide and Peter Jaszi have, for example, strongly advised in their book *Reclaiming Fair Use* that tinkering with fair use, in attempting to make it less ambiguous, would favor copyright owners over users.[29] Instead, they note that fair use's malleability to differing fact patterns can be viewed as indicative of its continuing strength and versatility.

Second, Section 107 sets forth that "fair use of a copyrighted work" does not constitute copyright infringement of that work "for purposes such as criticism, comment, news reporting, teaching (including multiple copies for classroom use), scholarship, or research," though the exclusive rights in Section 106 and a very narrow requirement that addresses the rights of visual artists via Section 106A may be applicable and consequently require assessment of their impacts on potential fair uses. These uses will not apply to all fair uses and are not to be construed as an exhaustive list of permissible uses. To that end, parody and transformativeness have also been frequently deemed acceptable fair uses of copyrighted works, though, again, specific facts for each instance of fair use will be dispositive. A review of fair use's largely prescribed uses yields the insight that these specified uses bear a strong nexus with democratic principles:

- freedom of the press
- free expression
- the right to petition one's government
- the right of peaceable assembly
- the advancement of knowledge
- informed citizenries

Third, the built-in, legislatively intended pliability of statutory fair use provides judges with an "escape hatch." Where strict, immutable application of fair use might yield an inequitable outcome, judges (and juries, where applicable) can utilize fair use as a means for providing a kind of escape hatch from a rigid finding of copyright infringement.

The scope and focus of this book do not permit a more detailed investigation and commentary on the vast corpus of fair-use-related cases that are both relevant to the work of information professionals and informative for any person wishing to be knowledgeable about copyright-law-related developments. Sources like the aforementioned Stanford University Copyright and Fair Use Center provide ready access to numerous summaries and analyses of fair use cases. However, several cases warrant signposting and thumbnail overviews in this book because of their especially noteworthy status and implications for information centers and larger societies:

- The so-called Google Books case offers useful insights regarding fair use and digitization, which is discussed at more length later in this chapter. A class action suit was brought by the Authors Guild against Google for its mass digitization of books written by the suing authors, alleging copyright infringement of their digitized books, with neither permission of the authors nor compensation to them. The case lasted more than a decade and was decided by the Second Circuit Court of Appeals in New York City. Judge Denny Chin found that Google's digitization of the author's books was permissible fair use under a rationale of transformativeness. *Authors Guild, Inc. v. Google, Inc.* (721 F.3d 132, 2nd Cir. 2015).
- In a 2021 6–2 decision, the US Supreme Court in *Google, LLC. v. Oracle America, Inc.* applied the four-part fair use test and found that Google's "limited use" of Oracle's Java application programming interface (API) was fair use that did not constitute copyright infringement. As in the Google Books case, transformativeness was also a factor in the high court's decision for Google. *Google, LLC. v. Oracle America, Inc.* (141 S.Ct. 1163, 2021).
- In May 2023, the US Supreme Court published its decision in another highly anticipated fair use case, *Andy Warhol Foundation for the Visual Arts, Inc., v. Lynn Goldsmith, et al.* Photographer Lynn Goldsmith sued the Andy Warhol Foundation, alleging copyright infringement. The late Andy Warhol had used a 1981 photo that Goldsmith had taken of the musical artist Prince in creating silkscreen prints and pencil illustrations. The foundation asserted that Warhol's "aesthetic changes" were transformative fair use. In a 7–2 ruling for the photographer, the court held that Warhol's changes were not transformative under fair use, but rather were "recognizably derived" from Goldsmith's original Prince photograph. *Andy Warhol Foundation for the Visual Arts, Inc., v. Lynn Goldsmith, et al.* (598 U.S. 508, 2023).
- The Internet Archive (IA), headed by Brewster Kahle, scanned physical books in its collection and made ebook copies of those books available for free online access from March 24, 2020, to June 16, 2020, during the COVID-19 pandemic lockdown. IA called this process controlled digital lending (CDL) and made the ebook copies available through its National Emergency Library program. Four publishers (Hachette Book Group, HarperCollins, John Wiley & Sons, and Penguin Random House) sued IA for "mass copyright infringement" of their works. As with the Google Books case discussed above, no compensation to or permission by the publishers for IA's lending of the ebook copies was attained. IA cited fair use as a defense to the publishers' infringement action. In March 2023, US District Judge John G. Koeltl, Southern District of New York, ruled that fair use was not a valid defense and found IA liable for copyright infringement. At the time of this writing, IA is appealing that decision. *Hachette Book Group, Inc. v. Internet Archive* (No. 20-cv-4160 [JGK], 2023 WL 2623787 [S.D.N.Y. 2023]).

Copyright, Ethics, and Equity: Issues, Questions, Dilemmas

With a solid, working foundation laid regarding the origins and fundamentals of copyright and fair use, we can now delve into some examples of ethically fraught copyright issues, questions, and dilemmas for consideration. Some examples of these include the following:

- Who gets to make money from copyrightable works? For example, authors/creators, government entities, corporations, publishers, licensors, vendors, users?
- Who can use or build on/transform copyrightable works?
- Who retains ownership and control of copyrightable works?
- How long do copyright protections last? How long *should* copyright protections last?
- What are the economic implications if a copyright owner isn't known or can't be found (i.e., orphan works)?
- What are the fiscal differences between how physical information objects and digital items (e.g., ebooks) are treated under copyright law, and what are the financial implications of those distinctions?
- What barriers exist for accessing copyrightable works? What are the economic impacts of those barriers on individuals and communities, particularly those facing ability challenges, financial insecurity, digital divides, and the like?
- Who enforces copyright law?
- Against whom is copyright law enforced and not enforced?
- Under what legal, ethical, and policy rationales and schemas is copyright law enforced?
- By what mechanisms is copyright law enforced?
- Who and what forums are available to adjudicate copyright disputes? Are there appeal rights to copyright judgments and adjudications?
- What are the economically related civil and criminal consequences of judicial/administrative findings of copyright law breaches?

Copyright, as a legal concept and system, intrinsically involves the creation of copyrightable works that have facets and relationships with existing cultures and, many times, the creation of new cultures. Copyright and copyrightable works, hence, raise and have connections with a host of issues that relate to and have influences on cultures. Customarily, these issues have kinship with important democratic and US constitutionally enshrined and protected principles, such as

- free speech
- freedom of expression, including commentary, dissent, satire, and petitioning of governments
- creativity and the creation of new culture (e.g., via the US Constitution's copyright and patent clause, Article 1, Section 1, Clause 8)
- free and independent presses

- access to information, data, and technologies
- informed citizenries
- advancement of knowledge
- archival functions like memory, the historical record, and preservation
- historically marginalized and underserved individuals and communities

Culturally anchored democratic principles are intertwined with complex ethics questions and policy concerns. Many of these culturally related precepts, not surprisingly, also possess economic dimensions and mirror and overlap with questions identified in the previous list of economics issues. Some instances of culturally centered copyright and ethics questions include the following:

- Who gets to make culture? Is it corporations, Big Tech and social media platforms, copyright content holders, users, government entities, other persons and organizations to which copyright rights are licensed, transferred, or assigned?
- Who can access and use copyrighted works for expression, scholarship, education, and other like uses?
- When does fair use apply and not apply to uses of copyrighted works?
- Who may quote from or sample copyrighted works?
- Who can appropriate, parody, or transform existing works?
- Who—or what—is an "author," according to existing copyright law?
- In what ways will emerging disruptive technologies like AI change and not change copyright law systems?
- In what ways will copyright law change or not change AI and other emerging disruptive technologies?

Copyright Status and Digitization

One of the biggest IP-related legal and ethical concerns, mentioned briefly earlier, is whether to digitize or not to digitize. For a large number of information professionals, whether they are working in libraries, archives, museums, government agencies, for-profit companies, or nonprofits organizations, determining the copyright status of works within their purview is a necessary but challenging, potentially minefield-strewn component of their jobs and responsibilities. Answering the question as to whether an information object is copyrighted, or whether its copyright status (if it ever had copyright status) has ended and the item has entered the public domain, is an important preliminary step to deciding whether digitization can or should occur. Additionally, an item may still be within the copyright protection time period, but copyright ownership may have been sold, assigned, or transferred by the copyright holder to the information center or another party.

Once copyright status and ownership has been uncovered and established, the information professional can then move on to considering other steps in the digitization process. If the information center definitively owns the work that is to be digitized, the information center can do so, without the fear of a claim of copyright infringement. It is important to keep in mind that physical or digital possession of an information object is not the same as copyright ownership of that object and does not guarantee valid copyright ownership. For example, one may purchase a book from, say, a bookseller, but that purchase and then subsequent ownership of that book does not confer *copyright ownership* of that book to the book buyer; rather, it is the author of that book that was purchased who (likely) retains the copyright (though in some case, it may be the publisher, or another entity, that is the copyright holder). In the next section on what are referred to as orphan works, we explore why information professionals may have legitimate fears of digitizing an item whose copyright status is unknown or uncertain.

Orphan Works

Before digitization of an information object occurs, it is a prudent rule of practice and due diligence for information professionals to perform a copyright status check on the item. Moreover, depending on the age of a work, its physical condition and provenance, and other factors, information professionals may need to do a more involved, well-documented analysis of a work to determine if it is still within the copyright law protection period. Currently, in the United States and much of the developed world, the copyright law protection period is the life of the author, plus 70 years from the date of the author's death. A copyright status analysis is conducted to determine whether digitization—the scanning of an analog hard-copy information object into a digital format of bits and bytes—of a work in question is legally permissible, pursuant to copyright law.

Information professionals also, as matter of good professional practice, frequently perform a risk/benefit analysis, weighing whether the risks in digitizing an item outweigh the benefits, or vice versa. Sometimes clear-cut answers are not possible. In such instances, information professionals face ethical dilemmas of deciding whether providing access to digitized versions of works that may still be under copyright protection is of greater priority than potential liability for copyright infringement. The main concern in these circumstances is whether a copyright holder may appear at some future time and raise objections to an unauthorized digitization of the copyright owner's work. Such digitization might violate a copyright holder's *exclusive rights* under Title 17's Section 106 of the US Code's Copyright Act.[30] At issue are so-called orphan works. Orphan works are works for whom the copyright holder may not be known or findable. The term *orphan work* denotes that ownership of a work is ambiguous and uncertain. Because of this uncertain copyright status of an orphan work, potential users of an orphan work may be "chilled" or reluctant to make use of the work due to fear of possible liability. Some chief explanations why a work may be an orphan work include the following:

- The copyright holder may be deceased.
- The copyright was transferred, licensed, or assigned to another person or company, or to a series of persons or companies.
- The copyright was never registered or renewed.

Julia Perry's Musical Legacy: Chilling Impacts of Orphan Works

The musical compositions of Julia Perry, a Black composer from Akron, Ohio, who died in 1979 at the age of 55, are a powerful example of the real-world implications of orphan works. Perry achieved a measure of success and recognition in the 1950s and 1960s before her career was later curtailed due to financial hardships and illness. As described in a November 2023 *New York Times* article, "Her Music Fell Into Obscurity. Now It's Back at the Philharmonic," "dozens [of her works] cannot be performed or recorded because there is no established copyright holder."[31] Perry's works are still protected by copyright. As a result, symphony organizations, like the Akron Symphony, that want to perform her works must hire lawyers to help resolve the copyright ambiguities. The hope is that these copyright barriers can be resolved in order to facilitate licensing of these works, particularly with 2024 marking the centennial of Perry's birth, and the desire to provide more platforms that can showcase her groundbreaking contributions as an African American female composer.

Orphan Works Remedial Strategies

Efforts to address and remedy the uncertain copyright status of orphan works, which are estimated to number millions of items in the United States, have not come to fruition in the US Congress, despite various attempts over the past two decades (e.g., the failed Shawn Bentley Orphan Works Bill in 2008). Indeed, I asked top US Copyright Office administrators about the status of orphan works remediation efforts during an April 2021 American Bar Association IP law conference panel discussion in Washington, DC, and was told that no orphan works measures were on the horizon. Contrast this with the European Union, which has developed and implemented a process for ameliorating the status of orphan works in the EU. This is a useful reminder why it is vital to know and stay updated on the copyright laws, exceptions, and work-around schema across domestic and international jurisdictions.

Judaica Sound Archives Case Study

The Judaica Sound Archives (JSA), located at Florida Atlantic University (FAU) in Boca Raton, is a powerful exemplar illustrating the real-world challenges of digitization and orphan works dilemmas. An insight-rich case study, it tangibly depicts an intricate mix of copyright law components intersecting with diverse stakeholders and colliding with complex ethics and policy issues like access, preservation, cultural heritage, the historical record, memory, and digitization. It also offers a revealing view of the often uncertain roads that information and cultural heritage professionals must navigate in delineating appropriate lines between applicable laws and written and unwritten codes of ethics, policy, and professional practice, as well as public interests and spaces.

A 2011 *Wall Street Journal* article about the *Golan v. Holder* US Supreme Court copyright case alluded to the unnamed JSA as an example of the ethical challenges faced by professionals committed to providing access to cultural heritage artifacts:

> Justice Stephen Breyer... said the government's position could undermine scholarship and preservation if researchers end up being forced to hire lawyers to track down owners or risk ignoring the law.
>
> He cited the example of a group that seeks to preserve and publish Jewish music from the early 20th century but because the Nazis destroyed Eastern Europe's Jewish communities cannot identify the copyright owners.[32]

The JSA was the unspecified group to which Justice Breyer was referring. It first started at FAU in 2002 with a different name—the Judaica Music Rescue Project. The rescue part of that name points to the endangered state of Jewish music—diaspora in musical form—and the need for saving and providing it with a new home. Putting out calls for "collectors around the world to donate their recordings of Judaic music and Jewish performers to FAU Libraries," the project swelled to a repository sanctuary for "thousands of recordings."[33] In 2005 the Rescue Project updated its name to the Judaica Sound Archives. Eventually, the JSA was folded into the larger Recorded Sound Archive (RSA) at FAU.

A 2012 phone conversation I had with Maxine Schackman, at that time director of the RSA, revealed some instructive insights and lessons learned:

- Many of the record companies that had recorded the Jewish music (at issue for the JSA) were out of business as of 2012.
- Records for these works were not findable, so many of the works were orphan works and the copyright status of many in the JSA's possession uncertain.
- From a collection comprising about 16,000 recordings, 8–9 percent had been digitized by the JSA.
- Some of the JSA's pre-1923 recordings (i.e., squarely in the US public domain) had been digitized by the JSA.
- Some post-1923 recordings had been digitized, for which express written consent from the performer, record producer, or heirs had been attained.
- One *exception* that the JSA had made to their requirement for "express written consent for a post-1923 recording" was for a group of culturally significant recordings called Vistas of Israel.
- The Vistas of Israel were recordings that were made after Israeli independence in 1948. The recordings were deemed to be of important cultural and historical significance. The JSA conducted an investigation to find the copyright owners, which did not prove fruitful. Consequently, the JSA decided to digitize the Vistas of Israel recordings in its possession because of their significance, the inability to locate the copyright owner (or owners), and a commitment to providing free access to the recordings for the scholarly community and public at large.
- No one, as of my 2012 conversation with Ms. Schackman, had challenged the JSA's digitization of a recording.

The most powerful lesson from this JSA case study is that sometimes taking a chance is the greater ethical good, to do something that one believes is a moral imperative, in order to save someone—or some *thing*—in danger of being lost forever. The Jewish music digitization, preservation, and access achievements that Ms. Schackman and the JSA realized are the living embodiment of that lesson. The mission-themed byline under the JSA's name in 2012 succinctly encapsulates their understanding of the significance, rationale, and need—the justification, as it were—for taking risks that can make endangered cultural heritage artifacts freely accessible to all: "Rescuing, Preserving & Sharing a Heritage."

Risk Assessment

The JSA case study compellingly illustrates the real-world ethical decision-making processes—and the human-centered dimensions—that institutions must engage in and be willing to live with. Often, the issues and "answers," or rather, the possible courses of action for addressing these issues, are neither clear nor readily apparent. The interests and perspectives of varied stakeholders must be considered. Benefits and risks must be identified and weighed. Each fact pattern is, in essence, singular. Each organization, therefore, will vary in terms of its risk assessment and the amount of risk that the organization is willing to undertake when deciding whether to pursue or not pursue a particular action, like the digitization actions that were performed by the JSA.

The bottom line: risk assessment is an essential step in weighing the applicability of IP rights vis-à-vis the upsides of digitization, preservation of valuable and irreplaceable materials, and greater access. Sometimes, as in the JSA case study, the rewards will more than justify and outweigh the potential risks. As a final cautionary note, too, it is vital for information professionals not only to assess their own comfort levels with risk but also to know where the institutions for whom they work stand on the continuum of comfort level with risk.

Trademarks and Ethical Issues

Of the four types of IP (patents, copyrights, trade secrets, trademarks), trademarks are arguably the most *visible* examples of IP. Trademarks can be slogans (Nike's "Just Do It"), logos (McDonald's M-suggesting Golden Arches), phrases (Paris Hilton's "That's Hot"), distinctive patterns (Adidas's three parallel stripes), and shapes (the conjoined pyramidal shape of Toblerone chocolate bars). To be granted and maintain a legally enforceable trademark, the mark has to be "used in commerce" or, put more plainly, used for selling goods and services.

Branding is a shorthand way of crystallizing an essential function of trademarked goods and services. Effective branding of goods and services is dependent on both marketing (i.e., the message[s]) *and* attracting the eyeballs and clicks of consumers (i.e., the audience[s] for the message[s]). At an imaginary party of made-up partygoers representing the four kinds of IP, trademarks would be the extroverted attention-grabbers, the guests saying "look at *me*," "notice *me*." Making it easy to readily identify the *source* of goods and services is vital for strong trademarks. Hence, returning to an earlier example, a consumer, already familiar with McDonald's and its products, seeing McDonald's Golden Arches, would likely be able to associate that famous trademark with aspects of the company: the foods and beverages it sells, its long-standing drive-through service, and so on. Quality, reputation, and goodwill are also inextricable elements of trademarks.

Trademarks, like copyrights, raise a number of ethical issues. In the United States, the Lanham Act (also called the Trademark Act, signed into law in 1946 by President Harry S. Truman and taking effect the following year) is the law that governs trademarks at the national level. (States also have their own individual trademark laws, so it's important to be aware that federal *and* state trademark laws may be applicable to a trademark.) In recent years, several provisions of the Lanham Act have raised thorny legal and ethical issues, spurring lawsuits that have risen all the way to the Supreme Court.

"Disparaging" Trademarks Versus Free Speech Clause

In *Matal v. Tam* (2017), a Pacific Northwest–based four-member Asian American rock band, calling themselves the Slants, applied for a federal trademark on that name.[34] Recognizing the word's deeply troubling history as a racially disparaging epithet wielded against people of Asian ancestry, the band members sought to reclaim and trademark the term as a manifestation of self-empowerment. Other marginalized groups have similarly reclaimed historically offensive terms for themselves (e.g., African Americans, lesbians) that have been used derisively against such groups. Before seeking federal trademark status for their name, the band had sold various kinds of merchandise utilizing the band name. Because federal trademark law provides federal trademark holders with some beneficial advantages (such as constructive notice of registration and prima facie evidence of a mark's validity), the band's intent was to secure trademark protection on their name at the federal level. The next stop on their road to federal trademark protection: the United States Patent and Trademark Office (USPTO).

The USPTO, founded in Washington, DC, in 1836 and in more recent years relocated downriver to Alexandria, Virginia, is the governmental entity that decides whether or not an application for a federal trademark is approved or denied. Citing the Section 2(a) disparagement clause of the Lanham Act, the USPTO turned down the band's trademark application. Section 2(a) forbids the granting of federal trademarks that disparage racial or ethnic group members. With band-member Simon Tam serving as the lead plaintiff, the band challenged the USPTO's denial of their trademark application, alleging that the Lanham Act's disparagement clause violated their First Amendment right of freedom of expression. After wending its way through the USPTO and the courts and garnering mixed decisions, the band's case was granted a writ of certiorari for a hearing by the US Supreme Court. In a 2017 unanimous 8–0 decision (Justice Neil Gorsuch did not participate in the case) in favor of the band, the high court struck down the disparagement clause as an unconstitutional restriction on the First Amendment's free speech clause.

"Immoral or Scandalous" Trademarks Versus Free Speech Clause

Two years after the decision in the Slants case, in 2019 the high court took up another trademark case that also raised impactful questions of law and ethics: *Iancu v. Brunetti*.[35] Where *Matal v. Tam* addressed the Lanham Act's Section 2(a) as it pertained to *disparaging* trademarks, *Iancu v. Brunetti*'s dispute arose from Section 2(a)'s prohibition on *immoral or scandalous* trademarks. In the latter case, Erik Brunetti, a West Coast–based artist and designer, had filed a federal trademark application for his clothing line named FUCT. Phonetically pronouncing the term as one syllable renders it as a conjugated expletive. Brunetti, however, asserted that FUCT is an acronym for "friends U can't trust" and that its pronunciation is based on saying each letter individually, one after another, in a four-syllable fashion.

After filing for a federal trademark, a USPTO trademark examiner denied Brunetti's application, on the grounds that, per Section 2(a), that term is impermissible under an "immoral or scandalous" trademark rationale. Again, as with *Matal v. Tam*, the *Brunetti* case eventually reached the US Supreme Court, and in 2019, in another unanimous decision, all nine justices sided with the designer. As with *Matal v. Tam*, the court found that applying an "immoral or scandalous" standard to federal trademark applications like Brunetti's violates the First Amendment's free speech clause.

Complicating the government's case was evidence that the "immoral or scandalous" standard was applied to applications in ways that often appeared viewpoint-based, in violation of the First Amendment's free speech clause (i.e., favoring some views on drug use, religion, and terrorism over others). For example, as noted by Justice Elena Kagan, who wrote the majority opinion, the USPTO "rejected marks conveying approval of drug use (YOU CAN'T SPELL HEALTHCARE WITHOUT THC for pain-relief medication, MARIJUANA COLA and KO KANE for beverages).... But at the same time, the PTO registered marks with such sayings as D.A.R.E. TO RESIST DRUGS AND VIOLENCE and SAY NO TO DRUGS—REALITY IS THE BEST TRIP IN LIFE."[36] In a nutshell, the court rejected the government's argument that its use of the "immoral or scandalous" trademarks clause was "viewpoint-neutral."

Ethics Implications of *Tam* and *Brunetti* Decisions

In the aftermath of the *Matal v. Tam* and *Iancu v. Brunetti*, US Supreme Court cases striking down the respective "disparaging" trademarks and "scandalous and immoral" trademarks sections of the Lanham Act, questions remain. If, for example, the federal government must be viewpoint-neutral and not disqualify a trademark application on the grounds, say, that the trademark examiner believes it to be disparaging, scandalous, or immoral, are there *any* ethical guardrails or prohibitions when it comes to applying for federal trademarks? Put more simply, is it anything goes now for federal trademark applicants? Can essentially anything that satisfies other federal trademark application requirements qualify for acceptance? After the Slants case holding, some questioned, for instance, whether merchandise with Nazi swastikas or adorned with racial epithets can now receive federal trademark protection. A Post-*Tam* 2017 Vox article, "People Are Filing Trademark Applications to 'Own' the Swastika and the N-word," identifies some legal hurdles to a would-be applicant simply slapping inflammatory symbols and slurs on merchandise and applying for a federal trademark: "'If a design or word is widely used by others (as the swastika is), the USPTO is likely to refuse to issue a trademark registration for that mark, since it won't be seen by consumers as source identifying,' Aaron Johnson, a trademark attorney with Lewis Roca Rothgerber Christie LLP, told Vox in an email."[37]

Additionally, some articles have identified post-*Tam* self-anointed de facto "disparaging trademark vigilantes" who are seeking to register slurs and symbols of hate in order to keep them from being used by hate groups, in essence embargoing those items and not making them available for use in commerce by bonafide supremacist groups. Following the Slants decision, a 2017 NPR piece, "After Supreme Court Decision, People Race to Trademark Racially Offensive Words," shed light on white knights "trying to trademark racial slurs" in preemptive efforts to keep them from being used to advance hate speech and discrimination.[38] As an example, a Mississipian named Curtis Bordenave filed a trademark on a racial epithet and explained his motivation to NPR: "I thought that I had a duty—you know what I'm saying?—and a responsibility to protect that word, to secure that word—you know what I'm saying?—and to make sure that it's used in a way that I think would not disparage people."[39]

Though decisions on some free-speech-related provisions of the Lanham Act have been rendered by the high court recently, additional trademark-related questions remain for judicial clarity. To that end, in October 2023, the US Supreme Court heard oral arguments in another federal trademark application case, *Vidal v. Elster*. In *Elster*, the defendant applied for a federal trademark on the phrase "Trump Too Small." In a unanimous ruling, the court rejected the Elster's efforts to attain the federal trademark.[40]

Just Because We Can... *Should* We?

Beyond the legality or lack thereof of federal trademark applications, nations like the United States continue to wrangle with trademark-related questions that fall squarely within the realm of ethical inquiry. Just because someone *can* legally file a federal trademark application that a preponderance of people would likely view as disparaging, immoral, and scandalous, *should* that trademark application be filed and approved? This question that may seem on its face to be abstract and theoretical becomes a real-life ethical dilemma when we look at the world of sports: should professional- and amateur-level sports teams be able to use mascots, logos, slogans, names, and merchandise that are derived from or associated with people and groups that have experienced systemic discrimination, displacement, and even genocide?

The highly offensive term *Redskins*—a stand-in for Native Americans—was first associated with the National Football League (NFL) team in Washington, DC, in 1933. Decades of advocacy by Native peoples, like Suzan Shown Harjo, and allies led the USPTO to revoke the federal trademark for that team in 2014. Team owner Daniel Snyder appealed the decision and repeatedly stated his intent to resist all calls for changing the team's name and mascot. With the US Supreme Court's striking down of the federal trademark statute's prohibition on disparaging trademarks with its 2017 decision in the *Tam* case, the USPTO's revocation of the Washington team's trademark was rendered moot. Under sustained pressure and advocacy, though, in 2020 Snyder finally agreed to a name change. Once that decision was announced, the team used the generic placeholder name "Washington Football Team," until implementing its new name, Washington Commanders, in 2022.

Similarly, Cleveland's Major League Baseball (MLB) team since 1915 had used the term *Indians*. In 1947, the team added a Native-themed mascot named Chief Wahoo, styled as a grinning, feather-wearing, cartoon caricature. Following decades of calls for retiring the mascot and name, in 2021 the team announced that it would begin using a new name, Cleveland Guardians, in the 2022 season. After resolving a trademark dispute with a local roller derby team that was already using the name Guardians, both teams continue to use the name.

Notwithstanding these changes, however, names derived from historically marginalized groups, most notably Native Americans, continue to be used by professional sports teams. Two teams continue to use Native-themed terms for their teams: the Atlanta Braves and the Kansas City Chiefs. At the college and high school sports levels, debates persist as to the ethics of using these kinds of controversial, racially charged names and mascots. A 2022 documentary, *Imagining the Indian: The Fight Against Native American Marketing*, focuses on communicating the voices and perspectives of Native people who work toward changing this kind of branding. The film advances three key points: (1) Native American-related "mascots and branding are offensive"; (2) these mascots, logos, and other forms of branding are just one component within a centuries-long practice of "systemic racism" in the United States; and (3) the continued use of these symbols has significant negative impacts on Native individuals and tribes.[41]

In some instances, professional and collegiate sports teams are changing names and mascots for their teams, often following lengthy and contentious battles among stakeholders. The University of North Dakota stands as one example. Following years of battles that the National Collegiate Athletic Association (NCAA) waged with a number of universities to change sports team names that the NCAA deemed offensive, voters in the state of North Dakota voted in 2012 to change the Fighting Sioux name that had been in place for the school since 1930. After a period of online voting, a new name—Fighting Hawks—was adopted.[42]

Yet other examples endure in which controversial names and mascots for university sports teams survive to this day. These decisions are often highly emotional, igniting the passions and lived experiences of people across life's spectrum. Whose voices in these ethically charged scenarios *should*—and, ultimately, *shall*—carry the most weight continues to be a question with which we will grapple for the foreseeable future. Advocacy, education, and information remain essential for those striving to change the use of these contested symbols.

Patents and Ethical Issues

The one-word stand-in for patents is *inventions*. Like copyrights, patents are specifically recognized in the US Constitution's patent and copyright clause (Article 1, Section 1, Clause 8). The World Intellectual Property Organization (WIPO) provides a useful definition: "A patent is an exclusive right granted for an invention, which is a product or a process that provides, in general, a new way of doing something, or offers a new technical solution to a problem. To get a patent, technical information about the invention must be disclosed to the public in a patent application."[43]

As the definition notes, a patent can be approved for something brand new or for something that has already been invented but which another inventor thinks can be improved to do something better. One of the best encapsulations of this latter idea is an oft-said expression by nineteenth-century philosopher and essay writer Ralph Waldo Emerson: "Build a better mousetrap, and the world will beat a path to your door." In other words, innovate and improve on something, and one can (potentially) reap the award of a patent. Regarding the parenthetical "potentially" in that preceding sentence, it's important to note that just because one gets a patent approved does not mean that that patent will actually come to the marketplace. Myriad inventions have garnered patents "on paper" that have never been brought to life and monetized.

Innovation is part and parcel of inventing and patent law and policy. Sometimes, as Emerson opined, transforming an existing idea can be as impactful, or even as revolutionary and evolutionary, as devising something de novo. As one example, earplugs that reduce if not eliminate noises—like sleep-disturbing snoring or battlefield gunfire—have been around for hundreds of years, since earning patent protection in the mid-nineteenth century. Despite the long existence and patentability of earplugs, inventors continue to come up with new ideas for "a better set of earplugs." In 2012, for instance, Pittsburgh-area inventor Marilyn Mance, working with her father, turned their idea for improved earplugs—called EarMuffers—into a patented and trademarked business.[44]

Patents Versus Copyrights and Trademarks

Patents, like their copyrights sibling, have a built-in expiration date. Both are conceptualized and systemically designed to provide time-limited monopolies to inventors. Patents come in three distinct varieties: utility patents, design patents, and plant patents. Utility and plant patents last for 20 years. Depending on when they are filed, design patents expire after 14 or 15 years.

One way that patents differ from copyrights is the office that manages them. Like trademarks, patents are administered by the USPTO. Copyrights are overseen by the US Copyright Office (USCO), located under the umbrella of the Library of Congress (LoC) in Washington, DC, near the US Capitol and Supreme Court. Those seeking federal patent, trademark, and copyright protections send applications for patents, trademarks, and copyrights to their respective offices. Once received, designated examiners within the USPTO and USCO are assigned to review applications to either approve or deny them. Applications that are denied can be appealed.

One of the somewhat counterintuitive aspects of patents is that the idea behind an invention is fully disclosed to the public at the very moment when a patent application is filed. This is borne out by the sketches that are required to accompany patent applications. In exchange for transparently disclosing the description of a patent in the application, as well as if and when a patent is granted, the patent seeker receives the assurance of protection from the US federal government against potential and actual infringers. As with copyright, once the patent protection period ends, the patented invention enters the public domain. Previously patented inventions can be used as building blocks or as inspiration to create new inventions. Hence, like copyright, the patent system uses a bargain system, whereby each party gains benefits and pluses and minuses are balanced.

Legal Issues and Ethical and Policy Challenges

USPTO-approved patents for Big Tech companies, as well as patents granted to small start-ups, have in some cases garnered almost unimaginable sources of revenue and licensing royalties for patent-rich inventors in recent decades. Commensurately, litigiousness in the patent arena has become more noticeable by the general public, as plaintiffs and defendants wage multi-billion-dollar high-stakes patent infringement wars, seeking *kaiju*-level damages. These *Monster Island* "patent battles" have been fueled by the proliferation of and revenue-generation from patents on digital devices and technologies, especially in the smartphone area. Key players in smartphone legal dustups include large tech company brand names that every consumer knows: Apple, Google, Motorola, Nokia, Samsung, Huawei, Hitachi, Intel, and Qualcomm, among others. The most frequent patent infringement scenario:

> Company A receives a patent on X technology.
>
> Rival company, Company B, releases a new product.
>
> Company A brings a patent infringement lawsuit against Company B, claiming that Company B is illegally using Company A's patent(s) in Company B's product(s), without permission or licensing royalties paid to Company A.
>
> Courts—and experts with highly technical engineering and scientific expertise—must then sort out (1) whether infringement has or has not occurred and (2) what remedies may or may not need to be meted out.

Rise of the Patent Trolls (AKA Nonpracticing Entities)

If members of the public think at all about patent issues, so-called patent trolls are likely one of the most widely recognized examples. Affordability and availability of patent-protected medicines is perhaps even higher profile, which is explored later below. Patent trolls are also referred to, euphemistically and arguably less pejoratively, as "nonpracticing entities" (NPEs) or "patent assertion entities" (PAEs). The glossary of the legal site Thomson Reuters offers a helpful definition of an NPE:

> A person or company that acquires a patent or patent rights but does not practice the patented invention. NPEs typically do not manufacture or sell any products or conduct any commercial processes. NPEs exploit acquired patents by:

- Enforcing the patent rights against alleged infringers.
- Obtaining royalties or other payments for licensing or transferring the patent rights to third parties.[45]

Teasing out aspects of these synonymous but varying terms yields insights into accompanying ethical controversies and issues. Like the proverbial troll in a fairy tale—an ogre-like creature residing under a bridge, shaking down would-be bridge crossers by extracting a toll and wreaking penance if said toll is unpaid—patent trolls are viewed in some quarters as extorting monies from would-be patent users for desired uses of a patent over which the patent troll exercises control and dominion. The more legalistic and less disparaging term *nonpracticing entity* emphasizes the *nonuse* of the patent that the NPE holds. In a similar vein, patent assertion entities affirmatively declare their ownership of a patent they own, while (frequently) foregoing use of that patent.

Patent Troll Examples and Harms

What are some examples of patent trolls? Electronic Frontier Foundation (EFF) offers Acacia Research Corporation as an example of a patent troll that "buys patents (or 'partners with'... patent owners) to sue companies, and splits the proceeds."[46] Some patent trolls, too, are awarded patents from the USPTO but never successfully bring their inventions to market. Indeed, just because the USPTO approves a would-be inventor's application for a patent does not mean that an invention will ever be commercially viable. Hence, some patent trolls are "companies that never got a product off the ground, and now focus instead on litigating patents in court."[47]

What are some of the alleged harms of patent trolls? Patent trolls can generate litigation that contributes to judicial logjams and backlogs. Trolls may also sue in jurisdictions where they believe they have better chances of winning, a practice known as "forum shopping." This can force defendants to incur greater legal and ancillary costs by having to defend themselves in those venues. Another potential harm is innovation. Because patent trolls often bring infringement lawsuits against parties who may seek to use patents acquired by the trolls, trolls can impede innovation. This can also have negative ripple effects on other stakeholders, regions, and nations. EFF explains that "some patent owners are simply individuals who acquire patents from the U.S. Patent Office for the purpose of making infringement accusations against others. It's all bad behavior that hurts innovation—whether they use the term 'inventor' or not."[48]

Is a Patent Troll by Any Other Name *Still* a Patent Troll?

Depending on one's self-interest as well as one's perspectives on the larger interests at stake, the conscious decision by a person or organization to use the terms *patent troll*, *NPE*, or *PAE* will necessarily vary. The term that *is* used by people and organizations may offer insights into that their tacit leanings, if not overt positions, on the issues at hand: namely, people who tend to talk about patent trolls are often proactively seeking, or at the least are interested in or sympathetic to, reform of what they see as a broken patent judicial system that facilitates patent trolls' abuse and extortion of more "well-intentioned" patent holders. In contrary fashion, those using the terms *NPEs* and *PAEs* can, potentially, be seen as more inclined to take a middle-ground position. Lawyers and people in regulatory positions, such as those within the USPTO, will often use *NPE* and *PAE* as less charged terms for what is seen as a controversial and complex landscape of competing stakeholders and interests.[49]

Though these kinds of debates over nomenclature and classification (i.e., what to call things) can seem nitpicky or of little consequence, "naming and framing" can confer considerable power and influence in shaping both perception and policy. This is evident in other areas of IP, where debates occur over whether to use the phrases *IP theft* or *IP piracy*, whereby the latter terminology can seem, by some, to be manipulatively "loaded language" and more biased toward necessitating—or justifying—harsher rhetoric and civil and criminal penalties. Similarly, this was seen in the trademark discussion sections earlier regarding controversies over words and phrases that can—or *should*—be afforded trademark protection status.

As one could reasonably expect, viewpoints on patent trolls/NPEs/PAEs differ depending on the stakeholder. Over the past decade, bills and policy proposals of varying hues have been championed by an array of competing stakeholders to curtail, maintain the status quo, or expand the powers and prevalence of patent trolls. Where some stakeholders see problems with so-called bad patents, others see efforts to limit patentability and opportunity.[50]

For the foreseeable future at least, no matter what name is used, stakeholders will continue to spar over the existence and effects of patent trolls/NPEs/PAEs on the patent system and societies.

Affordability and Accessibility of Medicines

Medicines are an issue that like patent trolls/NPEs/PAEs involves stakeholders with very different and competing interests. In the United States, affordability of and access to medications have been and continue to be important issues to American consumers. Politicians frequently talk about medication costs and ways to facilitate access to them. Lobbyists on Capitol Hill outlay enormous sums on persuading politicians on behalf of the pharmaceutical industry and consumer watchdog groups. Big academic research institutions over the past half century have increasingly become more involved in the issues of medicine affordability and access, conducting "translational research" and touting "precision medicine" targeted to individualized treatments. These higher education research institutions also have made monetization and commodification of research, in the form of IP rights, a top priority. IP rights can and do generate large amounts of monies for universities; "works made for hire" rules may limit how much the university-affiliated people making research discoveries and advances can share in the fiscal benefits of that research. Much of this research is also funded by the US government, so the public, arguably, has an interest in also reaping benefits from their taxpayer-funded contributions to this research. The issues of taxpayer-supported research are discussed in more detail later in this book.

The Roles of Big Pharma

Messaging by the large pharmaceutical companies (also called Big Pharma) that hold lucrative IP rights to many medications, notwithstanding meds that have lost their patent status at the end of their patent protection period, tends to be two-pronged. First, Big Pharma emphasizes the significant amounts of money that must be expended in the form of R&D (research and development) to discover, test, attain regulatory approval, and be able to make medications available. This yearslong process is shorthanded as "bench to bedside"; Duke University succinctly defines it as "the process of translating laboratory discovery into patient care."[51] Stanford (University) Medicine has a distinct internet section on "Bench-to-Bedside Research."[52]

Second, vast sums of money are spent by pharmaceutical companies every year on the marketing of drugs to consumers. This practice is not without its critics: a 2006 peer-reviewed article in the *Milbank Quarterly*, "A History of Drug Advertising: The Evolving Roles of Consumers and Consumer Protection," observed in its very first sentence that "direct-to-consumer advertising (DTCA) of prescription drugs in the United States is controversial."[53] DTCA was prohibited by the Food and Drug Administration (FDA) until 1985. As the author explains: "In September

1985, the FDA rescinded the moratorium on DTCA advertising and required the advertisements to meet the same legal requirements as those directed at physicians."[54]

As a related aside, attorneys were also disallowed from advertising their services until the US Supreme Court struck down this prohibition. In a 5–4 decision in *Bates v. State Bar of Arizona*, the high court found that the plaintiff attorney's First Amendment free speech rights had been violated when he and his firm "sought to provide low-cost legal services to people of moderate income who did not qualify for public legal aid" and needed to raise awareness of their "availability and low fees."[55] In this controversial decision, the majority ruled that "allowing attorneys to advertise would not harm the legal profession or the administration of justice, and, in fact, would supply consumers with valuable information about the availability and cost of legal services." At its core, the majority's decision can be read as an "access to information" ruling, in this case the right to access information relating to the practice of law and the provision of legal services. Moreover, the case has an equity component, as it endeavors to protect access to information that is targeted at and of relevance to a specific niche of marginalized community members.

Big Pharma spends billions of dollars every year on advertising. A 2021 US Government Accountability Office (GAO) study, "Prescription Drugs: Medicare Spending on Drugs With Direct-to-Consumer Advertising," looked at Medicare purchases of 553 medicines over a three-year period from 2016 to 2018. The study found that the makers of those drugs spent approximately six billion dollars per year, for a total of 17.8 billion dollars. The study further yielded some intriguing data points: "Almost half of this spending was for three therapeutic categories of drugs that treat chronic medical conditions, such as arthritis, diabetes, and depression. GAO also found that nearly all DTCA spending was on brand-name drugs, with about two-thirds concentrated on 39 drugs, about half of which entered the market from 2014 through 2017."[56]

It's important to note, too, that these figures do not represent the total pie chart of funds spent by pharmaceutical companies on ads. The 2021 GAO report's "Fast Facts" section states that "drug manufacturers advertise to encourage consumers to ask their doctors for specific medications."[57] That may be one reason they advertise, but it isn't the sole reason. Indeed, a slew of reasons can be identified. One is to expose the public, including medical providers, to these medications, to put these medicines on people's radar, as it were. Another reason is to normalize, in general, the ingestion of medications and, in specificity, Big Pharma's ad-targeted meds.

On that point, probably every person reading this has, at some point in the past few years, seen at least one of the ubiquitous plethora of Big Pharma ads on US television and social media sites. To make these medications more familiar, memorable, and even hummable and sing-along-able to consumers, many of these advertising campaigns are of cinematic production quality. As just one example, an ad for the diabetes medication Jardiance, owned by Boehringer Ingelheim, portrays an ebullient female protagonist walking onto a Hollywood-style set and being taken through "wardrobe, hair, and makeup" preparations before emerging in a canary yellow dress to dance with scores of similarly dancing extras, while belting out the med's ballyhooed benefits in the elaborate musical production. Other ads employ recognizable pop song melodies by retro artists, whose lyrics are altered to be applicable to the products the ads are selling. An exemplar of this strategy is Ozempic, another drug for treating diabetes (and also touted recently as a weight loss medicine), owned by Novo Nordisk, which reworks a 1974 song titled "Magic" by the Scottish band Pilot.

Given the huge amounts of DCTA money spent by Big Pharma then, one must assume that the companies and their shareholders are offsetting those outgoing sums with money coming back to the companies in profits. The data bears this out: the website Pharmaceutical Technology reported that the top five revenue-generating pharmaceutical companies in 2023 were

1 Johnson & Johnson—$85.2 billion
2 F. Hoffmann-La Roche Ltd—$66.4 billion
3 Merck & Co—$60.1 billion
4 Pfizer—$58.5 billion
5 AbbVie—$54.3 billion[58]

A February 2023 Reuters article, "Focus: Drug Companies Face Covid Cliff in 2023 as Sales Set to Plummet," provides additional illuminating quantitative data, reported in most instances by the companies themselves, about the scale of 2022 revenues that derived from COVID-19 vaccines and treatments.[59] Just as armed conflict is economically beneficial in filling the coffers of arms manufacturers, illness is similarly fiscally fortuitous for pharmaceutical companies.

The greatest means for pharmaceutical companies to generate revenue is through licensing of their IP-protected medications. Licensing of IP-protected medicines and vaccines requires nations to expend significant amounts of capital. Residents of nations and regions where Big Pharma companies are headquartered in the Global North have greater access to medications and vaccines than those living in the Global South. In large part, this is true because nations that are more developed have much greater economic resources to buy medications and vaccines than those that are more financially insecure. As a result, vast disparities in health-care metrics exist between these two classes of people on the planet. A coalition of organizations works to facilitate enhanced access to medicines and vaccines by the Global South. Some of the groups that work in this space include the World Health Organization's (WHO) Access to Medicines and Health Products, Association for Accessible Medicines (AAM), Access to Medicine Foundation, and Patients for Affordable Drugs.

Balancing Act

WIPO has the laudable, though unenviable, task of striving to balance IP rights with the stewarding of better access to health care and medicines in all parts of the world. In many ways, these two goals are mutually exclusive, raising ongoing, difficult, if not seemingly irreconcilable ethical questions and dilemmas. Traditional Western-founded IP systems and traditional knowledge (TK) of indigenous peoples are similarly, in key ways, at diametrical purposes. Later in this book, we examine both TK and vaccine access in greater detail.

An April 2023 WIPO web page spotlighting the theme for World Health Day 2023 (and the 75th anniversary of the WHO's founding in 1948) perhaps best captures the challenge that it—and the world—confronts: "IP and Health for All."[60] The site provides concrete examples of forums, partnerships, and research that afford "a good opportunity for WIPO's Global Health team to examine what we at WIPO do to accelerate health for all." It also encapsulates the continuing challenge: "Scientific and technological innovation contributes to important advances in public health. However, new technologies can take a long time to reach people in low- and middle-income countries. How does intellectual property (IP) help address this?"

Trade Secrets and ~~Right~~ Permission to Repair

The so-called right to repair offers a perfect intersection of multiple kinds of IP (including the neglected-stepchild IP subarea of trade secrets) and a mix of stakeholders with competing interests, all amid a bevy of law, ethics, and policy questions. The right to repair is exactly what its words imply: the right to be able to repair *things*, all kinds of things. Journey back in our minds for a moment to a time before the internet's birth in the 1990s, to the analog print era that had existed for centuries. In this (not so long ago) time, buying something came with the implicit understanding that if and when that thing broke, the purchaser-cum-owner of that thing could repair that thing themselves or enlist the services of someone else (i.e., a skilled family member or friend, a handyperson, a small repair shop, etc.) to repair that thing and get it back in working order. In the latter part of the twentieth century, libraries even collected and made available print repair manuals for varying kinds of automobiles, appliances, and other items in need of tinkering and fixing. (I recall, while working in a busy northern California public library that had these repair manuals, seeing grease-stained-overalls-clad auto mechanics periodically running in to the library to consult these invaluable print materials.)

Enter the digital age, in which virtually anything and everything has gotten "smarter" with infusions of electronic gadgetry, while commensurately getting infinitely more complex and expensive to repair when things, inevitably, go wrong. Probably the best-known example of this complexity and dependency is the cars that many of us own and drive. All it takes is one electronic component to "go on the fritz" and, for most people, a visit to the car repair service becomes necessary to get back on the road. Even fairly adept "car repair hounds" can find it challenging if not impossible to fix a broken-down vehicle because of these intricate internal electrical systems and components.

Now consider the example of farmers who own and depend on the good working condition of myriad kinds of farming equipment that they use to prepare fields and harvest crops that feed peoples around the world. We all recognize words like *tractor*, *harvester*, and *combine*. However, these seemingly ordinary-sounding agricultural machines are increasingly anything but ordinary in the brave new world of the internet of things. They are also extraordinarily expensive investments. Having made the decision to purchase high-cost farm equipment, many farmers want to have the option of repairing their machines themselves or, perhaps, having someone else other than the manufacturer troubleshoot and fix issues that may arise. Travel time and related expenses may also be issues, as a farmer may be located a significant distance from an authorized repair location. In addition to farm equipment and cars, other goods that may be subject to right-to-repair restrictions include, but are not limited to, cellphones, wheelchairs, ventilators, home appliances, and other electronic devices.

From the farm equipment manufacturer's standpoint, components within the equipment may be protected by patents, copyrights, and trade secrets. Trade secrets are a subarea of IP, along with copyrights, trademarks, and patents. They are sometimes called the hidden IP right, as they are not as well known as patents, trademarks, and copyrights. WIPO provides an informative definition of trade secrets:

> In general, any confidential business information which provides an enterprise a competitive edge and is unknown to others may be protected as a trade secret.
>
> Trade secrets encompass both *technical information*, such as information concerning manufacturing processes, pharmaceutical test data, designs and drawings of computer programs, and *commercial information*, such as distribution methods, list of suppliers and clients, and advertising strategies.
>
> A trade secret may be also made up of a *combination of elements*, each of which by itself is in the public domain, but where the combination, which is kept secret, provides a competitive advantage.
>
> Other examples of information that may be protected by trade secrets include financial information, formulas and recipes and source codes.[61]

As mentioned earlier, Coca-Cola's formula and the "11 herbs and spices" recipe for KFC's Original Recipe fried chicken, created by the late Colonel Harland Sanders, are two of the most famous and closely guarded trade secrets in the world. Unlike patents and copyrights, which are time-limited in duration, trade secret protection can last as long as the information continues to remain secret and known to only a select few; those who do know the secrets, as part of their employment agreements, are routinely bound by nondisclosure agreements (NDAs). Reverse engineering of trade secrets, however, is permissible under the law. One of the advantages for manufacturers who assert trade secret protection of their IP is the ability to use trade secret status as a justification for not being able to explain or provide information about the workings and designs of their IP-protected products. Trade secrets are protected by state statutes as well as a federal law called the Defend Trade Secrets Act (DTSA) of 2016. Entities whose trade secrets have been misappropriated or acquired through illegal means can bring suit in federal court under the DTSA.

Currently, several states have codified, and other states are considering, state-specific laws that protect consumers' right to repair. Advocacy groups are also supporting further legislation. Additionally, in July 2021, President Joseph R. Biden signed an executive order that in part addresses "unfair anticompetitive restrictions on third-party repair or self-repair of items."[62] Manufacturers continue to oppose the right to repair, asserting concerns about safety, data breaches, and cybersecurity.

Summary

Intellectual property—products of the mind—comprises four distinct types: copyrights, patents, trademarks, and trade secrets. IP has a profound influence, at the macro level, on the economies and cultures of the world. At the individual nation level, one out of every three Americans works directly in IP-related industries. IP generates more than 40 percent of US gross domestic product and significantly influences the creation of culture. IP is the legal area that most impacts the day-to-day lives of information professionals and centers. Information professionals need to cultivate the important skill set of IP literacy. In turn, greater awareness of IP can enable information professionals to be better prepared to recognize and address IP-related legal, ethics, and policy issues.

Questions for Reflection, Exploration, Action, Discussion (READ)

1. As research shows, many people need information that explains and demystifies intellectual property. Information professionals are in the business of providing access to information. What are some specific steps that information professionals can take to better help people understand legal and ethical issues of intellectual property?
2. The fair use defense (also sometimes called the fair use doctrine) provides would-be users with an "escape hatch" against copyright holders' claims of infringement. Identify some of the ethics-based reasons why copyright owners often believe that uses of their copyrighted works should not be permitted by people claiming that those uses are permitted under a fair use rationale. Then, identify some of the ethics-based reasons why users may feel that their uses of copyrighted works should be allowed under fair use.
3. Examining Supreme Court decisions.

a. In the aftermath of the *Matal v. Tam* (2017) and *Iancu v. Brunetti* (2019), what are some of the ethical ramifications of allowing people to attain federal trademarks on words and phrases that have historically been used as epithets and put-downs against ethnic and racial groups and gendered and sexual minorities?

b. Can you envision any unintended consequences of the *Tam* and *Brunetti* decisions?

4. The affordability of patented medicines and vaccines is a contentious IP rights and equity issue. Identify (a) some of the arguments that corporations frequently make for maintaining patents on vaccines that are desired by low- and middle-income countries and (b) some of the arguments that advocates make for free or lower-cost vaccines for low- and middle-income countries.
5. In pre-internet decades, some libraries collected print repair manuals for cars (e.g., Chilton car repair manuals), trucks, and appliances and made these available for library users to reference in the library or check out. Some libraries now license and provide digital access to these kinds of repair manuals. Hence, information professionals and centers have precedent for assisting users with the repair of their vehicles and appliances. Identify some of the chief ethical arguments and rationales for and against the right to repair.

Notes

1. Sciretta, P. (2016, October 17). *Jonathan Nolan responds to that* Westworld *location theory*. SlashFilm. https://www.slashfilm.com/547162/westworld-location-theory/
2. Legal Information Institute, Cornell Law School. (n.d.). *Intellectual property.* Retrieved February 20, 2025, from https://www.law.cornell.edu/wex/intellectual_property
3. Is the law playing catch-up with AI? (2025, January 16). *Harvard Law Today.* https://hls.harvard.edu/today/is-the-law-playing-catch-up-with-ai/
4. United States Intellectual Property Alliance. (2022, January). *Intellectual property awareness and attitudes.* https://static1.squarespace.com/static/5ed4066e709a614ffb39f898/t/634d92189b1e4d7e226418d1/1666028057285/USIPA_Survey_on_Intelectual_Property_REPORT_Q422.pdf
5. United States Patent and Trademark Office. (2022, March 17). *Latest USPTO report finds industries that intensively use intellectual property protection account for over 41% of U.S. gross domestic product, employ one-third of total workforce.* https://www.uspto.gov/about-us/news-updates/latest-uspto-report-finds-industries-intensively-use-intellectual-property-0
6. Harris County Public Library. (n.d.). *Gardening tools*. Retrieved February 20, 2025, from https://hcpl.net/gardening/
7. Boyer, K. (2014, January 15). *Library lends kitchen tools and appliances.* Public Libraries Online. https://publiclibrariesonline.org/2014/01/library-lends-kitchen-tools-and-appliances/
8. State of America's libraries 2020. (2020, April 20). *American Libraries*. https://americanlibrariesmagazine.org/blogs/the-scoop/state-of-americas-libraries-2020/
9. Watson, K. (2023, March 1). Libraries on call. *American Libraries*. https://americanlibrariesmagazine.org/2023/03/01/libraries-on-call/
10. Domen, J. (2023, October 16). *The Maryland library where you can get your blood pressure read*. WTOP News. https://wtop.com/prince-georges-county/2023/10/the-library-where-you-can-read-and-get-your-blood-pressure-read/
11. McDonald, R. (2023, October 28). *This robot cat offers purrfect companionship for Eugene Public Library patrons to check out*. OPB. https://www.opb.org/article/2023/10/28/eugene-public-library-robotic-cats/
12. Maxouris, C., & Griggs, B. (2018, September 18). *This library loans out ties and handbags so those with limited resources can go to job interviews*. CNN. https://www.cnn.com/2018/09/18/health/new-york-library-loans-ties-handbags-trnd/index.html
13. Minnesota Tool Library. (n.d.). *Access over ownership.* Retrieved February 20, 2025, from https://www.mntoollibrary.org
14. Kline, D. (2018, January 20). Every college student's dream: An 8 am class on patents. *Wired*. https://www.wired.com/story/every-college-students-dream-an-8-am-class-on-patents/
15. Segawa, N. (2018, May 14). Japan wants to teach kids how to patent their ideas. *Nikkei Asia*. https://asia.nikkei.com/Life-Arts/Education/Japan-wants-to-teach-kids-how-to-patent-their-ideas
16. United States Patent and Trademark Office. (n.d.). *Learn more about PTRCs.* Retrieved February 20, 2025, from https://www.uspto.gov/learning-and-resources/patent-trademark-resource-centers/learn-about-ptrcs
17. Ibid.
18. United States Patent and Trademark Office. (n.d.). *USPTO locations.* Retrieved February 20, 2025, from https://www.uspto.gov/about-us/uspto-office-locations
19. American Library Association. (2021). *ALA code of ethics.* https://www.ala.org/tools/ethics

20. Society of American Archivists. (2020). *Code of ethics for archivists.* https://www2.archivists.org/statements/saa-core-values-statement-and-code-of-ethics#code_of_ethics
21. Ibid.
22. Legal Information Institute, Cornell Law School. (n.d.). *17 U.S. Code § 108—limitations on exclusive rights: Reproduction by libraries and archives.* Retrieved February 20, 2025, from https://www.law.cornell.edu/uscode/text/17/108
23. Museums Association. (n.d.). *Code of ethics for museums.* Retrieved February 20, 2025, from https://www.museumsassociation.org/campaigns/ethics/code-of-ethics/
24. Jenkins, J., & Boyle, J. (n.d.). *Public domain day 2025.* Center for the Study of the Public Domain. Retrieved February 20, 2025, from https://web.law.duke.edu/cspd/publicdomainday/2025/
25. *Public domain day 2019.* (n.d.). Center for the Study of the Public Domain. Retrieved February 20, 2025, from https://web.law.duke.edu/cspd/publicdomainday/2019/
26. *Public domain day 2023.* (n.d.). Center for the Study of the Public Domain. Retrieved February 20, 2025, from https://web.law.duke.edu/cspd/publicdomainday/2023/
27. *U.S. Code, title 17, chapter 1.* (n.d.). Office of the Law Revision Counsel. Retrieved February 20, 2025, from https://uscode.house.gov/browse/prelim@title17/chapter1&edition=prelim
28. Stanford Libraries. (n.d.). *Copyright and fair use*. Retrieved February 20, 2025, from https://fairuse.stanford.edu
29. Aufderheide, P., & Jaszi, P. (2018). *Reclaiming fair use: How to put balance back in copyright* (2nded.). University of Chicago Press.
30. Legal Information Institute, Cornell Law School. (n.d.). *17 U.S. Code § 106—exclusive rights in copyrighted works.* Retrieved February 20, 2025, from https://www.law.cornell.edu/uscode/text/17/106
31. Schumann, G. (2023, November 20). Her music fell into obscurity. Now it's back at the philharmonic. *New York Times*. https://www.nytimes.com/2023/11/20/arts/music/julia-perry-stabat-mater.html
32. Bravin, J. (2011, October 6). Copyright law challenged. *Wall Street Journal*. https://www.wsj.com/articles/SB10001424052970204294504576613182462401072
33. Jewish Music Research Centre. (n.d.). *Judaica sound archives*. Retrieved February 20, 2025, from https://web.archive.org/web/20170427151652/https://jewish-music.huji.ac.il/content/judaica-sound-archives
34. *Matal v. Tam*. (2017). Oyez. https://www.oyez.org/cases/2016/15-1293
35. *Iancu v. Brunetti*. (2019). Oyez. https://www.oyez.org/cases/2018/18-302
36. *Iancu v. Brunetti*. (2019). Supreme Court. https://www.supremecourt.gov/opinions/18pdf/18-302_e29g.pdf
37. White, A. (2017, July 31). *People are filing trademark applications to "own" the swastika and the n-word.* Vox. https://www.vox.com/identities/2017/7/31/16068972/trademarking-n-word-swastika-slants-supreme-court
38. Chang, A. (2017, July 21). *After Supreme Court decision, people race to trademark racially offensive words*. NPR. https://www.npr.org/2017/07/21/538608404/after-supreme-court-decision-people-race-to-trademark-racially-offensive-words
39. Ibid.
40. Mann, R. (2024, June 13). Supreme Court rejects "Trump too small" trademark. *SCOTUSblog*. https://www.scotusblog.com/2024/06/supreme-court-rejects-trump-too-small-trademark/
41. Winkelman, N. (2023, March 30). "Imagining the Indian" review: Fighting offensive imagery. *New York Times*. https://www.nytimes.com/2023/03/30/movies/imagining-the-indian-review.html

42. Borzi, P. (2016, March 1). The Sioux nickname is gone, but North Dakota hockey fans haven't moved on. *New York Times*. https://www.nytimes.com/2016/03/03/sports/hockey/with-sioux-nickname-gone-north-dakota-hockey-fans-are-fighting-change.html
43. World Intellectual Property Organization. (n.d.). *Patents*. Retrieved February 20, 2025, from https://www.wipo.int/en/web/patents/
44. Twedt, S. (2012, October 14). Upper St. Clair woman develops ear plug to drown out bothersome noises. *Pittsburgh Post-Gazette*. https://www.post-gazette.com/business/finance/2012/10/14/Upper-St-Clair-woman-develops-ear-plug-to-drown-out-bothersome-noises/stories/201210140205
45. Thomson Reuters Practical Law. (n.d.). *Non-practicing entity (NPE).* Retrieved February 20, 2025, from https://uk.practicallaw.thomsonreuters.com/4-513-0928
46. Mullin, J. (2020, April 29). *Who's a patent troll, and who's an inventor?* Electronic Frontier Foundation. https://www.eff.org/deeplinks/2020/04/whos-patent-troll-and-whos-inventor
47. Ibid.
48. Ibid.
49. Mullin, J. (2023, November 6). *Congress shouldn't limit the public's right to fight bad patents*. Electronic Frontier Foundation. https://www.eff.org/deeplinks/2023/11/publics-right-fight-bad-patents-must-be-protected
50. Ibid.
51. Hart, D. (2014, July 8). *From bench to bedside and beyond*. Duke Today. https://today.duke.edu/2014/07/benchbedside
52. Stanford Medicine. (n.d.). *Bench-to-bedside research.* Retrieved February 20, 2025, from https://med.stanford.edu/research/bench-to-bedside-research.html
53. Donohue, J. (2006, December). A history of drug advertising: The evolving roles of consumers and consumer protection. *Milbank Quarterly,* 84(4), 659–699. https://pmc.ncbi.nlm.nih.gov/articles/PMC2690298
54. Ibid.
55. *Bates v. State Bar of Arizona*. (1977). Oyez. https://www.oyez.org/cases/1976/76-316
56. US Government Accountability Office. (2021, May 18). *Prescription drugs: Medicare spending on drugs with direct-to-consumer advertising.* https://www.gao.gov/products/gao-21-380
57. Ibid.
58. Maragkou, I. (n.d.). *The world's biggest pharmaceutical companies: Top ten by revenue*. Pharmaceutical Technology. https://www.pharmaceutical-technology.com/features/top-ten-pharma-companies-in-2023/
59. Erman, M., & Wingrove, P. (2023, February 6). *Focus: Drug companies face COVID cliff in 2023 as sales set to plummet*. Reuters. https://www.reuters.com/business/healthcare-pharmaceuticals/drug-companies-face-covid-cliff-2023-sales-set-plummet-2023-02-06/
60. World Intellectual Property Organization. (2023, April 4). *World health day 2023—IP and health for all.* https://www.wipo.int/policy/en/news/global_health/2023/news_0003.html
61. World Intellectual Property Organization. (n.d.). *Trade secrets.* Retrieved February 20, 2025, from https://www.wipo.int/en/web/trade-secrets
62. Cunningham, R., & Hobbs, D. (2023). The evolution of the right to repair. *Antitrust Magazine, 37*(3). https://www.americanbar.org/groups/antitrust_law/resources/magazine/2023-summer/evolution-of-right-to-repair/

Chapter 9

Open Movements and Traditional Knowledge

Introduction: Disrupting the System

This chapter expands our understanding of intellectual property (IP) issues and ethics questions raised in the preceding chapter, particularly with regard to copyrights and patents. We also investigate and consider the so-called upstart open movements: open access, open data, Creative Commons licenses, and open educational resources (OER), among others. These movements are challenging existing "legacy" IP legal systems and conceptualizations that arose and are predominant in Western industrialized nations. Legacy IP legal systems have expanded within the past century to other areas of the world, like the Asian IP powerhouses of China, Japan, and South Korea. Indeed, the world's five largest national patent offices are in the United States, the EU, China, Japan, and South Korea. Befitting their long histories and well-entrenched pedigrees, Western-originating IP systems have been the global norm for hundreds of years, dating back to the Statute of Anne, the world's first known copyright law, codified in Great Britain in 1710. This primordial copyright statute and operating schema was carried across the Atlantic from Britain: first as an idea in the Founding Fathers' fledgling American colonies, then eventually manifesting as an idea-cum-expression—the very essence of statutory copyright—in the patent and copyright clause of the US Constitution in 1789, which declares that Congress shall have power "To promote the progress of science and useful arts, by securing for limited times to authors and inventors the exclusive right to their respective writings and discoveries."[1]

Relevance of Open Movements to Information Professionals

Upstart movements like open access and open data are also of increasing relevance to libraries and information centers. For example, academic librarians must know the basics of open access (OA), which is prevalent in more and more academic settings; faculty submitting articles for publications in OA journals may turn to librarians for help and guidance. Open textbooks and OER are important to K–12 and higher education environments. Archivists may have responsibilities for curating open data. All information professionals need to have a minimum level of familiarity with Creative Commons licenses, the terms of which often implicate access to information objects by internal and external users.

Given that the information professions are committed to diversity, equity, and inclusion principles, increasingly included as explicit precepts in ethics codes like the American Library

Association's ninth principle added in 2021, it is incumbent on information professionals to be aware of movements like traditional knowledge (TK) that challenge systemic oppression, call out exploitation and cultural appropriation, and champion the rights of historically disadvantaged peoples. The character Dr. Grace Augustine, played by Sigourney Weaver in 2009's *Avatar* film, gives impassioned voice to the values of equity, inclusion, and autonomy when she calls out a character exhibiting a colonial attitude toward the native Na'vi people of the planet Pandora: "You need to wake up, Parker. The wealth of this world isn't in the ground—it's all around us. The Na'vi know that, and they're fighting to defend it. If you wanna share this world with them, you need to understand *them*."[2]

Access

In a prior chapter we explored the ethics of access-related issues. IP similarly has numerous linkages with access. Free access is a hallmark of most libraries and information centers. The carved-in-stone maxim "Free To The People" greets visitors at the stately entrance to Andrew Carnegie's flagship Carnegie Library of Pittsburgh, just as Toni Morrison's previously highlighted statement apotheosizes access at New York Public Library's main library. Libraries and other information centers house untold numbers of copyright-protected works. So too do they provide access to many items that are no longer protected by copyright. In the United States, as of January 1, 2025, any works that were created or published before 1929 are deemed to be in the "public domain"—a copyright-free "sandbox" or commons of digital and print "building blocks." Anyone can access and use public domain building blocks to create new works.

Send in the Zombies

One illustrative instance shows how copyright-free public domain works can be accessed and repurposed—an example pitting Regency England gentry characters against marauding, bloodthirsty zombie monsters. Seth Grahame-Smith's 2016 madcap novel *Pride and Prejudice and Zombies* "mashes up" zombies with celebrated author Jane Austen's now public domain characters from her classic and oft-adapted 1813 book *Pride and Prejudice*. Moreover, in 2019 a film based on Grahame-Smith's debuted. It's important to underscore that the zombie content Grahame-Smith uses in his "mashup" book was legally attained from the public domain.

The ability to access this public domain zombie content derived from a copyright registration flub arising from the late Pittsburgh filmmaker George Romero's seminal 1968 film *Night of the Living Dead*. *Night of the Living Dead* is seen as the progenitor of the zombie genre, spawning several Romero-helmed sequels, as well as film and television works made by other creators, such as *Shaun of the Dead* (2004), *Zombieland* (2009), and *The Walking Dead* (2010). Because of an error in how the copyright notice was displayed in the 1968 film, required at that time by the US Copyright Act before the statute was amended in 1976, Romero lost all copyright protection for his film and proto-zombies.[3] The very moment *Night of the Living Dead* was shown on movie theater screens, the doomed copyright notice's fatal flaw had the Draconian effect of nullifying the copyrightability of Romero's film and "living dead" zombies. Consequently, the film and

Romero's monsters were cast into the public domain, however unintentionally, becoming "fair game" fodder for subsequent creators.

The 1976 US Copyright Act remedied the unforgiving consequences of these kinds of pre-1976 copyright registration requirement errors by no longer requiring copyright registration. Instead, post-1976, copyright protection status attaches automatically at the moment of creation, for original works that are fixed in a tangible means of expression.[4] This chapter critically examines additional IP examples and areas that raise ethical issues like access, equity, and the cultural rights of indigenous peoples.

Rising Costs and the Push for Open Access

Since the 1990s, growing numbers of libraries and information centers have struggled to provide access to scholarly articles due to skyrocketing journal costs. These costs have been seen by some information center subscribers as price gouging by large academic publishers. The escalating prices of these journals have also been quantitatively demonstrated as wildly out of alignment with annual rates of inflation, especially when compared with other goods and services.[5] As a result, librarians coined this high-priced-journals environment the "serials crisis."

Over the past few decades, many libraries and information centers have also been subject to capriciously draconian price hikes and take-it-or-leave-it "bundled subscription" licensing contracts that provide access to suites of journals: some wanted, and some not. These budget-busting "bundled" contracts are akin to the controversial "bundled cable channels" practices notoriously engaged in by telecom cable companies that have driven some consumers to "cut the cable cord." Just as some cable subscribers (and nonsubscriber content seekers) have looked for alternative means for accessing content in ways they feel are cheaper and more equitable (albeit, often by illegally pirating streaming content via Torrent sites), the progenitors of what would come to be called open access were—and have been—seeking fairer and freer ways to access journal content.

Big academic publishers, like the Netherlands-headquartered Elsevier, have rejected such labeling and negative characterizations. The advent and rise of accessible internet technology in the mid-1990s disrupted the business models and decades-long monopolies on the dissemination of peer-reviewed research by these publishing companies. The growing presence and market pressures of free open-access publications, as well as increasing calls and nonnegotiable mandates for open access from researchers as well as nations and regions (e.g., the European Union's Plan S) have catalyzed these for-profit publishing corporations to compete and remarket themselves.

Consequently, large academic publishers, for the past decade or so, have been increasingly touting the "value-added services and features" that they provide to researchers, subscribers, and users. Examples of these value-added features and services include abstracting, indexing, enhanced search features, and the status-enhancing opportunities (i.e., promotion and tenure) and "crowing capabilities" of being published in top-tier peer-reviewed journals. Being published in these often well-established journals can also facilitate "impact factors" (i.e., citations). Additional value-added contributions, they argue, are elements like professional formatting, built-in preservation capacity and archiving repositories, and the marketing and selling advantages that are incumbent to long-established, well-regarded publishing entities.

With a web-driven sea change in scholarly communication beginning in the 1990s and advancing into the twenty-first century, academic and civil watchdog activists (inspired by earlier open-source efforts to open up software), created a countermovement against Big Publishing's monopolies: a movement christened "open access." This open access "new kid on the block" movement found its inspiration—and its rallying cri de coeur—in challenging the status quo of the academic publishing "factory model process" and self-congratulatory "new product roll-outs." Moreover, this upstart movement saw itself as standing up—in *Norma Rae* revolt-inspired fashion—to an academic publishing conglomerate that had come to be seen by many (most of all, the researchers supplying the publishers with product) as hegemonically and avariciously hoarding, silo-ing behind paywalls, and monetizing publicly funded research, research so highly sought and fought over for control and commodification as to be reminiscent of the allegorical spice drug (melange), extracted from the dune-dwelling sandworm denizens of sci-fi author Frank Herbert's desert planet Arrakis. ("He who controls the spice, controls the universe."[6]) In reaction to this quantitative and qualitative evidence-based perception of a monopolistic, oligarchical, large-academic-publishing-industrial-complex, open access disciples, a la *Dune*'s Fremen, evangelized and spread a gospel of "information access as human right."

Open Access and Equity

The most powerful argument *for* open access is equity-based, rooted in notions of basic fairness and common sense. Open access posits that because researchers (often) receive public money (i.e., taxpayer-funded research grants) to conduct research, the public that has funded that research should be entitled to free, unfettered—*unpaywalled*—access to any and all research that derives from those taxpayer-supported research investments. Open access then, as a conceptual model, anchors itself on the premise that most academic research to date is publicly funded. Accordingly, to require information centers and consumers to pay for access to research that the public has already fiscally enabled is viewed by many open access proponents as a kind of "double dipping." At its heart, open access seeks to democratize access to publicly funded research and to curb the gatekeeper and tolltaker roles, powers, and excesses of Big Publishing. With those aims at front of mind, each year for one week in October, International Open Access Week (openaccessweek.org) events throughout the world celebrate the advancements of open access and provide forums for knowledge sharing, research collaboration, and OA-focused coalition building.

Though aspiring to laudable values like free and equitable access, OA—as with all human endeavors—is not (yet) immune from ethics-related downsides and policy sticking points. Chief among these are "predatory journals," specifically defined and discussed in more detail in the next section. Other challenges and concerns include publishing costs, also called author processing charges (APCs), which are many times borne by the author(s); the quality, reputation, and status of the journal; impact factors (e.g., metrics regarding citation statistics); sustainability of the journal; and copyright restrictions and issues like reuse rights by authors. The Netherlands' open access site offers a useful pros and cons list as a resource.[7] A particularly crucial observation this site makes is that open access requires a commitment of time: a long-view perspective, dedicated commitment, and time to transition to an open-access publication framework. Additionally, time, perseverance, financial resources, open minds, and adaptive systems are necessary

- to attract more researchers to publishing in OA journals
- to improve the quality and long-term viability of OA journals
- to develop more financial support streams for researchers to be able to pay charges (APCs) that are required for publishing in some OA journals
- to persuade higher education promotion and tenure committees to better recognize the trail-blazing, paradigm-shifting contributions of tenure-seeking faculty, as well as all faculty members seeking promotion, who publish in OA journals (many of these OA journals may not (yet) be first- or second-tier journals, but they are still of quality and can serve as socially progressive vehicles for breaking down historic, systemic barriers to publishing and providing platforms for research voices that have been "gatekeepered" and marginalized)
- to equalize the scholarly publishing "playing fields" of the historically publishing-advantaged Global North and the underresourced Global South, by making more publishing avenues available and accessible for the latter via OA journals

Predatory Journals

Impressively expanding annual increases in the numbers of credible open access journals and articles over the past two decades have, sadly, been commensurately met by a burgeoning propagation of so-called predatory journals with dubious ethics standards. Elmore and Weston's 2020 *Toxicologic Pathology* article "Predatory Journals: What They Are and How to Avoid Them" provides a helpful definition: "Predatory journals—also called fraudulent, deceptive, or pseudo-journals—are publications that claim to be legitimate scholarly journals, but misrepresent their publishing practices."[8] Use of the word *predatory* is also a deliberate choice to convey the sense that these journals prey on researchers who may be especially vulnerable to unscrupulous enterprises that take advantage of their desire to be published. In their desire and need to have their research published, particularly for those on a tenure track, some researchers may not conduct due diligence to ensure that the publishers enticing them with entreaties to be forums for their work are in fact legitimate.

Journals that prey on would-be authors for submissions and publications routinely give off "red flags," including the following:

- nonexistent or low-grade peer review, despite claims of being peer-reviewed
- aggressive solicitation, often via email messages that have spam-like qualities (e.g., misspellings, grammatical errors)
- hidden/nontransparent APCs, frequently revealed after an author has been notified that a submission has been accepted for publication

- lack of indexing in major databases or publications in the same or related fields of study as the paper
- questionable editorial board members
- poor quality of the journals and articles themselves (e.g., misspellings, grammatical errors)
- pairing of publishing solicitations with invitations for participation in professional conferences that also raise "red flags"

As a May 2023 article on Researcher.Life points out, though, distinguishing between legitimate journals and predatory journals can be challenging and requires nuanced analysis.[9] Moreover, threats of legal action, such as claims of damage to reputations, as well as defamation and slander allegations can have chilling effects on commendable efforts to shed light on fraudulent scholarly publishing purveyors. Reputable sources, like DOAJ (Directory of Open Access Journals), can aid just-beginning as well as intermediate and more seasoned authors in finding and verifying journals that are of high quality and above board in their publishing processes and protocols. DOAJ, as of January 2025, contains information on more than 21,000 open access journals, representing 136 nations and eighty-nine languages. Additionally, more than 13,000 of those journals—roughly two-thirds—charge no fees for publishing in them. The nonprofit site Think. Check. Submit. "helps researchers identify trusted journals and publishers for their research."[10] Its 15-word vision statement concisely conveys its ambitious research integrity-centered mission: "Our vision is for all researchers to be aware of how to avoid predatory publishing."[11]

Efforts to promote research integrity that are coupled with research *publication* integrity are salutary for other stakeholder communities, most notably, users and consumers who access and rely on the accuracy of the information included in these publications and articles. However, scholarly publication authors themselves—not potentially unwary readers—must serve as the frontline "volunteer brigade" soldiers battling against the information ecosystem degradation, disinformation, and, ultimately, distrust that is unleashed and sown by predatory journal publishers. Perhaps the best reminder for publication-minded researchers, then, is a variation of the age-old admonition to purchasers: "let the author beware."

Creative Commons and Other Open Movements

For many researchers and stakeholders seeking and using research, Big Academic Publishing's scholarly communication monetization model is seen as fundamentally unfair and unethical. The US Constitution's patent and copyright clause (Article I, Section 8, Clause 8) states that Congress shall have power to "promote the progress of science and useful arts, by securing for limited times to authors and inventors the exclusive right to their respective writings and discoveries."[12] An argument can be made that the fencing off of publicly funded research for subscribers and licensees—not making that research freely accessible to all persons—impedes, rather than promotes, scientific progress. From a utilitarian analysis, too, Big Academic Publishing's model benefits "the few" rather than "the many."

As discussed previously, open access arose over the past couple of decades and has gained traction as a work-around reaction to the perceived excesses of Big Academic Publishing. In a similar vein, the pervasive sense among various stakeholder groups that present-day aspects of US copyright law have become unfair and unbalanced induced the formation of the Creative Commons movement in 2001.[13] One of the Creative Commons founders' principal concerns was, and remains today, that copyright protection periods have become longer and longer. Currently, the copyright protection period in the United States, as well as in the EU, UK, Canada, Australia, and Japan, is for the duration of the life of the author, plus 70 years. This protection period is demonstrably longer than the US Copyright Act of 1790, in which the copyright protection period was an initial 14 years, with a possible 14-year renewal. Longer copyright protection periods mean copyrighted works are covered by copyright law for many more years. This in turn diminishes the number of copyrighted works that can enter the public domain. Fewer copyright-free works in the public domain translates to fewer freely accessible works and building blocks that can be transformed into new copyrightable works.

Amid this increasingly copyright maximalist landscape, in 2001 the fledgling Creative Commons movement launched a number of free licenses. One of the aims of these licenses, specifically designed to be understandable and usable by anyone, was to make it easier to access, freely utilize, and protect copyrighted works. A key point of noteworthiness is that Creative Commons licenses are designed to *complement* copyright, not to supplant it. One can be a copyright holder and choose to retain some copyright rights, while at the same time electing to make some uses of a copyrighted work permissible through the flexibility afforded by a Creative Commons license. As Creative Commons explains, if copyright signifies "all rights reserved," Creative Commons enables "some rights reserved."[14] Going one step further some years later, Creative Commons released Creative Commons Zero, a tool making it possible for copyright holders to relinquish all exclusive rights under copyright law (i.e., "no rights reserved").[15]

Open Textbooks and Open Educational Resources

Earlier in this chapter, we looked at how the skyrocketing costs of academic journals in the 1990s—the "serials crisis"—was an impetus for the open access movement. Ever-increasingly expensive costs for academic textbooks over past decades have, comparatively, catalyzed the rise of the open textbooks and open educational resources (OER) movements.[16] Access, again, is the common denominator of all these movements. High costs for textbooks trigger equity and ethical issues. Students must, in most cases, shoulder the costs of textbooks that are not covered under grants or tuition assistance. Textbook prices deleteriously impact students from lower- and middle-class households to a greater degree than students in higher economic levels. The open textbooks and, more recently, the OER movements endeavor to level the unequal economic playing field by providing (often) free, high-quality textbooks. These textbooks span a wide range of disciplines and subject domains.

Rice University's OpenStax, operating as a 501(c)(3) nonprofit since 2012, is an example of one initiative to make textbook information more accessible, free when possible, and more affordable when not (charging "modest fees" for some print textbooks). The stated mission of OpenStax is "to improve educational access and learning for everyone."[17] Breaking down "the most common barriers to learning," of which high textbook prices are one, is central to its mission.[18] As of mid-2023, OpenStax offers approximately sixty openly licensable, peer-reviewed, free digital textbooks for courses in high schools, colleges, and universities. Subjects like accounting, world history, intellectual property, statistics, chemistry, marketing, precalculus, philosophy, and more are accessible to users.

OER movements, like open textbooks efforts, are focused on providing free textbooks. OER's mandate, however, encompasses many more types of educational resources, as telegraphed by the "educational resources" part of the name. Examples of the kinds of educational resources made available include lectures, quizzes, discussion questions, small and larger group activities, and study guides.

At the global level, the United Nations Educational, Scientific and Cultural Organization (UNESCO) recognizes the importance of facilitating access to learning and knowledge. Unsurprisingly, then, UNESCO sees great value in OER as a mechanism for promoting access and provides an expansive definition of these resources: "Learning, teaching and research materials in any format and medium that reside in the public domain or are under copyright that have been released under an open license, that permit no-cost access, re-use, re-purpose, adaptation and redistribution by others."[19] In 2019, UNESCO approved its "Recommendation on OER." The agency bills this recommendation as the "first international normative instrument to embrace the field of openly licensed educational materials and technologies in education."[20] UNESCO is continuing its work assisting participating nations in implementing the 2019 OER recommendation.

Open Data

Open data, like open access and OER, originates from the paramount goal of enabling access. For open data, this means access to diverse kinds of data. Underpinning open data is the belief that making it easier to access data can promote innovation, research, and societal advancements and enhancements that can in turn benefit the lives of individuals and communities.[21] Over the past few years, we have seen increasing numbers of organizations, cities, regional federations, and nations that are adopting open-data principles, hiring for varying open-data-related positions, and creating portals for people to access myriad types of open data for a host of uses.

Types of data made accessible via open-data portals run the gamut: climate and environmental, health, financial, census, and utilities, as examples. Drilling down into large data sets can yield granular observations of micro-level areas, such as wheelchair accessibility data. Data that provide insights into barriers and opportunities regarding wheelchair-friendly streets, sidewalks, ramps, and curb cutouts can be of particular usefulness to people with physical disabilities/challenges. One germane example: SkedGo, an Australia-based company, has developed an app that uses open data to help people who use wheelchairs to locate wheelchair-ready routes. As SkedGo's website describes,

> One in five people in [New South Wales] have [*sic*] some form of disability. Using transport can pose a significant challenge for this group, reducing their ability to get around, get to work, or participate actively in day-to-day life.

> Our TripGo app and API already had some wheelchair-friendly features integrated, for example showing wheelchair-friendly footpaths (providing we have access to the relevant data, it is integrated into the wheelchair-friendly routing. We rely on open data and community support. Users can contribute via OpenStreetMap).[22]

As with other open movements we've examined, open data's encouragement of free access to data can be of benefit to many stakeholder groups. However, barriers and legal, ethical, and policy issues can also be present that need to be considered. Some of these barriers and issues include the following:

- intellectual property rights in data (i.e., who owns the data at issue)
- licensing issues (e.g., what are the licensing costs, if any, and terms of use)
- privacy sensitivities and applicable privacy laws
- confidentiality laws and ethical guidelines
- social media terms of use and end-user license agreements
- data collection and use laws
- codes of conduct and ethics codes
- cloud computing issues and laws
- cybersecurity laws, ethical principles, and policies
- critical infrastructure laws and policies (e.g., utilities, elections and voting machines, banking institutions, social media)
- national security and law enforcement requirements

Ethical decision-making processes and models can be useful in conducting SWOT analyses of data. Analyses like these and others may enable data holders to better weigh the strengths, weaknesses, opportunities, and threats in determining when, where, why, and what data to make freely available through open data initiatives.

IP Theft/Piracy and IP Rights Enforcement

In shedding more light on the motivations for the open textbooks and OER movements, as well as ongoing initiatives to advance these movements at local and global levels, it is also important to acknowledge the prevalence and impacts of IP theft. IP theft is also often referred to as IP piracy, though the latter term *piracy* is controversial in some quarters because of the perception that it is a loaded term that demonizes uses and acquisitions of IP. Estimates on the specific value of theft of US IP vary widely but are estimated to number in the hundreds of billions of dollars. The US Trade Representative, the US Chamber of Commerce, and the National Intellectual Property Rights Coordination Center (IPR Center) provide some of the most current data on estimated theft values. In 2008, President George W. Bush signed new IP legislation that created the cabinet-level position of intellectual property enforcement coordinator (IPEC, also colloquially referred to as the IP czar). One of the chief responsibilities of the IPEC is "to ensure effective intellectual property protection and enforcement, domestically and abroad."[23] In making annual reports, such as its annual intellectual property report to Congress, IPEC works in coordination with the White House and a number of US government agencies and departments, including the "Departments of Agriculture, Commerce, Defense, Health and Human Services, Homeland Security, State, and Treasury, the Office of the U.S. Trade Representative, and the U.S. Copyright Office."[24]

Impacts of IP Theft

US items that account for the greatest amounts of IP theft are pirated software, trade secrets, and counterfeit goods. All three of these areas of IP theft have significant ramifications for cybersecurity, law enforcement, military readiness, and national security. Counterfeit goods also have very serious implications for public safety, particularly with regard to counterfeit medications. It's one thing to purchase, say, a pair of fake Uggs boots or a bogus Nike backpack. Imagine, though, someone taking what they believe is an antibiotic medication prescribed by a medical professional for an illness. Instead, the person taking that medicine may be ingesting substances that at best are worthless facsimiles (in which case, the illness remains untreated by medicine) and at worst may be injurious or fatal due to toxic ingredients or allergic reactions caused by the foreign matter, such as synthetic opioids. The US Food and Drug Administration (FDA) is a timely source for information and updates on counterfeit medicine.[25]

I vividly recall attending a December 2021 USPTO all-day webinar discussion ("E-Commerce and China, Strategies for Fighting Online Counterfeits, Part 2") reporting on law enforcement efforts against counterfeit goods manufactured in China.[26] One of the presenters, Andrew Love, head of brand security/investigations with elite-level cycling company Specialized Bicycle Components, provided a jarring example, via a split-screen PowerPoint slide with two photos, of the harm that fake products can exact on consumers: on the left side, a bogus Specialized bike helmet; to the right, a male cyclist's head in profile with a blood-speckled stitched-up wound at his right temple. The cyclist had suffered that head injury from wearing the fake Specialized helmet he had unsuspectingly relied on as being genuine. Two takeaway messages: (1) counterfeit goods can hurt people in serious physical ways, and (2) contrary to some characterizations by both the sellers and the buyers of counterfeits, counterfeiting goods is not a victimless crime. After the bike helmet example, Love proceeded to run through six additional knockoffs of his company's products (carbon frames, jerseys and shorts, seat saddles, handlebars, high-performing Roval carbon wheels, and socks and gloves), leaving attendees with the unforgettable impression that anything and everything can and is subject to counterfeiting by bad actors.

Another section of the USPTO panel highlighted raids and seizures in China of children's toys coated in lead and other hazardous substances that had been manufactured in China and were intended to be sent to the United States and other overseas markets before being seized. This portion of the webinar was perhaps even more chilling *and* angering because of the real and potential threats that such seemingly innocuous toys impose for the smallest and most vulnerable of humans. The mind can only imagine in dread how many of these dangerous toys were *not* intercepted before being shipped out and were eventually handled by parents and their children.

As Love concluded, *effectively* policing these kinds of dangerous counterfeit products requires a coordinated three-pronged approach. A first step is trademark-related enforcement by the brands whose products are counterfeited. Second, internet-centered enforcement by the platforms where many of these bogus goods are sold needs to take place. Finally, a multi-pronged third step is prosecutorial enforcement, with the potential imposition of civil and criminal penalties (1) by the national governments where the illicit activities occur or (2) by the national governments where the impacted companies are headquartered or whose companies are having their products counterfeited.

The Dark Web/Dark Net

No exploration of IP theft would be complete without including a brief foray into the Dark Web, also called the Dark Net. My first substantive encounter with the Dark Web was in 2018 at the American Bar Association (ABA) annual IP Law Conference in Arlington, Virginia, via a panel presentation ominously titled "DarkNet: Enter at Your Own Risk." Its subtitle, "Inside the Digital Underworld," further stoked a sense of foreboding danger. The program description's first sentences boldly stated that "intellectual property infringement on the Dark Net/Web is out of control" and that "knockoffs, grey market goods, brands and identities are being stolen at a rate like no other."[27] In addition to promising tips and strategies on protecting IP and sensitive data, the most tantalizing draw was "a real-time and first-hand demonstration of the Dark Net/Web... [and] what transpires there." True to the program's billing, Krista Valenzuela with New Jersey's Cybersecurity & Communications Integration Cell—assuming the role of law enforcement's version of the mythical River Styx-navigating ferryman Charon—transported the audience from the already-scary-enough real world into this shadowy digital underworld of illicit marketplaces populated by anonymous buyers and sellers.

So what exactly is the Dark Web? A 2018 CNBC video, "The Dark Web and How to Access It," succinctly describes it as "a series of encrypted networks that serve to anonymize peoples' use on the internet."[28] Further back, in 2014, Vox explains it as "a general term for the seedier corners of the web, where people can interact online without worrying about the watchful eye of the authorities."[29] A 2021 entry, "An Introduction to the Dark Web," written by two lawyers for the legal information website JDSupra, offers a less menacing interpretation: "The dark web is neither good nor bad. It is simply an anonymized virtual space where users of the internet can operate with greater anonymity."[30] They liken the Dark Web to "an alter-ego to the World Wide Web that we are all used to." Unlike the World Wide Web though, the Dark Web is neither indexed by web crawlers nor meant for "casual browsing" by "a general audience."

Distinguishing Dark Web and Deep Web

Access is another crucial aspect for understanding why the Dark Web is such an appealing realm for counterfeit goods. The general public is able to search the World Wide Web with commonly used search engines, like Chrome, Firefox, Bing, Yahoo!, and DuckDuckGo. Think of the World Wide Web as the *Surface Web*. As the 2018 *Scientific American* article "Illuminating the Dark Web" explicates, World Wide Web websites and Dark Web sites "can be viewed by a standard web browser."[31] Where Surface and Dark Web sites differ though is with regard to access: "The difference is that [Dark Web sites] can only be accessed through special network-routing software, which is designed to provide anonymity for both visitors to websites and publishers of these sites."[32] Tor, I2P, and Freenet are examples of the special software that provides access to the Dark Web.

Sometimes the Dark Web is confused with the Deep Web. The Deep Web refers to sites that are hidden. Reasons for why these sites are hidden include being behind cybersecurity walls, protected by encrypted passwords, and designed to block "web crawlers" from finding and indexing them. The analogy of an iceberg is helpful: the World Wide Web or Surface Web is the part of the iceberg that sticks above the water's surface and is readily visible. The Deep Web is the much larger part of the iceberg that is unseen below the water's surface. Below the entire iceberg is the zone of the Dark Web.

Benign and Malign Uses of the Dark Web

The Dark Web has an aptly deserved negative connotation. However, some uses of the Dark Web can be positive and benefit societies. For example, its anonymity can be a boon for investigative journalists and whistleblowers, facilitating safe places to communicate, share sensitive information and data, and more easily avoid detection by parties (e.g., repressive governments) who could harm reporters and their sources. The Society of Professional Journalists provides specific guidance for "leakers" on "how to share information with reputable news media" via "secure pathways."[33] Some examples of these news organizations are the *Guardian*, the *Intercept*, and the International Consortium of Investigative Journalists. The 2021 JDSupra article cited earlier also points out that human rights activists and people working in law enforcement are able to leverage the Dark Web's anonymity in ways that can further their respective objectives.[34] To this point, in December 2023, an undercover law enforcement official posing as a contract killer was able to arrest a Utah man who was on the Dark Web seeking a hit man to murder two individuals in New York and pay him with Bitcoin.[35]

The burgeoning rate of overdose deaths from widely accessible illegal drugs, particularly synthetic fentanyl opioid drugs, bought in the Dark Web's online markets is one of the most heart-breaking examples of its dangers. Over the past decade, the Federal Bureau of Investigation has engaged in efforts to disrupt and shut down Dark Web marketplaces. In 2021, as one example, the FBI partnered with European law enforcement in an opioid trafficking crackdown effort called Operation Dark HunTor. This initiative led to 150 arrests and the seizure of $31 million in crypto-currency and cash, as well as drugs and weapons.[36] In May 2023, US Attorney General Merrick Garland discussed an international crackdown that led to the seizure of a Dark Web market-place, arrest of almost 300 individuals, and confiscation of $53 million. Announcing the results of the collaborative operation, Merrick's message was one of both deterrence and consequence: "Our message to criminals on the dark web is this: You can try to hide in the furthest reaches of the Internet, but the Justice Department will find you and hold you accountable for your crimes."[37]

Ebooks and E-Journals Theft/Piracy

The advent and meteoric rise of ebooks (i.e., digital versions of books) over the past decade has brought about a sea change in how people access information. Ebooks, like the printing press of the fifteenth century and the internet of the late twentieth century, represent another kind of disruptive technology for societies. They also have impacted the business models of publishers and have provided new revenue streams for publishers, authors, and intermediaries, such as booksellers and licensors. Moreover, digital books have created opportunities for those who illicitly and without permission seek to make money from these works, as well as people who want to make these works freely accessible.

Napster-ization of Ebooks?

In the first decade of this century, as the market share of e-readers, like Amazon's Kindle and Barnes and Noble's Nook, and the ebooks read on them grew, some authors openly expressed concerns that their print works could and probably would be pirated if they granted permission for ebook versions of their works. Many contracts between authors and publishers that predated the internet of the 1990s did not expressly provide for ebooks. So authors were left with the choice of whether to greenlight digital formats for their books. Rapid rises in ebook purchases by the public each year applied commercial pressure for authors to authorize digital copies of their books. Eventually, even the authors who were holding out on "going digital" got on board the ebook train. A Government Printing Office website identifies a watershed moment for ebook market penetration; in 2012 "eBook sales in the United States surpassed all hardcover book sales for the first time, topping over $3 Billion in revenue nationwide."[38]

Fears that their digital books would be pirated was a major concern of print authors who were reluctant to embrace electronic formats of their works. Within publishing circles, many noted how digital music (i.e., the Napster-ization of music) had irrevocably altered the music industry's business models and dramatically reduced music royalties for a wide swath of music copyright stakeholders. Napster was, at its inception, a digital music file-sharing site that debuted in 1999. (Today, Napster exists as a reincarnated arm of Rhapsody, a subscription digital music service for licensed music.) Over time, Napster became a high-profile exemplar for the practice of downloading digital music. Downloading of digital music from sites like Napster constituted infringement under copyright law, as downloaders neither paid for these files nor had permission from the music rights holders to do so. Some downloaders received cease-and-desist orders. Others received demands for legal damages. A few, such as Jammie Thomas-Rasset, were sued by organizations like the Recording Industry Association of America, in widely reported prosecutions and trials that were intended to deter others from illegally downloading and sharing copyrighted music.[39]

Articles of this first decade and a half of the twenty-first century talked about whether the publishing industry and authors would suffer from the same levels of file sharing of books that had forever changed the ways that people access music. These articles were also seen as cautionary tales for publishers and authors: namely, to learn from the experiences of music creators—and the people sharing in and dependent on the food chains of those creators—who had seen their music-related livelihoods and income streams change, often for the worse. Unfortunately, for authors and publishers, worries about theft of their ebooks was not misplaced. Many of those who had been among the last authors to agree to allowing digital versions of their books learned that their ebooks had been uploaded to piracy sites within hours of their arrival on legitimate commercial sites.

E-Journal/Ebook Robber Barons and Robin Hoods

Open access and related issues like predatory journals were examined earlier in this chapter. A foundation has been laid then for understanding the high costs of journals and research access issues that catalyzed the open access movement over the past few decades. If open access can be seen as a kind of crusading white knight of the research world—a person or thing that comes to someone's aid—how shall we label and characterize the people who acquire or steal, depending on one's lens, copyright-protected journal articles and make those articles available for free, anywhere in the world, via unpaywalled internet sites? To some, these sites are seen as flagrant IP thieves and scofflaws. To others, they operate in stealth as digital "black ops" Robin Hoods: "stealing" journal articles from for-profit Big Publishing entities that are viewed as de facto research-hoarding and -monopolizing robber barons, then making their "acquisitions" or "plunder" (even those nouns shape the lenses through which these actions are viewed) accessible to the research-poor and those existing in scholarly communication deserts. In this analogy, law enforcement entities like the FBI are seen as the sheriffs of Sherwood Forest, well-meaning perhaps, in terms of the rule of law, but nevertheless instruments of systemic inequity and impediments to efforts to level the playing field and right the wrongs of economic imbalance and scarcity for billions of research-have-nots.

Sci-Hub

The digital site Sci-Hub is one of a number of these self-appointed internet Robin Hoods. It was started in 2011 by Alexandra Elbakyan, a graduate student studying science. She was born in 1988 in Kazakhstan, three years before the dissolution of the Soviet Union, of which her native country was an impoverished and underdeveloped part. Frustrated and dismayed by the inability to access research behind journal paywalls and armed with computer hacking skills, she proceeded to build what has become "the world's largest pirate site for academic papers."[40] A 2016 *Science* piece profiling Elbakyan crystallized the ethical conundrum of her creation: "[Sci-Hub is] an awe-inspiring act of altruism or a massive criminal enterprise, depending on whom you ask."[41]

In spite of warrants for Elbakyan's arrest and copyright infringement judgments against her by overseas court, users around the planet—including researchers and students in the Global North—continue to download Sci-Hub's trove of millions of articles. In September 2023 watchdog group Electronic Frontier Foundation (EFF) awarded Elbakyan its EFF Award for Access to Scientific Knowledge. EFF explains that her goal was "to provide free and unrestricted access to all scientific knowledge" and that Sci-Hub "has grown a database of more than 88.3 million research articles and books freely accessible for anyone to read and download."[42] The site is useful to diverse stakeholders, as EFF notes in the award announcement: "millions of students, researchers, medical professionals, journalists, inventors, and curious people all over the world, many of whom provide feedback saying they are grateful for this access to knowledge."[43]

Z-Library and Library Genesis

In addition to Sci-Hub, other ebook and journal piracy sites vie for slices of the book and scholarly communication pies. These sites, like Sci-Hub, vex authors and publishers, understandably frustrated at having their content made freely available, but delight users who, also understandably perhaps, seek free and accessible copies of readings and research. Z-Library's home page bills itself as "The world's largest ebook library."[44] However, copyright owners view the site in very different terms, namely, as a major digital piracy site for illegally accessing copyrighted works. A 2022 *Fast Company* article reported that at the height of its popularity and availability, Z-Library was making "more than 10 million e-books and 86 million articles" available at no cost.[45]

Around 2020, I examined the site. At that time, a notable feature of Z-Library was a display in the top left corner of its website: a blue box showing the number of books accessible through the site. That number was more than 22 million. Beside the books box was a green box that showed the number of available articles, numbering more than 284 million. This "running tally" facet was reminiscent of similar quantitative numbers showcased on the Directory of Open Access Journals (DOAJ) site, indicating the ever-increasing numbers of open access journals and articles. While DOAJ's numbers are within the law, each one of the books and articles listed in the numbers tally on Z-Library represented an instance of alleged copyright infringement.

In November 2022, the FBI shut down Z-Library. The US Attorney's Office for the Eastern District of New York announced in a November 16, 2022, press release that two Russian nationals had been arrested in Argentina and "charged with running a massive e-book piracy website."[46] By court order, the bureau seized Z-Library's network of internet domains and took them offline. The press release reports illuminating details about Z-Library in the indictment and court filings:

> Z-Library bills itself as "the world's largest library" and claims to offer more than 11 million e-books for download. Z-Library, which has been active since approximately 2009, offers e-book files in a variety of file formats, stripped of their copyright protections, and encourages users to upload and download titles. Many of the e-books offered by Z-Library are protected intellectual property for which authors hold copyrights and publishers hold exclusive distribution rights, and which Z-Library has no right or license to distribute, and which are available elsewhere only with anti-circumvention measures applied. As such, a central purpose of Z-Library is to allow users to download copyrighted books for free in violation of U.S. law. In addition to its homepage, Z-Library operates as a complex network of approximately 249 interrelated web domains.[47]

This FBI sting on Z-Library also had ripple effects on social media communities. As reported in a November 17, 2022, *Washington Post* article, pro and con views about the pirate site's takedown ignited impassioned interchanges among TikTok users who talk about books and authors via a forum called BookTok. The piece brilliantly captures the competing, diametrically oppositional points of view about these sites:

> Readers were mourning that their ability to download free textbooks, novels and academic papers had disappeared overnight. Some BookTokers compared the shutdown of the website to the mythical burning of the library of Alexandria in 48 B.C., [TikTok content creator Lexi] Hardesty said. "Some even said that shutting it down was an extension of slavery."...
>
> Yet authors across BookTok were relieved. "Piracy costs us our sales, specifically for marginalized authors; it adversely impacts public libraries; and it hurts the publishing industry," said Nisha Sharma, an author and BookToker. "Essentially when you mourn Z-Library, you are mourning the end of theft."[48]

As with most enterprises like Sci-Hub and Z-Library, law enforcement plays whack-a-mole, shutting down these sites until they emerge again on new hosting sites. In November 2023, Z-Library was once again back, this time, as pointed out by a site called Good E-Reader, with the updated domain name of z-lib.io. Interestingly, the post about Z-Library's resurrection advises users of the importance of being mindful of legal issues relating to the use of Z-Library: "While it provides a rich repository of resources, the legality of certain materials may pose copyright concerns."[49] The author of the piece admonishes users to follow copyright laws and be respectful of the IP rights of content holders, in order to coexist "with legal and ethical standards." Right below that admonition though, the reporter provides an update. Users are warned that the prior updated domain name turned out to be a scam. The author provides the correct URL and apologizes "for the error in reporting." This correction underscores the importance of having a healthy degree of caution and skepticism when accessing anything on the internet, as well as in print. Verification of the accuracy of information found anywhere is essential. It also shows that even controversial sites accused of improprieties and unorthodox activities are not immune from scams and opportunists.

Library Genesis is another operator accused of mass copyright infringement within the digital book and journal sharing space. Like Z-Library, the adjective *shadowy* is frequently employed in talking about Library Genesis. In September 2023, another group of stakeholders launched a legal salvo against these book and journal file-sharing sites: textbook publishers. Four publishers—Cengage, Macmillan Learning, McGraw Hill, and Pearson Education—brought suit against Library Genesis (also known as LibGen). The publishers alleged "copyright infringement on a 'staggering' scale," as reported by Reuters.[50] Intriguingly, shadow libraries like Z-Library and LibGen have been alleged in lawsuits against Big Tech companies like Meta and OpenAI to be the sources of illegally attained AI training data sets used as large language models for those companies to create controversial AI chatbots like OpenAI's ChatGPT. These companies have been sued in 2023 for copyright infringement in multiple lawsuits filed by authors, publishers, music copyright holders, and others. AI is discussed in more detail in Chapter 11.

Project Gutenberg: Legal Sharing of Ebooks and E-Journals

Ebook and journal file-sharing sites Sci-Hub, Z-Library, and Library Genesis represent the most daring and prominent rule-of-law-flouting outlaws challenging Western-based copyright law-and-order systems. Yet, for every law enforcement action taken against them, these sites, like Newton's third law of motion, have demonstrated a remarkable ability to execute equal and opposite reactions that enable them to adapt to new limits imposed on them and propagate in new digital spaces. Decades before these sites were even a thought in their progenitors' minds, however, Project Gutenberg was launched in 1971 as a global playing-inside-the-copyright-rules ebook file-sharing operation. Its founder, the late Michael S. Hart, somewhat accidentally began what would become Project Gutenberg when he created and then posted an electronic text document of the Declaration of Independence to the networks of which he was a part at the time at a University of Illinois lab.[51]

Hart died in 2011, but the mission and vision for his public domain ebook repository and library brainchild survive. As written and copyrighted by Hart in 1992 and accessible today via a "History and Philosophy of Project Gutenberg" web page, the overarching philosophy of the site is "is to make information, books and other materials available to the general public in forms a vast majority of the computers, programs and people can easily read, use, quote, and search."[52] Unlike Sci-Hub and its law-flouting kinfolk though, Project Gutenberg's ethos was and is one that hews closely within the bounds of copyright law. A guiding principle of the project is that only public domain copies of books are digitized and made accessible. Even then, the system is configured to be in compliance with the copyright laws of respective nations. As an example, a book that is in the public domain in America can be accessed within the United States but will not be accessible by someone in another country if that book is not in the public domain of the other country.

As of 2023, the corpus of Project Gutenberg, operating as a volunteer-based organization, is approximately 70,000 ebooks strong. This is a mere fraction of the hundreds of millions of ebooks and journal articles made freely available by Sci-Hub and others. Users and providers of Project Gutenberg, however, need not fear crackdowns or arrest warrants that disrupt their access of the site's law-abiding offerings.

Native Americans and IP Rights

Counterfeit goods have negative impacts on individual creators and rights holders. They also harm the groups and communities within which these creators are situated. Native Americans are one historically disadvantaged group experiencing the damaging consequences of arts and crafts that are falsely claimed as made by native peoples. Bloomberg reported in October 2023, for some context, that "Native Art is a $1.5 Billion Market Plagued With Problems," one of which is fakes.[53]

Native American tribes, pursuant to US law, are sovereign nations; indeed, many tribes officially use that designation: Navajo Nation, Cherokee Nation, Seminole Nation of Oklahoma.

The US Department of the Interior is the federal agency that has a "government-to-government relationship" with Native American tribes, which it refers to as "distinct, independent political communities."[54] The long history of conflict between the US government and native peoples and ongoing efforts toward "promoting and supporting tribal self-determination" is reflected in an illuminating statement included on the "Tribes" web page of the Interior Department: "Acknowledging the history of past mistreatment and destructive policies that have hurt tribal communities, the U.S. continues efforts to restore and heal relations with Native Americans, respect tribal self-determination, ensure meaningful consultation with tribes on matters affecting their interests, and work in partnership with tribal governments to support the health and development of native communities."[55] As of December 11, 2024, the US federal government recognizes 574 tribes.[56] Drawing from US Census data reported in November 2023, the "estimated American Indian and Alaska Native alone population was 4.3 million and the alone or in combination population" was 7.1 million on July 1, 2022.[57]

Within the Department of the Interior is the Indian Arts and Crafts Board (IACB). It was created in 1935 by the US Congress. The IACB's chief purpose is "to promote the economic development of federally recognized American Indians and Alaska Natives (Indians) through the expansion of the Indian arts and crafts market."[58] A federal statute is also an important driver of the IACB's work: "The Indian Arts and Crafts Act (IACA) of 1990 (P.L. 101-644) is a truth-in-advertising law that prohibits misrepresentation in the marketing of Indian art and craft products within the United States."[59] This law was beefed up in 2010 to augment its enforcement capabilities: "The Indian Arts and Crafts Enforcement Act of 2010 authorized all federal law enforcement officers, not just the FBI, to conduct investigations and revised the Act's criminal penalties."[60]

A May 17, 2023, Department of Justice press release offers a concrete example of the kinds of fraud that is prosecuted:

> A 67-year-old Western Washington artist was sentenced today to 18 months of federal probation for violations of the Indian Arts and Crafts Act (IACA) by representing himself as a Native American artist, when he had no tribal enrollment or heritage, announced U.S. Attorney Nick Brown. Jerry Chris Van Dyke aka Jerry Witten, of Seattle, pleaded guilty in March 2023. At the sentencing hearing, U.S. District Judge Tana Lin noted that this was not a one-time error, but a ten-year period of "undermining a community and identity."[61]

As demonstrated in this excerpt, law enforcement officials at state and federal levels work to hold counterfeiters of native arts accountable. Efforts at combatting these crimes focus on raising awareness and educating people about the impacts of counterfeiting. To that end, I attended a November 30, 2023, roundtable on "Tribes, Intellectual Property, and Consumer Protection" that was held at the USPTO.[62] The event brought together a diverse swath of practitioners, policy makers, artists, and thought leaders, including attorneys general from states such as Oklahoma and Kentucky, officials from federal agencies like the FBI's Intellectual Property Rights Center Coordination Center, the Department of Justice, and the Department of the Interior, native people working at state levels and within nonprofit organizations such as the Association on American Indian Affairs (founded in 1922), and academic researchers like those at Michigan State University's Center for Anti-Counterfeiting and Product Protection.

Participants, through a series of five panels, highlighted enforcement efforts and programs that authenticate native arts, as well as the domestic and global issue of traditional and indigenous knowledge, which is explored later in this chapter. One such State Council on the Arts–sanctioned effort is Alaska's Silver Hand Program, which, as of 2023, authenticates art created by approximately 1,500 Alaska Native artists. The program's website explains that

> the Silver Hand program helps Alaska Native artists promote their work in the marketplace and enables consumers to identify and purchase authentic Alaska Native art. The seal indicates that the artwork on which it appears is created by hand in Alaska by an individual Alaska Native artist. Only original contemporary and traditional Alaska Native artwork, not reproductions or manufactured work, may be identified and marketed with the Silver Hand seal.[63]

Regarding authentication of artists eligible for the program: "To be eligible for a two year Silver Hand permit an Alaska Native artist must be a full time resident of Alaska, at least 18 years old, and provide documentation of membership in a federally recognized Alaska Native tribe. The Silver Hand seal may only be attached to original work that is produced in the state of Alaska."[64]

One of the most significant takeaways from this roundtable was making the issue of counterfeit native arts more tangible and understandable at a human level. A law enforcement panelist, for example, pointed out that native arts are an integral economic lifeline for many tribal communities; he noted that, in some instances, arts and crafts account for more than 80 percent of all income generated for these communities.

A Native American panelist, Osceola Red Shirt, assistant director for the Law Enforcement Division of the Interior Department, noted that native cultures go through peaks and valleys of popularity vis-à-vis the majority culture in the United States. During peak times, such as the present due to popular television shows like *Reservation Dogs* and *Dark Winds*, native artists benefit from greater interest in their arts and crafts. However, this greater popularity also makes them more susceptible to artists who claim to be native and are selling art pieces not made by genuine native people. Red Shirt observed that Native Americans struggle with understanding their IP rights and who to turn to for help.

One of the roles of the IACB is to raise awareness about fake native arts and crafts and the impacts of those counterfeit goods on native people and communities. To that end, a visit to the IACB home page in late 2023 revealed a "special message from Secretary Debra Haaland in support of Indian artists and authentic Indian art."[65] Deb Haaland, a member of the Pueblo of Laguna in New Mexico, was the secretary of the Department of the Interior and the first Native American to serve in a cabinet position. She assumed the role of secretary in March 2021, after serving for two years as a Democratic member of the House of Representatives. Echoing the special message of Secretary Haaland as well as varied panelists in the November 2023 USPTO roundtable, education is key to better informing consumers about the pitfalls and consequences of purchasing fake native arts and crafts. Paul DelPonte, executive director for the National Crime Prevention Council, famous for its public service announcements featuring McGruff the Crime Dog ("Take a Bite Out of Crime") since 1980, talked at the roundtable about the need for more public education. He pragmatically opined that "we're not going to be able to arrest ourselves out of the problem of counterfeits."[66] DelPonte more broadly asserted that "fake products are the crime of the 21st century" and that mass online sellers like Amazon and Etsy as well as new technologies such as AI and 3D printing are enabling "criminals to make lots of money."

The negative impacts of counterfeit native arts and crafts are more than solely economic harms. Fake native goods also detrimentally impact the cultures of native peoples. We further examine issues of indigenous culture in the next section.

Traditional Knowledge/Indigenous Knowledge

Much of the world is part of a Western-based IP system of laws. Nearly all nations have their own legal systems that govern copyrights, patents, trademarks, and trade secrets at the individual nation-state level. Additionally, at the global level, treaties provide law, ethical guidance, policies, and mechanisms for addressing grievances, enforcement, remedies, and reform. The Berne Convention for Protection of Literary and Artistic Works is a world copyright treaty approved in 1886 in Berne, Switzerland. Its purpose is chiefly twofold, from the words of the treaty itself: "the protection of works and the rights of their authors."[67] At present, 181 nations out of 195 on the planet are members of this treaty. Treaties and global schema also exist in the areas of trademark (e.g., Madrid Agreement and Protocol) and patents (e.g., Patent Cooperation Treaty).

Amid this world-spanning sea of nations with Western-influenced IP systems are islands and oases of alternatives. These alternative IP conceptualizations and schema are varyingly referred to by categories such as traditional knowledge (TK), indigenous knowledge (IK), traditional cultural expressions (TCE), and communal copyright. The United Nations' World Intellectual Property Organization (WIPO) defines TK as "knowledge, know-how, skills and practices that are developed, sustained and passed on from generation to generation within a community, often forming part of its cultural or spiritual identity."[68] Some stakeholders include genetic resources and folklore within TK as well.

One of the biggest challenges facing TK is defining what it is: opinions vary widely on a specific, implementable, and potentially enforceable definition. Without a clear, boundaried definition, TK protections and rights remain elusive. Adding to this challenge of identity and nomenclature for TK is that the "owners"—or more accurately the progenitors and practitioners—of TK are native peoples who have been historically disadvantaged and disenfranchised. Examples include tribes of the Amazon rainforest in South America, Native Americans in the United States, First Nations in Canada, the San people of southern Africa, Aboriginal Australians, Maori of New Zealand, and indigenous groups throughout Southeast Asia and Asia, such as the Ainu of Japan's northernmost island, Hokkaido. For many of these groups, for whom Western rules of law are anathema and countercultural to millennia of practice, ethical principles and guidance can be a welcome alternative for dispute resolution, as well as an antidote to cultural appropriation. We will shortly look at one instance of remarkable cultural appropriation by Urban Outfitters. Ethical precepts and codes, as well as cultural competency and IP rights-related due diligence, can assist companies to avoid these kinds of missteps *before* the law is called to intervene.

Facets of TK are in key ways inverse to global IP. For one, TK is often oral in nature. The definition of US copyright includes the requirement that an "original" work be "fixed in a tangible medium of expression." How then to protect often-oral TK such as stories, lore, and creation myths that are not—from their nascent nature and amorphously emergent conception—in a "fixed medium"?

Adding to the challenges in harmonizing TK and IP is that TK is seen as being "shared by" or "communal" to the collective group, rather than "owned." Western-influenced conceptualizations of IP, in which one "owns" a copyright, trademark, or patent, are, at their core, antithetical to the very essence of TK. The pain-alleviating sap from a native plant or the medicinal properties of a native tree's bark are viewed and hailed as being open to use among group members. Indeed, native peoples throughout the world can be credited as the originators of a worldwide "proto-open knowledge" network writ large. TK's bedrock idea of unfettered, free access is embedded in all of the open movements today. Negotiations and policy formulation to "reconcile" IP and TK have been ongoing for many years at the international level, as well as in some nations where indigenous communities (e.g., Australian Aboriginals and southern Africa's San people) have been more successful in achieving some legal recognition and defensive mechanisms for TK (see, e.g., 2016 Intellectual Property Watch article "On Eve of WIPO Traditional Knowledge Negotiations, Nations Swap Experiences"[69]).

Batik is an ancient art, still practiced in Southeast Asia: fabric is first dipped in wax and then dyed. A story in 2009 reported that the Indonesian government was strongly encouraging Javanese batik makers to copyright their unique designs, to protect them from copycats. The batik practitioners balked at these efforts though, believing that "being grateful [to God] is sufficient for them."[70] Over the past decade though, the governments of Indonesia and Malaysia have instituted frameworks to protect batik as examples of copyrightable works and industrial design.

A major TK milestone occurred at WIPO's Swiss headquarters on May 24, 2024. WIPO's member states approved a groundbreaking new treaty, following years of diplomatic negotiations:

> This is the first WIPO Treaty to address the interface between intellectual property, genetic resources and traditional knowledge and the first WIPO Treaty to include provisions specifically for Indigenous Peoples as well as local communities.
>
> The Treaty, once it enters into force with 15 contracting parties, will establish in international law a new disclosure requirement for patent applicants whose inventions are based on genetic resources and/or associated traditional knowledge.[71]

WIPO members must now ratify the treaty in their respective nations in order for the treaty to come into force. In early December 2024, Malawi became the first country to ratify the historic treaty.[72]

Cultural Appropriation

An ethical issue that is also present in considering the IP rights of indigenous peoples is cultural appropriation. Cultural appropriation is defined by the *Cambridge Dictionary* as "the act of taking or using things from a culture that is not your own, especially without showing that you understand or respect this culture."[73] In a case that was reported widely, clothing seller Urban Outfitters in 2016 was accused of cultural appropriation of Navajo Nation designs. As the *Guardian* reported, "Urban Outfitters reached a settlement with the Navajo Nation after illegally using the tribe's name for a collection that included 'Navajo hipster panties' and a 'Navajo print flask.'"[74] Of note was the fact that the Navajo Nation had registered the term *Navajo* as a trademark in 1943, as indicated by the suit's records. Yet Urban Outfitters had used the term since 2001. The Navajo Nation sued the company for trademark infringement in 2012. Prior to this incident, Urban Outfitters attracted similar negative attention and outcry for utilizing Palestinean and Indian designs and themes in its commercial products. Incidents like this not only raise difficult ethical issues but also can inflict pain on individuals and communities that have historically experienced trauma, discrimination, and exploitation.

The World Intellectual Property Organization

WIPO, based in Geneva, Switzerland, is one of 16 specialized agencies of the United Nations. As of 2023, 193 countries (out of 195) are member states of WIPO (only the two observer states, the Holy See and the State of Palestine, do not belong to WIPO). Not surprisingly, owing to its near universal membership and signal platform, WIPO has the world's largest megaphone on all things relating to TK. In addition to administering the Berne Convention (the world's largest copyright treaty), discussed earlier in this chapter, one of WIPO's chief activities is to develop IP-related policies. In recent years, as a noteworthy example emanating from emerging technologies that intersect with IP, WIPO has been convening periodic forums to actively learn from AI experts, listen to stakeholder communities, and promote public dialogue as it works to develop ethics guidelines and policies for addressing AI and copyright. We examine these ongoing AI-focused efforts in more depth in a later chapter.

Since its establishment in 2000, the WIPO Intergovernmental Committee (IGC) has been working to better recognize and advance the IP interests of indigenous peoples in what are hoped to be meaningful measures, and not merely lofty, illusory rhetoric. Given how significant IP is to the GDPs of the world's economic powerhouse nations and the inherent incentives for these IP storehouses to amass and control as much IP as possible, as well as the pragmatic reality that Western-styled IP systems and TK/IK/TCE community-centered, predominantly nonlegally codified behavioral norms are in many ways in diametric opposition to Western IP, arriving at a mutually agreeable, viable, implementable, sustainable IP-TK framework that can function as a kind of "two-systems solution" has been elusive. In 2023, a WIPO background brief provided some encouraging updates and reawakened hopes for progress.

> In July 2022, WIPO Member States decided to convene a Diplomatic 1 Conference to conclude an International Legal Instrument Relating to Intellectual Property, Genetic Resources and Traditional Knowledge Associated with Genetic Resources no later than 2024.
>
> This brief describes the origins and rationale of the IGC, the participation of members and observers including Indigenous Peoples as well as local communities, its achievements and the state of ongoing negotiations under its mandate.[75]

The brief's introduction also explains that the WIPO IGC "is currently undertaking text-based negotiations with the objective of finalizing an agreement on one or more international legal instruments that would ensure the balanced and effective protection of genetic resources, traditional knowledge and traditional cultural expressions (the terms 'traditional cultural expressions' and 'expressions of folklore' are used interchangeably in WIPO discussions)."[76] Illustrating the considerable challenges that exist in securing a substantive agreement, the introduction goes on to observe that "such an instrument or instruments could range from a recommendation to WIPO members to a formal treaty that would legally bind countries choosing to ratify it." The sizeable distance between a "recommendation" on the one hand and a legally binding ratifiable treaty on the other lays bare the diplomatic efforts and hurdles that remain to be surmounted to achieve greater protections for TK, IK, and TCE.

IP Nexus to Gene Patents, Vaccines, and Medicines

Genes are the architectural blueprints of life on earth. In humans, as for all animals, genes encode instructions on the color of our eyes, the height of our bodies, and the pigments of our skin. They also influence our health and wellness, whether we have the capacity to develop various gene-carried diseases, like Parkinson's, Huntington's, and breast cancer. Over the course of decades of research, the human DNA genome was first fully mapped in 2003 by the Human Genome Project. Since that time, one of the goals of human gene sequencing has been to apply understanding of human genes to medical care and treatment. In the wake of gene sequencing, pharmaceutical companies are already leveraging "precision medicine"—individually tailored, bespoke drugs and procedures—in ways that are revolutionizing health care. This bleeding-edge research and development entails billions of dollars in investment, research, testing, government approval, manufacturing, licensing, and delivery of these treatment regimens from the research lab to patients' bodies, or what is more colloquially shorthanded "from bench to bedside." Unsurprisingly, IP plays a big role in this life cycle. In this section, we delve into ethical issues and questions of three topics with a nexus of IP and medicine: gene patents, access to vaccines, and affordability of medicines. High research and development costs for medicines translate to a commensurately high incentive for investors and companies to reap dividends on those investments.

Gene Patents and DNA Data

Question: Can a *human* gene be patented?
Answer: That depends on *where* you want to apply to patent that human gene.

Whether a human gene patent can be patented was argued before the US Supreme Court in 2012 in *Association for Molecular Pathology v. Myriad Genetics*. At issue: whether a section of the US Patent Act allows patents on genes. In a landmark unanimous 9–0 decision in June 2013, the justices answered yes and no to the human gene patentability question.[77] Yes, the court said: patent protection can be granted for *synthetic* gene patents—"gene sequences refined by synthetic processes to create molecules that do not occur naturally." But no, the court decided as to whether "naturally occurring gene sequences, and their natural derivative products" are eligible for patent protection under the US Patent Act. "Discovery of natural products" does not fall within US patent protection eligibility because DNA is a "product of nature."[78] The landmark ruling had significant implications: "Prior to this ruling, more than 4,300 human genes were patented. The Supreme Court's decision invalidated those gene patents, making the genes accessible for research and for commercial genetic testing."[79]

In some other jurisdictions like the European Union, however, human gene patenting *has* been deemed legally permissible. The European Patent Office (EPO) provides some illustrative insight into its rationale and policy:

> The EPO grants patents in strict accordance with its legal basis, the European Patent Convention (EPC) and takes ethical considerations into account when granting patents. Inventions whose exploitation is deemed to be contrary to Article 53 EPC are not patentable. The law lists exceptions to patentability for ethical reasons, among them human cloning, modification of the human genome and commercial uses of human embryos.
>
> For patents in biotechnology the rules of the EU's Directive on the legal protection of biotechnological inventions ("Biopatent Directive") also apply. The Directive for instance clarifies that human genes, plants and animals are patentable when all conditions for a patent are fulfilled. The Directive became part of the EPC in 1999.[80]

Despite some currently controlling case law (e.g., *Myriad Genetics* in the United States and EPO policy in the EU), the issue of gene patents is not universally agreed on and remains open to debate. A June 2023 ACLU article revisits the landmark *Myriad Genetics* ruling, somewhat surprisingly including Myriad Genetic's June 2023 post to Twitter stating that the company no longer is in support of patents for products of nature.[81] However, positions on these genetics-related issues remain unsettled. Indeed, as the ACLU article observes, "[A] number of Congressmembers have supported legislation to restore patenting of laws of nature and abstract ideas, including human genes, and have made clear they will push this issue in the future."[82]

The ACLU article additionally cautions that the public's right to control their own genetic information is imperiled. DNA testing kits are increasingly popular with consumers and have been marketed through television advertising as birthday, Father's Day/Mother's Day, and holiday gifts. The Federal Trade Commission in 2023 offered some advisory notes to weigh, though, like data privacy implications, when thinking about giving family and friends these kits as gifts.[83] Saliva-based kits like 23AndMe, whose parent company is Alphabet/Google, and Ancestry.com, recently acquired by the private equity firm the Blackstone Group, are owned and leveraged as lucrative portfolio assets by companies. On the one hand, consumers' ability to gain insights into their own genetic makeup can have numerous potentially positive and negative health and medical care implications. Alternatively though, complex privacy, bodily autonomy, informed consent, and data collection and use issues, both known and as yet unknown, are present with the retention and possible uses of consumers' genetic personally identifiable information by these large corporate entities.

Rapid advances in biotechnology and genetic science portend that legal, ethical, and policy understandings and positions on gene-related patents may change over time. Recent patent decisions related to gene editing technologies (e.g., CRISPR-Cas9) are also providing fertile ground for discussion of the need for ethical guidelines and codes of conduct. More cases and case law in these emerging areas are likely to arise.

COVID-19 and Open COVID Pledges

Hollywood films have—for almost a century—both stoked and mirrored humankind's gnawing muscle-memory fears of extinction-level epidemics (e.g., fourteenth-century Black Plague, 1918 influenza pandemic). Movies depicting run-amok infestations and doomsday pandemics evoke virologically fueled, survival-of-the-fittest terror. Some films on the subject include *Faust* (1928), *Panic in the Streets* (1950), *Invasion of the Body Snatchers* (1956; 1978), *The Last Man on Earth* (1964), *The Andromeda Strain* (1971), *The Omega Man* (1971), *Dawn of the Dead* (1979), *12 Monkeys* (1995), *Outbreak* (1995), *28 Days Later* (2002), *Children of Men* (2006), *I Am Legend* (2007), *The Happening* (2008), *Blindness* (2008), *Carriers* (2009), *Contagion* (2011), *Black Death* (2011), *Rise of the Planet of the Apes* (2011), *World War Z* (2013), and *Bird Box* (2018).

Recent comic book and video game TV show adaptations, like *The Walking Dead* (2010), its ever-expanding spin-off progeny (e.g., *Fear The Walking Dead*, 2015; *The Walking Dead: Dead City*, 2023), and *The Last of Us* (2023), have similarly channeled gruesome plagues and quixotic quests for may-or-may-not-exist vaccine cures. These genre works serve purposes other than scaring and entertaining us, though. They're also replete with morally ambiguous antihero characters grappling with ethical questions and issues of life-or-death consequence. Though fictitious—thankfully—these characters, plotlines, and settings function, too, as pragmatic stand-ins and allegories for *real-world* people, challenges, and places.

Fears of a real-world human epidemic came to life in 2019 in China, reaching pandemic status the following year. To remove stigmatizing impacts of naming and associating this emergent disease with the Wuhan area of China where the disease first appeared, the scientific community called the disease COVID-19, shorthand for the longer official designation coronavirus disease 2019. As of February 2025, the World Health Organization (WHO) has logged worldwide COVID-19 deaths of over seven million people and more than three-quarters of a billion cases. WHO also notes that more than 13.5 billion vaccine doses have been given to recipients.[84]

Amid this exigent world health crisis, ethical issues abounded. Once vaccines for COVID-19 were developed and ready to be rolled out in the spring of 2021, two critical considerations were at the top of the list. The first, which demographic groups would be deemed most in need to receive the initial injections. The second, whether and how poorer nations and peoples would be afforded equitable opportunities to receive the life-saving patent-protectable vaccines. Little controversy ensued with regard to the first question: those involved in health care, as well as persons living in long-term care and nursing home facilities, were indeed the first groups to get their COVID-19 vaccine shots. Ethically, both of these groups' priority needs for the vaccines were readily apparent and on philosophically firm ground: namely, inoculating the caregiver and health-care professionals on the front lines of the pandemic and persons residing in facilities who were most at risk of succumbing to the highly contagious impacts of the virus.

At the outset of the COVID-19 outbreak, some companies and entrepreneurs agreed to temporarily suspend their IP rights for a whole range of IP-protected goods and services in an all-hands-on-deck kind of approach to the health emergency. Creative Commons led this effort, called the Open COVID Pledge:

> Immediate action is required to halt the COVID-19 Pandemic and treat those it has affected. It is a practical and moral imperative that every tool we have at our disposal be applied to develop and deploy technologies on a massive scale without impediment.
>
> We therefore pledge to make our intellectual property available free of charge for use in ending the COVID-19 pandemic and minimizing the impact of the disease.
>
> We will implement this pledge through a license that details the terms and conditions under which our intellectual property is made available.[85]

Founding adopters of the pledge included Big Tech companies: Amazon, Apple, Facebook, Hewlett Packard Enterprise, IBM, Intel, and Microsoft. As one signatory of the pledge, Hewlett Packard Enterprise granted "free access to all of its patented technologies for the purpose of diagnosing, preventing, containing and treating COVID-19."[86] Other diverse sector participators comprised financial giant Morgan Stanley, telecom AT&T, rideshare gig economy maverick Uber, and consumer and electronics manufacturers Fujitsu and Mitsubishi Electric.

What exactly are the purposes of Open COVID Pledge licenses? Open COVID License—Patent (OCL-P) 1.1 spells out that "pledgors" license (i.e., allow) others to use their IP-protected property "to speed the development and dissemination of the technologies needed to end the COVID-19 Pandemic and mitigate the effects of the disease."[87] Put simply, in layperson language: pledge participants allow others to make, use, copy, and modify their *stuff*, whatever that IP stuff may be. The pledgors individually and collectively can be viewed as a kind of cavalry-to-the-rescue inventors brigade, an Avengers team wielding patent tools rather than star-spangled shields, trick arrows, and Mjölnirs. Life-saving respirators—in short supply and in high demand—were just one kind of desperately needed patent-protected technology for treating and attempting to save the lives of lung-impacted patients afflicted by COVID-19. So, as an example of how pledgors stepped up to the plate, the NASA-funded Jet Propulsion Laboratory (JPL) at the California Institute of Technology provided testing data and designs for people to make 3D-printed COVID-19 respirators.[88]

Another notable aspect of these licenses is that they are time-limited: Open Covid Pledge licenses are *temporary* in duration, not meant to last forever (i.e., in perpetuity). For example, Morgan Stanley's license, which has since expired (and so is no longer viewable online), stated, "This license is effective as of December 1, 2019 and lasts until one year after the World Health Organization declares the COVID-19 Pandemic to have ended, but in any event not beyond January 1, 2023, unless otherwise extended by the Pledgor." Another feature is that parties using the Patent (OCL-P) 1.1. license must accept these licenses without any warranties and "as is." Additionally, in the Creative Commons section of this chapter, I noted that Creative Commons Zero (CC0) is a tool enabling copyright holders to give up all rights in their copyrights (i.e., no rights reserved). As an example of the Creative Commons Zero option, the Patent (OCL-P) 1.1. license stipulates that "all copyright and related rights in the Open COVID License are waived via CC0."[89]

In addition to mega companies like Apple and Microsoft that signed on to the Open COVID Pledge, some smaller-size companies also got on board. Anyone who experienced the COVID-19 pandemic will recall that face masks of various kinds were in fever-pitch demand. Coupled with unprecedented demand, masks were often difficult, if not impossible, to attain. So Tom Bihn, a small Seattle-based company that makes "backpacks, travel bags, and briefcases," took the Open COVID Pledge and agreed to "make our face mask designs available to all."[90] To that end, the company "posted the pattern + instructions of our V1 Reusable Cloth Face Mask—for *free use by all* under the Creative Commons Attribution-ShareAlike 4.0 International License."[91] Like health-care workers, people working in grocery stores and food and restaurant service were also designated as "essential workers," toiling—at high personal risks for infection from the coronavirus—on the front lines of service. Massachusetts' Bow Market, "to reduce undue pressure on our frontline workers by providing safe access to essential goods," provided a downloadable "Field Guide" as a "blueprint for safely connecting food, people, and public space."[92]

Difficult soul-searching questions arise from these kinds of good Samaritan efforts though: Are these Open COVID Pledges materially significant and praiseworthy? Or do efforts like this represent a kind of performative virtue signaling, what could be called *IP-washing*: a public relations effort to absolve an IP content holder of other morally questionable practices by performing an act that has negligible benefits, particularly to historically disadvantaged communities?

COVID-19 Vaccines 1.0 and mRNA Vaccine Technology Transfer Hub

With exponentially expanding COVID-19 deaths and illnesses throughout the world in 2020, President Donald J. Trump's administration spearheaded Operation Warp Speed to accelerate development of vaccines to combat the disease.[93] In the United States, three vaccines were developed by three separate US Big Pharma corporations—Pfizer, Moderna, and Johnson & Johnson (J&J). All of these vaccines are based on messenger RNA (mRNA) scientific research and technologies. Because of the emergent and exigent nature of the COVID-19 crisis, these vaccines were given emergency use authorization (EUA) without going through formal approval by the FDA.

Early in the pandemic, one of the chief ethical concerns was how to procure and deliver vaccine doses for poorer, economically developing nations. Enter COVAX: a collaborative initiative to get COVID-19 vaccine shots in arms around the world. Its tagline: "No one is safe, until everyone is safe."[94] COVAX's partners are the WHO, the United Nations Children's Fund (UNICEF), the World Bank, the Bill and Melinda Gates Foundation (Gates Foundation), the Coalition for Epidemic Preparedness Innovations (CEPI), and Gavi, the Vaccine Alliance. *Gavi* stands for Global Alliance for Vaccines and Immunisation. It was founded in 2000, building on the work of the Gates Foundation and other collaborators to "encourage manufacturers to lower vaccines prices for the poorest countries in return for long-term, high-volume and predictable demand from those countries."[95] In the words of Gavi today, the organization "helps vaccinate almost half the world's children against deadly and debilitating infectious diseases."[96]

Opinions vary, however, on the effectiveness of COVAX in providing COVID-19 vaccine doses to Global South nations. COVAX had some success in getting vaccines to poorer people but fell short in reaching goals and aspirations. A March 2022 *Washington Post* article's headline sums up the latter assessment: "Why Covax, the Best Hope for Vaccinating the World, Was Doomed to Fall Short": "The idea: to coax wealthy and poor countries to pool their money to place advance orders for vaccine doses. Participating countries would then share doses equitably to protect their most vulnerable people first. The initiative's backers badly misjudged the desperation and myopia of wealthier countries, which raced to manufacturers to snatch up doses for their own people."[97]

Patent holders on the eponymous Pfizer, Moderna, and J&J vaccines have also largely resisted calls for waiving their IP rights that would enable low- and middle-income nations to manufacture the vaccines within their own countries. A search for information on the topic of whether or not Big Tech companies should waive IP rights for their vaccines reveals myriad articles, studies, and opinion pieces spanning the continuum of thoughts on this ethical question. It is noteworthy that the inventor of the polio vaccine, the late Jonas Salk, relinquished all patent ownership rights in his vaccine. When asked who owned the patent, Salk famously responded, "The people, I would say. There is no patent. Could you patent the sun?"[98] The Salk vaccine is used worldwide to this day.

Flash-forward to mid-2023: many of the same Global South nations that were seeking Western vaccine doses in the throes of the pandemic have been channeling their disappointment and frustration with Western-controlled vaccine access into proactive, in-country efforts of their own. The intended solution to this enduring inequity: create R&D hubs within Global South countries that can jump-start the discovery and deployment of biomedical weapons for fighting future epidemics and diseases. In short, to replace the United States and other rich nations as the intermediaries and white knights for vaccines sent to poorer countries.

To that end, countries including Argentina, Bangladesh, Brazil, India, Indonesia, South Africa, and Vietnam have been pursuing creation of their own mRNA vaccines, as well as other kinds of vaccines and disease treatments. This WHO-supported initiative is called the mRNA Vaccine Technology Transfer Hub. A group called the Medicines Patent Pool (MPP) is a copartner with WHO. As a WHO FAQ about the initiative explains, research conducted in-country will be done at designated research "hubs."[99] "Spokes"—selected manufacturers radiating from a research hub—will produce vaccines discovered by a hub. IP that is developed through this research will be owned by the inventors. MPP will assist in negotiating licensing and use terms for any IP developed through the initiative. Perhaps a fitting bookend to the earlier-cited March 2022 *Washington Post* article is this July 2023 one: "As Pandemic Raged, Global South Lacked Vaccines. Never Again, Vow Researchers."[100]

Summary

Intellectual property—"products of the mind"—is comprised of four distinct types: copyrights, patents, trademarks, and trade secrets. Maverick open movements are challenging Western-based "legacy" IP systems. Movements for open access, Creative Commons licenses, open data, and open educational resources (OER), among others, strive to promote free, unfettered access to information and knowledge. Examples of such works include scholarly communication, data sets, and copyright-free textbooks. These movements serve as alternative approaches to the perceived imbalances and restrictions created by IP systems. WIPO is a UN agency that provides leadership on IP and related issues, such as counterfeit goods and traditional knowledge. The COVID-19 pandemic catalyzed a number of pharmaceutical companies to develop IP-protected vaccines. These companies agreed to make their vaccines available, under certain exceptions and for limited times, under an agreement called the Open COVID Pledge. Vaccines like these raise thorny ethical questions for marginalized, resource-scarce communities and Global South citizenries.

Questions for Reflection, Exploration, Action, Discussion (READ)

1. Theft of Native American IP has ripple effects beyond the artisans that create IP-protected indigenous art. Make a list of some of these impacts. Then, compose a script for a 15-second social media ad campaign that can educate people about these IP theft impacts on Native American artists and communities.
2. Imagine you are on a WIPO task force that is working to raise awareness of the IP interests of indigenous peoples around the globe. Identify three to five of the biggest ethics issues involving IP and indigenous peoples. Then, brainstorm some strategies for communicating those issues to the public.
3. Open access has given rise to predatory journals that are blurring the lines of high-quality peer-reviewed scholarly communication. Develop a list of ethics issues involving predatory journals. Then, briefly highlight what information professionals can do to mitigate the impacts of those ethics issues on authors and users.

Notes

1. Constitution Annotated. (n.d.). *Article I, Section 8, Clause 8*. Retrieved February 21, 2025, from https://constitution.congress.gov/browse/essay/artI-S8-C8-1/ALDE_00013060/
2. Cameron, J. (Director). (2009). *Avatar*. 20th Century Fox.
3. Kennedy, M. (2019, November 16). *How* Night of the Living Dead *accidentally became public domain*. Screen Rant. https://screenrant.com/night-living-dead-movie-public-domain-copyright-accident/
4. Legal Information Institute, Cornell Law School. (n.d.). *17 U.S. Code § 102—subject matter of copyright: In general*. Retrieved February 21, 2025, from https://www.law.cornell.edu/uscode/text/17/102
5. Piled Higher and Deeper (PHD Comics). (n.d.). *Open access explained!* Retrieved February 21, 2025, from https://www.youtube.com/watch?v=L5rVH1KGBCY
6. Herbert, F. (1965). *Dune*. Chilton Books.
7. OpenAccess.nl. (n.d.). *Pros and cons*. Retrieved February 21, 2025, from https://web.archive.org/web/20240930161401/https://www.openaccess.nl/en/what-is-open-access/pros-and-cons
8. Elmore, S. A., & Weston, E. H. (2020, April 22). Predatory journals: What they are and how to avoid them. *Toxicologic Pathology, 48*(4). https://journals.sagepub.com/doi/10.1177/0192623320920209

9. George, E. (2023, May 31). *What is Beall's List and why was it shut down?* Researcher.Life. https://researcher.life/blog/article/what-is-bealls-list-and-why-was-it-shut-down/
10. Think. Check. Submit. (n.d.). *About.* Retrieved February 21, 2025, from https://thinkchecksubmit.org/about/
11. Ibid.
12. Constitution Annotated. (n.d.). *Article I, Section 8, Clause 8.* Retrieved February 21, 2025, from https://constitution.congress.gov/browse/article-1/section-8/clause-8/
13. Creative Commons. (n.d.). *A history of Creative Commons.* Retrieved February 21, 2025, from https://creativecommons.org/timeline/
14. Creative Commons. (n.d.). *About CC licenses.* Retrieved February 21, 2025, from https://creativecommons.org/share-your-work/cclicenses/
15. Creative Commons. (n.d.). *CC0.* Retrieved February 21, 2025, from https://creativecommons.org/public-domain/cc0/
16. EBSCO. (2019, December 16). *The history and future of open educational resources in academic libraries*. https://www.ebsco.com/blogs/ebscopost/history-and-future-open-educational-resources-academic-libraries
17. OpenStax. (n.d.). *Preface*. Retrieved February 21, 2025, from https://openstax.org/books/college-success/pages/preface
18. Ibid.
19. UNESCO. (n.d.). *Open educational resources.* Retrieved February 21, 2025, from https://www.unesco.org/en/open-educational-resources
20. UNESCO. (n.d.). *UNESCO's mandate in OER.* Retrieved February 21, 2025, from https://www.unesco.org/en/open-educational-resources/mandate
21. Data.gov. (2013, April 4). *Open data: A history.* https://data.gov/blog/open-data-history/
22. SkedGo. (2018, February 1). *Improved trip planning for wheelchair users*. https://skedgo.com/improved-trip-planning-wheelchair-users/
23. White House. (n.d.). *Office of the Intellectual Property Enforcement Coordinator*. Retrieved February 21, 2025, from https://bidenwhitehouse.archives.gov/ipec/
24. White House. (2025, January 13). *IPEC Annual Intellectual Property Report to Congress.* https://bidenwhitehouse.archives.gov/briefing-room/statements-releases/2025/01/13/ipec-annual-intellectual-property-report-to-congress-4/
25. US Food and Drug Administration. (n.d.). *Counterfeit medicine.* Retrieved February 21, 2025, https://www.fda.gov/drugs/buying-using-medicine-safely/counterfeit-medicine
26. US Patent and Trademark Office. (2021, December 2). *E-commerce and China: Strategies for fighting online counterfeits, part 2*. Webinar.
27. American Bar Association. (2018). *Intellectual property law conference session.* Arlington, Virginia.
28. Sigalos, M. (2018, April 14). *The dark web and how to access it*. CNBC. https://www.cnbc.com/2018/04/13/the-dark-web-and-how-to-access-it.html
29. Lee, T. B. (2014, December 31). *The dark web: What it is, how it works, and why it's not going away*. Vox. https://www.vox.com/2014/12/31/7470965/dark-web-explained
30. Handa, S., & Lenz, J. (2021, October 14). *An introduction to the dark web*. JDSupra. https://www.jdsupra.com/legalnews/an-introduction-to-the-dark-web-3939932/
31. Gehl, R., & the Conversation US. (2018, October 31). Illuminating the dark web. *Scientific American*. https://www.scientificamerican.com/article/illuminating-the-dark-web/
32. Ibid.

33. Society for Professional Journalists. (n.d.). *SPJ leakseeker*. Retrieved February 21, 2025, from https://www.spj.org/leak-seeker/
34. Handa, S., & Lenz, J. (2021, October 14). *An introduction to the dark web*. JDSupra. https://www.jdsupra.com/legalnews/an-introduction-to-the-dark-web-3939932/
35. Chavez, P. (2023, December 7). Man paid hitman in Bitcoin to make planned killings look like botched robbery, feds say. *Sacramento Bee*. https://www.sacbee.com/news/nation-world/national/article282800258.html
36. Archives, US Department of Justice. (2021, October 26). *International law enforcement operation targeting opioid trafficers on the darknet results in 150 arrests worldwide and the seizure of weapons, drugs, and over $31 million*. https://www.justice.gov/archives/opa/pr/international-law-enforcement-operation-targeting-opioid-traffickers-darknet-results-150
37. Corder, M., & Whitehurst, L. (2023, May 2). *Watch: Garland announces results of operation targeting dark web opioid, fentanyl traffickers*. PBS News. https://www.pbs.org/newshour/politics/watch-live-garland-announces-results-of-operation-targeting-dark-web-opioid-fentanyl-traffickers
38. Bartram, M. (n.d.). *The history of ebooks from 1930's "readies" to today's GPO ebook services*. Government Book Talk, Government Printing Office. Retrieved February 21, 2025, from https://govbooktalk.gpo.gov/2014/03/10/the-history-of-ebooks-from-1930s-readies-to-todays-gpo-ebook-services/
39. Electronic Frontier Foundation. (n.d.). *Capitol v. Thomas*. Retrieved February 21, 2025, from https://www.eff.org/cases/capitol-v-thomas
40. Bohannon, J. (2016, April 28). The frustrated science student behind Sci-Hub. *Science*. https://www.science.org/content/article/frustrated-science-student-behind-sci-hub
41. Ibid.
42. Romero, C. (2023, September 7). *EFF award winner: Alexandra Asanovna Elbakyan*. EFF. https://www.eff.org/deeplinks/2023/09/eff-award-winner-alexandra-asanova-elbakyan
43. Ibid.
44. Z-Library. (n.d.). *Home page*. Retrieved February 21, 2025, from https://z-lib.io/s
45. Cramer, J. (2022, November 4). Pirate e-book site Z-Library vanishes—sending college students into a panic. *Fast Company*. https://www.fastcompany.com/90806657/z-library-ebook-piracy-shut-down-alternatives
46. US Attorney's Office. (2022, November 16). *Two Russian nationals charged with running massive e-book piracy website*. https://www.justice.gov/usao-edny/pr/two-russian-nationals-charged-running-massive-e-book-piracy-website
47. Ibid.
48. Javaid, M. (2022, November 17). The FBI closed the book on Z-library, and readers and authors clashed. *Washington Post*. https://www.washingtonpost.com/nation/2022/11/17/fbi-takeover-zlibrary-booktok-impacted/
49. Mandal, S. (2023, November 2). *Z-Library is back with a new domain name*. Good E-Reader. https://goodereader.com/blog/digital-library-news/z-library-is-back-with-a-new-domain-name
50. Brittain, B. (2023, September 14). *Textbook publishers sue "shadow library" Library Genesis over pirated books*. Reuters. https://www.reuters.com/legal/litigation/textbook-publishers-sue-shadow-library-library-genesis-over-pirated-books-2023-09-14/
51. Hart, M. (1992, August). *The history and philosophy of Project Gutenberg*. Project Gutenberg. https://www.gutenberg.org/about/background/history_and_philosophy.html
52. Ibid.
53. Riley, E. V. (2023, October 6). Native art is a $1.5 billion market plagued with problems. *Bloomberg*. https://www.bloomberg.com/news/features/2023-10-06/native-art-and-jewelry-is-plagued-by-fraud-and-exploitation

54. US Department of the Interior. (n.d.). *Tribes*. Retrieved February 21, 2025, from https://www.doi.gov/international/what-we-do/tribes
55. Ibid.
56. US Department of the Interior. (n.d.). *Tribal leaders directory.* Retrieved February 21, 2025, from https://www.bia.gov/service/tribal-leaders-directory
57. US Census Bureau. (2023, November 24). *Related statistics: Stats for stories.* https://www.census.gov/library/stories/2023/10/2020-census-dhc-a-aian-population.html
58. Indian Arts and Crafts Board, US Department of the Interior. (n.d.). *Who we are—IACB*. Retrieved February 21, 2025, from https://www.doi.gov/iacb/who-we-are
59. Indian Arts and Crafts Board, US Department of the Interior. (n.d.). *The Indian Arts and Crafts Act of 1990.* Retrieved February 21, 2025, from https://www.doi.gov/iacb/act
60. *Testimony of Meridith Stanton before the Senate Committee on Indian Affairs*. (2017, July 7). US Department of the Interior. https://www.doi.gov/ocl/indian-arts-crafts-0
61. US Attorney's Office. (2023, May 17). *Artist who falsely claimed Native American heritage sentenced for violations of the Indian Arts and Crafts Act.* https://www.justice.gov/usao-wdwa/pr/artist-who-falsely-claimed-native-american-heritage-sentenced-violations-indian-arts
62. US Patent and Trademark Office. (2023, November 30). *Roundtable discussion: Tribes, intellectual property, and consumer protection.* Alexandria, Virginia.
63. Alaska State Council on the Arts. (n.d.). *Silver Hand program.* Retrieved February 21, 2025, from https://arts.alaska.gov/silverhand
64. Ibid.
65. Indian Arts and Crafts Board, US Department of the Interior. (n.d.). *Special message from Secretary Debra Haaland.* Retrieved February 21, 2025, from https://web.archive.org/web/20231207002656/https://www.doi.gov/IACB
66. US Patent and Trademark Office. (2023, November 30). *Roundtable discussion: Tribes, intellectual property, and consumer protection. Author's notes.* Alexandria, Virginia.
67. World Intellectual Property Organization. (n.d.). *Berne Convention for the Protection of Literary and Artistic Works.* Retrieved February 21, 2025, from https://www.wipo.int/treaties/en/ip/berne/
68. World Intellectual Property Organization. (n.d.). *Traditional knowledge*. Retrieved February 21, 2025, from http://www.wipo.int/tk/en/tk/
69. Saez, C. (2016, November 28). *On eve of WIPO traditional knowledge negotiations, nations swap experiences*. Intellectual Property Watch. https://web.archive.org/web/20220128162920/https://www.ip-watch.org/2016/11/28/eve-wipo-traditional-knowledge-negotiation-nations-swap-national-experiences/
70. Doctorow, C. (2009, July 13). *Javanese batikers say thanks but no thanks to copyright*. Boing Boing. https://boingboing.net/2009/07/13/javanese-batikers-sa.html
71 World Intellectual Property Organization. (2024, May 24). *WIPO member states adopt historic new treaty on intellectual property, genetic resources and associated traditional knowledge.* https://www.wipo.int/pressroom/en/articles/2024/article_0007.html
72. World Intellectual Property Organization. (2024, December 5). *Malawi is first to ratify WIPO treaty on IP, genetic resources and associated traditional knowledge.* https://www.wipo.int/pressroom/en/articles/2024/article_0019.html
73. Cambridge Dictionary. (n.d.). *Cultural appropriation.* Retrieved February 21, 2025, from https://dictionary.cambridge.org/us/dictionary/english/cultural-appropriation
74. Woolf, N. (2016, November 18). *Urban Outfitters settles with Navajo Nation after illegally using tribe's name*. https://www.theguardian.com/us-news/2016/nov/18/urban-outfitters-navajo-nation-settlement

75. World Intellectual Property Organization. (2024). *The WIPO Intergovernmental Committee on Intellectual Property and Genetic Resources, Traditional Knowledge and Folklore. Background brief No. 2.* https://www.wipo.int/edocs/pubdocs/en/wipo-pub-rn2023-5-2-en-the-wipo-intergovernmental-committee-on-intellectual-property-and-genetic-resources-traditional-knowledge-and-folklore.pdf
76. Ibid.
77. *Association for Molecular Pathology v. Myriad Genetics*. (n.d.). Oyez. Retrieved February 21, 2025, from https://www.oyez.org/cases/2012/12-398
78. National Library of Medicine. (n.d.). *Can genes be patented?* MedlinePlus. Retrieved February 21, 2025, from https://medlineplus.gov/genetics/understanding/testing/genepatents/
79. Ibid.
80. European Patent Office. (n.d.). *Biotechnology patents at the EPO.* Retrieved February 21, 2025, from https://web.archive.org/web/20190405194520/https://www.epo.org/news-issues/issues/biotechnology-patents.html
81. Strum, L. (2023, June 13). *AMP v. Myriad: The fight to take back our genes*. ACLU. https://www.aclu.org/news/privacy-technology/the-fight-to-take-back-our-genes
82. ACLU. (2022, August 3). *ACLU opposes legislation to enable patents on human genes*. https://www.aclu.org/press-releases/aclu-opposes-legislation-enable-patents-human-genes
83. Fair, L. (2023, December 19). *If DNA test kits are on your holiday shopping list*. Federal Trade Commission. https://consumer.ftc.gov/consumer-alerts/2023/12/if-dna-test-kits-are-your-holiday-shopping-list
84. World Health Organization. (n.d.). *WHO COVID-19 dashboard*. Retrieved February 21, 2025, from https://data.who.int/dashboards/covid19/
85. *Open COVID Pledge.* (n.d.). Retrieved February 21, 2025, from https://opencovidpledge.org/
86. Hewlett Packard Enterprise. (2020, April 20). *HPE opens its patents to fight COVID-19*. https://www.hpe.com/us/en/newsroom/blog-post/2020/04/hpe-opens-its-patents-to-fight-covid-19.html
87. Open COVID Pledge. (2020, April 17). *OCL-P v.1.1.* https://opencovidpledge.org/v1-1-ocl-p/
88. *JPL designed 3D printable COVID-19 respirators.* (n.d.). Retrieved February 21, 2025, from https://github.com/nasa-jpl/COVID-19-respirators
89. Open COVID Pledge. (2020, April 17). *OCL-P v.1.1.* https://opencovidpledge.org/v1-1-ocl-p/
90. Tom Bihn. (2020, August 30). *Free pattern! V3 reusable cloth face mask*. https://www.tombihn.com/blogs/main/threelayermask
91. Ibid.
92. Safe Supply Field Guide. (2020, April 22). *Bow Market takes Open COVID Pledge*. https://www.safesupplyfieldguide.org/open-covid-pledge
93. US Government Accountability Office. (2021, February 11). *Operation Warp Speed: Accelerated COVID-19 vaccine development status and efforts to address manufacturing challenges.* https://www.gao.gov/products/gao-21-319
94. World Health Organization. (n.d.). *COVAX.* Retrieved February 21, 2025, from https://www.who.int/initiatives/act-accelerator/covax
95. Gavi. (n.d.). *About our alliance.* Retrieved February 21, 2025, from https://www.gavi.org/our-alliance/about
96. Ibid.
97. Taylor, A. (2022, March 22). Why Covax, the best hope for vaccinating the world, was doomed to fall short. *Washington Post*. https://www.washingtonpost.com/world/2022/03/22/covax-problems-coronavirus-vaccines-next-pandemic/

98. Intellectual Property Expert Group. (n.d.). *Jonas Salk, inventor of the polio vaccine, "Could you patent the sun?"* Retrieved February 21, 2025, from https://www.ipeg.com/jonas-salk-inventor-of-the-polio-vaccine-could-you-patent-the-sun/
99. World Health Organization. (n.d.). *FAQ—the mRNA vaccine technology transfer hub.* Retrieved February 21, 2025, from https://www.who.int/initiatives/the-mrna-vaccine-technology-transfer-hub/faq.
100. Maxmen, A. (2023, July 16). As pandemic raged, Global South lacked vaccines. Never again, researchers vow. *Washington Post*. https://www.washingtonpost.com/health/2023/07/16/vaccines-mrna-global-south/

Part 4

Technologies and Regulatory Oversight

Chapter 10

The Internet and Social Media

Introduction: Open Ranges and Fences on the Internet Frontier

Modern life in America is replete with legal contradictions, rooted in ethics and policy determinations. Consider the following:

- To drive an automotive vehicle, one has to have a legal license, proof of registration, and valid insurance, which everyone seemingly accepts; however, requirements on gun ownership, registration, and background checks are mired in vituperative disputes and a mishmash of state and federal laws.
- Anyone who goes for a medical appointment encounters the privacy and confidentiality requirements of the Health Insurance Portability and Accountability Act of 1996 (HIPAA), but internet users have virtually no rights over the data that websites collect on them and sell to third parties, outside of economic powerhouse California and a slowly expanding handful of states.
- To legally fly a drone, one must be at least 13 years old and be registered with the Federal Aviation Administration, but requirements for minors to surf the internet and use social media platforms are still few and far between.
- A librarian may face legal threats of imprisonment and civil fines if a minor reads a contested print book in a public or school library, while that same minor's access to sites that revel in hate speech and graphic violence are virtually unrestricted.

All of these examples include known and unknown risks and dangers that involve technologies. Difficult questions arise:

- Who makes the decisions about what risks and dangers are acceptable?
- What are the bases for how decisions about regulation and nonregulation of technologies are made, even when majorities of people are, in some cases, demanding that their elected officials implement safeguards and regulations?
- What ethics-based judgments tacitly or overtly undergird some technology-related issues and the priority that these issues are accorded, while other areas of life touched by technologies are ignored and unaddressed?
- Utilitarianism, in part, weighs the needs of the one versus the few or the many. Why, in some of the tech-related examples described above, do the needs of one person outweigh the needs of

the few or the many, while in other examples determinations based on the needs of a few people are deemed to outweigh the needs of the individual or a community? Who is afforded the power to make these determinations, and why? Why, for example, is it deemed permissible for one person, or even just a few, to influence decisions about, say, an entire school community's ability to access a book in a school library, while in most instances one person has no power to tangibly influence who can access a firearm?

Many factors are at play, certainly (e.g., monied interests; lobbying; executive, legislative, and judicial priorities), but each of these decisions again can be traced back to ethics judgments and valuation of what is and is not important. This chapter explores ethics issues involving a technological development that has done more to change the world in the last three decades than all other developments over the previous three centuries: the internet. The internet, in turn, enabled the rise of a new way for people to socialize and exchange information: social media. Though the term *social media* is widely used today, let's first define what it is and what it encompasses. *Cambridge Dictionary* defines social media as "websites and computer programs that allow people to communicate and share information on the internet using a computer or cell phone."[1]

Drilling into the internet age, we delve into issues stemming from the proliferation of social media and smartphones. We also look at cyberbullying and organizational efforts to diminish its incidence, raise awareness, and provide support. This chapter has links with a prior chapter

Figure 10.1 Bizarro comic. (© 2014 Bizarro Studios, distributed by King Features Syndicate, Inc. Used with permission)

on accuracy of information, misinformation, and disinformation, as social media is an influential means for the dissemination of information. Rapid adoption of the internet by users in the mid-1990s and into the twenty-first century, without education on "internet dos and don'ts," created deficits of what would eventually come to be called "digital citizenship." *Cambridge Dictionary* defines digital citizenship as "the state of being skilled in using the internet in order to communicate with others, buy and sell things, and take part in politics, and understanding how to do this in a safe and responsible way."[2] Unlike a 2014 *Bizarro* cartoon's depiction of the biblical figure Moses being given internet ethics "commandments" (see Figure 10.1), societies and internet communities have had to develop their own norms and guidelines.

Internet Frontier Meets Civilization

This chapter's introduction identified some examples in which ethics influences law and policy in ways that favor a regulatory stance and other instances that take a more hands-off, nonregulatory, free-market approach. Latter examples reflect the early ethos, and in many ways now myth, of the internet. In the 1990s and into the early years of the twenty-first century, the World Wide Web was frequently touted and even lionized as a space that could, should, and would be free, unfenced, and open. This view, or philosophy as it were, tapped into elements of the country's American West frontier mythos. At its core, this mythos homilizes that *anything* is possible with hard work, grit, and determination.

Little discussed regarding that frontier ethos, though, are some of the systemic impediments and competing interests of stakeholders and societies and the roles that the natural world plays. For example, the interests of sheep farmers and cattle ranchers sharing the same or adjacent lands were prone to clashing, with violent results. Sheep farmers did not want to have lands fenced, so their sheep could roam, while many cattle ranchers wanted fenced lands to keep their animals within protected boundaries. Water rights, also called riparian rights, were frequently contested by a wide range of stakeholders as well. The fluctuating availability of water was and remains a resource that is influenced by the unpredictable vagaries of Mother Nature and complex, interconnected systems of the natural world. The interests of indigenous peoples and settlers were also often at odds, with the latter benefiting from their larger numbers and the supportive structures of legal systems weighted to their advantage. The often romanticized discourse of the American West also reflects a kind of paradoxical duality: one of wide, open, free-range spaces (i.e., a kind of comforting vision as immortalized in Kansas's 1871 state song "Home on the Range") versus another of homesteader-settled plots of bounded, fenced-in land, where each landholder knows what's theirs and will fight to protect their hard-earned property from any who would encroach on it. Many American Western films have also reinforced these themes over the years.

In ways that are similar to the mythologizing of the "free and open" American West of the nineteenth century, the internet, in its early youth, was idealized and propagandized as that kind of new frontier, where anything and everything imaginable was seemingly possible through this new technology, where the "little guy" could compete with the "big guy" in those unfenced digital spaces, and where all information just wanted to be free. In more recent years, we have learned that much of that is not the case. Just as in the nineteenth-century frontier West, some want fences and some don't. Some tear down fences while others build them up, and then rebuild them when they are damaged or torn down. Some benefit from fences and some do not. In this digital age, in order to fundamentally understand the most substantive civil and constitutional rights challenges of our time—privacy, intellectual freedom, free speech, truth, the roles of new technologies in our societies—we must always ask and include in our analyses an ethics-anchored question: who *benefits* from the systems that do not have fences and who benefits from the systems that do have fences? Similarly, we must look at the *costs* that are created by having and not having fences.

Social Media: Like Driving With No Speed Limits

"Social media is like driving with no speed limits."[3] Those were the words of the US surgeon general, Dr. Vivek Murthy, in March 2024. Murthy has been one of the principal voices from the medical community sounding alarms about society's ever-growing reliance on smartphones. He has been especially concerned about the deleterious impacts of smartphones on the development and well-being of children and youth. Empirical studies continue to show escalating rates of depression, anxiety, suicidality, and other mental health issues in boys and girls, with rates for girls even higher than boys. One of the most troubling issues for Murthy is the assumptions that have ostensibly been made, or acquiesced to, that young peoples' social media and smartphone use is safe. As the *Guardian* reported in March 2024, Murthy served as surgeon general in the administrations of Barack Obama and, following the intervening Donald Trump administration, Joe Biden. In 2023, he disseminated a formal warning that social media constituted "a profound risk of harm" to children's mental health and wellness.[4] Murthy contends that there is insufficient evidence to conclude whether social media is safe enough for children to use.

Data released in March 2024 has concerning implications for the health of young people: "For the first time since the data was first collected in 2012, 15- to 24-year-olds in North America say they are less happy than older generations."[5] That data comes from the 2024 annual World Happiness Report conducted by researchers at the University of British Columbia and Simon Fraser University (Canada), the London School of Economics (UK), Columbia University (US), and the international joint-venture university Xi'an Jiaotong-Liverpool University (China and UK). In the report's summary section, the researchers discuss some of their findings on happiness, broken down by age groups:

> In the West, the received wisdom was that the young are the happiest and that happiness thereafter declines until middle age, followed by substantial recovery. But since 2006–10, as we shall see, happiness among the young (aged 15–24) has fallen sharply in North America—to a point where the young are less happy than the old. Youth happiness has also fallen (but less sharply) in Western Europe.[6]

In addition to worries for North American youth, Murthy is also concerned about the well-being of youths around the globe, including young people in European and Asian countries that are being studied. The growing substitution of hours and hours each day spent on smartphones and social media instead of in face-to-face social interaction between people is exceptionally worrisome to him.[7] Murthy is an advocate for governmental regulation of social media. He has urged technology corporations to provide data backing up their claims that their social media platforms are safe. Murthy has also likened the conduct of technology companies to twentieth-century automobile manufacturers that made vehicles without safety features like seatbelts and airbags. Only when those companies were required by law to provide these now common public safety measures did they do so. Murthy employed driving as a metaphor for the state of social media use today:

> "What's happening in social media is the equivalent of having children in cars that have no safety features and driving on roads with no speed limits," he said. "No traffic lights and no rules whatsoever. And we're telling them: 'you know what, do your best—figure out how to manage it.' It is insane if you think about it."[8]

Like Murthy, other experts share deep concerns about unregulated social media use and unrestricted access to the addictive qualities of smartphones. Social psychologist Jonathan Haidt is a professor of ethical leadership at New York University's Stern School of Business. His 2024 book is titled *The Anxious Generation: How the Great Rewiring of Childhood Is Causing an Epidemic of Mental Illness*. As one book review noted, Haidt "believes this mental health crisis has been driven by the mass adoption of smartphones, along with the advent of social media and addictive online gaming."[9] A fascinating distinction Haidt draws is the tendency for parents today to be overprotective with regard to less unsupervised play, compared to prior generations. At the same time, he observes that those same parents, in the words of the book reviewer, "have allowed the young too much freedom to roam the internet, where they are at risk of being bullied and harassed or encountering harmful content, from graphic violence to sites that glorify suicide and self-harm."[10] To remedy this urgent health crisis, Haidt recommends some strong measures:

- banning smartphone use in schools
- requiring tech companies to exercise a duty of care for young people
- raising the age of adulthood vis-à-vis the internet to 16
- necessitating that tech companies implement age verifications

Social Media Issues

Given the state of evidence-based concerns expressed by some researchers, academics, and health-care professionals, what are some of the other issues that are arising from social media use, and what can be done to address these issues, particularly from legal, ethics, and policy standpoints? In this section, we identify some of the principal issues and examine various strategies and approaches for leveraging positive elements and mitigating harmful impacts.

Social-media-related issues present complex and novel questions and challenges arising from the nexus between digital technologies and changes in how people communicate and interact with one another as they use that digital tech. These issues are interconnected with many thorny ethics-related issues with which societies are grappling, and which we are critically examining in this book. As with all technologies, social media has positive and negative aspects and uses.

Probably the biggest advantage of social media is that it makes it possible for people to communicate more easily and in different ways. The headline of a February 2011 *New York Times* article about internet use by enlisted military members stationed abroad to stay in touch with their families effectively illuminates both the promise and peril of this technology: "Staying in Touch With Home, for Better or Worse." Moreover, the first two sentences of the article deftly convey the life-changing impacts of the internet and social media on the everyday lives of the human beings who serve their country: "Forget the drones, laser-guided bombs and eye-popping satellite imagery. For the average soldier, the most significant change to modern warfare might just boil down to instant chatting."[11] Social media, the piece describes, makes it possible for military members to stay in more frequent touch with their families back home, to be a part of "births and birthdays in real time." Long gone are the days of service members having to wait weeks or months for their letters to be received back home; indeed, a 1917 letter from my great uncle serving in World War I in France took months to be received by his mother in Pennsylvania, per a surviving newspaper article describing that letter.

However, the *New York Times* article notes that this can also be a downside, as enlistees can get more easily drawn into everyday issues involving broken appliances, stressed relationships, and bills that need to be paid. For military leaders at the top of the food chain, the article identifies competing interests: the desire to lift the morale of troops by enabling "instant communication," while also balancing cybersecurity and safety concerns should enemy combatants be able to breach internet communications and home in on signals that can reveal secret troop locations and put them in danger.

Social Media Statistics

Social media usage data can be very helpful to get a richer understanding of social media's prevalence in societies. It is also beneficial to parse the data and better understand social media use by various demographic groups. The Pew Research Center (PRC) has studied tech adoption, such as the internet and social media use, for years. Hence, PRC is a valuable source for accessing research data on social media. For example, in June 2019, Pew identified that about seven in 10 Americans were using social media "to connect with one another, engage with news content, share information and entertain themselves."[12] That same 2019 report also provided rather startling comparative data points showing the steady rise in social media penetration among the American public: "When Pew Research Center began tracking social media adoption in 2005, just 5% of American adults used at least one of these platforms. By 2011 that share had risen to half of all Americans, and today [in 2019] 72% of the public uses some type of social media."[13]

Additional PRC research involving social media offers insights, such as an April 2023 report, "Teens, Social Media, and Technology 2023." That report points out the changes that young people are confronting today, as indicated earlier in this chapter by social media commentators like Murthy and Haidt. "Today's teens are navigating a digital landscape unlike the one experienced by their predecessors, particularly when it comes to the pervasive presence of social media. In 2022, Pew Research Center fielded an in-depth survey asking American teens—and their parents—about their experiences with and views toward social media."[14] These are some key findings from the study:

- "YouTube is the platform most commonly used by teens, with 95% of those ages 13 to 17 saying they have ever used it, according to a Center survey conducted April 14–May 4, 2022, that asked about 10 online platforms."
- "About three-quarters of teens (77%) say they use YouTube daily, while a smaller majority of teens (58%) say the same about TikTok. About half of teens use Instagram (50%) or Snapchat (51%) at least once a day, while 19% report daily use of Facebook."
- "About a third of teens (36%) say they spend too much time on social media, while 55% say they spend about the right amount of time there and just 8% say they spend too little time."
- "Teens are relatively divided over whether it would be hard or easy for them to give up social media. Some 54% say it would be very or somewhat hard, while 46% say it would be very or somewhat easy."
- "Some 32% say social media has had a mostly negative effect on people their age, while 9% say this about social media's effect on themselves."
- "Half of teens think criminal charges or permanent bans for users who bully or harass others on social media would help a lot to reduce harassment and bullying on these platforms."
- "Six-in-ten teens say they think they have little (40%) or no control (20%) over the personal information that social media companies collect about them. Another 26% aren't sure how much control they have. Just 14% of teens think they have a lot of control."[15]

Cyberbullying

In 2010, I was asked to relaunch an information ethics course for the upcoming 2011 spring term. That course had last been offered several years earlier at Pitt's School of Information Sciences (SIS) by the former dean of the school, Dr. Toni Carbo, one of the principal contributors to this field of inquiry, as mentioned earlier in the acknowledgments. I was given freedom to determine what should be in the relaunched course. At that time, cyberbullying was emerging as a tech issue that was receiving more mainstream press attention and gaining greater public awareness. The sad fact was that this was due to a number of tragic losses of life by young victims of cyberbullying. As a result of those tragic deaths, and because cyberbullying represented a kind of novel *bad act* perpetrated through the newish technology of the internet, I believed that cyberbullying needed to be a part of the revamped course from its first offering in early 2011.

Unfortunately, though efforts have been implemented to diminish cyberbullying and online harassment in the ensuing years, these behaviors continue to plague and harm individuals and societies. In this section of the chapter, we define cyberbullying, learn how it is being studied, examine some examples, and identify some initiatives to address it.

Defining Cyberbullying and Related Terms

As with many scientific and research fields, securing agreement on how to define even the most fundamental terms that are the very basis of those fields can be fraught with disagreement and unresolved debate. How to define, say, *information* or *data* are well-known examples. Defining *cyberbullying* and other terms with links to cyberbullying has proven similarly challenging over the past couple of decades. One reason is because cyberbullying intersects with constitutional rights like free speech. A concern voiced by some is that if the definition of what constitutes cyberbullying is made too broad, that definition may chill or even criminalize free expression. Others argue, too, however, that if the definition is too narrow or rife with caveats, it may be meaningless and unenforceable.

Additionally, it is often said that it sometimes takes time for the law, as well as ethics and policy, to catch up with science and technology developments that may outpace rule-based institutions. This is certainly the case with rapid-fire developments in AI-related advances that are compelling nations to update their legal systems to address these new emerging technologies, as we will explore later in this book. In a like manner, societies have had to grapple with how to categorize digital bad acts via terms such as *cyberbullying*, *cyberharassment*, *cyberstalking*, and more. Adding to this nomenclature challenge as well are other kinds of undesirable tech-related behaviors that relate to cyberharassment and cyberbullying but which in some ways are also distinguishable. Examples include *doxing*, *swatting*, and *revenge porn*, along with its legally preferable term *nonconsensual pornography*. We define and explore these terms later in this chapter.

The Cyberbullying Research Center, which we learn more about later in this chapter, defines cyberbullying as "willful and repeated harm inflicted through the use of computers, cell phones, and other electronic devices."[16] It is also known as "cyber bullying," "electronic bullying," "e-bullying," "sms bullying," "mobile bullying," "online bullying," "digital bullying," and "internet bullying."[17]

Cyberbullying Data

A PRC study, "The State of Online Harassment," conducted in September 2020 and reported in January 2021, opened an illuminating window on how Americans felt about online harassment and technology industry responsibility: "79% of Americans think social media companies are doing an only fair to poor job when it comes to addressing online harassment or bullying."[18] To reinforce that point, nearly eight of every 10 Americans viewed this kind of activity as something that tech companies were doing a mediocre or bad job in handling. For any industry, that degree of user dissatisfaction with performance would be remarkable, and one could reasonably think that the leadership of that industry would be impelled to make dramatic changes. Regrettaby, few substantive, meaningful, or lasting changes have been made by the major tech companies.

Indeed, this was apparent at a January 31, 2024, US Senate hearing on tech platforms and child safety, where the heads of some of the world's biggest tech companies—Meta/Facebook, TikTok, X/Twitter—appeared for questioning (more aptly *grilling*) by a bipartisan group of senators. One of the viral moments at that hearing occurred when the anguished parents of children who had died because of the relentless online bullying they had experienced on social media angrily confronted Meta/Facebook founder and CEO Mark Zuckerberg. His apology to those families fell flat, as this AP news story "Mark Zuckerberg's Long Apology Tour: A Brief History" recounted:

> "I'm sorry for everything you've been through," the Meta CEO said Wednesday. "No one should go through what you and your families have suffered." Then he returned to corporate mode, noting Meta's continued investments in "industrywide" efforts to protect children.
>
> Zuckerberg has accumulated a long history of public apologies, often issued in the wake of crisis or when Facebook users rose up against unannounced—and frequently unappreciated—changes in its service.[19]

Absent from Zuckerberg's statement was any acknowledgment of personal responsibility for his company's pivotal role in the deaths of the children.

Returning to PRC's 2021 report "The State of Online Harassment," what to even call these types of behaviors can be debatable. (This is discussed in more detail later, too, regarding other digital bad acts, like doxing, swatting, and nonconsensual pornography.) For its 2021 report, PRC set out a "Defining Online Harassment" statement. The statement explains that PRC employed "six distinct behaviors" as metrics for assessing online harassment:

- offensive name-calling
- purposeful embarrassment
- stalking
- physical threats
- harassment over a sustained period of time
- sexual harassment[20]

The statement also delineates persons who have experienced these behaviors from those who have not and explains how they derived gradations of online harassment:

> Respondents who indicate they have personally experienced any of these behaviors online are considered targets of online harassment in this report. Further, this report distinguishes between "more severe" and "less severe" forms of online harassment. Those who have only experienced name-calling or efforts to embarrass them are categorized in the "less severe" group, while those who have experienced any stalking, physical threats, sustained harassment or sexual harassment are categorized in the "more severe" group.[21]

In the report, the researchers observed a number of noteworthy findings. They found that the "biggest surprise" is that "the nature of the harassment has changed over time."[22] Specifically, they found "that the level of harassment is increasing in two key ways: (1) People are more likely to have encountered multiple forms of harassment online, and (2) severe encounters have become more common." One of the additional striking findings was that online harassment had "intensified" since PRC's study in 2017. In 2014, the number of people who were subjected to online harassment-related issues was 15 percent, in 2017 the number rose to 18 percent, and in the 2021 report that number had risen again to 25 percent, or one out of every four. Social media continued to be the overwhelming source of and space for online harassment, per the report, in keeping with prior PRC surveys: "As was true in previous Center surveys about online harassment, social media continue to be the most commonly cited online venues where harassment takes place. When asked where their most recent experience with online harassment occurred, 75% of targets of this type of abuse say it happened on social media."[23]

Cyberbullying Advocacy and Responses

In the wake of increases in cyberbullying and the suicide deaths of cyberbullying victims, advocacy and support organizations have been created to combat bullying and cyberbullying through outreach, education, and legislative lobbying. Many of these groups specifically address marginalized populations, such as LGBTQ+ people and communities of color. These initiatives and foundations include the following:

- StopBullying.gov
- Trevor Project
- Tyler Clementi Foundation
- It Gets Better
- Born This Way Foundation

Entrepreneurs and technology designers have also stepped into the cyberbullying response space by creating apps. Though these apps can alert parents and guardians to bullying and cyberbullying-related situations involving children, these apps also raise some privacy, cyber-tracking, and data ethics concerns. Bark is an app that enables parents and guardians to monitor their childrens' online and smartphone activities. The company's website advertises a variety of activities the app can perform:

- manage screen time
- get alerts 24/7
- monitor content
- track location
- block websites[24]

In addition to bullying-related activities that Bark purports to address, it can also raise concerns about self-harm that can enable adults to aid children. The app provides some data points about usage and incidents:

- 7.4 million children covered
- 3.6 million severe self-harm situations detected
- 5.7 million severe bullying situations detected[25]

Legal Actions

Sometimes a necessary response to cyberbullying is in the courts. Amanda Jones, discussed in the prior intellectual freedom and censorship chapter, is the Louisiana librarian and author of *That Librarian* who has been harassed and cyberbullied relentlessly for simply sharing her opposition to censorship in libraries. As mentioned previously, Jones has sued her harassers for defamation. As part of the June 2024 *Guardian* headline about Jones's suit proclaims: "I Decided to Fight Back." The article notes that Jones is "one of the first in the US to file a lawsuit for defamation against her detractors."[26]

Naming and Classifying

Nomenclature (i.e., what to call things) and taxonomy (i.e., how to classify things) are information-organizing areas that sit squarely within the bailiwicks and work activities of librarians, archivists, and other information professionals. Moreover, identifying, naming, and classifying are activities that all sciences and scientists perform. These practices add to existing foundations of knowledge and development of new avenues for scientific inquiry and research. Disciplines and scientists can then build on these foundations to advance our collective understandings—and questions yet to be answered—about the world and universe.

One of the challenges for any new legal area, such as cybercrimes, is how to identify, name, and classify various acts. Cyberbullying has presented just such a challenge. It has also generated questions that don't always have clear answers:

- What exactly is "cyberbullying" and how should it be defined?
- What are the parameters of cyberbullying?
- When does free speech activity (e.g., criticism, commentary, expression) cross the line and become cyberbullying? Who gets to make such a determination, which can arguably be subjective? Nations vary, for example, as to what constitutes hate speech and what is protected free speech.
- How does cyberbullying differ from hate speech?
- How is cyberbullying different, or not different, from "cyberharassment"?
- Where does a term like *cyberstalking* fit into conceptions of cyberbullying and cyberharassment?
- How do other kinds of internet-facilitated problematic or harmful acts, such as revenge porn and doxing, intersect with cyberbullying? More will be discussed about this below.

At the time of this writing, no federal cyberbullying or bullying law exists, despite various legislative efforts to enact such a law at the congressional level. Nevertheless, some cyberbullying-related actions that occur in schools can trigger the application of federal law. For example, StopBullying.gov explains that "in some cases, when bullying is based on race or ethnicity, color, national origin, sex, disability, or religion, bullying overlaps with harassment and schools are legally obligated to address it."[27]

Technologies like the internet enable new types of activities that legal, ethics, and policy systems and practitioners (e.g., lawmakers, lawyers, judges, etc.) need to first assess and then create frameworks and mechanisms for addressing, criminalizing as warranted, adjudicating, and enforcing. As an example, when cyberbullying first manifested and a number of cyberbullying victims tragically took their own lives, legal systems struggled to decide which existing laws might encompass the actions of alleged cyberbullying perpetrators.

In most jurisdictions, as incidents of cyberbullying began to gain broader awareness throughout the United States and world, state laws didn't exist that specifically applied to cyberbullying. In some cases, prosecutors chose to try to "shoehorn" alleged cyberbullying crimes into already existing laws. Such laws may have a tenuous relationship to cyberbullying, the internet, and the social media platforms that enabled alleged transgressors to avoid liability for their actions. Hence, in the past decade and a half, US state legislatures have needed to evaluate their laws and, in most states, amend existing laws or enact new laws that deal with cyberbullying-related activities. The Cyberbullying Research Center has an informative and regularly updated webpage, "Bullying Laws Across America." This information page includes whether the state law provides for criminal sanctions. It also indicates whether schools have disciplinary sanctions or policies that address cyberbullying. Additionally, information is provided regarding cyberbullying that occurs off of the school grounds/campus: "Federal case law allows schools to discipline students for off-campus behavior that results in a substantial disruption of the learning environment at school."[28] Some states have opted to incorporate that federal law into their state statute.

Holistically oriented enforcement systems also build in components to address the needs of victims, as well as preventative and rehabilitative facets. This chapter has previously talked about some of the educational and preventive efforts that nonprofits and government entities are providing. Research shows, too, that victims of cyberbullying can sometimes become perpetrators, or the reverse. Information centers have roles they can assume in educating people about cyberbullying-related acts and helping to lower incidence rates of these harmful activities. This is even more persuasive when we consider that most libraries and other information centers today offer digital technologies in their physical sites or via devices that can be checked out and which may potentially be used in cyberbullying incidents.

Other Kinds of Internet- and Social-Media-Enabled "Bad Acts"

As mentioned previously, other types of internet- and social-media-enabled injurious and detrimental actions can intertwine with cyberbullying. Examples include the following:

Catfishing. This term "refers to the practice of setting up a fictitious online persona, most often for the purpose of luring another into a fraudulent romantic relationship."[29] The Cyberbullying Research Center (CRC) views catfishing "as a form of cyberbullying" and provides examples of some high-profile incidents, like 2012's catfishing against Notre Dame University football athlete Manti Te'o.[30]

Doxing. CRC notes that this term (also spelled *doxxing*) comes from the phrase "dropping documents." It "typically occurs when someone collects another's private personal information, such as a home address, contact information or social security information, and subsequently broadcasts or 'outs' that information to the public without permission."[31]

Swatting. The US Department of Education explains that swatting "is a false report of an ongoing emergency or threat of violence intended to prompt an immediate tactical law enforcement response."[32] Two of the most common kinds of swatting are false bomb threats and active shooter claims. Swatting has resulted in losses of life. In 2017, a video gamer in California swatted an innocent man, with no relationship to the gamer, in Wichita, Kansas, who was tragically killed when local police went to his home. The gamer, 26-year-old Tyler Barriss, was sentenced to 20 years in federal prison. US Attorney Stephen McAllister emphasized that "swatting is no prank."[33] He further observed that "sending police and emergency responders to anyone's home based on utterly false information as some kind of joke shows an incredible disregard for the safety of other people." Swatting can be perpetrated on both public and private figures.

Nonconsensual Pornography. This term is more commonly known as "revenge porn." The National Association of Attorneys Generals defines nonconsensual pornography as "a sexually graphic image of video of an individual distributed without the consent of the person depicted in the media."[34] Often, this practice occurs between partners who were previously intimate. At present, no federal nonconsensual porn law exists. However, as of 2023, nearly all US states have enacted laws against this invasive practice.

Dr. Mary Anne Franks, a professor at George Washington University Law School, has studied, published extensively, and spoken publicly on this subject. I first became acquainted with her work via a panel presentation at an ABA intellectual property law conference because of the intersection of nonconsensual pornography with intellectual property and civil rights interests. Her August 2024 article "The Criminalization of Non-consensual Pornography in the United States" highlights successes and challenges that remain in combating this transgression against both individual privacy and human dignity.[35] On the plus side of the ledger, Dr. Franks points out that forty-eight states now criminalize nonconsensual porn, compared with just three states that did so a decade ago. However, on the debit side, despite efforts to pass a federal law, no such law has yet been achieved. Additionally, "the definition and classification of nonconsensual pornography varies by jurisdiction."[36] This "legal patchwork" makes it more difficult for victims to receive justice. Yet Dr. Franks observes that "courageous victims and the civil society advocacy organizations that support them" have achieved hard-won gains.[37] One of these is the creation of the Cyber Civil Rights Initiative in 2013 by Dr. Holly Jacobs. As with other cybercrimes discussed in this section, nonconsensual pornography can sometimes overlap with other malicious cyber acts. One example is "sextortion," detailed next.

Sextortion. An FBI special agent, in an educational video about sextortion, describes this unlawful act: "Sextortion is a serious crime that occurs when someone threatens to distribute your private and sensitive material if you don't provide them images of a sexual nature, sexual favors, or money."[38] The agent further states that "the perpetrator may also threaten to harm your friends or relatives by using information they have obtained from your electronic devices unless you comply with their demands."

Internet- and Social-Media-Related Moderation and Censorship

Technology, the internet, and social media have exponentially multiplied the amounts of data and information, as well as the means by which the production, communication, and dissemination of data and information occurs. Though there are clearly many benefits of digital technology, the speed, volume, and ease by which data and information are produced and disseminated through the use of electronic technologies today makes the management of intellectual freedom and censorship issues from the analog area seem quaint and halcyon by comparison. The proliferation of digital technology has propagated many more contexts in which issues of intellectual freedom and censorship are implicated.

Examples

The following are summaries involving censorship within social media and internet contexts since approximately 2011. Innumerable examples exist in this intersectional space. These instances represent a select sampling of some noteworthy instances, all of which implicate other ethics issues examined in this book, such as access, freedom of speech, intellectual freedom, privacy, surveillance, accuracy of information, transparency, and accountability.

The Great Firewall of China

Over the past decade, China's government under President Xi Jinping has implemented the world's most formidable censorship system: "The Great Firewall." As the *Washington Post* reported in 2016, "First there was the Berlin Wall. Now there is the Great Firewall of China, not a physical barrier preventing people from leaving, but a virtual one, preventing information harmful to the Communist Party from entering the country."[39] Western social media sites like Twitter and Facebook, search engines such as Google, news networks like CNN, and newspapers like the *New York Times* are frequently blocked within China. In fact, as an example, China was so spooked by the fear that the 2011 Arab Spring protests in Egypt could spread to China, it blocked searches of the word *Egypt*.[40]

The *Atlantic* describes some of the additional impacts of the Great Firewall in a 2016 report:

> Virtual private networks and other work-arounds, tolerated a decade ago—the academic who invented China's 'Great Firewall' system of censorship even bragged about the six VPNs he used to keep up on foreign developments—are now under governmental assault. When you find a network that works, you dare not mention its name on social media or on a website that could alert the government to its existence. "It's an endless cat-and-mouse," the founder of a California-based VPN company, which I'm deliberately not identifying, recently told me. "We figure out a new route or patch, and then they notice that people are using us and they figure out how to block it. Eventually they wear most users down."[41]

The Great Firewall is one example of an extraterritorial technology that has created thorny transnational ethical dilemmas for US technology companies. This was described in a 2017 *Slate* article about pressure applied to Apple and Amazon by China's vast state-mandated censorship apparatus.[42] The next sections explore some examples of censorship involving US tech companies and other nations.

Microsoft's Bing Search Engine

In 2014, Microsoft's Bing search engine engendered considerable controversy when it was discovered that Bing was censoring search terms used by Chinese-language users in the United States. Chinese-language searches for "Dalai Lama," "June 4" (the Chinese means of denoting the 1989 Tianenmen Square protests), and "Falun Gong" rendered very different results than their English-language counterpart terms. The Bing-censored Chinese-language terms in the United States mirrored similar kinds of censored searching in China. Microsoft claimed that "a system fault" had resulted in the elimination of some of the search results conducted by users in countries outside of China.[43] Microsoft has been criticized previously for censoring China's version of Skype. Tech companies have often been caught between Scylla and Charybdis when it comes to balancing ethical standards and the pragmatic realities of doing business in China: "This censorship often means foreign Internet companies must tread a careful path in China to exploit business opportunities without compromising a carefully nurtured image as champions of open societies and free speech."[44]

Google "Blacklisting" Search Rankings and Sources

Another "dirty little secret" revealed several years ago is the internet search engine practice of "blacklisting" websites and news. Facebook has admitted censoring conservative news sites, for example. Google has also acknowledged using human reviewers and algorithms to determine sites to include and exclude. A 2016 *U.S. News* article stated:

> When Google's employees or algorithms decide to block our access to information about a news item, political candidate or business, opinions and votes can shift, reputations can be ruined and businesses can crash and burn. Because online censorship is entirely unregulated at the moment, victims have little or no recourse when they have been harmed. Eventually, authorities will almost certainly have to step in, just as they did when credit bureaus were regulated in 1970. The alternative would be to allow a large corporation to wield an especially destructive kind of power that should be exercised with great restraint and should belong only to the public: the power to shame or exclude.[45]

Flickr "Community Rules"

Other US tech companies confront thorny censorship-related ethical decisions. In 2011, the photo-sharing social media site Flickr ignited controversy when it removed headshot photos that a number of Egyptians had found on police headquarters websites and uploaded to the website. The headshots were of Egyptian police who were alleged to have perpetrated abuse on Arab Spring Egyptian protesters. The identities of the police were unknown. The hope was that they could be identified through crowdsourcing via the Flickr photographs. Flickr removed the photos, though, claiming that the persons posting them did not own the rights to the photos. The company claimed that this violated the terms of use and community rules when someone signs up for a Flickr account. On the one hand, the activists found a novel use for Flickr. However, from Flickr's standpoint, this kind of use was never envisioned when Flickr rolled out its platform. Flickr isn't the only tech company to face these kinds of ethical dilemmas, as the *New York Times* explains:

> Built as a platform for amateur and professional photographers to share their work, Flickr is among the social media networks, like Facebook, Twitter and YouTube, that are increasingly being used by activists and pro-democracy forces especially in the Middle East and North Africa.
>
> That new role for social media has put these companies in a difficult position: how to accommodate the growing use for political purposes while appearing neutral and maintaining the practices and policies that made these services popular in the first place.[46]

For users of these sites, the actions of technology companies can seem arbitrary and insensitive. As one Egyptian protester expressed after the police headshots he had posted were taken down by Flickr, "I thought it was a good platform," he said of Flickr. "Now, for me, it is an unethical platform, and I would not recommend it to anyone."[47] This real-life ethical scenario is the kind of exercise I've used in my information ethics course as a capstone discussion, precisely because it presents tough questions without easy answers or solutions. Questions like,

- What are the information ethics issues raised by this story?
- Who are the stakeholders?
- Is Flickr "an unethical platform"? If so, why?
- What policies would *you* implement to address the issues raised?
- What might be some of the impacts of *your* policies?

Twitter (Renamed X in July 2023) Blocking Tweets

In 2023, businessman Elon Musk, well-known for his car company Tesla and space corporation SpaceX, bought Twitter. Subsequently, Musk rebranded Twitter as X. This section of the chapter looks at some practices of the company that predate Musk's acquisition.

Earlier, I alluded to the difficulty in deciding which laws to apply when dealing with issues involving more than one state or country. Bodies of case law exist that wrestle with these kinds of questions of "conflicts of law" and "choice of forum." Some of the key questions for consideration are these:

- Whose laws should be paramount?
- Which jurisdiction should be deemed to have greater weight to adjudicate a dispute?

Companies like Twitter/X have increasingly faced these kinds of issues in terms of social media use. Though Twitter/X is a US company, its users are located throughout the world. As such, other countries assert that Twitter/X is subject to the rules, laws, and cultural mores of their respective nations. As the *New York Times* notes, "Twitter, like other Internet companies, has always had to remove content that is illegal in one country or another, whether it is a copyright violation, child pornography or something else."[48]

In 2012 though, Twitter/X catalyzed controversy:

> Twitter announced that upon request, it would block certain messages in countries where they were deemed illegal. The move immediately prompted outcry, argument and even calls for a boycott from some users.
>
> Twitter in turn sought to explain that this was the best way to comply with the laws of different countries. And the whole episode, swiftly amplified worldwide through Twitter itself, offered a telling glimpse into what happens when a scrappy Internet start-up tries to become a multinational business.[49]

As with the previous Flickr dilemma, Twitter's/X's technology has generated and continues to create challenging ethical issues and novel uses since its founding in 2006. In a 2008 CNN article, Twitter cofounder Biz Stone and others foresaw "wide-reaching effects" of the technology.[50] For example, people used Twitter to communicate when earthquakes and other natural disasters happened. It also was used for activism and journalism purposes.

Like all existing and emerging technologies, Twitter/X has its upshots and downsides. For users, Twitter/X is a free speech tool, a conduit for communication, and much more. Sometimes it has also proven to be a lifeline for users who find themselves in life-and-death situations in conflict-riven places. For US graduate student James Karl Buck, Twitter was a literal SOS beacon for help while reporting in Egypt in 2008:

> Buck, a graduate student from the University of California-Berkeley, was in Mahalla, Egypt, covering an anti-government protest when he and his translator, Mohammed Maree, were arrested April 10. On his way to the police station, Buck took out his cell phone and sent a message to his friends and contacts using the micro-blogging site Twitter. The message only had one word. "Arrested."
>
> Within seconds, colleagues in the United States and his blogger-friends in Egypt—the same ones who had taught him the tool only a week earlier—were alerted that he was being held.[51]

Ethics of Content Moderation and Moderators

Post-2016 US presidential election revelations of unprecedented disinformation and "fake news" disseminated by Russian-sanctioned troll farms and bots catalyzed calls in some political quarters for remedial action by tech companies. As just one example, whistleblower disclosures emanating from the Facebook/Cambridge Analytica data harvesting scandal and Facebook CEO Mark Zuckerberg's subsequent April 2018 testimony before the US Congress compounded

Figure 10.2 Non Sequitur comic. (© 2018. Used with permission from Andrews McMeel Syndication)

pressure on tech companies to better manage user-generated content and promote safer digital hygiene practices. For more than a decade, Big Tech companies have been lobbied by some stakeholders (e.g., watchdog groups, politicians, parents, and other users) to moderate "hate speech" and better address cyberbullying. Newer challenges have also arisen in recent years, like AI-generated deep-fake images.

Moderation of content that is uploaded to the internet and social media is one strategy for addressing these concerns. What then is content moderation? Consider the untold trillions of pieces of content that are uploaded every day: a certain percentage includes inappropriate, dangerous, false, or illegal material. Someone (or some things, as we discuss in a bit) must make decisions about that content. UK-based nonprofit the Bureau of Investigative Journalism (TBIJ), founded in 2010, explains what content moderation is and, even more importantly from a humane-consciousness perspective, whom the people are who carry out this unrelenting work:

> Content moderators are paid by tech companies to sift through the text, images and videos posted on the platforms they run, such as social media sites. They determine whether that content should be deleted, restricted or left up. They will also often make decisions on whether the behaviour of a particular account or profile requires action, like suspension or banning for posting inappropriate content or harassing other users. Their decisions are meant to be based on both the laws of the country where the content is being read and the rules set by the platform.
>
> Working to strict targets, content moderators often spend their shifts reviewing a constant feed of potentially harmful content flagged by automated systems or reported by individual users.[52]

As this reference lays out, human beings, with minds and emotions, are employed by tech companies to review questionable content and make decisions as to whether we (i.e., everyone who isn't a content moderator) see that content. Companies that have relied on these unsung social media first responders/moderators include content-rich platforms, like Facebook and Instagram (rebranded under the corporate umbrella Meta in 2021), Twitter-cum-X, Alphabet/Google/YouTube, and TikTok.

Who are content moderators and where do they reside? Most of these workers, per TBIJ, are based in "middle- and low-income countries."[53] Why is this advantageous to tech companies? Big Tech companies that hire these outsourced workers can pay them lower wages, thereby increasing profits for the corporations and their shareholders. In what other way does this type of outsourcing benefit high-income countries? As TBIJ observes, "Users in rich regions such as the US and Europe are therefore protected from harmful content by people working hard thousands of miles away."[54] Tech companies and proponents of this practice can argue that they are providing these workers with jobs in markets where such opportunities are often scarce.

TBIJ points out that content moderation can be traumatic because moderators may be "exposed to extreme violence, images of sexual abuse and other disturbing content."[55] Viewing traumatic content can have impacts on both mental and physical health for these workers. The potential health-related consequences of content moderation work is dramatically illustrated in the *Washington Post* September 2024 piece "I Quit My Job as a Content Moderator. I Can Never Go Back to Who I Was Before."[56] Through the reporting of the *Post*'s Beatrix Lockwood and as depicted through the artistry of Maya Scarpa (who illustrates the article with a comic strip), readers learn of the first-person experiences of content moderator Alberto Cuadra. Having to look at disturbing content day in and day out takes a toll on the mental health of him and many of his fellow content moderators. As an indicator of how unhealthy the content is, the unnamed tech company in the article/comic strip offers employees "a certain number of free sessions with a therapist." Indeed, the article even includes an italicized "*Warning: The following illustrations contain references to disturbing content.*" Alberto eventually decides he must leave his job for the sake of his mental health, after having performed content moderation there for almost one year. Even after quitting, he notices that he still experiences feelings of anger and anxiousness, such as when he sleeps and travels. During one depicted trip, he wonders, "What if I'm in the next video?" and asks himself whether his former coworkers will "have to watch me die." In the article's final comic panel, Alberto poignantly pronounces that "if I ever have children, I won't let them on any type of media until they're 18."

It's important to underscore that tech companies aren't the only parties to employ content moderation as a strategy for lessening the presence of hate speech, misinformation and disinformation, and deep-fake materials on Big Tech platforms. The Santa Clara Principles on Transparency and Accountability in Content Moderation reflect a collaborative stakeholder effort to create and implement guardrails and guideliness for technology companies and users. In 2018, a group composed of civil watchdog organizations/advocates and academic subject area experts, identified three principles for internet platforms to use in managing and moderating user-generated content. The three principles created by the group are transparency about (1) numbers/data on user posts that have been removed and accounts that have been impacted by violations, (2) notice provided to users who have been impacted by content moderation actions, and (3) appeal rights having been communicated to persons whose content has been removed or whose accounts are suspended.[57] The gathering's aim was that these principles would

- promote more transparency and accountability by tech companies,

- balance due process considerations, and
- respect users' rights.[58]

Moreover, the group stressed that the principles were to be seen as a "starting point," with the goal of "more in-depth dialogue" moving forward.

Flash-forward to 2025 and the record on how tech companies have performed on content moderation efforts is mixed. Following the January 6, 2021, insurrection at the US Capitol, Big Tech leaned into instituting more stringent content moderation efforts. For example, some public figures were banned from platforms like Twitter and Facebook under a rationale of posting fallacious information and fueling conspiracy theories. As this book prepares to go to press in early 2025, the ground on content moderation has shifted again. Many public figures who were banned from platforms like Facebook and Twitter after January 6 have had their accounts reinstated. In general, Big Tech companies are now decreasing their content moderation practices or eliminating them entirely, as described in more detail in the next section.

Social media, free speech, and content moderation issues present potentially irreconcilable interests and unresolvable questions. One multifaceted question, for example, is how to promote free and open internet and social media cultures and respect users' freedom to exercise free speech by posting and accessing content of their choice, while ensuring that platforms maintain some degree of mutually agreed-on minimal standards for public safety, truthfulness, and lawful activities.

Social Media Developments and Potential Regulatory Actions

As examined in the previous sections, social media continues to be a highly contested area catalyzing ethical concerns and impassioned debates among diverse stakeholders. Amid concerns about the impacts of increasing smart phone usage on young people, the US Congress and states face growing calls for regulation of social media and restrictions on its use by children. Additionally, in the past few years Congress has held public hearings to consider revising the safe harbor provision (i.e., Section 512) of the Digital Millennium Copyright Act (DMCA) that some believe enables tech companies to avoid responsibility for harms caused by content posted on their sites. Opponents stress that revision or revocation of Section 512 would be the death knell of an open internet and would stifle free expression and creativity.

On January 7, 2025, Mark Zuckerberg announced that Meta would no longer be using fact checkers and content moderators: "Meta is abandoning the use of independent fact checkers on Facebook and Instagram, replacing them with X-style 'community notes' where commenting on the accuracy of posts is left to users."[59] Some politicians have argued that content moderation censors right-wing voices. In an article and video by Zuckerberg explaining the policy change as an effort to promote greater free speech, Meta's headline stated, "More Speech and Fewer Mistakes." The article further described the company's stance: "Meta's platforms are built to be places where people can express themselves freely. That can be messy. On platforms where billions of people can have a voice, all the good, bad and ugly is on display. But that's free expression."[60]

It remains to be seen what the consequences of Zuckerberg's new policy of "community notes" crowdsourced commenting and discontinuation of content moderation will be. Reporter Sarah Chatta expressed her concerns about the policy change in a January 14, 2025, *New York Times* op-ed, "Meta's Decision to End Fact-Checking Could Have Disastrous Consequences." She experienced firsthand "the importance of fact-checking while working as a reporter in Sri Lanka in 2018 when an episode of violence tied to Meta's platforms rocked the country."[61]

The fate of the phenomenally popular China-based social media company TikTok, regarding its access by US citizens, remains to be decided in 2025. Though owned by a Chinese company named ByteDance, the company has long raised concerns among US intelligence officials about the security implications of the Chinese government potentially being able to access the personal data of millions of US citizens via the TikTok app that they have and use on their smartphones. In the spring of 2024, the US Congress and Biden took combined legislative and executive branch actions to order a ban on TikTok in the United States unless it was sold to an entity outside of China by January 19, 2025.[62] This action was appealed through the US courts. On January 17, 2025, the court upheld the Biden administration's decision in a unanimous 9–0 ruling.[63] TikTok briefly went dark just before the ban was set to take effect on January 19. On that same day, president-elect Donald Trump declared that he would restore TikTok through executive branch action.[64]

Information Centers and Social Media

Information professionals and organizations have key positions of opportunity in educating users about responsible social media use and supporting programming and advocacy efforts that combat cyberbullying. It's crucial to point out, though, that libraries and information centers are not required to participate in social media themselves, as ALA underscores: "Libraries are under no legal obligation to participate in social media, nor are they required to host public conversations."[65]

Guidelines and resources like ALA's "Social Media Guidelines for Public and Academic Libraries" can aid information centers in identifying options and determining their institutional stances on social-media-related issues and ethical challenges.[66] Information professionals as individuals are also encouraged to assess the roles that they can perform as stewards of principled information and technology access and use. Moreover, librarians and other information professionals have decades of experience in promoting intellectual freedom, standing up to censorship, and helping users discern the quality of information sources. These skill sets, among others, can empower information centers to confidently take lead positions in helping users to face cyberbully-related and other social media challenges that are still unfortunately present in all too many internet and social media venues.

Summary

Social media use via smartphones has become a routine part of the lives of citizens around the globe. It can promote communication and meaningful connections, among many other positive uses. However, its potential to be used as a vehicle for hate speech, misinformation, disinformation, cyberbullying, and other worrisome and harmful activities has raised concerns. Some researchers and medical experts have called for smartphone and social media access to be restricted for children of certain ages. Discussions of these restrictions have included consideration of bans on smartphone and social media access in schools. Big Tech companies like Facebook/Meta, Twitter/X, Alphabet/Google/YouTube, TikTok, and others provide many of the most popular apps that children and adults access. Some stakeholders have advocated for government officials to institute legislative restrictions on technology companies to ensure that standards for online safety and well-being are implemented and enforced. Social media overlaps with other fundamental ethics areas and civil liberty interests, including access to information and technology, free speech and expression, intellectual freedom, privacy, accuracy of information, data autonomy, privacy, and equity. Information professionals and organizations have important voices to share on issues like social media use and access because of their influential and privileged societal roles as information stewards and curators.

Questions for Reflection, Exploration, Action, Discussion (READ)

1. Go to the Cyberbullying Research Center website (https://cyberbullying.org) and familiarize yourself with the website and its resources. Identify a resource or takeaway insight you could potentially apply in an information center setting or activity.
2. In the following hypothetical scenario, you are a young adult librarian working in a public library. A 12-year-old user whom you have casually become acquainted with from several programs that the user has attended at the library over the past year is near your desk and appears visibly distressed as they look at their phone. When you ask the young person if they are okay and if they need any help, they tell you that some other students have been calling them names, posting things about them, and threatening to beat them up.
    - What additional information, if any, do you want to know?
    - What do you do?

    Explain your thought process and decision(s).
3. Examining social media content moderation.

a. Identify three to six ethical issues that are implicated by having content moderation on social media sites like Facebook and Twitter/X.

b. Identify three to six ethical issues that are implicated by not having content moderation on social media sites like Facebook and Twitter/X.

4. Human beings currently conduct the majority of content moderation on internet and social media sites. Many of these workers reside in the Global South and are unknown to users in the Global North.

a. Make a list of the pros and cons of tech companies employing outsourced workers in middle- and low-income countries to perform content moderation.

b. Now, imagine you serve on the board of a Big Tech company. Discuss your positon on whether your company should employ or not employ content moderators in Global South nations.

[*Note:* You may want to return for part C of this question after reading the next chapter on AI.]

c. The CEO of the Big Tech company for which you serve as a board member presents a proposal to terminate all human content moderators and have content moderation performed by AI chatbots. (1) Make a list of the pros and cons of this proposal. (2) Articulate your decision and the rationale for your position.

5. Identify some of the key ethical issues involved with banning smartphone and social media use in schools. Then, imagine you are an information professional serving on a public school board that is determining whether to ban smartphone use in public schools. When the time for the board's public vote on the issue arrives, do you ultimately vote for the ban, against the ban, or abstain? Explain your decision.

6. Congratulations, you've just been elected to the US Senate! Members of Congress routinely serve on committees, and you've been assigned to the Judiciary Commmittee. Next week, the leaders of a number of Big Tech companies will appear before your committee for a live public hearing.

a. Distill your position on social media into a 60-second introductory statement, after which you will then pose some questions to the Big Tech CEOs for your Q&A allotment of five minutes.

b. Identify two questions you will ask the company leaders. Explain why you are asking those questions and what you hope to learn from the responses.

7. The US Supreme Court has upheld legislative and executive branch actions banning TikTok.

a. Do you agree or disagree with this decision?

b. Identify the ethics issues implicated by the question of whether or not to ban TikTok within the United States.

Notes

1. Cambridge Dictionary. (n.d.). *Social media.* https://dictionary.cambridge.org/us/dictionary/english/social-media
2. Cambridge Dictionary. (n.d.). *Digital citizenship.* Retrieved February 21, 2025, from https://dictionary.cambridge.org/us/dictionary/english/digital-citizenship#google_vignette
3. Booth, R. (2024, March 19). "Social media is like driving with no speed limits": The US surgeon general fighting for youngsters' happiness. *Guardian*. https://www.theguardian.com/media/2024/mar/20/vivek-murthy-us-surgeon-general-social-media-youth-happiness
4. Ibid.
5. Ibid.
6. World Happiness Report. (2024). *Happiness and age: Summary.* https://worldhappiness.report/ed/2024/happiness-and-age-summary/
7. Booth, R. (2024, March 19). "Social media is like driving with no speed limits": The US surgeon general fighting for youngsters' happiness. *Guardian*. https://www.theguardian.com/media/2024/mar/20/vivek-murthy-us-surgeon-general-social-media-youth-happiness
8. Ibid.

9. McBain, S. (2024, March 21). *The Anxious Generation* by Jonathan Haidt—A pocket full of poison. *Guardian*. https://www.theguardian.com/books/2024/mar/21/the-anxious-generation-by-jonathan-haidt-a-pocket-full-of-poison
10. Ibid.
11. Dao, J. (2011, February 16). Staying in touch with home, for better or worse. *New York Times*. https://www.nytimes.com/2011/02/17/us/17soldiers.html
12. Pew Research Center. (2024, November 13). *Social media fact sheet*. https://www.pewresearch.org/internet/fact-sheet/social-media/
13. Ibid.
14. Vogels, E. A., & Gelles-Watnick, R. (2023, April 24). *Teens and social media: Key findings from Pew Research Center study*. Pew Research Center. https://www.pewresearch.org/short-reads/2023/04/24/teens-and-social-media-key-findings-from-pew-research-center-surveys/
15. Ibid.
16. Cyberbullying Research Center. (n.d.). *About us*. Retrieved February 21, 2025, from https://cyberbullying.org/about-us
17. Ibid.
18. DeSilver, D. (2021, January 13). *Q&A: What we've learned about online harassment*. Pew Research Center. https://www.pewresearch.org/fact-tank/2021/01/13/qa-what-weve-learned-about-online-harassment/
19. Hamilton, D. (2024, February 1). *Mark Zuckerberg's long apology tour: A brief history*. AP News. https://apnews.com/article/mark-zuckerberg-facebook-apology-c7055b654f63a23d09b6a96388dfa2b4
20. Vogels, E. A. (2021, January 13). *The state of online harassment*. Pew Research Center. https://www.pewresearch.org/internet/2021/01/13/the-state-of-online-harassment/
21. Ibid.
22. Ibid.
23. Ibid.
24. Bark. (n.d.). *Home page*. Retrieved February 21, 2025, from https://www.bark.us/
25. Ibid.
26. Empson, O. (2024, June 2). The US librarian who sued book ban harassers: "I decided to fight back." *Guardian*. https://www.theguardian.com/books/article/2024/jun/02/librarian-book-ban-interview
27. StopBullying.gov. (n.d.). *Laws, policies & regulations*. Retrieved February 21, 2025, from https://www.stopbullying.gov/resources/laws
28. Cyberbullying Research Center. (n.d.). *Bullying laws across America*. Retrieved February 21, 2025, from https://cyberbullying.org/bullying-laws
29. Patchin, J. W. (2013, February 7). *Catfishing as a form of cyberbullying*. Cyberbullying Research Center. https://cyberbullying.org/catfishing-as-a-form-of-cyberbullying
30. Ibid.
31. Hinduja, S. (n.d.). *Doxing and cyberbullying*. Cyberbullying Research Center. Retrieved February 21, 2025, from https://cyberbullying.org/doxing-and-cyberbullying
32. *Swatting: Mitigation strategies and reporting policies*. (n.d.). School Safety Center, Washington. Retrieved February 21, 2025, from https://rems.ed.gov/docs/WA_Swatting.pdf
33. US Attorney's Office, District of Kansas. (2019, September 13). *Ohio gamer sentenced in deadly swatting case*. https://www.justice.gov/usao-ks/pr/ohio-gamer-sentenced-deadly-swatting-case
34. Carter, C. (2021, November 16). *An update on the legal landscape of revenge porn*. National Association of Attorney Generals. https://www.naag.org/attorney-general-journal/an-update-on-the-legal-landscape-of-revenge-porn/

35. Franks, M. A. (2023, August 23). *The criminalization of non-consensual pornography in the United States*. Social Science Research Network. https://papers.ssrn.com/sol3/papers.cfm?abstract_id=4939776
36. Ibid.
37. Ibid.
38. FBI. (n.d.). *What is sextortion?* Retrieved February 21, 2025, from https://www.fbi.gov/video-repository/what-is-sextortion/view
39. Denyer, S. (2016, May 23). China's scary lesson to the world: Censoring the internet works. *Washington Post*. https://www.washingtonpost.com/world/asia_pacific/chinas-scary-lesson-to-the-world-censoring-the-internet-works/2016/05/23/413afe78-fff3-11e5-8bb1-f124a43f84dc_story.html
40. Wong, E., & Barboza, D. (2011, January 31). Wary of Egypt unrest, China censors web. *New York Times*. https://www.nytimes.com/2011/02/01/world/asia/01beijing.html
41. Fallows, J. (2016, December). China's great leap backward. *Atlantic*. https://www.theatlantic.com/magazine/archive/2016/12/chinas-great-leap-backward/505817/
42. Glaser, A. (2017, August 2). How Apple and Amazon are aiding Chinese censors. *Slate*. http://www.slate.com/blogs/future_tense/2017/08/02/apple_and_amazon_are_helping_china_censor_the_internet.html
43. Carsten, P. (2014, February 12). *Microsoft denies global censorship of China-related searches*. Reuters. https://www.reuters.com/article/microsoft-bing-censorship/microsoft-denies-global-censorship-of-china-related-searches-idUSL3N0LH21C20140212
44. Ibid.
45. Epstein, R. (2016, January 22). The new censorship: How did Google become the internet's censor and master manipulator, blocking access to millions of websites? *U.S. News*. https://www.usnews.com/opinion/articles/2016-06-22/google-is-the-worlds-biggest-censor-and-its-power-must-be-regulated
46. Preston, J. (2011, March 27). Ethical quandary for social sites. *New York Times*. https://www.nytimes.com/2011/03/28/business/media/28social.html
47. Ibid.
48. Sengupta, S. (2012, January 27). Censoring of tweets sets off #outrage. *New York Times*. https://www.nytimes.com/2012/01/28/technology/when-twitter-blocks-tweets-its-outrage.html
49. Ibid.
50. Simon, M. (April 25, 2008). *Student "Twitters" his way out of Egyptian jail*. CNN. https://web.archive.org/web/20101129104755/http://articles.cnn.com/2008-04-25/tech/twitter.buck_1_cell-phone-blog-anti-government-protest?_s=PM:TECH
51. Ibid.
52. Bureau of Investigative Journalism. (2024, May 17). *What is content moderation?* https://www.thebureauinvestigates.com/stories/2024-05-17/explainer-what-is-content-moderation/
53. Ibid.
54. Ibid.
55. Ibid.
56. Lockwood, B. (2024, September 9). "I quit my job as a content moderator. I can never go back to who I was before." *Washington Post*. https://www.washingtonpost.com/opinions/2024/09/09/shifts-content-moderator-comic/
57. *Santa Clara principles on transparency and accountability in content moderation.* (n.d.). Retrieved February 21, 2025, from https://newamericadotorg.s3.amazonaws.com/documents/Santa_Clara_Principles.pdf
58. Ibid.

59. McMahon, L., Kleinman, Z., & Subramanian, C. (2025, January 7). *Facebook and Instagram get rid of fact checkers*. BBC News. https://www.bbc.com/news/articles/cly74mpy8klo
60. Kaplan, J. (2025, January 7). *More speech and fewer mistakes*. Meta. https://about.fb.com/news/2025/01/meta-more-speech-fewer-mistakes
61. Chatta, S. (2025, January 14). Meta's decision to end fact-checking could have disastrous consequences. *New York Times*. https://www.nytimes.com/2025/01/14/opinion/meta-fact-checking-policy.html
62. Fung, B. (2024, April 24). *Biden just signed a potential TikTok ban into law. Here's what happens next.* CNN. https://www.cnn.com/2024/04/23/tech/congress-tiktok-ban-what-next/index.html
63. Schonfeld, Z., & Shapero, J. (2025, January 17). Supreme Court upholds TikTok ban. *Hill*. https://thehill.com/regulation/court-battles/5083305-supreme-court-upholds-tiktok-ban/
64. Cho, J. H., Palumbo, A., & Alpert, B. (2025, January 19). TikTok is back up after Trump says he'll delay the ban. What comes next. *Barrons*. https://www.barrons.com/articles/tiktok-ban-7ff3c00e
65. American Library Association. (2018, June). *Social media guidelines for public and academic libraries.* https://www.ala.org/advocacy/intfreedom/socialmediaguidelines
66. Ibid.

Chapter 11

Artificial Intelligence

Introduction: Attitudes Toward Technology

Humans have had love-hate relationships with technologies throughout recorded history. Take, for example, the printing press. Johannes Gutenberg invented and perfected his revolutionary printing press in mid-fifteenth-century Germany. His innovative system, employing movable metal type for faster printing, made mass printing a reality. His printing press, and its continuously perfected printer progeny, eventually enabled the creation of the "incunabula": books printed by applying ink to wood block letters and metal keys.[1]

Mass printing, made manifest with the emerging technological advances of Gutenberg's press and his system for printing, represented a paradigmatic shift from the laborious, time-consuming work of scribes and scriveners who printed and copied works by hand. Monks fastidiously hand-copied texts and hand-painted works like illuminated manuscripts. *The Book of Kells*, circa AD 800 and stewarded by Trinity College Library in Dublin, Ireland, is arguably the best-known exemplar of this ancient visual art form.

The introduction of the printing press provided the means for mass-producing print works for largely commercial ends. Greater supply exponentially increased access to books and knowledge on a scale never before seen. Demand for this supply inevitably followed. For the first time, members of the lower and middle classes, not just the upper and aristocratic classes, were more easily able to attain reading matter. With greater access to print, literacy rates rose. New forms of print materials were created and distributed. One example was the so-called penny dreadfuls: cheap serialized booklets mass-produced in mid-nineteenth-century Britain to sate the desires of increasing numbers of working-class readers. A 2016 *Guardian* article referred to these often salacious and gruesome page-turners as "the Victorian equivalent of video games."[2]

Over decades, the cumulative effect of more easily attainable books and newspapers presented more opportunities for entertainment, education, and awareness of current events. Moreover, these information sources created new pathways for people of lower and middle economic strata to become more informed citizens. However, more informed citizens may become members of rebellions and revolutions, as well as democracies. Through the printed word and other means, citizen-readers within all societal classes became more aware of the unchosen hierarchical strata within which they lived, worked, paid taxes, worshiped, prayed, and died. Today, we would say that these lower- and middle-class people lacked agency and autonomy over their own lives. Mass-printed books and newspapers enabled them to become more conscious of the inherent inequities of monarchical ecosystems, systems made up of and dependent on strict adherence to the disparate roles of the hereditarily privileged and the unprivileged masses. As book and information access grew, unsurprisingly, monarchies were not as thrilled as readers with mass printings' unintended implications for the now-better-informed subjects inhabiting the realms over which these kings and queens ruled. As Marshall Leaffer, Distinguished Scholar in Intellectual Property Law, University of Indiana at Bloomington School of Law, explains, in England the new technology of the printing press "threatened the Crown"; rulers feared "widespread dissemination of works advocating religious heresy and political upheaval," even referring to mass printing as this "dangerous art."[3]

Other historical examples of technophobia abound. Perhaps most notoriously and lastingly, nineteenth-century Luddites in Great Britain—so incensed that a novel technology could render their work as hand loom weavers obsolete—destroyed the new loom machines and burned down the factories. Their unwitting legacy: giving the world an eponymous term, Luddite, for anyone opposed to new technology or ways of working.[4] At the dawn of the twentieth century, musicians and composers like John Philip Sousa—while testifying before Congress about the US Copyright Act of 1909—railed against another new mechanical technology, the "player piano," which they perceived as a dire threat to their livelihoods. More recently, video gaming, streaming music, augmented reality handhelds, online gambling, and various social media apps and platforms have incited their own unique reactions and recycled spins on the recurrent theme of technology-stoked moral panics. As the Bible's Book of Ecclesiastes (1:9) (and variations of this verse, seen, for example, in the Book of Mormon and twenty-first-century science-fiction show *Battlestar Galactica*) observes: "What has happened before will happen again. What has been done before will be done again. There is nothing new in the whole world."[5]

Centuries of clashes over contested issues such as speech, religion, petitioning of governments, and rights of assembly mirror many of the same struggles that exist around new technologies today. Everything old *is* indeed new again.

Disruptive Emerging Technologies

Britannica defines technology as "the application of scientific knowledge to the practical aims of human life or, as it is sometimes phrased, to the change and manipulation of the human environment."[6] Emerging technologies are new applications of scientific knowledge that gradually gain traction within societies until greater market penetration and mass adoption of the technology has been reached. By their very nature, these kinds of tech innovations disrupt human behaviors and economic business models. Like a dinosaur-killing asteroid, new technologies can be "extinction events" that wipe out dominant life-forms. They can also catalyze existing technologies to adapt and evolve, in Darwinian fashion, in order to survive.

While dispassionately serving as a death knell for some extant technologies, emerging tech can also enable new life-forms to emerge. Brick-and-mortar Blockbuster video stores dominated the home entertainment business from 1985 until their eventual disestablishment in 2013. (As of early 2025, the last remaining Blockbuster, of the more than 8,000 storefronts that existed during their heyday, is in Bend, Oregon.[7]) Anyone alive during that era remembers going to one of the thousands of Blockbuster video stores (or smaller companies like Hollywood Video and mom-and-pop video outlets) to browse and rent VHS tapes, and then DVDs, of movies. I called this Blockbuster in Bend on January 24, 2025, to verify that the store is still open and ask what formats they rent. The clerk with whom I spoke said they rent Blu-Rays, DVDs, and some 4K, like the latest *Transformers* movie, but no VHS any longer.

The arrival in 1997 of a new company named Netflix delivered an eventual one-two punch to Blockbuster and its popular business model. First, Netflix made it possible for consumers to more easily rent and obtain video rentals. It did this through home deliveries of movies, protected in square-shaped red Netflix-labeled envelope sleeves, sent through the US Postal Service. Once a movie was watched, the viewer merely needed to reenclose the film in its Netflix sleeve and drop it in the mailbox for return to the nearest Netflix distributor. No longer did a consumer need to physically travel to Blockbuster stores.

Second, and more consequentially, Netflix bet the house on a novel way for users to consume content, without having to venture out to a rental store or have movies delivered through the mail. Cofounder Reed Hastings heavily invested in the promise and eventual realization of an innovative technology called streaming. Streaming enabled consumers to log into digital content platforms like Netflix to watch TV shows and movies from the comfort of their own homes. They could also do this on smartphones and laptop and tablet computer devices while, for example, commuting on the bus or waiting in an airport terminal. Consumers eagerly embraced the time savings and viewing options afforded to them via streaming technology. Blockbuster, married to its more restrictive model, could not compete with the advantages of streaming, bringing about the end of the VHS- and DVD-technology era for users.

In this chapter and the next, we explore the major emerging technologies that are most influencing individuals and societies today, focusing on how ethical issues intersect with these technologies. Chapter 11's spotlight is on artificial intelligence (AI). Chapter 12 looks at technologies that are not as attention-grabbing as AI but which are still making impacts and raising many ethical considerations:

- the internet of things (IoT)
- drones
- robots
- autonomous vehicles
- immersive technologies (i.e., VR, AR, MR, XR)

All of these technologies share a nexus with the overarching disruptive effects and innovative impacts of the internet that emerged in the mid-1990s. These two chapters also call back to many of the key themes, pressing ethics issues, and principal stakeholders from preceding chapters.

A Tale of Two Technologies: AI and Nonfungible Tokens

No technology topic has engendered more pontificating and prognosticating in the recent past and present than AI. The veritable avalanche of AI-related spin-offs, subtopics, and applications that are proliferating means that *this* technology is and will be a global game-changer with staying power. It's also in many ways the best of financial times for the entities that own the eye-poppingly profitable rights to AI technologies, notwithstanding the mushrooming number of generative AI-related copyright infringement lawsuits some of these tech companies are defending themselves against, which are examined later in this chapter.

It's worth noting that the rise of AI followed the colossal crash-and-burn of another previously rapidly emerging technology, nonfungible tokens (NFTs). A 2024 case study of NFTs describes what they are: "A non-fungible token (NFT) is a unique cryptographic asset representing the ownership of digital content, such as an image, an audio file, a video, or any other digital asset."[8] NFTs have also been referred to as "monetized graphics."[9] They are bought and sold and accrue value through "cryptocurrencies": "decentralized virtual currencies that rely on anonymous peer-to-peer networks that maintain a public ledger of transactions."[10] NFTs rely on blockchain technology as their "infrastructure."[11] Blockchains are "anonymous peer-to-peer networks" that enable "irreversible" transactions.[12]

No images are probably more associated with NFTs than the so-called Bored Ape images. Bored Ape Yacht Club (BAYC) is a collection of 10,000 NFT headshots of cartoon apes. *Rolling Stone* described the primates as looking like "hipsters"; examples include apes that smoke, eat pizza, and wear fashionable glasses and hats.[13] Viewed as status symbols[14] and owned by celebrities like Enimen and Snoop Dogg,[15] the value of Bored Apes, first owned by Yuga Labs, soared after their launch in April 2021.[16] In 2022, Yuga Labs was valued at $4 billion.[17] By 2023, though, Bored Apes, along with other NFTs, saw precipitous valulation drops in billions of dollars.[18]

I vividly recall attending a continuing legal education (CLE) session—"NFTs and Intellectual Property: The Wild West of Open Marketplaces"—at the American Bar Association (ABA) IP law conference in Washington, DC, in April 2022. A Bored Ape image adorned the title page of the presenters' PowerPoint slides. The presenting attorneys gushed about the NFTs that they had acquired, while exhorting those of us in attendance to buy our own NFTs. *Caveat emptor* (Let the buyer beware): as Shakespeare's *Merchant of Venice* reminds us, "All that glitters is not gold." Indeed, NFTs turned out to be just the latest self-proclaimed "next best thing" in a long line of overhyped technologies whose shortcomings are all too often underacknowledged by their evangelists. The unfortunate but unsurprising truth for many was that most NFTs were every bit as worthless and illusory as they appeared to be. In January 2023, an *Observer* article's headline noted Bored Apes' reversal of fortune from the stratospheric "Center of the NFT Boom" to a more earth-bound reality of lawsuits and plummeting value.[19] As a September 2023 report by dappGambl observed, "NFTs had a bull run then crashed."[20] More strikingly and quantitatively, dappGambl found that "95% of people holding NFT collections are currently holding onto worthless investments."

If the smoke-and-mirrors NFTs scams can teach us anything, it's that technologies are almost always more complex, more nuanced, and more ethically fraught than their inventors, owners, and financial-interest-holding promoters and shareholders (often) make them out to be. Certainly, the Theranos "blood-testing lab" debacle—with its convicted founder Elizabeth

Holmes currently serving a more-than-11-year sentence in federal prison for wire fraud—has also taught us to be dubious about things that seem, well, too good to be true.[21] When it comes to meaningful responsibility for technologies, much more is required than checking off and cursorily rubber-stamping items on a list; rather, it is critical to

- Ask the tough—*and* the right—questions.
- Maintain healthy skepticism about what is claimed and professed.
- Look "under the hood," "kick the tires," and perform periodic inspections and fulsome audits.
- Implement and uphold legal, ethical, policy, and regulatory guidelines and guardrails.
- Seek transparency and truthfulness.
- Expect competent oversight.
- Demand accountability and—when necessary—justice.

Public-safety-centered steps like these are vital action-oriented, measurable imperatives to ensure that the public interest and trust are *always* among the highest priorities in managing technologies.

One silver lining of the NFTs crash-and-burn fad is that blockchain technology, on which NFTs are based, may prove to be more lasting and leverageable than NFTs. For example, some have suggested that blockchain may offer ways to protect copyrighted images from opportunistic, impermissible, and potentially copyright-infringing AI bot scraping and generative AI training data use. A June 2024 piece in the Wrap, discussing generative AI and copyright law challenges facing the entertainment industry, explained that blockchain technologies "might be used to track original content" and changes made to that content.[22]

NFTs were, in fairly short order, unmasked as nothing more than overhyped piffle, like the notorious Wizard of Oz hiding behind a curtain. AI, rather, has already shown itself to be world-changingly substantive, consequential, and highly profitable. In June 2024, Silicon Valley–based Nvidia, founded in 1991, surpassed Microsoft and Apple as the world's most valuable public company. What juiced the company's skyrocketing growth? "Nvidia has about 80% of the market for AI chips used in data centers, a business that's ballooned as OpenAI, Microsoft, Alphabet, Amazon, Meta and others have raced to snap up the processors needed to build AI models and run increasingly large workloads."[23]

AI algorithms, chatbots, image generators, and deep fakes, as examples of AI-related technologies, are catalyzing societal changes and soul-searching about how to deal with and live with, or without, these welcomed and unwelcomed developments. As with many prior inventions, thinkers and technologists of myriad stripes ruminate and expound on AI's benign and malign possibilities. Writing for *MIT Technology Review*, Will Douglas Heaven poetically conveys peoples' often diametrically opposite feelings and perceptions of AI in his July 2024 article "What Is AI?": "AI is sexy, AI is cool. AI is entrenching inequality, upending the job market, and wrecking education. AI is super-smart and as dumb as dirt. AI will increase abundance and empower humanity to maximally flourish in the universe. AI will kill us all. What the hell is everybody talking about?"[24]

As an example of the Armageddon-level fears that AI is fomenting in some circles, at various points in 2023 writers of numerous pieces in mainstream publications all pondered the issue of whether the development of AI is akin to the invention of the atomic bomb, including

- the *Atlantic* ("AI Is Like... Nuclear Weapons?")[25]
- *Fortune* ("The Atom Bomb or the Airliner: The Battle Over A.I. Analogies Will Determine the Fate of AI Regulation")[26]
- the *New York Times* ("A.I. or Nuclear Weapons: Can You Tell These Quotes Apart?")[27]
- Vox ("AI Is Supposedly the New Nuclear Weapons—but How Similar Are They, Really?")[28]

In July 2024, *New York Times* op-ed writer Nicholas Kristof continued the AI-for-good versus AI-for-bad debates with a piece titled "AI May Save Us, or May Construct Viruses to Kill Us."[29] As his conclusion discerns, "Managing A.I. without stifling it will be one of our great challenges as we adopt perhaps the most revolutionary technology since Prometheus brought us fire." Other voices, like David Brooks, another longtime *New York Times* opinion columnist, originally titled his July 2024 piece "Many People Fear A.I. They Shouldn't."[30]

Amid this lack of consensus on AI's true threat level by experts and nonexpert commentators, *is* the nuclear weapon analogy—or virus threat—an apt one for AI? Will we harness the power of AI for the purposes of good, or will our technology-stoked hubris in daring to tame and wield AI, akin to Prometheus's bold but doomed act of "stealing the sun," eventually be humanity's undoing? In this chapter, we weigh those presently unknowable questions and more:

- What are the tool (i.e., plowshare) and weapon (i.e., sword) capabilities of AI?
- What are the ethics issues sparked by AI?
- In what ways is AI influencing broader aspects of societies?
- How can humans maintain control over AI?
- What safeguards and preparedness plans need to be put in place to address and respond to AI issues?
- What accountability frameworks need to be created, implemented, and enforced for human actors and systems that are responsible for AI?

We also critically examine how AI is impacting information centers and the ways that information professionals are striving to exercise agency over this revolutionary technology to benefit people and their communities.

Defining AI

There are many ways to define AI. Indeed, a United Nations site, referencing the World Intellectual Property Organization website on artificial intelligence and intellectual property, states that "there is no universal definition of artificial intelligence (AI)."[31] This site further explains that

> AI is generally considered to be a discipline of computer science that is aimed at developing machines and systems that can carry out tasks considered to require human intelligence. Machine learning and deep learning are two subsets of AI. In recent years, with the development of new neural networks techniques and hardware, AI is usually perceived as a synonym for "deep supervised machine learning."[32]

Carnegie Mellon University (CMU) is one of the world's foremost AI research institutions. As such, it is a good source for explanatory materials on AI (including a timeline, a glossary of common terms, and explanations like "What AI is *not*").[33] In just eight sentences CMU's Heinz College delivers a clear definition of AI—with a dash of Bill Nye the Science Guy, complemented by a digestible portion of historical development explaining why AI is the most popular *plat du jour* on the emerging tech menu, and finishes up with a couple of relatable, illustrative "AI technologies in everyday life" for dessert:

> AI refers to the ability of machines and computers to perform tasks that would normally require human intelligence. These tasks include things like recognizing patterns and making predictions. Ultimately, that's not magic; it's math.
>
> To understand what's going on with AI today, it's helpful to think of AI in phases of development. Early AI systems were machines that received an input—the data they were fed by humans—and then produced a recommendation. That response is based on the way the system was trained, and the algorithms (the math!) that tell the system what to do with the data. It's computers that can play checkers or chess. It's Netflix knowing that you loved "Karate Kid" and suggesting that you watch "Cobra Kai."[34]

The definition of AI adopted by the European Union for its EU AI Act is somewhat lengthy and legalistic, yet it also captures some of the complexity and breadth that definitions of AI can struggle to encompass at present, especially pertaining to emerging technologies. It differs, too, from prior examples cited because the EU AI Act is representative of the challenges faced by legislative bodies in attempting to codify the essential meaning(s) of AI (and other technologies) in the form of statutory language. The act defines an "AI system" as a "machine-based system designed to operate with varying levels of autonomy and that may exhibit adaptiveness after deployment and that, for explicit or implicit objectives, infers, from the input it receives, how to generate outputs such as predictions, content, recommendations, or decisions that can influence physical or virtual environments."[35]

Given decisions by, for example, the US Supreme Court at various times throughout its history in striking down laws that are deemed to be fatally flawed due to "vagueness and overbreadth," getting the statutory language "just right" is no small feat. It is important to underscore, too, that the signing of a law isn't the end of the road. Rather, once a law has been enacted, the policies that lay out the nuts and bolts for implementation of those laws must be created, executed, enforced, and regularly revisited and revised as necessary. This is also true for the ethical guidelines that must be sown and planted to provide guardrails for laws and policies, as well as to address the conduct of people.

AI: Apocalypse, Salvation, or Something in Between?

AI technologies are an exponentially growing topic of interest and debate, spawning exacting ethical questions that defy easy answers:

- Like the Four Horsemen from Christianity's Bible, will AI usher in humanity's demise?
- Will AI technologies save humankind in ways known now and those yet to be engineered and applied?
- Will emerging applications of AI be a mix of benefits and costs, like all technologies?

These are just some of the conundrums that are especially roiling the minds of philosophers, scientists, technologists, ethicists, clergy members, and futurists, among others. In this chapter, we investigate several case studies. These will enable us to scrutinize positive and negative aspects of AI and critically evaluate the knowledge and opinions of sundry experts who are, depending on their respective stances, sounding the alarm bells, heralding the advent, or encouraging more nuanced considerations of the promises and perils of AI.

Any substantive discussion of the key personages who are central to AI would be remiss without including Ray Kurzweil. Kurzweil is the author of the 2005 highly read book *The Singularity Is Near*. He is also unapologetically of a techno-optimistic strain. As a June 2024 *Guardian* article described, Kurzweil's 2005 book "sparked imaginations with sci-fi like predictions that computers would reach human-level intelligence by 2029 and that we would merge with computers and become superhuman around 2045, which he called 'the Singularity.'"[36] The prognostications in his 2024 follow-up book *The Singularity Is Nearer*, as the *Guardian* noted, "no longer seem so wacky."[37] In fact, Kurzweil predicts that by 2045 humans will have made AI "brain-computer interfaces" manifest, that these "will go noninvasively into our brains through the capillaries," and that these hybrid-interfaces-cum-nanobots will "deepen our awareness and consciousness."[38] He also believes that, eventually, a kind of human-computer immortality will be achieved. Kurzweil is unabashed about his plans "to stay alive" long enough to enter into his own consciousness-infused replicant body, akin to the 2002 *Altered Carbon* cyberpunk book and TV series.

On the dystopic side, in the same *Guardian* piece, Kurzweil acknowledges that "we do have to work through certain types of issues": namely, AI's capacity to conjure up deep fakes, disseminate disinformation, displace workers, increase wealth disparities and tech inequities, facilitate unjust uses of data, and perpetuate human biases.[39] Ultimately, though, Kurzweil is buoyant that AI's rewards will outnumber the drawbacks: "It is not going to be us versus AI: AI is going inside ourselves. It will allow us to create new things that weren't feasible before. It'll be a pretty fantastic future." Time will tell whose vision is more accurate.

AI Concerns and Vulnerable People and Communities

One of the most prominent academic researchers to effectively sound the alarm about AI's unchecked dangers is Cathy O'Neil. Her 2016 book *Weapons of Math Destruction: How Big Data Increases Inequality and Threatens Democracy* asserts two main points: first, that AI algorithms are increasingly infiltrating all aspects of everyday life, and second, that AI algorithms have unfair and disparate impacts on some individuals and groups.[40] Later in this chapter, we look at some collaborative efforts to address AI impacts on vulnerable people and communities.

Joy Buolamwini is another academic researcher working to educate people about the perils and pitfalls of AI algorithms, particularly design aspects. She first came to prominence in a 2016 TED Talk, "How I'm Fighting Bias in Algorithms," when she was a Massachusetts Institute of Technology graduate student. Buolamwini was studying facial recognition software when she made a shocking discovery involving herself: the software could not detect her face. Delving into this problem, she learned that the software designers "who coded the algorithm hadn't taught it to identify a broad range of skin tones and facial structures."[41] As a woman of color, Buolamwini was essentially invisible to the software.

That firsthand experience with machine learning bias galvanized Buolamwini to speak out about this phenomenon she termed the "coded gaze." It further emboldened her to found the Algorithmic Justice League (AJL) in 2016. Its mission: "to raise public awareness about the impacts of AI, equip advocates with resources to bolster campaigns, build the voice and choice of the most impacted communities, and galvanize researchers, policymakers, and industry practitioners to prevent AI harms."[42] Buolamwini and other members of AJL continue to advocate for more equity, transparency, and accountability in technologies like AI. Her 2023 book *Unmasking AI: My Mission to Protect What Is Human in a World of Machines* "looks at the social implications of the technology and warns that biases in facial analysis systems could harm millions of people."[43] While some speaking out about AI have the tendency to present it as a matter of concern and potential catastrophe for future generations, in a 2023 NPR interview Buolamwini underscored her concern "with the way in which AI systems can kill us slowly already."[44]

Burgeoning reliance on algorithms for making decisions that materially affect the lives of real people—often in inequitable ways for disadvantaged and vulnerable communities—is of great concern to researchers like O'Neil, Buolamwini, and others discussed later in this chapter. Raising awareness about how these types of AI-influenced decisions create tangible impacts and lobbying for equitable improvements and remedies is of vital importance. Many of us still do not know the extent to which AI algorithms are invisibly running behind the scenes and making unappealable judgments about everyday aspects of human lives, like

- who gets a mortgage and who doesn't
- who gets a shorter prison sentence or probation and who doesn't
- who gets a job and who doesn't
- who gets admitted to colleges and universities and who doesn't

These kinds of questions are far from simply thought problems for philosophers and theoretical university classroom discussions, as the following real-world examples demonstrate. Workday Inc., the maker of AI algorithmic hiring tools, was sued in 2024 in California for allegedly violating "federal anti-discrimination laws that protect [job applicants] based on race, age, and disability."[45] Although housing discrimination based on race, religion, national origin, and color is prohibited under the 1968 Fair Housing Act, illegal exclusionary actions persist. One way this behavior continues is through the use of AI algorithmic tenant screening technologies by property owners. These systems can be employed to "mine data such as criminal records, credit reports, and civil court records to make predictions about prospective tenants."[46] Law enforcement officials in Spain are using AI algorithms, such as one tool called VioGén, to make literal life-and-death decisions involving domestic abusers and their victims. Data plugged into VioGén is used to assist police and judges in making determinations as to whether a known domestic abuser is likely to commit abuses again and what kinds of protective measures to implement. As the headline of a July 2024 *New York Times* article chillingly conveyed, "An Algorithm Told Police She Was Safe. Then Her Husband Killed Her." The article goes on to note the alarming degree to which these emerging technologies have penetrated into everyday use in Spain: "Spain has become dependent on an algorithm to combat gender violence, with the software so woven into law enforcement that it is hard to know where its recommendations end and human decision-making begins."[47]

In another example, a company called Ultrasound.AI received a patent for an invention using AI to analyze ultrasound pictures of fetuses in wombs. The AI software utilizes data analytics to make predictions about whether a birth is likely to be premature. Moreover, the company provides a quantitative assessment asserting that it can narrow down the estimated time of birth to a specific time frame, such as, for example, a premature birth two to four weeks or four to six weeks before the due date. This example raises a number of ethics issues. On the one hand, knowing with reasonable certainty that a birth may be premature can enable medical professionals to prenatally take steps to aid the mother and unborn child and to be better prepared if the child is born prematurely. On another hand, one wonders whether some people might use quantitative assessment of a projected premature birth to make decisions about a fetus that could raise bioethics concerns for some stakeholders. Many additional ethics and policy questions arise:

- Who will have access to these kinds of technologies? Will insurance plans (ever) cover services like this? Health-care access divides may make this technology unavailable to financially insecure people.
- How may legislatures and governments respond to these kinds of medical advancements?
- Could pregnant mothers ever be required to submit to testing of their fetuses by these kinds of AI technologies as part of their medical insurance coverage? One can imagine an argument being made by some parties from a fiscal standpoint that having information about a premature birth likelihood should inform medical care costs and insurance coverage. Such considerations may raise equity issues.
- What policies and ethics guidelines, if any, need to be created and implemented to address these types of AI technologies?

Some of the most disconcerting facets of these examples (though not for the Ultrasound.AI instance) are nonexistent or inconsistent consent and agency for those whose lives are directly impacted by these AI tools. No one asked job applicants applying to institutions using Workday AI job screening products whether they agreed to or were comfortable with giving the applicant "a thumbs up or thumbs down" when applying for a position. Prospective tenants are not asked whether they consent to having their rental applications reviewed and scored by AI algorithmic applications that harvest and rank publicly available data. Victims of domestic and gender violence are not asked how they feel about machine learning tools deciding whether their abusers are likely to harm them again. Such calculations are rife with ethical issues and concerns. In turn, decisions to green-light these calculated assessments are being weighted, sloughed off, or decided by people and entities in positions of privilege and power, eschewing any input from the very people whose existences are influenced by the answers to these data-driven calculations.

AI algorithm products are also being used to make decisions about some of the most vulnerable members of any society. One especially vulnerable subset is people residing in nursing and assisted living residences. Most of us probably have or have had a loved one or friend residing in an assisted living facility at some time. Most likely we also know someone outside of our family and friend circle for whom that has been the case. Now imagine how AI might impact the life of that person living in an assisted living facility, or, more directly, consider how AI might affect those of us who may very well find ourselves living in one of these facilities at some point in the future. In April 2024, a *Washington Post* article, "Assisted Living Mangers Say an Algorithm Prevented Hiring Enough Staff," revealed that the United States' largest assisted living company, Brookdale, was utilizing AI algorithms. This AI system, called Service Alignment, was used to make decisions about how many staff to hire. Moreover, these AI algorithms were relied on to decide how many minutes staff could spend with individual residents. "Brookdale, based in Brentwood, Tenn., pioneered staffing systems based on algorithmic formulas, an approach experts say is ill-suited to caring for the elderly, who are growing more frail and are more likely to suffer from chronic conditions than previous generations."[48] The prediction- and order-based, impersonal, seemingly infallible nature of algorithmic systems such as Service Alignment finds itself running headlong into the unpredictable, chaos-riven personal reality of human beings whose aging bodies are inevitably failing them. One of the things the Brookdale algorithms did not account for vis-à-vis a facility in Chicago was the "tiny elevators."[49] This detail consequently slowed down staff and residents as they moved residents to and from dining areas.

Very thought-provoking value-laden questions emerge from this example: Is the use of AI algorithms like this to care for vulnerable aging people, many of whom are not able to effectively use their own voices to advocate for themselves, an acceptable practice and "new normal" that each of us is ethically comfortable with? If yes, why? If no, why not? In thinking about the care of older human beings in need, consider an additional tweak to this real-life thought experiment: How does each of us feel about robots substituting for human staff persons to help the people residing in assisted living facilities? Would we want a robot to provide care in an assisted care or nursing facility, even in part, for ourselves or a loved one? In some facilities, this is already occurring. We investigate thorny issues and developments involving additional emerging technologies like helper robots in the next chapter.

Beneficial Uses of AI: Vesuvius Challenge Case Study

AI-related news stories on almost a daily basis are increasingly raising awareness about the already-realized and potential ways that AI technologies may be used to aid people and societies. As a case in point, the *Guardian* reported two such stories on March 29, 2024:

- "Sepsis Blood Test Combined With AI Could Offer Early Detection Tool"[50]
- "New York City to Test AI-Enabled Gun Scanners in Subway System"[51]

The sepsis story is an example of the many ways that AI can serve as a critical medical diagnostic and treatment tool for health-care providers and patients. Because early detection of conditions like sepsis is so important, in order to increase the odds that patients can be treated as expediently and precisely as possible, anything that can give medical providers an edge against ticking time clocks is an innovative game changer. The gun scanner story showcases ways that AI can aid law enforcement and communities in responding to crime and public safety issues. At the same time, though, privacy and Second Amendment right-to-bear-arms issues may be raised by those concerned about what might happen to those accurately or mistakenly identified by scanners as carrying guns. False positive results can occur, as we know sometimes happens with analog metal detectors, which can implicate civil-liberties-related concerns.

A defining feature of AI is that it can accomplish tasks humans have been unable to achieve. One example of this is the way that AI is being used to decipher artifacts that are approximately 2,000 years old. Ancient papyrus scrolls recovered in the eighteenth century were charred and rendered indecipherable by the volcanic eruption of Mount Vesuvius and the subsequent destruction of Pompeii in AD 79. Until the emergence of AI, humans had been unable to unlock the secrets contained within these damaged and fragile historic records. As recounted in an October 2023 *Scientific American* article, "AI Reads Ancient Scroll Charred Mount Vesuvius in Tech First," a 21-year-old US computer science student named Luke Farritor achieved the heretofore unthinkable with the assistance of AI. Farritor "developed a machine-learning algorithm that has detected Greek letters on several lines of the rolled-up papyrus, including πορφυρας (*porphyras*), meaning 'purple.' Farritor used subtle, small-scale differences in surface texture to train his neural network and highlight the ink."[52]

This imaginative effort was part of an initiative called the Vesuvius Challenge, which provides monetary prizes to those who can decipher passages from the scrolls. Farritor was awarded "$40,000 for reading more than 10 characters in a 4-square-centimetre area" from one of the papyrus artifacts.[53] In February 2024, the *Guardian* updated readers that "a team of three computer-savvy students, Youssef Nader in Germany, Luke Farritor in the US, and Julian Schilliger in Switzerland, had won the $700,000 (£554,000) grand prize after reading more than 2,000 Greek letters from the scroll."[54] One Greek scholar, Richard Fowler, referred to this latest success as "a complete gamechanger."[55]

The Vesuvius Challenge was started in 2023 by a University of Kentucky computer science researcher named Brent Seales, who was joined by a number of Silicon Valley financiers. As the *Guardian* article points out, Seales and his fellow researchers "spent years developing algorithms to digitally unwrap the scrolls and detect the presence of ink from the changes it produced in the papyrus fibres."[56] What is particularly noteworthy, though, is that after having laid the groundwork for deciphering these elusive scrolls by creating the algorithms, Seales "released the algorithms for contestants to build on in the challenge."[57] His decision ties back to previous chapters on IP and open movements, where we looked at the benefits of making data, inventions, and other aspects of science open and accessible for others to learn from and build on.

Unsettling Uses of AI: Online Gambling Case Study

George Orwell, celebrated twentieth-century author of the classic satirical works *Animal Farm* (1945) and *1984* (1949), was suspicious of governments. Through his works of fiction, he presciently opined about the dangers of authoritarian regimes and their tacit and overt control of information and people. In *1984*, Orwell warned of the ways that those in positions of power can subtly and insidiously shape perceptions and behaviors:

> So long as they [the Proles] continued to work and breed, their other activities were without importance. Left to themselves, like cattle turned loose upon the plains of Argentina, they had reverted to a style of life that appeared to be natural to them, a sort of ancestral pattern.... Heavy physical work, the care of home and children, petty quarrels with neighbors, films, football, beer and above all, gambling filled up the horizon of their minds. To keep them in control was not difficult.[58]

An example of the unsettling use of AI to influence behavior involves online gambling. Gambling is big business in the United States and throughout the world. A 2022 Pew Research Center survey on all types of sports-related gambling found that 19 percent of US adults had "personally bet money on sports in some way."[59] Statista is a well-regarded German data analytics company that offers helpful background and insights on US online gambling. Statista reported in October 2024 that gross gaming revenue in 2023 was $66.65 billion US dollars.[60] As noted by Statista, in the ensuing years since the 2018 *Murphy* holding, rates of online gambling have continued to rise significantly. Reinforcing this new reality, a February 12, 2024, *Fortune* magazine article's headline, after the Super Bowl overtime win by the Kansas City Chiefs over the San Francisco 49ers, trumpeted that "So Many Americans Were Gambling on the Super Bowl Online That Transactions Hit Nearly 15,000 per Second."[61]

Turn on any television now or spend a few minutes online and one is likely to encounter frequent online gambling ads. These advertisements feature celebrities such as actor Jamie Foxx (for BetMGM), football player Rob Gronkowski and the late actor Carl Weathers (for FanDuel), and comedian/actor Kevin Hart (for DraftKings) exhorting viewers to place bets via their respective companies. Contrast this with the province of Ontario, Canada (where Toronto is situated), with a population of more than 14 million, which in August 2023 banned "gambling commercials featuring athletes and celebrities," effective as of February 28, 2024.[62] The *New York Times* reported in March 2024 that the Canadian Ban Ads for Gambling coalition "wants the federal government to step in and go much further by banning all gambling advertising."[63] Their rationale derives from "severe restrictions on cigarette advertising in Canada, which studies have shown to be an effective tool for reducing smoking."[64] The group also asserts that the downsides created by sports betting, such as gambling addictions that put added stress on public healthcare systems, "are higher than the tax revenue it generates."[65] Bruce Kidd, the chairperson of the coalition, sports and public policy academic, and member of Canada's 1964 Olympic track team, identifies another important ripple effect of sports gambling. Some gamblers are so focused on winning their bets on sporting events that they are resorting to abusive behavior: "Athletes were now finding themselves targeted by sometimes racist online abuse—not for losing a game but for not scoring enough, or too much, to fit some gamblers' bets."[66]

The fuel that makes money for online gambling companies is AI algorithms, as revealed in the February 4, 2024, CBS News investigative story, "Technology Has Fueled a Sports Betting Boom and a Spike in Problem Gambling, Addiction Therapist Warns." The premise is that sports and gambling are markedly connected. Phones present new ways for Americans to gamble on sports events. However, there is a "mismatch, pitting man against machine: gamblers—overwhelmingly young men; versus gambling companies, armed with sophisticated AI, data, and engineering."[67] Moreover, CBS asserts that sporting fans are enticed to not just bet on games but to gamble on every play that occurs within games. As the piece explains, the largest demographic of online gambling users and addicts is men aged 25–34. Recovering gambling addict Joe Ruscillo, 26 years old, started gambling in high school. His words describe the simplicity and seductive attraction of online betting: "You know, you can wake up in the middle of the night and take your phone out, set an alarm for a match, maybe overseas or something like that. I would place a bet on anything, anywhere, at any time."[68] Gambling addiction therapist Harry Levant sheds light on the AI-enabled advantages that online gaming companies wield. Gambling companies like DraftKings can employ machine algorithms to continually tweak betting odds in ways that the average human bettor can't replicate. Additionally, the companies are able to access statistical data, combine that with AI, and then make predictions about what will entice people to gamble.[69]

What makes online gambling even more insidious, though, is the algorithms that create individual profiles of users and track them, that know when they tend to bet and when they don't, and that send them prompts to induce them to return when they've been away from betting or are in addiction recovery. Indeed, UK gambling reform advocate and recovering gambling addict Matt Zarb-Cousin offers a chilling portrait of one gambler who was targeted and tracked by a Big Gambling company:

> Recently, Zarb-Cousin was able to use Britain's public information laws to access data the betting company Flutter—owner of FanDuel—had on a U.K. customer. That data was used to tailor offers and push notifications to keep the guy in action....
>
> "So, about 93 different data points they—they had on this individual were when they bet, what offers worked, what inducements worked. On this particular one, he played slots for three to four days straight. They knew the life stage—the customer life stage he was at. So, 'win back,' they described it: so people that have given up gambling for a while, and they're trying to get them to come back."[70]

Zarb-Cousin encourages people to see online gambling companies for what they are and what they do: "They are big data companies that are extractive."[71] Moreover, gambling therapist Levant admonishes, "This is a public health emergency happening. And we're not talking about it yet."[72] The *1984* quotation that began this section observed how "gambling filled up the horizon of [the Proles'] minds." Similarly, unchecked online gambling AI algorithms threaten to contribute to less individual autonomy, more negative health outcomes, and greater rates of addiction, particularly among younger men.

This online gambling case study and cautionary tale is included in the AI chapter of this book; however, as has been illustrated, AI has links to many of the technology ethics issues explored in earlier chapters, including privacy, cybertracking, surveillance, data collection and use, social media usage, the right to be forgotten, intellectual property, public health and safety, and more. It is important, thus, to see issues like AI-facilitated online gambling as intersectional. Strategies for addressing and ameliorating such multifaceted issues will likely also need to be multipronged. To this end, information professionals are well suited to taking on a variety of vital roles—educators, legal/ethics/policy advocates, change agents, technology experts, helpers, and more—in preparing to come to the aid of users who are subjected to these and other types of technology-related predatory behaviors. As the late Pittsburgh-based children's public television personality Mister (Fred) Rogers famously and empathetically reminded, "Look for the Helpers." Thankfully, information professionals are singularly positioned at the vanguard of those who can and do assist others. The growing trend of libraries employing social workers to work in analog and digital capacities with at-risk and in-need people offers other promising avenues of tangible help and hope.

Mixed Uses of AI

Crisis Text Line Case Study

Another use of AI, this one health-related, illustrates how AI can be leveraged in ways that may initially appear positive and benign but later can raise concerns about possible harms and misuses. Even if legal, however, these kinds of uses can feel like they have an air of impropriety about them, and that they aren't on the "up and up." The nonprofit organization Crisis Text Line (CTL), in its own words, "provides free, 24/7, high-quality text-based mental health support and crisis intervention by empowering a community of trained volunteers to support people in their moments of need."[73] Its mission and vision are similarly well-intentioned and laudatory: "Our Mission: At the intersection of empathy and innovation—we promote mental well-being for people wherever they are. Our Vision: An empathetic world where nobody feels alone."[74]

CTL provides services in the United States, Canada, all parts of the UK (i.e., England, Northern Ireland, Scotland, Wales), and Ireland. In addition to English, it began offering Spanish-language services in 2021. The organization's data philosophy explains that it is centered on "data privacy, security, and ethics."[75] To that end, the philosophical statement further states that CTL's conversations with diverse users are "de-identified, anonymized and aggregated conversations." The sheer numbers of conversations that are managed by the organization, now in the millions, are supported by humans and AI working in tandem. A June 2018 Vox article, "How Data Scientists Are Using AI for Suicide Prevention," describes how CTL is able to prioritize those who are most at imminent risk:

> Deluges have the potential to overwhelm the human staff.
>
> So data scientists at Crisis Text Line are using machine learning, a type of artificial intelligence, to pull out the words and emojis that can signal a person at higher risk of suicide ideation or self-harm. The computer tells them who on hold needs to jump to the front of the line to be helped.[76]

The Vox piece explains that the organization amasses huge amounts of data from its text conversations with people in crisis. It then draws a distinction on the kinds of data that it collects and the kinds of data that big for-profit companies scoop up: "While Netflix and Amazon are collecting data on tastes and shopping habits, the Crisis Text Line is collecting data on despair."[77]

A few years later, in January 2022, a *Politico* article's headline shed light on an eyebrow-raising application of that vast cache of so-called despair data: "Suicide Hotline Shares Data With For-Profit Spinoff, Raising Ethical Questions." CTL had entered into a relationship with Loris.ai, a "for-profit spinoff [that] uses a sliced and repackaged version of that information to create and market customer service software."[78] Loris.ai asserted that the data from CTL had been anonymized. However, experts expressed concerns that users could still potentially be identified, "citing past examples involving health records, genetics data, and even passengers in New York City taxis."[79] Numerous authorities interviewed by *Politico* identified ethics issues involving the paucity of meaningful informed consent from CTL users. They also expressed skepticism about the long-term integrity of the data anonymization's ability to protect privacy and stave off confidentiality breaches. Jennifer King, a privacy and data policy fellow at the Stanford University Institute for Human-Centered Artificial Intelligence, crystallized the crux of the ethical dilemma at hand: "Using that data to help other people is one thing, but commercializing it just seems like a real ethical line for a nonprofit to cross."[80]

The scrutiny and outspokenness of these data privacy experts, coupled with greater awareness through reporting, led CTL to make significant changes within the next few weeks. A February 1, 2022, headline on the Verge succinctly summed up the about-face: "Crisis Text Line Stops Sharing Conversation Data With AI Company."[81] CTL communicated the dissolution of its relationship with Loris.ai via a January 31, 2022, post:

> During these past days, we have listened closely to our community's concerns. We hear you. Crisis Text Line has had an open and public relationship with Loris AI. We understand that you don't want Crisis Text Line to share any data with Loris, even though the data is handled securely, anonymized and scrubbed of personally identifiable information. As a result, we have ended our data-sharing relationship with Loris. This change includes our request that Loris delete the data it has received from Crisis Text Line. We have updated our Terms of Service & Privacy Policy accordingly to reflect this change.[82]

Identification and Surveillance of Homeless Encampments

Homelessness, also known as housing insecurity, is a global problem. Nowhere in the United States has greater housing scarcity per capita than California. The US Department of Housing and Urban Development's 2023 Annual Homelessness Assessment Report (AHAR) to Congress indicated that California had the highest rate of homelessness in the country. California's homeless people numbered 181,399, while 123,423 individuals were unsheltered.[83] Axios, unpacking the report, notes that "unsheltered homelessness is defined as those in places not meant for human habitation, as opposed to emergency shelters, transitional housing or safe haven programs."[84]

The report highlights various data breakout sections offering views of homelessness for "a single night in January 2023," including this one regarding the staggering level of continuums of care (CoC) in California: "In five major city CoCs, all located in California, more than 70 percent of people experiencing homelessness were unsheltered: San Jose (75%), Los Angeles (73%), Oakland (73%), Long Beach (72%), and Sacramento (72%)."[85] Axios also observes that "28% of all people experiencing homelessness in the country," more than one out of every four people, are located in California. Additionally, the Golden State is home to "nearly half of all unsheltered people in the U.S."[86] California has a higher rate of youth homelessness than any other state: 20 percent, or one out of every five homeless youths.[87]

Against this alarming statistical backdrop of flesh-and-blood human beings in need, in March 2024 the world learned that San Jose, California, was believed to be the first city to utilize AI to identify homeless encampments. As reported by the *Guardian*, from interviews and publicly accessible records:

> For the last several months, a city at the heart of Silicon Valley has been training artificial intelligence to recognize tents and cars with people living inside in what experts believe is the first experiment of its kind in the United States.
>
> Last July, San Jose issued an open invitation to technology companies to mount cameras on a municipal vehicle that began periodically driving through the city's district 10 in December, collecting footage of the streets and public spaces. The images are fed into computer vision software and used to train the companies' algorithms to detect the unwanted objects. . . .
>
> Some of the capabilities the pilot project is pursuing—such as identifying potholes and cars parked in bus lanes—are already in place in other cities. . . . Local outreach workers, who were previously not aware of the experiment, worry the technology will be used to punish and push out San Jose's unhoused residents.[88]

Proponents of this effort in San Jose assert that the technology can be used to identify people who are in need of help and steer them to community services. However, this case study implicates many often competing interests and contentious civil liberties issues. Some of these interests and issues include the following:

- data analytics
- data ethics
- data collection and use
- privacy
- surveillance
- access to information
- accuracy of information
- historically disadvantaged persons and communities
- technological innovation and entrepreneurship
- public safety and general welfare
- protection of minors

Public Awareness and Feelings About AI

What does empirical research show us about how aware people are of AI and how people feel about AI? Data points from the Pew Research Center (PRC) offer some insights into these queries. A February 2023 PRC study, "Public Awareness of Artificial Intelligence in Everyday Activities," found that at least half of Americans were able to identify that AI was present in some of the six examples they were presented with, though lesser numbers of people were able to correctly identify that AI was present in all six examples. The six examples were the following:

- Wearable fitness trackers that analyze exercise and sleeping patterns
- A chatbot that immediately answers customer questions
- Product recommendations based on previous purchases
- A security camera that sends an alert when there is an unrecognized person at the door
- A music playlist recommendation
- The email service categorizing an email as spam[89]

Sixty-eight percent of the survey respondents were able to identify that AI was present in fitness-tracking technology, compared with only 51 percent who were able to recognize that AI plays a role in the email service example. These findings indicate that more awareness-raising efforts are needed to increase understanding of AI for significant numbers of Americans.

PRC survey research has also provided beneficial insights into how people feel about AI. For example, an August 2023 study, "Growing Public Concern About the Role of Artificial Intelligence in Daily Life," found that only 10 percent of the survey participants were "more excited than concerned" about AI.[90] In contrast, 52 percent reported that they felt "more concerned than excited about the increased use of" AI, and 36 percent said they felt "an equal mix of these emotions." A November 2023 PRC survey study provided further evidence of people's emotional responses to AI. That study, "What the Data Says About Americans' Views of Artificial Intelligence," found that "Americans are increasingly cautious about the growing role of AI in their lives generally. Today, 52% of Americans are more concerned than excited about AI in daily life, compared with just 10% who say they are more excited than concerned; 36% feel a mix of excitement and concern."[91] This study noted that Americans do "see some specific uses of AI positively" and that their feelings hinge "on the context of how and why AI is being used."

Later in this chapter, we look in detail at some of the chief stakeholders and contentious issues involving generative AI, copyright law, and ethics. A March 2024 PRC study reported that 54 percent of Americans think that generative AI technologies "need to credit the sources they rely on to produce their responses."[92] Only 14 percent stated that the generative AI didn't need to provide attribution for the source of the training data that had been used in creating the AI tools. Approximately one in three weren't sure how they felt about the issue.

Poynter Institute is a source of journalistic ethics and guidance that was examined in the prior chapter on accuracy of information. In May 2024, Poynter, in conjunction with the University of Minnesota, conducted focus groups on the topic of AI being used in various ways, such as to create news articles. The focus group findings revealed a range of emotions about AI being used in these ways, as telegraphed by an explanatory article's headline about the research: "We Asked People About Using AI to Make the News. They're Anxious and Annoyed."[93] The research was discussed at a June 2024 Poynter Summit on AI, ethics, and journalism focusing on "the ethical implications for journalists using AI tools in their work."[94] One of the main findings was peoples' fears about losing human connections with the news, such as when an AI voice can be clicked on to read a news article.

Synthesizing these polls, more people expressed caution than excitement about AI. More people, too, felt that AI technologies need to be more transparent about what sources they are built on. Mirroring the diversity of human beings, perhaps unsurprisingly, the Poynter study indicated a whole slew of feelings about AI, from "no big deal" to "big deal." How attitudes about AI may or may not change over the next few years will be interesting to watch. Attitudinal changes may also have implications for how segments of societies choose to address or not address AI from ethics, law, and policy standpoints.

Balancing Act: AI Guardrails, Innovation, and Regulatory Oversight

In light of how people feel about AI, as well as their varying degrees of awareness and understanding of AI technologies, what can and is being done to assuage concerns, promote greater proficiency, and provide safety measures? The Crisis Text Line case study convincingly shows how information and advocacy can be highly effective in addressing—and remedying—concerns about how emerging technologies like AI are used. It also demonstrates the vital roles that researchers and reporters, respectively, can play in raising greater awareness of technologies and their societal ramifications. Indeed, the critical watchdog functions that research integrity and free and independent presses perform cannot be overstated. We know from recent experience, for example, that the two prongs of sound peer-reviewed research and investigative journalism eventually revealed the decades-long duplicity and dangers of tobacco-industry-funded pseudo-research and public relations that suppressed knowledge about tobacco's harms. Respected voices like those of Cathy O'Neil, Joy Buolamwini, and countless others are underscoring the importance of having necessary and nuanced conversations about new technologies, in order to inform prudent regulatory oversight and more ethical decision-making and equitable policy making.

In recent years, corporate entities like Google, Microsoft, and Facebook have made headlines announcing the creation of various AI ethics advisory boards and frameworks. More often than not, these efforts have had mixed results. National governments and corporations have also taken steps to instill AI guidelines and guardrails, while still endeavoring to balance innovation and oversight. The following sections highlight some noteworthy examples. Later in this chapter, we learn more about the various AI leadership roles and policy positions that the world's religions are taking. We also look at how religious observers and nonreligious people are meeting the challenges and opportunities of the AI age.

Asilomar AI Principles (2017)

The "Asilomar Conference on Beneficial AI" was one of the world's first conferences to bring together diverse participants working to develop principles for both harnessing the benefits of AI and keeping AI harms in check. The gathering was sponsored by the Future of Life Institute (FLI) and was held in Asilomar, California, a ninety-minute drive south of Silicon Valley. Fruitfully, the conference generated twenty-three so-called Asilomar AI Principles, which FLI describes as "one of the earliest and most influential sets of AI governance principles."[95]

Principles 6–18 are grouped under the category "Ethics and Values." One principle notes that safety and security of AI systems should be maintained for the duration of their life cycles. Another indicates that the creators of AI systems are stakeholders who have both the responsibility and opportunity to influence how those systems are used and misused.[96]

Rome Call for AI Ethics (2020)

Pope Francis, the former head of the Roman Catholic Church who died in April 2025, made AI ethics a top priority. In 2020, at a gathering at the Vatican in Rome, the late pope and several corporate and governmental partners signed what became known as the Rome Call for AI Ethics:

> The Call for AI Ethics is a document signed by the Pontifical Academy for Life, Microsoft, IBM, FAO and the Ministry of Innovation, a part of the Italian Government in Rome on February 28th 2020 to promote an ethical approach to artificial intelligence. The idea behind it is to promote a sense of shared responsibility among international organizations, governments, institutions and the private sector in an effort to create a future in which digital innovation and technological progress grant mankind its centrality. Pointing to a new algorethics, the signatories committed to request the development of an artificial intelligence that serves every person and humanity as a whole; that respects the dignity of the human person, so that every individual can benefit from the advances of technology; and that does not have as its sole goal greater profit or the gradual replacement of people in the workplace.[97]

The agreement set out three impact areas: ethics, education, and rights. Additionally, the call identifies six principles:

- transparency
- inclusion
- accountability
- impartiality
- reliability
- security and privacy

In January 2023, at an event referred to as the "Abrahamic Commitment to the Rome Call for Ethics," leaders representing Christianity, Islam, and Judaism all pledged their fidelity to the principles and components of the 2020 Rome Call. Companies like IBM and Microsoft were also present. Archbishop Vincenzo Paglia, president of the Pontifical Academy for Life and the RenAIssance Foundation, observed the collaborative underpinnings of the assemblage:

> We have gathered with our Jewish and Muslim brothers in an event of great importance to call upon the world to think and act in the name of brotherhood and peace—even in the field of technology. The signing of the Rome Call by Jewish and Muslim religious leaders and the joint call for algorethics to guide the design of artificial intelligence refer precisely to the increasingly urgent need to build paths of peace, mutual respect, dialogue and community.[98]

On the other side of the globe, on July 9–10, 2024, religious leaders and technology company leaders once again came together: this time in Hiroshima, Japan. The event was titled "AI Ethics for Peace: World Religions Commit to the Rome Call."[99] On August 6, 1945, in the midst of World War II, the United States dropped an atomic bomb on Hiroshima, Japan, a city of more than 300,000 people, making it the site of the world's first use of an atomic bomb on a human population. Hiroshima was purposefully chosen as the site for the "AI Ethics for Peace" event because of its symbolism and significance as

> a city that stands as a powerful testament to the consequences of destructive technology and the enduring quest for peace. In this symbolic location, leaders of major world religions will gather to sign the Rome Call for AI Ethics, emphasizing the vital importance of guiding the development of artificial intelligence with ethical principles to ensure it serves the good of humanity.[100]

The organizers also explained that the Hiroshima gathering was intended to emphasize "peaceful coexistence and shared values" and "a multi-religious approach to vital questions such as AI ethics."[101] Moreover, the event participants accentuated the rationale and need for all religions to lead on AI ethics: "Religions play a crucial role in shaping a world in which the concept of development proceeds hand in hand with protecting the dignity of each individual human being and preserving the planet, our common home. Coming together to call for the development of an AI ethic is a step that all religious traditions must take."[102]

The "AI Ethics for Peace" event was groundbreaking in bringing together 11 world religions, 16 new signatories to the Rome Call for Ethics, and 13 attending countries.[103] Father Paolo Benanti, the late Pope Francis's AI ethics advisor, "presented the Hiroshima *Addendum on Generative AI*. This document focuses on the need for ethical governance of generative AI—an ongoing and iterative process that requires a sustained commitment from all stakeholders so that its potential is used for the good of humanity."[104] While speaking at the Hiroshima summit, Father Paolo strikingly remarked that "ethics need to contaminate computer science."[105] Leaders from Big Tech also attended and shared their thoughts. One of these, Microsoft President Brad Smith, among the first signatories of the Rome Call, stated that "with its profound place in human history, Hiroshima serves as a compelling backdrop to help ensure a technology created by humanity serves all of humanity and our common home."[106]

US AI Executive Order (2023)

On October 30, 2023, President Joe Biden, responding to a wide spectrum of concerns about AI in American society, issued a groundbreaking executive order on safe, secure, and trustworthy artificial intelligence. The official White House fact sheet announcing the order conveys its desired balance: not overregulating AI and thereby stifling innovation, while instilling sufficient guardrails and fail-safes to earn the "safe, secure, and trustworthy" verbiage in the order's name.[107] This balancing of interests is seen in the way that the order "establishes new standards for AI safety and security, protects Americans' privacy, advances equity and civil rights, stands up for consumers and workers, promotes innovation and competition, advances American leadership around the world, and more."[108]

The order isn't just lofty rhetoric though; rather, its design expressly includes some "teeth" and enforcement powers. Specifically, as a 2023 *Wired* article observed, the order's draft rules "would require federal agencies to assess AI systems currently in use in law enforcement, health care, and other areas—and to shut down any algorithms doing harm."[109] Agencies, in a nutshell, must conduct cost-benefit analyses and not use algorithms whose risks outweigh their benefits. It is important to point out, too, that the order does contain some carve-out exemptions. For example, one exemption is for algorithms related to national security. Another example is for algorithms that "would create an impediment to critical agency operations."[110]

Bletchley Declaration by Countries Attending the AI Safety Summit (2023)

In November 2023, the EU and twenty-eight additional countries, including the United States and China, participated in a two-day event called the "AI Safety Summit." Symbolically, as with the "AI Ethics for Peace" event held in Hiroshima in 2024, the site of the event, Bletchley Park in England, holds enduring significance because of its nexus with technology. Bletchley Park is a nineteenth-century mansion in Buckinghamshire, roughly an hour's drive north of London. Famously, it was the site of World War II–era Allied code-breaking efforts by Alan Turing, Joan Clarke, and other cryptographers and scientists. Their successful efforts to decrypt Axis codes, depicted in the 2015 Oscar-winning film *The Imitation Game* (Best Adapted Screenplay), are credited with shortening the war by years.

The fifty-five countries in attendance at the "AI Safety Summit" signed a historic agreement acknowledging both the opportunities and risks presented by AI. In pertinent part, this agreement recognized that

> AI systems are already deployed across many domains of daily life including housing, employment, transport, education, health, accessibility, and justice, and their use is likely to increase. We recognise that this is therefore a unique moment to act and affirm the need for the safe development of AI and for the transformative opportunities of AI to be used for good and for all, in an inclusive manner in our countries and globally. This includes for public services such as health and education, food security, in science, clean energy, biodiversity, and climate, to

> realise the enjoyment of human rights, and to strengthen efforts towards the achievement of the United Nations Sustainable Development Goals.
>
> Alongside these opportunities, AI also poses significant risks, including in those domains of daily life. To that end, we welcome relevant international efforts to examine and address the potential impact of AI systems in existing fora and other relevant initiatives, and the recognition that the protection of human rights, transparency and explainability, fairness, accountability, regulation, safety, appropriate human oversight, ethics, bias mitigation, privacy and data protection needs to be addressed. We also note the potential for unforeseen risks stemming from the capability to manipulate content or generate deceptive content. All of these issues are critically important and we affirm the necessity and urgency of addressing them.[111]

In this stronger-together-than-apart esprit-de-corps statement, the signatory nations also called for multinational joint action:

> Many risks arising from AI are inherently international in nature, and so are best addressed through international cooperation. We resolve to work together in an inclusive manner to ensure human-centric, trustworthy and responsible AI that is safe, and supports the good of all through existing international fora and other relevant initiatives, to promote cooperation to address the broad range of risks posed by AI.[112]

The participating countries pledged to meet in 2024 and "to continue research on frontier AI safety to ensure that the benefits of the technology can be harnessed responsibly for good and for all."[113]

European Union AI Act (2024)

The earlier privacy chapter demonstrated that the EU, with its 2018-launched General Data Protection Regulation (GDPR), is far ahead of the United States when it comes to a legally enforceable privacy and data collection and use framework. Similarly, the EU is outpacing the United States in the AI legislation arena, though President Biden's 2023 AI executive order is a proactive step forward. In February 2024, the EU's twenty-seven member nations moved another step closer to implementing the world's first comprehensive AI law: the EU's AI Act. As a February 2024 *The National Law Review* article noted, "The AI Act is, as it was intended to be, a comprehensive framework for AI, regulating AI in everything from policing to commercial products to education and employment, and everything in between. The legal community and policy scholars agree that it is likely to be to AI what the GDPR was to data privacy."[114] A February 2024 JDSupra article observed that the EU AI Act seeks "to set the new global standard for AI regulation."[115]

This AI legislation was passed in December 2023. Member states since that time have been hammering out various areas of contention and negotiating aspects of the law, such as how to address AI vis-à-vis medical devices.[116] The law entered "into force" on August 1, 2024, per the EU's website.[117] As a February 2023 *Scientific American* article suggested, the EU AI Act has global implications.[118] According to the European Commission, one of the principal features of the law is that it assesses AI based on four levels of risk: minimal risk, limited risk, high risk, and unacceptable risk.[119] Work on a "General-Purpose AI Code of Practice" for the EU AI Act is ongoing as of April 2025, with a first draft having been made available in November 2024.[120] A goal of the code of practice is "to facilitate the proper implementation" of the law's rules.[121] The code is also seen as a vital instrument that will shape the EU's "future development and deployment of trustworthy and safe general-purpose AI models."[122]

California AI Bills (2024)

In late August 2024, the California legislature passed SB 1047, a bill known as the Safe and Secure Innovation for Frontier Artificial Intelligence Models Act. As NPR noted, this bill "was seen as a potential blueprint for national AI legislation."[123] The bill proved divisive. Opponents, such as many tech companies and Democrats in Washington State, where Microsoft is located, feared the bill would stifle innovation and creativity. Proponents argued that the bill would provide necessary AI safety guardrails and rules. Companies could also be held legally liable for harms caused by large AI models.

Several weeks of speculation in August and September surrounded whether California Governor Gavin Newsom would sign or veto the bill. The answer came on September 29, 2024. Newsom vetoed SB 1047 and provided his reasons for doing so, indicating that the legislation

> does not take into account whether an AI system is deployed in high-risk environments, involves critical decision-making or the use of sensitive data. . . .
>
> Instead, the bill applies stringent standards to even the most basic functions—so long as a large system deploys it. . . . I do not believe this is the best approach to protecting the public from real threats posed by the technology.[124]

Newsom's website post on the day he vetoed SB 1047 also touted the AI measures and bills that he had signed:

> Last year, Governor Newsom signed an executive order laying out how California's measured approach will focus on shaping the future of ethical, transparent, and trustworthy GenAI, while remaining the world's GenAI leader.
>
> Within the past two weeks, Governor Newsom signed a series of bills to crack down on sexually explicit deepfakes and require AI watermarking, protect performers' digital likenesses and combat deepfake election content.
>
> AI is already changing the world, and California will play a pivotal role in defining that future. The state is home to 32 of the world's 50 leading GenAI companies, high-impact research and education institutions, and a quarter of the technology's patents and conference papers.[125]

United Nations

The United Nations has made AI a focal point of its planning and discussions the past few years. In the latter part of 2023, the UN launched its plan: "Roadmap—Towards Globally Inclusive

AI Governance." Following a call for experts in August of that year, an AI advisory body was announced in October. In December 2023, the AI advisory body released its interim report, "Governing AI for Humanity," with its final report published in September 2024.[126] The report's guiding principles include the following:

- Inclusivity: all citizens, including those in the Global South, should be able to access and meaningfully use AI tools.
- Public interest: governance should go beyond the do no harm principle and define a broader accountability framework for companies that build, deploy and control AI, as well as downstream users.
- Centrality of data governance: AI governance cannot be divorced from the governance of data and the promotion of data commons.
- Universal, networked, and multistakeholder: AI governance should prioritize universal buy-in by countries and stakeholders. It should leverage existing institutions through a networked approach.
- International law: AI governance needs to be anchored in the UN Charter, international human rights law, and the Sustainable Development Goals.[127]

The first guiding principle, inclusivity, is intended to promote AI access for the Global South and developing nations, as well as development of AI by these countries. In essence, the UN seeks to lessen the probability of a bifurcated world of AI haves and AI have-nots, in the vein of efforts to close the gap seen with prior digital divides. Moreover, UN policymakers are mindful of the current IP reality in which the vast majority of IP rights are controlled by first-world countries and multinational corporations. An example of efforts to make AI more accessible to the Global South was seen at the UNESCO-sponsored "2nd Global Forum on the Ethics of AI" in February 2024. *Fast Company* highlighted that "the meeting is aimed at broadening the conversation around AI risks and the need to consider AI's impacts beyond those discussed by first-world countries and business leaders."[128] Hearing from the Global South regarding what its citizens think about AI is vital in order to be more inclusive.

Boon and Bane of AI

The emergence of AI presents dueling polarities: exhilarating breakthroughs and troubling perils. As discussed, AI tools are already making extraordinary scientific advances practicable—like the deciphering of the charred Pompeii scrolls. Moreover, in terms of the quality of human life, AI is yielding major improvements in health care and medical treatments and augurs more leaps forward. A 2023 peer-reviewed study, as just one example, found that

> machine learning (ML), a subset of AI that enables computers to learn from training data, has been highly effective at predicting various types of cancer, including breast, brain, lung, liver, and prostate cancer. In fact, AI and ML have demonstrated greater accuracy in predicting cancer than clinicians. These technologies also have the potential to improve the diagnosis, prognosis, and quality of life of patients with various illnesses, not just cancer.[129]

At the same time, AI raises the specter of future AI-precipitated existential threats to humanity. Such threats have been depicted in cinema (often adapted from books) for decades, in movies like *2001: A Space Odyssey* (1968), *Blade Runner* (1982), *The Terminator* (1984), *The Matrix* (1999), and more recently *The Creator* (2023). In the real world, AI experts have shared their unvarnished fears about unchecked AI and the need for safeguards. In a May 2023 *New York Times* article, "A.I. Poses 'Risk of Extinction,' Industry Leaders Warn," more than 350 AI authorities, including leading technology experts such as Geoffrey Hinton (who resigned from Google in the same month), Sam Altman (cofounder of OpenAI in 2015), and numerous others, strongly advised that AI "should be considered a societal risk on a par with pandemics and nuclear wars."[130]

Others, like Joy Buolamwini, Cathy O'Neil, and Timnit Gebru, have called needed attention to present-day examples of actual and potential biases that are already evident in the ways AI systems are being utilized to make decisions influencing an array of life matters. These include issues related to housing, health care, employment, education, access to justice, incarceration, social programs, and finance, among others.

The unseen, unknowable "black box" nature of AI, frequently protected under the rationale of trade secrets, is also a concern. Using a rationale of IP-protected trade secrets, AI designers and owners can assert that their AI cannot be made available for review and scrutiny by unauthorized third parties and, more importantly, members of the public. The public, thus, is expected to simply trust the integrity of any AI in use.

Yet AI experts and watchdog groups have identified copious verifiable examples that buttress legitimate concerns about the oft-ballyhooed trustworthiness and superiority of AI programs. Anything made by humans or powered by machines can fail, though. The historical record is replete with examples: from the sinking of the iceberg-gouged "unsinkable" *Titanic* in 1912 to the 1937 hydrogen-gas-powered *Hindenburg* dirigible calamity and, more recently, from the faulty O-ring seals that caused the 1986 *Challenger* space shuttle explosion to the loose insulation tiles leading to the 2003 *Columbia* space shuttle disaster.

So too are emerging technologies not immune from errors, overhype, and hubris. The technologies and the people underlying self-driving vehicles reminded us of this in 2018 with a shocking, but not altogether surprising, pedestrian's death at the hands of an Uber test car in suburban Phoenix, Arizona. We examine this example and autonomous vehicles in Chapter 12. The bottom line is that oversight and healthy skepticism of all technologies will remain requisite for the foreseeable future. If we've learned anything from AI films—whether it's a malfunctioning HAL 9000 computer in *2001: A Space Odyssey* or AI-amped avatars gone rogue in *The Matrix*—anything technological can and will likely go wrong at some point. Ethics requires that we recognize, mitigate, and prepare for such inevitabilities.

Information Centers and AI Ethics

The gray nature of many AI tools and issues is a compelling impetus for providing societies with as much clarity and guidance as possible to better manage and contain these powerful technologies. The ever-looming presence of order and chaos—the yin and yang principles of the known universe—instruct us, too, to prepare as much as is possible for the worst, while hoping for the best. (Indeed, anyone who has seen any of the *Jurassic Park* movies knows the roles that chaos and seemingly random but predictable potentialities can exact on both mundane and lofty human endeavors.) With this vein of AI-centered strategic planning and disaster prevention and preparedness in mind, developing AI ethics-based structures has been a central focus in the past few years. Ethics codes, advisory bodies, policies, white paper reports, and more have been created and publicized by many nonprofit and for-profit organizations. As highlighted in this chapter, individual nations, cooperative regions like the EU, and collective global entities such as the UN have also been working in this space to craft and implement AI laws, ethical frameworks, and workable policies.

Libraries and information centers are among the institutions that are serving as conduits to information relating to AI ethics. The American Library Association (ALA) initiative Library of the Future, for instance, has a digital section on AI as one of its emerging trends.[131] The section includes discussion of how AI is developing and why AI matters. ALA has also been hosting various AI webinars. For example, an October 2023 two-day digital forum on generative AI and libraries examines "how such large language models might impact libraries in internal workflows, patron services, access to collections, information literacy training, and ethical practices; how it might impact society and what libraries can do to respond; and how libraries can best use this new availability of generative AI."[132] Workshops and courses are another means for information professionals to increase their AI acumen. For example, as part of the *Library Journal* and *School Library Journal* Professional Development series, in March 2024 information professionals could take a three-week live online course titled "Adopting Artificial Intelligence in Libraries" covering the impact on and practical applications of AI for libraries.

Panels at library professional conferences as well as trade publication articles are furthering conversations about AI as well. A January 2024 *American Libraries* magazine write-up, "Using AI Responsibly," reports on an AI panel, "AI and Libraries: A Discussion on the Future," that was held at ALA's 2024 LibLearnX Conference in Baltimore, Maryland. One of the panelists, Juan Rubio, a digital media learning program manager with Seattle Public Library (SPL), conveyed "excitement" about AI tech and the ways it can be used, while also identifying the need for more education so users can utilize AI responsibly.[133] Rubio also provides a window into how libraries like Seattle Public Library are addressing AI internally and externally: "'I think ethical use of AI comes with creating ethical people,' he said, adding that SPL has been thinking about implementing guidelines for using AI."[134] Two other panelists shared firsthand examples demonstrating privacy and accuracy of information concerns vis-à-vis AI usages.

Academic libraries, like public libraries, are also contributing information and education on AI. A 2022 book, *The Rise of AI: Implications and Applications of Artificial Intelligence in Academic Libraries*, "collects projects, collaborations, and future uses from academic librarians who have begun to embrace artificial intelligence (AI) in their work."[135] More specifically, academic libraries are weighing in on how to navigate ethical issues and AI. A May 2023 research study's survey reported in the January 2024 *Inside Higher Ed* article "University Librarians See Urgent Need for AI Ethics" found that "nearly three-quarters of university librarians say there's an urgent need to address artificial intelligence's ethical and privacy concerns."[136] The study also revealed that of the 605 participants, almost three out of four people noted a need for ethics training, particularly on issues like privacy, data misusages, and inaccurate citations generated by AI.

Grant-funded initiatives are another way that academic librarians are contributing to development of AI ethics awareness-raising. Montana State University has a three-year (2022–2025) Institute of Museum and Library Services (IMLS) grant project called Responsible AI in Libraries and Archives. The project's goals are "to support ethical decision-making for AI projects in libraries and archives," to ready information professionals "for the methodical consideration of potential harms of AI projects," with "hope that libraries and archives can provide models that resonate beyond the profession."[137] One of the deliverables of the effort is "to create case studies exploring AI-related ethical challenges in libraries and archives."

Within the academic milieu, other university entities are examining AI ethics-related thematic areas. The Sara Fine Institute, headed by Dr. Eleanor "Nora" Mattern, at the University of Pittsburgh School of Computing and Information (SCI), hosted a series of AI-centered panels during the 2023–2024 academic year. These panels explored such topics as "Generative AI and Research: Policies, Practices, and Ethics in Publishing" and "Generative AI and Pedagogy: Values and Opportunities."

Santa Clara University's Markkula Center for Applied Ethics is another venue for guidance and discourse on AI ethics. In addition to hosting AI-related panels and webinars, the Markkula Center periodically publishes materials, such as a 2023 handbook, *Ethics in the Age of Disruptive Technologies: An Operational Roadmap*. A Voice of America article, "New Handbook Highlights Ways to Develop Tech Ethically," explains that this resource is "the first product of the Institute for Technology, Ethics and Culture, or ITEC."[138] ITEC is the result of a collaborative effort between Markkula Center and the Vatican's Dicastery for Culture and Education. Its website states that ITEC "convenes leaders from business, civil society, academia, government, and all faith and belief traditions, to promote deeper thought on technology's impact on humanity."[139] The Voice of America article points out that the book was specifically designed "in the language of business and engineers so that it's familiar to them" and that the authors have striven "to make it as usable and practical as possible."[140] This tailoring of information to specific stakeholders and communities is an important point for information professionals to continue to prioritize.

Research institutions like the University of Pittsburgh and the Markkula Center are vital sources for data on AI and other emerging technologies. Such data can be of great utility to information professionals in determining the nature of information gaps and how they can satisfy those information needs. Markkula Center and ITEC contracted with Pollfish for an AI-related online survey that was conducted on November 7, 2023. The study, "Ethics in the Age of AI," looked at "how Americans feel about artificial intelligence (AI) from an ethical standpoint."[141] Three thousand Americans, aged 18 or older, were surveyed. The most striking finding from the study derived from this question: "Do you care whether AI is ethical or not?" Eighty-two percent of the respondents answered yes.

AI Standards and Transparency

Information professionals are wrestling with and taking steps to set standards on how to handle the use of AI tools, like ChatGPT, in the creation of content. A January 2024 statement on generative AI in academic publishing by the editor of the *American Archivist* magazine, a publication of the Society of American Archivists (SAA), provides a window into these thorny issues. The introduction explains the impetus for the statement:

> In Spring of 2023, *American Archivist* editors learned that authors had used artificial intelligence (AI) platforms such as ChatGPT in submissions to the Reviews Portal. This sparked a conversation within the Editorial Board about the ethical use of AI in *American Archivist* content. As ChatGPT and other AI programs are becoming ubiquitous, the *American Archivist* Editorial Board is endorsing the following Norms and Recommendations for AI use within out publications.[142]

The statement endeavors to strike a measured stance, while still establishing some boundaries for the use of AI in submissions. The editorial board, for instance, says that it understands the "distinction between using AI for assistance with spelling and grammar, and using AI to generate content."[143] This articulation raises the critical issue of the degree to which tools like AI are used. Referencing an older technology as an illustrative example, a Texas Instruments calculator is a tool for aiding one in solving a mathematical problem. However, the calculator cannot complete, say, a whole series of math problems and spit out the completed tasks. (As an aside, I recall vigorous societal debates about the long-term implications and teaching merits of permitting students to use calculators, rather than doing math without these kinds of aids.) AI can, at least in some instances, perform a whole series of tasks in ways that a "dumb" calculator cannot. The caveat is that one must still check AI output for accuracy. Myriad examples exist of AI creating content with errors and inaccuracies, such as two New York attorneys in 2023 who were sanctioned and roundly rebuked by a judge for using AI to write their legal brief. The brief was replete with fake cases and citations, and the lawyers admitted that they had not understood the AI technology they elected to try out.[144] The term of art for inaccurate or false content generated by AI is *hallucination*.

The norms and recommendations crafted by the SAA are intended to raise awareness of other issues that can arise from the use of AI tools for content submissions. These include issues such as copyright law and potential copyright infringement, attribution, reproduction of "biases from existing online content," misinformation, and nonexistent information.[145] The *American Archivist* editor's statement explains that "transparency in the use of AI in the research and publication process" is the intent of the SAA norms and recommendations.[146] Interestingly, the statement communicates that these guidelines are exactly that, without enforceability, and "do not purport to be a legally binding agreement." Rather, the norms and recommendations "represent an ethical agreement between authors submitting content to the journal and portal, and the Editorial Board." It is also noteworthy that the guidelines will be reviewed annually.

All organizations can benefit from the kinds of broad-minded, ethically centered thinking and concrete guidelines that the SAA has manifested through its norms and recommendations.

Moreover, it is highly advisable for organizations to take proactive positions on AI-related issues, like stances on the use of AI tools, in order to avoid and mitigate future potential legal and ethical problems and minefields that can harm individuals, communities, and institutions.

Standards and guidelines from other fields that can be applied to AI may already exist. In some cases, these standards may be able to be applied to AI in their already existing forms or modified to best fit with AI. The February 2024 issue of *Computer* (which, notably, is devoted specifically to ethics and computing), a peer-reviewed journal of the Institute of Electrical and Electronics Engineers (IEEE), describes a novel ethics-based approach from a team of researchers with the National Institute of Standards and Technology (NIST). This team's findings and recommendations are included in their study, titled "Avoiding Past Mistakes in Unethical Human Subjects Research: Moving From Artificial Intelligence Principles to Practice." The researchers identify bias in AI training data as a significant problem and reiterate an ongoing AI design challenge: "How can we make sure that humans train AI systems on data that reflects sound ethical principles?"[147] A NIST article about the research discusses a viable response to that question:

> "We looked at existing principles of human subjects research and explored how they could apply to AI," said Kristen Greene, a NIST social scientist and one of the paper's authors. "There's no need to reinvent the wheel. We can apply an established paradigm to make sure we are being transparent with research participants, as their data may be used to train AI."[148]

The NIST article explains that their recommended framework draws on the decades-long human subjects research protections that evolved from the landmark 1978 *Belmont Report: Ethical Principles and Guidelines for the Protection of Human Subjects of Research*.[149] The article lays out their recommendation: "We should apply the same basic principles that scientists have used for decades to safeguard human subjects research. These three principles—summarized as 'respect for persons, beneficence and justice'—are the core ideas of 1979's watershed Belmont Report, a document that has influenced U.S. government policy on conducting research on human subjects."[150]

One of the noteworthy challenges in applying the Belmont Report and common rule to AI contexts though is that "the Belmont Report's principles apply only to government research."[151] As the NIST researchers go on to note, those principles do not apply to entities like industries and for-profit endeavors. Hence, as one of the researchers acknowledges, "For the private sector, it is a choice whether or not to adopt ethical review principles."[152] Ultimately, the researcher observes that "the paper is best seen as a starting point for a discussion about AI and our data, one that will help companies and the people who use their products alike."

AI Education Pipelines: AI Curricula, Partnerships, and Resources

School districts throughout the United States, like other segments of societies, are increasingly coming face-to-face with AI and other emerging technologies. How school districts choose or don't choose to deal with emerging technologies, however, varies from venue to venue. As part of their educational missions to impart to children the knowledge and skill sets they will need to live fulfilling and productive lives (as children and their parent[s]/guardian[s] envision their lives), some schools and school districts *are* incorporating AI-related resources. Often, these resources include ethics components.

As one example from the recent past, *Education Week*, in a 2018 article, "How One District Is Preparing Students for an AI-Powered World," reported that a suburban Pittsburgh middle school was among the first in the country to offer a mentored course in AI for its students. Nearby tech powerhouse Carnegie Mellon University (CMU) was in the process of launching its first undergraduate degree in AI. CMU decided to collaborate with the David E. Williams (DEW) Middle School in Coraopolis on testing a new kind of curriculum for teaching students about AI fundamentals. Along with providing the school with the expertise of CMU faculty, the program incorporates field trips to places like Pittsburgh's "Robotics Row" where students can see start-up companies that are working on autonomous vehicles. One of those involved in the effort, Justin Aglio, a school administrator and a visiting fellow at Carnegie Mellon's Human-Computer Interaction Institute, emphasized that the program was envisioned for every student, in order to prepare them for innovations like machine learning.[153] One aspect the curriculum specifically focuses on is AI ethics. The aim is to assist all fifth-to-eighth-grade DEW students in thinking about questions like, "How can we build a better society? How can we use technology and AI for the greater good?"[154] Aglio observed that there is fear that technology like robots "will take over the world.... 'But it's the people who make the technology and program it who are really important. AI is scary only if the programmers don't get this training.'"[155]

More recently than the prior 2018 example, some US states are recognizing the need for better preparing their denizens for a world of ubiquitous AI, especially their youngest citizens. An *Axios* November 2023 article, "How States Are Guiding Schools to Think About AI," observed that only California and Oregon had issued AI "policy guidance for schools," per an October 2023 research report.[156] The research and subsequent report were conducted by the Center on Reinventing Public Education (CRPE), a research unit within Arizona State University's Mary Lou Fulton Teachers College. The report, titled "AI Is Already Disrupting Education, but Only 13 States Are Offering Guidance for Schools," describes the illuminating response rates from the 50 US states and the District of Columbia:

> CRPE reached out to all 51 State Departments of Education, requesting updates on their approach to AI guidance. Just two states—California and Oregon—have offered official guidance to school districts on using AI in schools this fall. Another 11 states are in the process of developing guidance, and the other 21 states who have provided details on their approach do not plan to provide guidance on AI for the foreseeable future. The remaining states—17, or one-third—did not respond to CRPE's requests for information and do not have official guidance publicly available.[157]

The researchers also noted, however, that although only thirteen states as of October 2023 had either published or begun efforts to offer AI guidance to their respective schools (translating to 25.5 percent of the subjects), this was an improvement from two months earlier when they "found that no states or territories had released official communication to help schools navigate the use of AI in teaching and learning."[158]

Ohio was one of the 11 states that CRPE had found was developing AI guidance for its school districts. In February 2024, Cleveland.com, a digital affiliate of Cleveland's newspaper the *Plain Dealer*, reported that the state of Ohio had launched an AI guide whose purpose is "to help schools and parents navigate generative artificial intelligence in an ethical manner."[159] This AI tool kit effort, which includes private-sector groups, is another example of a collaborative initiative, like the Pittsburgh-area one cited earlier.[160]

The tool kit is designed to further a number of objectives. A principal aim is advancing "AI readiness in Ohio schools."[161] To that readiness end, rather than being just a kind of playful exercise, this resource is intended to be a catalyst for conversations involving serious-minded topics such as ethics, student safety, data security, and privacy. It's also aimed at being useful to a variety of stakeholders: students, parents, families, teachers, administrators, legislators, policymakers, and more. The tool kit is being made available for voluntary usage rather than compulsory adoption by Ohio school districts. Another goal of the AI tool kit is to demystify AI, to show that societies have previously confronted new technologies and have adapted and thrived. In the words of an Ohio legislative proponent, then Lt. Gov. Jon Husted: "When you use the term AI, I know in some people's minds, it can sound scary.... Every technology that's come into society has been like that."[162] But Husted goes on to say that AI is the future and "it's important that students are exposed to it."

Circling back to the October 2023 CRPE study, that report suggested some reasons why some states are lagging when it comes to AI-related recommendations for their schools. One of their conclusions: most states "are cautious to influence AI decisions" at the individual school level and at the school district level.[163] The approaches form a patchwork quilt of stances: several examples include deferring to districts (Iowa, Montana, North Dakota, Wyoming), surveying school districts while giving wide latitude to individual school administrators to create AI policies "as deemed needed" (Vermont), and deferring to state legislatures or state boards until direction is provided by those entities (Colorado and Montana). Puzzlingly, some states like Kansas indicated to the researchers "that they have no plans to explore this topic in the foreseeable future."[164] As AI continues to become more prevalent throughout the world, and given the considerable revenue- and job-generating possibilities of AI, it will be interesting to watch how the policy positions of US states develop—or do not develop—in the coming years. In the call-to-action words of the prior CRPE study, though, "states have a responsibility to lead—and lead quickly."[165]

AI Uses in Education

AI technologies are already being used in educational settings in a number of ways that are impacting the creation of content. One topic sure to ignite strong opinions, especially among educators, is the issue of students who choose to use AI tech in the course of composing their assignments. Educators continue to discuss and debate the upsides and downsides of this practice along a continuum of acceptability and unacceptability. Some feel it is okay to use AI tech for brainstorming, for example, while others believe that these technologies are antithetical to the learning process and decrease individual human responsibility.

As discussed in an earlier section of this chapter, AI technologies are also being used at times by some academics. For example, some faculty members are using AI tech at various stages for research and scholarly communication creation, as evinced by the SAA guidelines on the use of AI for conference proposal submissions. Other uses are already giving rise to unintended consequences that have serious implications for science and societies. A March 2024 *New York Times* piece by neuroscientist Erik Hoel argues, as the title of the article indicates, that "AI-Generated Garbage Is Polluting Our Culture."[166] Highlighting a study of the impacts of chatbots on peer-reviewed studies, Hoel observes that even "the language of scientific research" shows evidence of transmutation, with certain words generated by chatbots showing up repeatedly in peer reviews. Hoel discusses a 2024 study, "Monitoring AI-Modified Content at Scale: A Case Study on the Impact of ChatGPT on AI Conference Peer Reviews," which found results suggesting "that between 6.5% and 16.9% of text submitted as peer reviews to these conferences could have been substantially modified by LLMs, i.e. beyond spell-checking or minor writing updates."[167] The researchers in this study discussed the implications of chatbot use on peer review, while calling "for future interdisciplinary work to examine how LLM use is changing our information and knowledge practices."[168] Hoel's article zeroes in on one of the study's big takeaways, namely, that "significant numbers of researchers at A.I. conferences were caught handing their peer review of others' work over to A.I.—or, at minimum, writing them with lots of A.I. assistance."[169] He also points out that more AI was found in the peer reviews as the deadline for the peer review submissions got closer and closer.

These practices raise difficult ethics questions and concerns:

- To further transparency, is there an ethical duty for faculty members to disclose when and how they use AI in reviewing research?
- Should faculty simply acknowledge their general use of AI, or should they specifically reveal which kinds of AI they are using, as some AI tech may be perceived as more reliable (or more problematic) than others?
- How accurate are the peer reviews that are conducted and created wholly or in part through the use of AI?
- Even when faculty disclose that they are using AI when reviewing research, does the use of AI tech in these ways create an air of impropriety and undermine trust and confidence in scholarly communication and science? Does AI-enabled peer review contribute to peer review being seen as a pro forma exercise that diminishes a view of research as a manifestation of scientific rigor and standards?
- What kinds of messaging does the use of AI for peer reviews send to students, budding researchers, and the public?
- How do grant providers, corporate interests, and governments feel about the use of AI to review research that they fund in part or in total?

Many if not all of these questions regarding the use of AI technology in peer reviews can also apply to AI tech used for grading, which is another area of potential AI usage that involves pluses and minuses. Other questions arise as well:

- Are there differences between using AI to grade, say, multiple choice quizzes and tests as opposed to grading essay exams?
- What are the duties to disclose the use of AI in grading?
- What are the rights of students to appeal grades and comments made by AI systems that students or their families believe are inaccurate or unjust?
- How will AI technologies effectively manage ADA accommodations that have been approved for student assignments?
- What will the explanations be for why faculty members are using AI to grade assignments rather than having a human grade submissions?

While writing this section, I serendipitously chanced upon an April 4, 2024, piece in the *Guardian* titled "Why I Wrote an AI Transparency Statement for My Book, and Think Other Authors Should Too."[170] The writer of the piece, Kester Brewin, is a British educator and author who researches AI's effects on the UK employment market. Brewin observes that AI is already impacting virtually all occupational sectors, including the writing profession. Because the maintenance of trust between authors and their readers is so important, he believes that until reliable tests have been developed that can detect AI in works, writers should proactively be transparent about whether they have or have not used AI tools in their writing processes. To build trust that he had not used AI technologies in writing his own book on AI, Brewin decided that he needed to include an "AI transparency statement." Not finding a template for such a statement, he created his own version. He determined that an effective AI transparency statement needed to address four issues: whether AI had generated any text, improved any text, suggested any text, or corrected any text.[171]

Because the act of reading a writer's work involves an investment of trust on the part of readers, Brewin asserts that the ability of AI machines to now write "profoundly undermines that historic trust."[172] Given that AI, like the atomic bomb, is "vastly powerful" and not going away, he concludes that "we have no choice now but to learn to survive alongside" this human invention in order to avoid a divisive "writing arms race." Being transparent about whether and how much AI has been used is a step that writers can proactively take to explain their own processes of writing and knowledge generation. Similarly, faculty members (as well as anyone who is writing and evaluating) would be well advised to disclose the roles of AI, if any, in their writing and grading processes, in order to promote transparency, accountability, and trust. I have used no aspect of AI at any stage of this book's writing, from beginning idea to end product. That was fortuitous because during a meeting with my book editor in 2024 when I anecdotally raised the issue of ways I had seen AI being used more and more in some professions, I learned that this book's publisher has an explicit prohibition on the use of AI by its contracted authors.

Educational venues that allow students, staff, and faculty to use AI tools are increasingly creating rules and guidelines for the responsible and transparent use of such tools. For example, instructors who permit students to use AI products in the course of creating assignments may specify that a student must disclose how and when they used a product. Some instructors and institutions also may set a limit on how much of an assignment can be composed by AI. Others may state that AI can only be used at the prompt stage. As AI sophistication develops, it can be expected that pedagogies for instructors and academic guidelines and requirements will need to be promulgated, regularly reviewed, and revised and adapted as necessary. The use of AI tools also has significant implications for academic integrity, as well as, in the converse, for plagiarism, copyright infringement, and research misconduct.

Another place where AI ethics issues are arising is in the college admissions context. A July 2024 *Education Week* piece looked at how students and other stakeholders are weighing ethical considerations of using AI tools when applying to universities and colleges. The article indicates that research conducted by a learning organization called foundry10 found that "about a third of high school seniors who applied to college in the 2023–24 school year acknowledged using an AI tool for help in writing admissions essays."[173]

Some students report using AI tools for brainstorming purposes to generate ideas and then write the rest of the essays themselves. Is that cheating or acceptable? Others utilize these tools to write parts or all of the actual essays. Do these purposes constitute cheating? Before answering, now consider the issue of college admissions coaches. People with money to do so can contract the services of individuals and companies that will not only write college admissions essays for students but also coach them through all levels of the college application process. These services are not cheap, as the *Education Week* article illuminates:

> College admissions coaching services typically cost from $60 to $349 per hour, according to data cited in Rubin's report from Prepmaven, an admissions-preparation company.
>
> The website of one such service, Ivy College Essay, advertises its Harvard connections. For between $1,500 and $4,800, depending on the number of applications, students receive help in brainstorming topics and "extensive written notes, comments, and guidance, focusing on both content and structure," according to the site.[174]

All of these admissions-related matters generate complicated questions that intersect with ethical principles, considerations of equal opportunity, transparency, and fundamental fairness:

- Is the hiring of college admissions coaches, as well as the provision of the services they can provide, cheating?
- Where are the lines between acceptable practices and unfair advantages?
- Who will decide the ethical guidelines for the use of AI tools and college admissions coaches in the higher education application process?
- Can the use of AI tools for writing college admissions essays by lower- and middle-income students be justified under a "leveling the playing field" rationale?

As the *Education Week* article points out, colleges and universities "permit students to use pricey admissions coaches... [b]ut they are mostly silent on how AI can be used in crafting essays."[175] Responses, or lack thereof, to the kinds of questions cited here will have consequences for students at all rungs of the socioeconomic ladder.

AI Advisors

The speed with which AI is advancing, coupled with the seemingly infinite ways that AI is influencing virtually every aspect of life, can feel unsettling and ungovernable at times. On top of this, the opaqueness of AI systems—shielded from transparency and accountability under the rationale of proprietary IP protection—contributes to what can feel like the purposeful mystification and inscrutability of these technologies. This sense of AI as being both "too complex to understand" and "too valuable to rein in and regulate" benefits Big Tech and other tech companies that control and leverage the IP rights to these technologies. The tech-driven laissez-faire message about AI props up a framework that comes at the expense of the general public, a framework that imbues and reinforces a psychological structure of inescapable "technology helplessness" (i.e., that nothing can be done to confront and modulate AI-driven forces and inequities).

Determining what we can do now to push back against this sense of helplessness and inevitability can lead to meaningful changes. The sobering reality of this thus-far-unequal power arrangement between Big Tech and the public gives rise to probing questions that can impel people to take action; namely, rather than solely or principally focusing on "how did we get here?" (which is still important to understand), expending time and energies on "what do we do now?" can be empowering and mobilizing:

- How can individuals themselves become more knowledgeable about AI?
- How can those who *are* knowledgeable about AI (e.g., information professionals) help others to become more knowledgeable about AI?
- What are some ways that citizens can become more capable of exercising agency over AI when so much about AI remains invisible and intentionally unknowable (to all but the chosen tech illuminati) by design?
- Who are trusted people and organizations that we can learn from and who are already speaking up about technologies like AI? What kinds of proactive activities are these individuals and groups engaging in to promote positive, substantive changes?

As shown in previous discussions, information professionals and centers of many kinds are stepping up to fulfill influential roles as advisors, guides, and teachers on all things related to AI. Others, too, are taking leading roles and responsibilities as activists, advocates, and watchdogs. Insights and lessons can be learned from a variety of fields and disciplines that are confronting the AI revolution. This in turn can inform human-centered strategic thinking and scenario planning by information professionals. In this section, we explore a few examples of organizations and individuals that are providing leadership roles on AI.

Because lawyers are members of a licensed profession, ethics rules mandate that they stay abreast of the latest developments in new technologies, in order to effectively represent and zealously advocate for their clients' interests. As mentioned previously, lawyers are also required not only to think about ethics but to comport themselves ethically as a requisite component of their licenses. The intersection between legal ethics and AI, hence, is catalyzing trailblazing work. Underscoring this reality, ABA President Mary Smith designated AI as a high priority for her 2023–2024 tenure. As a result, an ABA AI task force, with an ambitious, multipronged agenda, was created: "The mission of the AI Task Force is to: (1) address the impact of AI on the legal profession and the practice of law, and related ethical implications; (2) provide insights on developing and using AI in a trustworthy and responsible manner; and (3) identify ways to address AI risks."[176] One of the goals of the task force is to help users find information on the following issues: "(1) AI and the Legal Profession, (2) AI and Justice, (3) AI Governance, (4) AI Challenges: Generative AI, (5) AI in Legal Education, (6) AI Risk Management and (7) AI and the Courts."[177]

These seven critical areas of ABA focus can serve as a useful advisory source and framework for thinking about how information professionals and centers can proactively be prepared to respond to AI-related challenges in the near future of five, 10, and 20 years from now. The following are five critical issues of focus for the information professions.

Impacts of AI on Information Professions and Centers

All fields will ostensibly and eventually be enveloped by the tendrils of AI and its branching offshoots. The pivotal issue for every field to plan for now is potential obsolescence, namely, will the existence of AI make its human counterparts obsolete and replaceable by cheaper, faster, and more reliable AI? AI doesn't require health-care benefits or Family and Medical Leave Act (FMLA) coverage. AI doesn't need lunch breaks or weekends off. AI doesn't get the flu or need to leave work early to pick up a child from daycare or arrive at work late because of a freeway traffic jam or inadvertently delete the access codes for a departmental Zoom meeting. Returning to a cinematic lens, at some near or far future time,

- Will the increasing emergence and expansion of AI in our workplaces and societies resemble apocalyptic survival-of-the-fittest "alien replacement" plots, akin to that of films like John Carpenter's *The Thing* (1982) or the respective Eisenhower- and Carter-era versions of *Invasion of the Body Snatchers* (1956, 1978)? (These themes also have resonances with real-life disinformation conspiracy theories.)
- Is AI fated to more naturally and evolutionarily outgrow and outlive its human creators—in a humans-are-their-own-worst-enemies fashion—as shown in Steven Spielberg's *A.I.: Artificial Intelligence* (2001)?
- Is the vision of AI as a tool that aids and adds to human life, but doesn't supplant it, more likely, as depicted in the elegiacal indie film *After Yang* (2022)? In that inward-looking film, a grieving middle-class family faces unexpected feelings of legacy and loss following the unfixable mechanical breakdown and eventual death of its AI android child, named Yang.

What *is* clear is that ethics-minded thinkers from all compass points will continue to be needed to engage with some of the most consequential present *and* future decisions to ever face *Homo sapiens*. Examples of questions and issues regarding this topic include the following:

- How will AI impact the numbers and varieties of human staffing at information centers?
- How will AI influence supply and demand for information professionals?
- How will AI affect user perceptions of the reliability and availability of information available through information centers?
- In what ways will AI impact the kinds of information objects that information centers collect and offer and the types of experiences that they provide?
- How will AI influence decision-making about whether and when to provide and maintain brick-and-mortar libraries, archives, and museums?
- How will AI influence pricing structures for fee-based information services, like those provided by specialized libraries? Legal professions are already considering how AI will impact, for example, billable hours and the salary revenues of people working at all points in legal settings.

AI Influence on Access to Information, Technologies, and Human Information Professionals

Some people will likely always prefer interacting with humans, while others will prefer interfacing with AI. Some will want a mix of options. Examples of questions and issues in this area include the following:

- How will user preferences regarding AI be addressed by information professionals?
- What responsibilities, if any, do information professionals owe to those who never want to use or interact with AI?
- What new digital divides will be created via the use of AI?
- What are the ethical considerations to weigh in making decisions about using AI to carry out various functions of the information professions, such as reference services, collection development, programming, and classification of materials?

Generative AI Challenges for Information Professions

Some of this section overlaps with the two preceding sections, which identified some questions and issues that represent generative AI challenges for the information professions, particularly the existential issue of how AI may displace human information workers. Many other challenges remain, though. Some examples of other generative AI challenges for the information professions include the following:

- How will the information professions navigate present uncertainties involving copyright law and AI? For example, imagine a scenario in which a librarian wants to use an AI tool to generate images for use in a library program or exhibit. Legal issues may arise if those generated images implicate potential copyright infringement issues, particularly if they are used in marketing efforts. This scenario can also occur in an archive or a museum setting. Until case law provides some "bright lines" on what is and is not legally permissible, uncertainty will continue to exist for all fields and professions.

- Information professionals often work with data, whether using data, providing access to data, or stewarding data. AI tools occasionally hallucinate, generating false data. What then do information professionals need to do to put safeguards in place to ensure information professionals and information center users can rely on the integrity of data?
- It is well established that AI tools can perpetrate and perpetuate biases. These biases are invariably a result of variables such as the tools' technical components, the AI training data sets that may contain biases that are replicated in the AI tools, and the conscious or unconscious biases of the tools' human creators. What ethics guidelines will information professionals use to make decisions about, say, the selection of AI tools that do not subject themselves and external users to biases?

Roles of AI in Information/Data/Computing Education

Just as the legal profession is thinking about and will likely make curricular changes to address ways that AI is covered and included in legal education, the information professions are also considering and acting on these issues, though much more work is needed. Some of the questions and issues that are implicated are these:

- What kinds of courses need to include AI in their curricula and assessment outcomes? All or only some required courses? Some electives or all electives?
- Who will teach these courses?
- Where will instructors attain AI-related expertise?
- What professional development will be required or strongly encouraged?
- How will the role of ethics be addressed in courses that include exploration of AI?
- What are some of the chief student learning outcomes related to AI?
- In what ways, if any, will student learning outcomes vary for undergraduate and graduate students?
- What role will accrediting bodies play in setting out how AI will or should be addressed in degree programs?
- What courses may include or require student use of AI tools within courses to provide exposure and guidance?
- In what ways will academic integrity policies need to be updated to be current and keep pace with emerging technologies like AI?
- What policies and ethics guidelines are necessary to provide guardrails and parameters for how instructors use AI in their pedagogies, such as for lecture creation and delivery, grading of assignments, and creation of course materials?
- How will works-made-for-hire laws and policies need to be revised to address the use of AI in higher education settings and endeavors?

- Should degree programs require AI-related lab experience(s) or other types of experiential learning?
- What AI-related higher education partnerships can be created to foster learning, mentoring, and career opportunities?

Strategies for Managing AI Risks

Risks are a part of life. Some risks can be avoided or mitigated; others cannot. Managing risks is a vital activity for any individual and any organization, whether for-profit or not-for-profit. AI presents exciting opportunities. It also carries a number of known and unknown risks. A February 2024 NIST report encapsulates the potential risks and rewards of AI in its two-sentence abstract: "Artificial intelligence (AI) is increasingly affecting many aspects of our day-to-day lives. Although the benefits of AI to society are potentially transformative, many fear the cost to human rights may be too great without focused attention."[178]

What strategies then will information professionals devise and implement to avoid, mitigate, and manage risks? Some risk-reduction strategies have already been identified and discussed in this chapter. Others include

- crafting risk management checklists
- composing ethics codes and guidelines
- creating codes of conduct
- utilizing advisory boards that include diverse members to lessen blind spots and maximize a breadth of perspectives and lived experiences
- conferring with trusted legal counsel
- drafting and regularly reviewing and updating policies that address risks and responses to undesirable contingencies
- promoting transparency and accountability within and outside of information organizations

Overarching all of the AI focus areas above is this central question: what are the ethical considerations to weigh in making these kinds of decisions?

AI and Religion

Religious traditions and institutions have provided moral and ethical guidance and enacted strictures on human behavior throughout the millennia-long record of human history. Unsurprisingly, then, perhaps, religious orders today are also thinking about and advising on emergent technologies like AI. The Vatican, led from 2013 to 2025 by the 266th pontiff, Pope Francis, took a prominent role on AI and ethics, and with new technologies in general, as evinced by its collaboration with ITEC. At the Jesuit Santa Clara University, the Markkula Center's Framework for Ethical Decision Making, discussed earlier in this book, sets out important steps for grappling with ethical questions.[179] These include identifying ethics issues, obtaining the facts, and evaluating alternative actions.

A February 2024 article, "The Friar Who Became the Vatican's Go-to Guy on AI," reveals behind-the-scenes insights into how both the Roman Catholic Church and the Italian government get up to speed on "tricky questions."[180] They turn to Father Paolo Benanti, who was mentioned earlier in this chapter. He is an engineering-school-dropout-cum-ethics-professor at Rome's Pontifical Gregorian University, where he "spends his days thinking about the Holy Ghost and the ghosts in the machines."[181] In addition to advising, Father Benanti is a frequent participant on global AI panels, an author and lecturer, and a teacher of ethics and moral theology. In June 2023, he gave a globally accessible lecture titled "Algorethics: Opportunities and Challenges in the Age of AI."[182] He is one of an august cadre of several dozen AI experts invited to serve on the UN Advisory Body on AI. His most vital role, though, remains providing trusted counsel:

> For a church and a country looking to harness, and survive, the coming A.I. revolution, his job is to provide advice from an ethical and spiritual perspective. He shares his insights with Pope Francis, who in his annual World Day of Peace message on Jan. 1 called for a global treaty to ensure the ethical development and use of AI to prevent a world devoid of human mercy, where inscrutable algorithms decide who is granted asylum, who gets a mortgage, or who, on the battlefield, lives or dies.[183]

He is not persuaded that technology companies can regulate themselves; rather, his belief is that "some rules of the road" are necessary to protect democratic systems and ideals.[184] Far from a Luddite, he "also sees the potential of A.I." Encapsulating the ethical dilemmas with which he is wrangling, the article observes that "Father Benanti is concerned that A.I. could increase inequality, but he is also thinking about how it can help keep productivity afloat in Italy, which has one of the world's most aged populations."

One wonders, too, what Father Benanti's views may be of the intersection of AI algorithmic use with, for example, online gambling, given the historically fraught relationship between religions and industries that have been viewed as promoting moral vices. In the coming years, the ways positions on AI of the Vatican and other religious bodies develop, with the contributions of people like Father Benanti, will be instructive to watch. Recalling the Catholic Church's contentious history with matters of science, such as the seventeenth-century heresy trial of famed Copernican astronomer and polymath Galileo Galilei, the church's evolving relationship with technologies like AI is particularly noteworthy.

Although the Roman Catholic Church has taken the most visible religious faith role in talking about AI and endeavoring to influence policy on AI, other faith traditions are also engaging in conversations about AI. A March 2024 article, "Generative AI: Its Innovation, Implications and Use in the Christian World," on the *Presbyterian Outlook* web page shares seven brief interviews with "people in ministry who are finding innovative ways to use AI faithfully in the Christian world."[185] Biola University, a private Christian higher education institution in California, announced its launch of an AI lab "grounded in theological values" in June 2024.[186] In a Fox News article,

Dr. Michael J. Arena, the dean of the Crowell School of Business at Biola University, explained the focus of the AI lab, which is meant to be a place of discovery for students, faculty, and staff:

> The AI Lab at Biola University is unique in that it "places emphasis on moral and ethical discernment," rather than technical skills, Arena said....
>
> "By prioritizing ethical discernment, the lab strives to better facilitate the mitigation of potential biases, promote greater transparency and uphold human dignity within AI," Arena continued....
>
> Part of the work that will be done at the AI Lab at Biola University will involve the creation of AI applications for Christians.[187]

Arena explained that their goal is to respond to people's specific needs through "faith-contextualized tools" that can have concrete impacts.[188] He observed that AI is controversial but that educational institutions have to "embrace AI or risk falling behind."

Another example of religious faiths proactively engaging with ways to spiritually and socially engage with AI is an August 2024 collaborative summit, "AI & the Church Summit: Navigating the Ethical Frontier." This is a collaborative effort led by three Protestant Christian denominations: the Episcopal Church, the Presbyterian Church (USA), and the Evangelical Lutheran Church in America (ELCA). Participants included theologians, practitioners, bishops, professors, and laypeople; I was a participant. The initiative examined "AI's potential to address global challenges while critically examining ethical dilemmas like exacerbating inequality and threats to human dignity. We simply cannot shrink from the church's role in advocating for ethical, human-centered AI development that protects the vulnerable."[189] The late Pope Francis's AI ethics advisor, Father Paolo Benanti, who is also an advisor to the Italian government on AI ethics, served as the keynote speaker and guide for the group's conversations. A defining feature of this initiative is that the summit was not solely a one-off meeting; rather, the initiative was designed specifically to enable all of the participants to continue to work virtually over the course of a calendar year to advance the identified goals. This example illustrates efforts to bring together people from a variety of faiths to dialogue on AI ethics, shape policy, and work toward positive change, particularly for AI-insecure persons.[190]

In May 2024, the Islamic Center of Southern California hosted a one-day conference, "Islam and Artificial Intelligence: Ethical Horizons in a Digital Age." The conference's aim was to bring "together a diverse group of scholars, technologists, and community leaders to explore the intersections of artificial intelligence, ethics, and Islamic values."[191] Community is also a central tenet of the organizers: "The conference will delve into the ethical, social, and economic implications of artificial intelligence as seen through the eyes of the Muslim community and other conscientious thinkers. This event is not just an academic exploration but a community-building exercise where the wisdom of Islamic tradition meets modern technological innovation."[192] US Congressman Ro Khanna (D-California) delivered the keynote address. Examples of some of the ethics-related panel discussions include the following:

- "Policy and Ethics in AI": Exploring how biases in algorithms can impact society, with insights from Dina Chehata, Esq. (CAIR), and Sireen Sawaf, Esq. (Deputy City Attorney).

- "Muslim Women in AI": Focusing on transparency, trust, and legal risks, with panelists like Shabnam Tai Hofer, Esq., Leslie Hood (Microsoft), and Sujude Dalieh (Jazz Venture Partners).[193]

Judaism, too, is actively exploring and discussing AI's implications. A March 2024 *Mosaic* magazine article titled "What Artificial Intelligence Has in Store for Judaism," written by Moshe Koppel, a Computer Science Department faculty member at Bar-Ilan University, captures some of the nuanced and conflicted questions and feelings about AI: "AI has the potential to change the way Jews study Torah, observe Jewish law, work with rabbis, and teach their children. Will Jews resist those changes or welcome them?"[194] Koppel notes that AI will "offer great benefits, but will present even more serious challenges than the industrial revolution." The article is focused on two areas:

> First, drawing on work being done in my own AI lab in Israel, I will show how AI can provide tools that benefit Judaism by making Jewish texts and ideas more accessible. Second, I will suggest ways in which Judaism might, in return, offer models for purposeful and meaningful living, even as ubiquitous AI threatens to attenuate some of our deepest social and moral attachments.[195]

Through the lens of religion, scholars and commentators are also analyzing ways that religious texts and faiths can provide ethical and legal guidance on contemporary issues. One example is a 2023 piece by scholar and researcher Rabbi Yehuda Shurpin, "What Is the Jewish View on AI and Intellectual Property?"[196]

Buddhist ethics, drawn from the religious teachings of Buddhism, is an area that other thinkers and scholars are developing. A 2021 MIT *Technology Review* article, "What Buddhism Can Do for AI Ethics," identifies a Western-centric emphasis in the creation and adoption of AI ethics guidelines.[197] The author, Soraj Hongladarom, is a philosophy professor at Bangkok's Chulalongkorn University and critically analyzes the paucity of non-Western voices that have been included in the creation of many AI ethics guidelines and frameworks. Hongladarom points out that IEEE, the international professional organization for the engineering professions, and the European Union have created guidelines and frameworks on AI ethics. Other organizations have also explored and generated ethics principles addressing the creation and use of AI, though Hongladarom points out that "most of these guidelines are developed by groups or organizations concentrated in North America and Europe."[198]

Hongladarom acknowledges that there are areas of commonality and shared values among Western systems and Buddhism. Values like accountability, justice, and compassion are mutual ones. However, Hongladarom states that Buddhism looks at these values in some different ways and with different emphases that can be helpful in considering ethical uses of technologies. The author suggests that "the same is also true of many other non-Western value systems. AI ethics guidelines should draw on the rich diversity of thought from the world's many cultures to reflect a wider variety of traditions and ideas about how to approach ethical problems. The technology's future will only be brighter for it."[199]

The Conversation, in August 2023, looked at a number of thought-provoking ethical dilemmas in an article titled "Navigating the Intersection Between AI, Automation and Religion—3 Essential

Reads." One of the areas profiled is the use of robots to perform Hindu rituals in South Asia. Feelings are mixed about the role of robots and technological automation of religious rituals. Many people think that using robots to perform Hindu rites may cause departures from the religion. People also are uncertain about utilizing robotic technology "to embody religious and divine figures."[200] Many believe that the growth of robots within Hindu practices could lead to an increase in people leaving the religion.

> But there is another concern: whether robots could eventually replace Hindu worshippers. Automated robots would be able to perform rituals without a single error. This is significant because religions like Hinduism and Buddhism emphasize the correct execution of rituals and ceremonies as a means to connect with the divine rather than emphasizing correct belief.[201]

The examples described thus far in this section demonstrate ways that religions are grappling with challenging ethics-based questions about how to align new technologies like AI with religious teachings and cultural traditions. It is important to point out that those who do not follow faith-based practices, such as atheists, are also contributing to AI-centered discussions and scholarship. Writer Frank S. Robinson in a May 2024 *Medium* article, "Artificial Intelligence Explains There's No God," directed AI chatbot ChatGPT "to write a blog post on whether God exists."[202] He then included the unedited results of that prompt. This excerpt is ChatGPT's introductory statement:

> In a world filled with diverse beliefs and spiritual ideologies, the question of the existence of God has been a source of profound contemplation for centuries. Atheism, the rejection of the belief in gods or a higher power, has gained prominence as a rational and empirical alternative to religious doctrines. In this blog post, we will delve into the arguments that challenge the existence of God, exploring the philosophical and scientific perspectives that have led many to adopt atheism.[203]

Robinson appeared pleased overall with how ChatGPT composed the requested blog post, observing that "AI is faulted for occasionally 'hallucinating' false information. But on this subject, at least, it seems pretty clear-headed. It's human religious believers who are hallucinating."[204]

A March 2024 article in the *Harvard Gazette*, "It May Be Neither Higher nor Intelligence: Religious Scholars Examine Value, Limits of AI," is an apt encapsulation of the inchoate and evolving state of opinions on AI for many religious leaders and authorities at present. The article's author recounts an example that illustrates the range of reactions to AI. In a small German town in 2023, a sizeable group of people attended a Protestant church service that was largely conducted through AI. The religious ceremony included sermons, prayers, and music. This religious content was conveyed to the attendees via OpenAI's ChatGPT chatbot tool and four avatar humans visible on a screen.

> The experiment drew mixed reviews: Some detected a lack of authenticity, while others were enthusiastic about AI's potential to make services accessible and inclusive.

> Religious leaders and scholars are starting to wrestle with the ways in which AI may change religion—much as other emerging technologies over the centuries, from the printing press to radio to television to YouTube, have altered religious beliefs and practices.[205]

The article notes that Greg Epstein, Harvard University's humanist chaplain, is concerned that "people are treating AI itself as a higher power."[206] According to the article, Epstein asserts that Big Tech leaders are promoting "the glorification of AI," referencing examples of their grandiose language and evangelistic zeal. Epstein's 2024 book is titled *Tech Agnostic: How Technology Became the World's Most Powerful Religion, and Why It Desperately Needs a Reformation*. He encourages "people to be cautious when using AI without 'trying to eliminate it and go back to the days of no tech.'"[207]

AI Through Humanities and Interdisciplinary Lenses

Like Father Benanti and additional religious figures, other voices believe that protective guardrails are needed to fill in gaps in the world's AI laws, ethics, and policy infrastructure. Father Benanti's expertise straddles more than one domain, intersecting his positionality in academic and religious spheres. Academic researchers around the globe are also contributing ideas and study from diverse fields in furtherance of the development of interdisciplinary frameworks for addressing AI. Philosophy is just one field that is offering new thinking drawn from centuries of moral and spiritually grounded traditions and ways of living. Evan Selinger is a professor in the Philosophy Department at Rochester Institute of Technology (RIT) who believes that humanities scholars and disciplines have essential roles to contribute vis-à-vis emerging technologies like AI. He employs a humanities perspective and an interdisciplinary approach to pose questions about the ways AI can cause harm and what governments and AI tech companies can do to respond to and manage those harms. Sellinger also says that AI ethics is not simply about fixing technical aspects of the technology: "Philosophers and other humanities experts are uniquely skilled to address the nuanced principles, value conflicts, and power dynamics. These skills aren't just crucial for addressing current issues. We desperately need them to foster anticipatory governance."[208]

Selinger also shares his voice in ongoing efforts to define and provide guidance on what is referred to as "responsible AI." What that term means, though, in terms of measurable standards and accountability, is not yet a matter of consensus. The US government is actively engaged in research and development of AI for military and national security objectives, and Selinger collaborates on AI initiatives that are funded by the Defense Advanced Research Projects Agency (DARPA). Selinger explains the important skill sets he provides as a philosophy and humanities expert: "One of the skills I bring to the table is identifying core ethical issues in emerging technologies that haven't been built or used by the public."[209] He notes that after identifying the things that can go wrong with AI, it's then possible to put preventative steps in place to mitigate risks, like altering the technology's design. To further meaningful, substantive oversight and accountability, Selinger and others advocate the implementation of "AI audits" as a facet of responsible AI and AI ethics. Just as a financial audit is a procedure to ensure fiscal compliance with legal, ethical, and policy standards, AI audits are a process tool to periodically review and appropriately and proactively respond to AI-related activities within an organization.

More recent graduates of higher education humanities programs are also bringing valuable perspectives to AI ethics. Tess Buckley completed Northeastern University's graduate program in philosophy and AI in London in 2022. She articulates the role that philosophy-grounded lenses can provide for addressing AI ethics issues:

> Buckley feels philosophy is needed when confronting questions about the kind of role AI should play in society because these are age-old conundrums faced by people for centuries. "AI and generative AI are great but they are not new issues," she adds.
>
> They are just escalating issues that already exist. So this question of "Is this moral?"—that has been discussed in philosophy since forever ago.[210]

Another graduate of the same program in 2023, Tracy Woods, had observed how the General Data Protection Regulation (GDPR), implemented in the EU in 2018 and discussed in this book's earlier chapter on privacy and surveillance, "was going to create upheaval in how data was managed and governed" and "knew she needed additional skills to understand how AI might pose some difficult questions for developers and businesses."[211] Her tech employer, Cognizant, made many technically oriented learning opportunities available to people like herself, but she wanted to get "a handle on what AI's 'social impact' is likely to be." Woods refers to herself as a "pebble thrower"—someone who isn't afraid to ask questions that generate ripple effects in critically thinking about things like AI and the GDPR. To that end, she "helped found the workshop series 'Conversations on AI,' which looks at the technology from different perspectives, including regulation attempts and its impact on the creative industries."[212]

Whether it is the ability to spot ethics issues of emerging tech before something has been built, applying time-tested lessons and epistemologies of philosophy to new societal challenges, or taking a risk to be a pebble thrower and see where the ripples lead, humanities-oriented voices are making important contributions to AI ethics and the world. At a time when some are questioning the value of programs in humanities and the arts, these examples present persuasive arguments for those with these voices and points of view to be included in matters involving emerging technologies like AI.

AI Ethics Washing

In recent years, we have seen terms like *greenwashing* and *sportwashing* enter the lexicon. The *washing* part of the term in combination with whatever predicate signifies a kind of deceptive, obfuscatory effort by those who perpetrate the "washing." For instance, the term *greenwashing* refers to activities that are contrary to climate change mitigation efforts. As the UN describes, "By misleading the public to believe that a company or other entity is doing more to protect the environment than it is, greenwashing promotes false solutions to the climate crisis that distract from and delay concrete and credible action."[213] Similarly, sportwashing refers to efforts by entities and nations to use sports figures and events to distract from human rights concerns and abuses. Soccer (also called football), golf, and the Olympic Games are three areas that have all been at the center of various sportwashing controversies in recent years.

RIT's Selinger and others are speaking out about ethics washing, particularly related to AI. As Selinger observes, "AI can be co-opted into 'ethics washing'—weak checklists, flowery mission statements, and empty rhetoric that covers over abuses of power."[214] As examples of AI ethics washing, some companies, like Google/Alphabet and Facebook/Meta, have engendered criticism for showcasing splashy, headlines-garnering AI ethics-related initiatives, such as internal AI ethics boards. Some of these boards and individual members working on these boards and teams have run afoul of leadership within Big Tech companies when they have raised concerns about AI use by their organizations. Moreover, in some cases, individuals have even become whistleblowers and been terminated from these companies. Some of these terminated people have gone on to form their own start-up companies or have assumed advocacy roles calling for authentically responsible AI and AI ethics. One such person is Timnit Gebru, a computer scientist with globally respected expertise in AI and other tech areas like algorithmic bias. As a respected AI ethics expert, Gebru was employed with and prominently involved with AI ethics-based projects within Google until she was controversially fired in 2020. She subsequently founded the Distributed AI Research Institute (DAIR). The first and only sentence on DAIR's landing page succinctly communicates its reason for being: "The Distributed AI Research Institute is a space for independent, community-rooted AI research, free from Big Tech's pervasive influence."[215]

Initiatives like Joy Buolamwini's Algorithmic Justice League (AJL) and Timnit Gebru's DAIR strive to act as counterbalances to the massive influence and resources that are marshalled and distributed by Big Tech companies every day. AJL and DAIR are also examples of efforts to provide platforms for the voices and talents of BIPOC and other historically unwelcomed, unheard, and uninvited persons and communities in these incredibly impactful emerging-technology-based conversations and contested spaces.

AI and IP

Our perceptions of the seemingly unstoppable advance of AI can sometimes feel like standing in the path of the Great Wave depicted in Hokusai's celebrated 1831 woodblock print *Under the Wave off Kanagawa*: a mixture of awe and fear in the presence of an untamed, uncontainable force. Indeed, Pope Francis expressed exactly these kinds of conflicted emotions about AI in an unprecedented June 2024 speech to the G7 leaders "concerning the effects of artificial intelligence on the future of humanity."[216] As a point of fact, the speech's italicized byline reads "*An exciting and fearsome tool*."

This Janus-like view of AI's benefits and harms is evident, too, when Pope Francis stated that "perhaps we could start from the observation that artificial intelligence is above all else *a tool*. And it goes without saying that the benefits or harm it will bring will depend on its use. This is surely the case, for it has been this way with every tool fashioned by human beings since the dawn of time."[217] Of note at the very outset of the speech is the pope's exegetical grounding of AI from the Bible itself: "Sacred Scripture attests that God bestowed his Spirit upon human beings so that they might have 'skill and understanding and knowledge in every craft' (*Ex* 35:31). Science and technology are therefore brilliant products of the creative potential of human beings. Indeed, artificial intelligence arises precisely from the use of this God-given creative potential of human beings."[218]

The fear of AI identified by Pope Francis is palpable in the words expressed by stakeholders whose professional livelihoods are threatened by AI's advent and the potential extinguishing of their respective roles and outputs. Chief among these are visual artists and authors. In 2023 and 2024, a variety of authors, visual artists, publishers, and actors, including Sarah Silverman, filed more than a half dozen copyright infringement lawsuits against Big Tech companies, such as Microsoft's OpenAI, Meta/Facebook, Alphabet/Google, and more. The underlying basis (in legal parlance, *gravamen*) of all of these lawsuits is the assertion that tech companies have impermissibly used the plaintiffs' copyrighted content as AI training data in creating and continuing to perfect various AI products, with no remuneration to the parties whose content has been scraped and repurposed.

Commentators from diverse fields have suggested that copyright law may have a significant impact on the trajectory of AI and generative AI. A 2023 *New York Times* article quotes Fred Havemeyer, analyst at the financial research firm Macquarie: "Copyright will be one of the key points that shapes the generative A.I. industry."[219] The 2023 *Axios* article "Copyright Law Will Shape How We Use Generative AI" also emphasizes the importance of copyright on the future of AI.[220] Others, like IP scholar Jane Ginsburg at Columbia Law School, have observed that copyright infringement lawsuits, like the one that the *New York Times* filed against Microsoft and OpenAI in late December 2023 for the unauthorized use of *New York Times* content as generative AI training data, may hinge on whether such suits are decided on the basis of copyright law and fair use exceptions or the parties settle their disputes and enter into voluntary licensing agreements.[221]

The year 2024 saw a number of companies such as Reddit and Google enter into these kinds of licensing contracts. What is the upshot for Reddit in entering these kinds of quid pro quo legal arrangements? Money, in exchange for its users' social media data. It's important to remember that generative AI requires massive amounts of data that tech companies must secure in order to initially create and then continue to refine their products. Hence, we are likely to see tech companies seeking to strike deals like the one between Google and Reddit. In many cases, though, users are not on board with these deals. As a case in point, Meta/Facebook ignited vociferous outcries from privacy-concerned users of the Meta-owned Instagram social media platform when Meta announced on June 10, 2024, its intent to "train [an] AI model on European data."[222] Within days, Meta had done an about-face as TechCrunch summed up in its headline on June 14: "Meta Pauses Plans to Train AI Using European Users' Data, Bowing to Regulatory Pressure."[223] The EU's GDPR, discussed in this book's privacy chapter, is a bulwark against efforts by companies like Meta to harvest and monetize user data. However, no nationwide GDPR-like framework exists in the United States, where companies like Meta have been using the data of US users to train their AI products for years.[224]

Such efforts raise troubling ethics issues involving privacy, informed consent, IP ownership rights, and data agency and autonomy. A July 2024 research article, "The Great Scrape: The

Clash Between Scraping and Privacy," by Daniel J. Solove (George Washington University Law School) and Woodrow Hartzog (Boston University School of Law; Stanford Law School Center for Internet and Society) gives literary corporeal form to these formidable ethics disputes. The authors argue that those whom they name "scrapers" (i.e., tech companies) have blatantly ignored data protection and privacy laws in the unchecked creation of their AI systems that "depend on massive quantities of data, often gathered by 'scraping'—the automated extraction of large amounts of data from the internet."[225] This scraping of data has resulted in what Solove and Hartzog call the "great scrape." Consequently, to right this unjust taking of personally identifiable information that serves as uncompensated "grist for AI tools such as facial recognition, deep fakes, and generative AI," the researchers assert that "there must now be a great reconciliation"—namely, the scrapers' overdue reckoning with privacy law and the individuals and groups who believe their privacy rights have been violated.

Key AI and IP issues in the United States remain unsettled and uncertain. In the following sections, we examine developments in AI training-data-related copyright infringement lawsuits and the lack of clear lines of delineation on AI authorship and AI inventorship in copyright law and patent law, respectively. We also investigate some of the chief points of contention among the affected stakeholders.

Authorship and Inventorship

Probably the thorniest of all issues stemming from IP and AI is authorship and inventorship. A hallmark of the well-established laws and policies of copyrights and patents is the requirement of human authorship and inventorship. That is to say, under copyright law and policy, in order for a copyright to be granted by the US Copyright Office (USCO), the creator/author of that copyright must be a human being. (Efforts to have nonhumans be legally recognized as authors under US copyright law have been unsuccessful to date. In a case that garnered global attention, a Celebes crested macaque monkey named Naruto was held by the Ninth Circuit Court of Appeals to not be eligible for copyrights of the selfies he took after picking up a camera purposefully placed near him by British wildlife photographer David J. Slater in an Indonesian jungle.[226]) Similarly, patent law and policy require that an inventor be human in order to be granted a patent by the United States Patent and Trademark Office (USPTO).

Before approximately 2024, the USCO and USPTO both took rather firm stances on the uses of AI vis-à-vis copyright registrations and the granting of patents. Efforts by copyright applicants like Steven Thaler and Kristina Kashtanova to individually register works that they had created through the use of AI were turned down by the USCO. Thaler had attempted to secure a federal copyright on an AI-generated image titled "A Recent Entrance to Paradise." In August 2023, a federal judge dismissed Thaler's suit against the USCO for disallowing a copyright on this image created by an image generator in 2012, on the grounds that there was no human author.[227] Kashtanova used generative AI developed and owned by tech company Midjourney to create a comic book she named *Zarya of the Dawn*.

Kashtanova describes the comic as "an adventure of a non-binary person (Zarya) in different worlds to gather mental health tools to be able to handle their emotions and thoughts and to find connection with other people and creatures."[228] The comic is free to read and download after providing a first name and an email address. A donation option is present on the same access site for the comic. Kashtanova's initial application for a federal copyright was approved by the USCO on September 15, 2022. Following that award, however, the USCO learned of the use of AI in creating the comic's images. Subsequently, a February 21, 2023 USCO letter informed Kashtanova that its policy was to recognize Kashtanova as the author of the comic's textual elements "as well as the selection, coordination, and arrangement of the Work's written and visual elements," but that "the images in the Work that were generated by the Midjourney technology are not the product of human authorship."[229] The novel nature of this decision faced by both the applicant Kashtanova and the USCO makes the language of the body's decision particularly instructive:

> Because the current registration for the Work does not disclaim its Midjourney-generated content, we intend to cancel the original certificate issued to Ms. Kashtanova and issue a new one covering only the expressive material that she created.
>
> The Office's reissuance of the registration certificate will not change its effective date—the new registration will have the same effective date as the original: September 15, 2022. The public record will be updated to cross-reference the cancellation and the new registration, and it will briefly explain that the cancelled registration was replaced with the new, more limited registration.[230]

Interestingly, shifting to patents, Thaler has been equally unsuccessful in persuading the USPTO to grant him two patents for inventions created by a machine he calls DABUS: an acronym that stands for Device for the Autonomous Bootstrapping of Unified Sentience. The USPTO and lower courts all dismissed Thaler's patent-seeking efforts, and this decision was upheld by the higher US Court of Appeals for the Federal Circuit in August 2022.[231] A last-ditch writ of certiorari appeal by Thaler to the US Supreme Court sought to have the lower court rulings overturned. Thaler's Hail Mary bid to the nine justices was denied by that court in April 2023.[232] Hence, the US Court of Appeals' ruling stands "that patents can be issued only to human inventors and that his AI system could not be considered the legal creator of two inventions that he has said it generated."[233] Thaler also applied for patents on his DABUS inventions in the UK courts, appealing his losses all the way to the highest court in the land. As in the United States, a final decision came down for Thaler in December 2023 when the UK Supreme Court rendered judgment: "Now five Supreme Court judges have dismissed a bid to reverse those decisions, concluding that 'an inventor must be a person,' and that an AI cannot be named as an inventor to secure patent rights."[234]

Despite these rulings, in 2023 and 2024 the heretofore "hardball" positions of the USCO and USPTO in holding to strict interpretations of the human authorship and inventorship requirements shifted in some notable ways, while still adhering to the human author/inventor standard, for now. Before explaining how those positions have moderated and adapted to changing global circumstances, what accounts for these adaptive and evolving shifts by these two US IP entities? One reason is economic. AI-related technologies stand to generate almost unimaginable amounts of revenue. As a result, rhetoric from a range of corporate, financial, and governmental stakeholders indicates a desire to find some kind of middle ground that can assuage the concerns of artists, authors, and others whose livelihoods are directly threatened by AI encroachments on the products that they produce, while still enabling societies to foster and capitalize on the benefits that AI can provide. In other words, if AI is akin to the goose that laid the golden eggs in Aesop's famous fable, many parties have clear self-interests in not wanting the demise of the progenitor of those riches. To the contrary, indeed, myriad parties want a say *and* a stake in the riches and rewards that AI is projected to make manifest.

Another related reason why the USCO and USPTO are somewhat softening their positions on recognizing some roles for AI in author and inventor spaces in 2023 and 2024 is because of competition from other AI-developing nations. No country is battling the United States for supremacy in the global AI race more than China. Moreover, unlike the United States, China has recognized some narrow instances in which AI can be recognized as an author and inventor under Chinese copyright and patent laws and policies. China's first copyright infringement ruling in a dispute over AI-generated images was decided in November 2023.[235]

Given China's evolving stance on recognizing some roles for AI in the granting of copyrights and patents, it follows that this incentivizes the United States and other economic powers to adapt their legal, ethical, and policy frameworks, in order to remain competitive and not cede exclusive occupancy of the AI ground to China. Economic forces, thus, dictate that US laws, ethics, and policies on AI authorship and inventorship are likely to moderate and lean in toward global harmonization because it is in the interests of all parties to do so. The bottom line: no one wants to be the party left without a chair in an AI game of musical chairs.

A final explanation for the USCO and USPTO's graduated but perceptible movement toward greater acceptance of AI in IP schema is reality-based: now that AI has been unleashed on the world, like Pandora's proverbial box, it isn't disappearing anytime soon, and likely never will. The questions then aren't *whether* entities like the USCO and USPTO will ever modulate their laws and policies to recognize AI as authors and inventors, but rather

- *when* and *in what scenarios* they will make allowances, such as legal exceptions like fair use, for AI to be accorded sole legal status, or co-legal status as authors and inventors along with human collaborators,
- or, at the very least, when they will recognize AI as a legitimate tool that authors and inventors may use as a part of the author and inventor process.

Spoiler alert: that last part already occurred in 2023. On March 16, 2023, the USCO (and Library of Congress) published a policy statement entitled Copyright Registration Guidance: Works Containing Material Generated by Artificial Intelligence. The USCO one-sentence summary explains that "the Copyright Office issues this statement of policy to clarify its practices for examining and registering works that contain material generated by the use of artificial intelligence technology."[236]

This guidance document makes clear that human authorship is still requisite for the USCO to grant copyright status to a work submitted for review. However, what this document strives to do is to identify some instances in which the use of AI by a human author can result in copyright status being granted and some instances in which the opposite is the case. For example, the document specifies that

> if a work's traditional elements of authorship were produced by a machine, the work lacks human authorship and the Office will not register it. For example, when an AI technology receives solely a prompt from a human and produces complex written, visual, or musical works in response, the "traditional elements of authorship" are determined and executed by the technology—not the human user.[237]

Compare this with the converse fact patterns:

> In other cases, however, a work containing AI-generated material will also contain sufficient human authorship to support a copyright claim. For example, a human may select or arrange AI-generated material in a sufficiently creative way that "the resulting work as a whole constitutes an original work of authorship." Or an artist may modify material originally generated by AI technology to such a degree that the modifications meet the standard for copyright protection. In these cases, copyright will only protect the human-authored aspects of the work, which are "independent of" and do "not affect" the copyright status of the AI-generated material itself.[238]

As with the guidance issued by the USCO in 2023, on February 16, 2024, the USPTO published a policy document titled "Inventorship Guidance for AI-Assisted Inventions."[239] The summary states that

> Pursuant to the "Executive Order on the Safe, Secure, and Trustworthy Development and Use of Artificial Intelligence" (October 30, 2023), the United States Patent and Trademark Office (USPTO or Office) is issuing inventorship guidance for inventions assisted by artificial intelligence (AI). The guidance provides clarity for USPTO stakeholders and personnel, including the Central Reexamination Unit and the Patent Trial and Appeal Board (PTAB or Board), on how the USPTO will analyze inventorship issues as AI systems, including generative AI, play a greater role in the innovation process. This guidance explains that while AI-assisted inventions are not categorically unpatentable, the inventorship analysis should focus on human contributions, as patents function to incentivize and reward human ingenuity. Patent protection may be sought for inventions for which a natural person provided a significant contribution to the invention, and the guidance provides procedures for determining the same.[240]

These documents in total evince two government agencies that are responding in real time to changing domestic and global perceptions of AI. Information professionals need to stay attuned to these fast-changing developments to best understand how these issues are progressing and how they will impact users and communities. The following is a list of some of the types of ethics questions and issues that are implicated by the AI authorship and inventorship areas:

- What ethics issues are implicated and which ones are not addressed by guidelines on AI authorship and inventorship?

- How will users make sense of evolving guidelines on AI authorship and inventorship? What tools and resources are available for educational purposes?
- Are there channels for diverse input into the formulation and periodic revisiting and updating of guidelines like those for AI authorship and inventorship?
- Do these kinds of guidelines break down barriers for authors and inventors or do they reinforce existing barriers or create new barriers to IP creation and ownership?
- In what ways can information professionals aid users in better understanding and working with guidelines on AI authorship and inventorship?

AI Inputs and Outputs

Other as-yet-unresolved AI and IP areas that will shape the future of AI are AI inputs and outputs. The term *AI inputs* refers to AI training data, referenced in earlier parts of this chapter. Examples of AI training data include images, texts, books, poems, memes, prompts, posts, song lyrics, and other types of content. Tech companies claim that they need data to create their AI products, like ChatGPT. Unfortunately for those whose data was acquired by these tech companies, neither permission was obtained nor compensation given. Some of these individuals and institutions have turned to courts for redress, as described in *Marketplace*'s January 2024 story "The Economy and Ethics of AI Training Data": "Writer Michael Chabon is suing Meta. Getty Images is suing Stability AI. And both The New York Times and The Authors Guild have filed separate lawsuits against OpenAI and Microsoft. At the heart of these cases is the allegation that tech companies illegally used copyrighted works as part of their AI training data."[241]

Why do companies need AI training data? What are the alleged threats to their generative AI business models if they lose the copyright infringement suits facing them? In a June 2024 article about one of the copyright infringement lawsuits filed against Google by a proposed class action group of nine persons, Reuters reported that "Google responded in October [2023] that the use of publicly available data is necessary to train AI systems, and that the lawsuit would 'take a sledgehammer' to 'the very idea of generative AI.'"[242] As we know from this book's prior chapters on IP, the key defense to a claim of copyright infringement is fair use. Tech companies assert that "taking copyrighted material to train AI is legally 'fair use'—AI systems should be able to read and learn from the internet, just like humans do."[243]

On the flip side, artists, songwriters, actors, publishers, social media influencers, and many more forcefully rebut the fair use contention raised by Big Tech. Instead, they argue that the harvesting of their copyrighted content is out and out uncompensated theft. Strident opposition to the nonconsensual appropriation of their copyrighted works by technology companies is illustrated by the headlines from just a sampling of articles over the past couple of years:

- "Thousands of Authors Urge AI Companies to Stop Using Work Without Permission"[244]
- "The New York Times' AI Copyright Lawsuit Shows That Forgiveness Might Not Be Better Than Permission"[245]
- "How to Stop Your Data From Being Used to Train AI"[246]

- "Generative AI Is a Marvel. Is It also Built on Theft?"[247]
- "Generative AI Is Creating a Copyright Crisis for Artists"[248]

Embedded within these battles between content creators and AI tech companies are many ethics considerations. As examples:

- Even if the law were to ultimately side with tech companies under, say, a fair use analysis, as in the Google Books mass digitization case discussed in the IP chapter, is that the right thing to do morally?
- What about the rights of creators to not have their creative outputs vacuumed up, spit back out as new copyrightable works, and then commoditized, with no remuneration or attribution given to those who arguably made those new works possible?
- Indeed, the prior example is analogous to how indigenous persons have often expressed feeling when their traditional cultural expressions (TCE) and traditional knowledge (TK) are appropriated by nonindigenous people, repurposed, and monetized, without providing compensation or credit to the peoples and lands from where those raw materials were, in essence, harvested, mined, plundered, and profited on.
- Some tech companies like Meta/Facebook have claimed that terms of use and end user licensing agreements that users consent to when they sign up for accounts give Meta the right to make use of their data. Even if this is legally correct, though, is this an ethical practice for companies like Meta to engage in? Is there an unequal power differential between billion-dollar companies and individual users? Are users truly exercising "informed consent" when they click on terms-of-use boxes?

How these copyright infringement lawsuits will eventually be determined and resolved is anyone's guess at present. Legislative bodies, such as the US Congress, may also have pivotal roles to play in shaping these issues, regardless of or in addition to decisions made by judges. Additionally, executive branch actions can be taken, as in the executive order on AI signed by President Biden in late 2023.

Most societal discussions of AI to date have centered on AI inputs like training data. Yet AI outputs also present thorny legal, ethical, and policy issues. Such outputs may take the form of new images, books, plays, TV shows, web videos, songs, orchestral compositions, choreography, comic books, anime, manga, and many more creative works and genres. Now consider AI outputs created/generated by AI products that are owned by OpenAI, Google, Meta, Twitter/X, and other companies. Many questions arise with regard to these AI output works, as with AI inputs:

- Do tech companies own these new works outright, or should they share IP ownership with content owners whose works are hybridized within the new works created through generative AI?
- Who will make such determinations?
- What compensation, if any, is owed to those whose copyrighted content contributed to the manifestation of new AI-generated works?
- The copyright law of some nations, like France, recognizes moral rights, as discussed in the IP

chapter. Do copyright holders have moral rights to not have their works distorted or misused by AI products if they object to such practices?

- Do copyright owners have a moral right to have their original content attributed to them, if and when it is used to create new AI-generated works?
- What guidelines can be created to protect copyright creators from having their data used as AI training data without the creators' consent and provide them with more agency?

Regarding that last question, efforts are being taken to give creators more power and information to decide if and in what circumstances they agree to have their content used as AI training data. Fairly Trained is one nonprofit organization that is "trying to get a fairer deal for human creators."[249] The mission of the organization explains the need and significance of its advocacy:

> There is a divide emerging between two types of generative AI companies: those who get the consent of training data providers, and those who don't, claiming they have no legal obligation to do so.
>
> We believe there are many consumers and companies who would prefer to work with generative AI companies who train on data provided with the consent of its creators.
>
> Fairly Trained exists to make it clear which companies take a more consent-based approach to training, and are therefore treating creators more fairly.[250]

Fairly Trained has developed an initial certification, called an L certification, as a first step in encouraging parties to seek permission for using the data of copyright holders:

> We've launched the L certification as our first certification because we believe that AI companies getting licences for training data usage, rather than claiming fair use, should be applauded. We hope it will help reinforce the principle that rights-holder consent is needed for generative AI training. We don't propose this as an end to the debate over what creator consent should look like. But we do think there is a key difference between AI companies who scrape data and claim fair use, and AI companies who license training data. The Licensed Model certification is intended to highlight this difference.[251]

AI and Reliability

Of all the AI-related issues we've explored in this chapter and book to this point, reliability is arguably the single most consequential issue influencing AI's present and future. In short, the ability of people to be able to confidently rely on AI working accurately and consistently. AI technologies *can* clearly, in some cases, enable the discovery and accomplishing of great and novel things, as documented in this chapter. Yet, for all the successes of AI to date, copious examples can also easily be identified where AI technologies don't work, sometimes in jaw-dropping ways. As of mid-2024, empirical research and case study examples from numerous sectors demonstrate the continuing unreliability of AI. The unreliability of AI also has a direct correlation to whether individuals and groups can trust AI. Reliability and trust are indispensable to technologies. For example, people buy certain brands of cars because some cars are known to be more reliable than others. Based on reputation, track record, industry ratings, and owner reviews, people also develop trust or lack thereof in various car brands and models. Similarly, reliability and trust are integral to the aviation industry. For instance, public trust in Boeing-manufactured airplanes has been diminished over the past few years and recently because of high-profile public safety incidents with some Boeing jets. Statista's March 2024 article "Trust in Boeing Plummets After Safety Incidents" discussed survey research revealing that business travelers' trust in Boeing had diminished between the fourth quarter of 2023 and the first two months of 2024 by twenty-six points.[252]

How Boeing will restore reliability and trust in its jets is an open question. What this example underscores is that trust is an essential and highly variable factor in human activities and relationships. Research as well as lived experience tell us that once trust is damaged it can be difficult to repair and restore. A 2023 *Journal of Management* study, "How Can It Be Made Right Again? A Review of Trust Repair Research," provides an encapsulation of some fundamental aspects of trust and ways that it can be repaired:

> Trust is an integral part of a relationship and provides both interpersonal and organizational benefits. *Trust* is generally defined as involving risk and vulnerability. The presence of risk in such relationships increases the possibility of betrayal for the trustor. A betrayal invalidates the "trustworthy until proven otherwise" assumption of initial trust development and induces concern about additional harm in the trustor. Thus, trust repair is different from general trust development. Many scholars have examined the consequences of a betrayal and the process by which broken trust may be repaired. The intent of this paper is to review and integrate this research and suggest avenues to bolster the literature further.[253]

Reliability and trust are inextricably connected with the three foci of this book—ethics, information, and technology. Reliability and trust are also integral components of many of the topics investigated in this book, including ethical foundations and guardrails, accuracy of and access to information, social media use, human subjects research, AI, and other emerging technologies explored in the next chapter. Using a framework of reliability and trust, in this section we identify some of the most prominent reliability and trust issues related to AI. We also look at what is being done presently to address AI reliability and what can be done in the future.

Augmenting the reliability of AI may in turn potentially elevate public trust in AI. Voting integrity is a major concern of people and organizations dedicated to upholding free, fair, and accurate elections. How AI technologies may affect the integrity of upcoming elections remains a wild card. In the first part of 2024 alone, well-documented instances of AI already being used in ways that imperil election integrity, media accuracy and truthfulness, information quality, informed citizenries, and well-functioning democracies include

- fake social media accounts with false candidate disinformation
- use of AI voice technologies to replicate the voice of President Biden through bogus robocalls
- altered video clips, such as misleading edits of President Biden's appearances at the 80th anniversary of D-Day in Normandy, France

- inaccurate voting information pushed by AI chatbots, such as repeatedly providing the incorrect date of the November 2024 US presidential election

Some of these bulleted examples show AI's unreliability from the standpoint of functionality and design. Others represent examples of AI tech unreliability due to intentional human malfeasance and bad actors. Significant numbers of Americans express unease with and distrust of AI vis-à-vis elections. In a March 2024 survey, Pew Research found that "Americans' use of ChatGPT is ticking up, but few trust its election information."[254] The study strikingly found that "about four-in-ten adults have not too much or no trust in the election information that comes from ChatGPT. By comparison, just 2% have a great deal or quite a bit of trust."

Turning to the business sector, in June 2024 McDonald's, the largest fast-food chain in the world, announced that it had decided to halt its use of AI in drive-thru settings, in partnership with IBM. This decision was at least in part based on the current unreliability and lability of AI technology. Kotaku described the decision with the tongue-in-cheek headline: "McDonald's Drive-Thru AI Is McCanceled." The article highlighted how tech companies like Nvidia have ridden the AI wave in becoming some of the largest companies on the planet, while problems with AI tech still persist: "For example, McDonald's has just announced it's abandoning its disastrous experiment with AI, after it turned out people didn't want bacon on their ice cream."[255] McDonald's continues to experiment with incorporating AI into its businesses.

AI Hallucinations: The Good, the Bad, and the Ugly

Hallucinations by AI technologies are one of the principal factors influencing the reliability and trust of AI products and tools. IBM, one of the world's leaders in AI technologies, offers a helpful definition of AI hallucinations: "When a large language model (LLM) perceives patterns or objects that are nonexistent, creating nonsensical or inaccurate outputs."[256] AI hallucinations have led to embarrassing situations even for Big Tech companies, as IBM notes:

> Preventing issues with generative, open-source technologies can prove challenging. Some notable examples of AI hallucination include:

- Google's Bard chatbot incorrectly claiming that the James Webb Space Telescope had captured the world's first images of a planet outside our solar system.
- Microsoft's chat AI, Sydney, admitting to falling in love with users and spying on Bing employees.
- Meta pulling its Galactica LLM demo in 2022, after it provided users inaccurate information, sometimes rooted in prejudice.[257]

Before we explore some additional examples of AI hallucinations, it's important to acknowledge that these hallucinations are yielding some unexpected and beneficial scientific benefits. A December 2024 *New York Times* article discusses "How Hallucinatory A.I. Helps Science Dream Up Big Breakthroughs."[258] The article explains that "innovators are finding that A.I. hallucinations can be remarkably useful." Moreover, what we humans see as inaccurate or crazy made-up things that AI is generating also represent "new ideas" and "riots of unrealities that help scientists track cancer, design drugs, invent medical devices, uncover weather phenomena and even win the Nobel Prize." This kind of AI dreaming is giving new meaning to influential science fiction writer Philip K. Dick's musing—"I wonder what the machines will think"—from his short story *The Variable Man* (1953).[259]

Individuals, like companies, are not immune from AI hallucinations. Two New York attorneys, in an infamous high-profile example, submitted a legal brief to a Manhattan court that had been composed using the AI chatbot ChatGPT. Opposing counsel in reviewing the attorneys' filed brief "questioned the existence of the cases" cited.[260] The attorneys later admitted that they did not understand the technology they had employed. The court was not amused:

> A federal judge in New York City has ordered two lawyers and their law firm to pay $5,000 for submitting a brief with fake cases made up by ChatGPT and then standing by the research.
>
> In a June 22 decision, U.S. District Judge P. Kevin Castel of the Southern District of New York imposed the penalty on Peter LoDuca, Steven A. Schwartz and the firm Levidow, Levidow & Oberman.[261]

The trust that one of the attorneys had mistakenly and imprudently placed in ChatGPT continues to be an important lesson for anyone relying on emerging technologies: "At the sanctions hearing, Schwartz testified that he was 'operating under the false perception that [ChatGPT] could not possibly be fabricating cases on its own.'"[262]

Ethically, the decisions these two attorneys made in using a new technology, particularly for something as consequential as the representation of a client, were not in alignment with the Model Rules of Professional Conduct that guide and govern attorney conduct. Let's briefly look at the rules that are applicable to this fact pattern. Though not binding in any way on information professionals, there may be insights that can be applied to their contexts.

Rule 5.1 addresses the responsibilities of lawyers in managing roles within firms, such as partners, managers, and supervisory lawyers. It states that managing lawyers are responsible for all lawyers in the firm to "conform to the Rules of Professional Conduct."[263] Recall from earlier in this book that these rules equate to ethical standards and requirements for attorneys. Lawyers must maintain requisite ethical standards to ensure that they can retain their licenses to practice law and remain in good standing with the legal oversight entities that enforce ethical requirements. An implicit rationale for this, too, is that as a public policy Bar Associations, as well as societies in general, expect and require lawyers to possess and continuously maintain baseline levels of competency to effectively represent their clients. We look more at competency when we examine Rule 1.1 below.

Though we don't know if the lawyers in the prior example were being managed by anyone else, it's still important to underscore the supervisory and ethical responsibilities that attorneys owe to other attorneys in a firm. Managing attorneys are also responsible, under this rule, for other lawyers who violate the Model Rules of Professional Conduct in some instances.[264] Considering specifically, when AI tools are used in a firm, lawyers must ensure that use of those AI tools is in compliance with the Model Rules of Professional Conduct and that members of the law firm are being supervised in the use of such tools.

The takeaway from Rule 5.1 for information centers is that it is good ethical practice for all information professionals to adhere to their respective ethics code's principles in using technologies like AI. Additionally, information professionals should exercise appropriate managerial oversight over those who are using tools like these, as deemed necessary or relevant to their job responsibilities. Information professionals also have an ethical duty to maintain minimum levels of competency to effectively serve users and organizations. Evidence of this is borne out by competency standards that are set out by various information professional associations, like ALA.

Rule 1.1 for lawyers specifically addresses competence in the relationship between a lawyer and their client. An attorney is required to competently represent a client. The rule defines competent representation as including skill sets of "legal knowledge, skill, [and] thoroughness."[265] It also includes a level of preparation by the lawyers that is deemed "reasonably necessary for the representation." This rule is arguably the most important ethical guideline for lawyers, which likely explains why it is the very first rule, as all legal and ethical duties and responsibilities emanate from this standard. If a lawyer is not competent in exercising the duties owed to a client, the other model rules are essentially superfluous, as competence or lack thereof will influence all other matters. Of note, too, is the fact that the comment on Rule 1.1 specifices that lawyers have a responsibility for "maintaining competence."[266] Hence, attorneys choosing to use technologies like AI are required to possess the necessary knowledge and skill regarding AI, be thoroughly prepared to use AI, and maintain competence in these areas. Ethics codes for information professionals, such as ALA's, similarly address the need for professional development and competence. Specifically, ALA's principle 8 talks about striving for excellence by sustaining and honing one's knowledge and skills.[267] Moreover, the same principle delineates the need for fostering the professional development of coworkers.

It's important to highlight that lawyers are expected and required to be competent and maintain competence with respect to technology. Indeed, attorneys cannot avoid responsibility by simply saying they do not understand a technology. Comment 8 on Rule 1.1 specifically speaks to this issue. Lawyers are instructed that they "should keep abreast of changes" in laws and legal practice.[268] This means that attorneys should stay current with issues involving risks and benefits of using technologies. Lawyers should also ensure that they participate in "continuing study and education" and are in compliance with any continuing legal education requirements that are applicable to them. Applying Rule 1.1 and Rule 1.1's comment 8 to the two New York attorneys who negligently used ChatGPT in submitting a brief with nonexistent cases and citations, we can see that the attorneys breached their duties of competence. Additionally, they did not meet their duties to possess and "maintain the requisite knowledge and skill" of the ChatGPT technology that they chose to utilize in composing and then submitting their brief.[269]

Again, though these rules for attorneys are not directly applicable, it is ethically incumbent on information professionals to be competent in the duties that are owed to their patrons/customers/users. In addition, as with lawyers, information professionals should commit to securing and staying abreast of technologies that are relevant to the work of information centers and to the clients that they are pledged to serve. As for attorneys, thoroughness is a key aspect of competence.

One of the issues that concerns me in attending webinars in 2024 is that the presentations are not always as thorough as they should be regarding the risks aspect of assessing technologies (in conjunction with assessing the benefits), as required under the Model Rules examined here. For example, in some legal webinars I have attended, no mention has been made about the ongoing issue of AI hallucinations, which has yet to be decisively fixed. Similarly, in nonlegal webinars in which AI is discussed as a tool for educational purposes, such as assignment preparation, lecture creation, and even grading, it is essential to always explicitly acknowledge the risks and limitations of using AI to ensure that legitimate expectations are understood by all parties.

Growing numbers of voices in legal circles also remind lawyers that they need to be transparent in disclosing to clients how and when AI products are used in legal matters. This would include legal research as well as preparation of any legal documents like briefs, affidavits, and the like. Additionally, lawyers are advised to consider a variety of issues in setting their legal fees when using AI tools. For example, how much should a law firm charge if an AI legal research tool is used in the course of performing that research? Should a legal services charge be discounted because an AI legal research tool is used, or kept at the same rate that research would have been charged at if that tool had not been used? This is a gray area for lawyers in setting billable hours rates, which will continue to be refined as the efficacy of AI advances. Though many information professionals provide services for free, some information professionals are fee-based and work in for-profit settings. Examples of such professionals include competitive intelligence specialists, knowledge managers, chief information officers, data brokers, vendors, web developers, and more. As with attorneys, any information professional working with AI tools has an ethical responsibility to consider how the use of AI products influences the fees that they charge for their services.

It's crucial, as a side point of these discussions about technologies, to point out that information professionals have a duty to not engage in the unauthorized practice of law or medicine. To that end, information professionals can assist users in searching for information and accessing information, but they should never interpret legal and medical information, for fear of violating proscriptions that they not engage in legal and medical practice. Information centers often have guidelines explaining these distinctions. Information professionals who perform reference and other information services will often receive training in order to be mindful of these important bright lines.

As mentioned, the issue of AI hallucinations remains an area that has yet to be solved. To that point, it is essential for all members of societies to maintain healthy skepticism about the purported reliability of AI products. Illustrating this point through empirical research, a sextet of Stanford University and Yale University researchers in June 2024 published a study on the efficacy of AI legal research tools. The study, "Hallucination-Free? Assessing the Reliability of Leading AI Legal Research Tools," looked at the claims and accuracy of AI legal research tools of the companies Thomson Reuters and LexisNexis. The novel quality of this research is asserted by the researchers: "we design and report on the first pre-registered empirical evaluation of AI-driven legal research tools."[270] Often, companies will use a trade secrets rationale (i.e., we can't show you X because it is trade-secret-protected information) when asked about the workings of their products. The researchers point to this in their study, noting that "because of the closed nature of these systems, systematically assessing these claims is challenging." With that caveat, nevertheless, the study's findings are striking:

> We demonstrate that the providers' claims are overstated. While hallucinations are reduced relative to general-purpose chatbots (GPT-4), we find that the AI research tools made by LexisNexis (Lexis+ AI) and Thomson Reuters (Westlaw AI-Assisted Research and Ask Practical Law AI) each hallucinate between 17% and 33% of the time. We also document substantial differences between systems in responsiveness and accuracy.[271]

The findings are significant enough to warrant some additional parsing. Namely, the Thomson Reuters legal AI research product that the researchers examined generated results that were inaccurate one out of every three times. The LexisNexis product performed better than the Thomson Reuters product but was still wrong 17 percent of the time. These results necessitate a number of cautionary notes:

- First, anyone using these products must be wary of the outputs.
- Second, they must thoroughly and carefully review any outputs and correct any inaccuracies.
- Third, arguably there is a duty to disclose use of these products to relevant parties, which certainly includes clients.

Ethical Imperatives and AI Use Best Practices

A takeaway lesson from the previous section is that all people using *any* AI products should be wary of the outputs, thoroughly and carefully review any outputs and correct any inaccuracies, and disclose use of these products to relevant parties. As indicated by ethics codes and codes of conduct examined earlier in this book, information professionals across diverse sectors prize the accuracy of information and the provision of the most accurate information possible to users. To that end, information professionals have an ethical duty to assist users in thoroughly evaluating the use of any AI product. Information professionals also have an ethical duty to use AI products themselves with the highest degree of integrity and to be transparent and accountable in their use of all AI products. Users and societies deserve nothing less. Information professionals are uniquely trained and motivated to create best practices in the use of AI and should take leadership roles in stepping into these roles, as many already are.

Summary

The jury is still out on how the revolutionary powers of AI will impact the world's societies in the near term, let alone the far-flung future. AI technologies are already ushering in revolutionary discoveries and innovations in science and medicine. Yet AI is also raising concerns about its well-documented capabilities for promoting disinformation, bias, discrimination, job displacements, identity theft, nonconsensual image generation, copyright infringement, and privacy breaches, among other worrisome ills. Groups, religious orders, nations, and even the UN are dialoguing and creating ethics frameworks and guardrails in the attempt to nurture the public good that AI can do on the one hand while endeavoring to rein in its potential for unleashing known harms and unintended consequences on the other. Emblematic of AI's upsides and downsides, AI hallucinations can produce inaccurate, nonsensical information at times, while also providing scientists with innovative ideas that can spark scientific breakthroughs. Information professionals and centers have important leadership and helper roles to play vis-à-vis AI technologies, amid an uncertain but dynamically innovative environment.

Questions for Reflection, Exploration, Action, Discussion (READ)

1. Many information centers are adopting AI technologies for uses within their organizations. Imagine you are the manager of a library department that is increasingly using AI applications in your children's and young adult programming activities. A local reporter preparing a newspaper article on AI use in libraries has contacted you to get your comments, which your director has green-lit you to provide. Draft a brief set of talking points you can use to explain ways the library is working to prevent possible harms from use of AI technology by children and young adults, while also highlighting advantages of its use.
2. Online gambling addiction is an increasing public health emergency that is facilitated by AI algorithms. One of the chief demographic groups at risk for online gaming addiction are men aged 25–34. Develop an action plan that an information center can use to create a library program focused on raising awareness about the links between AI algorithms and online gambling addiction. Highlight resources that are available to combat addiction and support recovery. Include information on how users can take steps to protect their privacy from online gaming providers.
3. The use of AI tools in academic publishing is a growing concern. As a thought experiment, imagine you are a staff member in an archive, library, or museum. At an upcoming all-staff inservice for your institution, the small group to which you have been randomly assigned has been tasked to discuss the use of AI tools for organizational activities, like programming and exhibitions, and generate a one-page list of do's and don'ts. Create that list.
4. Imagine you are a digital services librarian working in a children's and young adult department of a public library. You've been contacted by a local sixth grade teacher who is teaching a course on new technologies like AI and would like to include you in a collaborative project involving the school, a large technology company based in Silicon Valley, and a state education agency. Compose a list of 5–10 questions you have about the collaboration before you can give an answer as to whether you and your library will or will not participate.
5. Some higher education faculty members are using (or advocating for) AI technologies not only to grade assignments but also to provide AI-generated comments on those assignments.

a. Identify some of the key positive and negative aspects of using AI to evaluate student assignments.
b. Discuss concerns that stakeholder groups, such as faculty members, university students, and the parents/guardians of university students, may have about the use of AI to grade and comment on student assignments.
c. Imagine you are a university professor:

1. Share your decision about whether you would or would not use AI to grade and comment on your students' assignments.
2. Articulate what you would tell students about your decision to use or not use AI to grade and comment on your students' assignments.

6. AI transparency statements are a way for writers to build trust with readers by revealing whether AI has been used by the writer in the process of writing.

a. Identify some reasons for and against AI transparency statements.
b. Talk about whether you would or would not include an AI transparency statement in your own work, if you use AI.

7. AI technologies are being deployed in ways that can ostensibly identify people who are experiencing homelessness.

a. Create a list of stakeholders and ethics issues that are implicated by the use of AI tech to identify homeless encampments.
b. If you were the mayor of a city considering whether or not to use AI technologies in this way, would you approve or not approve its use? Explain why or why not and discuss the basis for your rationale.

8. Whether AI can or should be a creator/author under US copyright law or an inventor under US patent law is a thorny area that involves aspects of ethics as well as law.

a. Identify the main stakeholder groups that are impacted by the decision as to whether AI can or cannot be an author or inventor.
b. Make a list of the ethics issues.
c. From what you currently know and understand about these complex issues, do you think AI should or should not be recognized as an author or inventor when copyright and patent applications are considered? Explain why or why not. What additional information would you like to know to make the best possible decisions?

9. Companies like Microsoft, Google/Alphabet, and Meta/Facebook have admittedly "scraped" and "harvested" copyrighted material from books and other sources to use as training data for generative AI systems like ChatGPT. Authors and other groups have filed copyright infringement lawsuits against these companies, alleging that the authors' permission was not attained and compensation from the companies was not provided.

a. Aside from the legal issues, what are the ethics issues involved in companies engaging in these kinds of "takings" of copyrighted materials to build proprietary technologies?
b. What do you think courts should do in deciding copyright infringement cases against tech companies that took copyrighted materials without permission from or compensation to authors?

10. A user asks a reference librarian to show them how to use ChatGPT to write a paper for their university course.

a. Analyze the ethical issues.
b. What do you think the reference librarian should do? Explain your reasoning.

11. Imagine you are a curator working in an art museum. Your board wants you to use generative AI to create new art from some of the historic and valuable artwork within your collection.

a. What are the ethics issues you need to consider?

b. Do you comply or not comply with the board's directive? Explain your thinking and decision.

12. Religious leaders from the world's major religions have taken proactive stances in calling for AI to be used ethically, in ways that ensure the dignity of all human beings and the protection of marginalized persons. Nonreligious leaders and organizations are also sharing their voices on AI ethics issues.

a. Do you think that religious leaders and religions or nonreligious leaders and organizations have roles to play in AI ethics? Explain.

b. What are some ways that religious or nonreligious leaders can influence the ethical development, dissemination, and use of AI products?

13. Consider the following, regarding college admissions.

a. Is it ethical for students to use AI tools when writing their college admissions essays?

b. Is it ethical for students to hire college admissions coaches to aid them in applying to colleges and universities?

c. What ethical guidelines, if any, would you recommend for guiding these kinds of college admissions considerations?

14. Create a best practices tool kit of ethical decision-making approaches, codes, guidelines, and resources that can be used to make policy decisions about the kinds of AI issues and tools profiled in this chapter.

Notes

1. Stillo, S. (n.d.). *Incunabula: The art & history of printing in Western Europe, c. 1450–1500.* Library of Congress. Retrieved February 24, 2025, from https://www.loc.gov/ghe/cascade/index.html?appid=580edae150234258a49a3eeb58d9121c
2. Summerscale, K. (2016, April 30). Penny dreadfuls: The Victorian equivalent of video games. *Guardian*. https://www.theguardian.com/books/2016/apr/30/penny-dreadfuls-victorian-equivalent-video-games-kate-summerscale-wicked-boy
3. Leaffer, M. (2005). *Understanding copyright law* (4th ed.). LexisNexis.
4. Merriam-Webster. (n.d.). *Luddite.* Retrieved February 24, 2025, from https://www.merriam-webster.com/word-of-the-day/luddite-2018-10-10
5. *Ecclesiastes 1:9.* (n.d.). Good News Translation (US Version). Retrieved February 24, 2025, from https://www.bible.com/bible/69/ECC.1.9.GNTD#
6. Britannica. (n.d.). *Technology*. Retrieved February 24, 2025, from https://www.britannica.com/technology/technology
7. Enebrad, C. (2024, October 28). *No, the last Blockbuster Video in Bend is not closing, despite viral social media post.* https://www.centraloregondaily.com/news/local/bend-blockbuster-video-closing-hoax/article_2d4896d2-957c-11ef-a0bd-ff1d5cf53a5d.html
8. Sinnott, A., & Zhou, K. Z. (2024, August 2). Governance implications toward safer online blockchain communities: A case study of Bored Ape NFTs. *Proceedings of the TPRC2024 The Research Conference on Communications, Information and Internet Policy*. https://papers.ssrn.com/sol3/papers.cfm?abstract_id=4913319&download=yes

9. Ibid.
10. Ibid.
11. Ibid.
12. Ibid.
13. Hissong, S. (2021, November 1). How four NFT novices created a billion-dollar ecosystem of cartoon apes. *Rolling Stone*. https://www.rollingstone.com/culture/culture-news/bayc-bored-ape-yacht-club-nft-interview-1250461/
14. Van Boom, D. (2022, August 11). Bored Ape Yacht Club NFTs explained. *CNET*. https://www.cnet.com/culture/internet/bored-ape-yacht-club-nfts-explained/
15. Sinnott, A., & Zhou, K. Z. (2024, August 2). Governance implications toward safer online blockchain communities: A case study of Bored Ape NFTs. *Proceedings of the TPRC2024 The Research Conference on Communications, Information and Internet Policy*. https://papers.ssrn.com/sol3/papers.cfm?abstract_id=4913319&download=yes
16. de Best, R. (2025, January 30). *Daily sales volume of NFT collection Bored Ape Yacht Club to January 30, 2025*. Statista. https://www.statista.com/statistics/1265505/bored-ape-yacht-club-nft-development/
17. Melinek, J. (2022, October 27). *Yuga Labs' Nicole Muniz to talk about NFTs and Bored Apes at TC Sessions: Crypto*. TechCrunch. https://techcrunch.com/2022/10/27/yuga-labs-nicole-muniz-to-talk-about-nfts-and-bored-apes-at-tc-sessions-crypto/
18. Rapoza, K. (2023, October 10). Bored Ape Yacht Club owners restructuring emblematic of NFT shakeup; billions lost in value. *Forbes*. https://www.forbes.com/sites/digital-assets/2023/10/10/bored-ape-yacht-club-owners-restructuring-emblematic-of-nft-shakeup-billions-lost-in-value/
19. Tremayne-Pengelly, A. (2023, January 18). Bored Apes were at the center of the NFT Boom. Now they're mired in lawsuits and plunging in value. *Observer*. https://observer.com/2023/01/bored-apes-were-at-the-center-of-the-nft-boom-now-theyre-mired-in-lawsuits-and-plunging-in-value/
20. NFTs*: The evolving landscape of the NFT market.* (n.d.). dappGambl. Retrieved February 24, 2025, from https://dappgambl.com/nfts/dead-nfts/
21. Deliso, M. (2025, February 24). *Theranos founder Elizabeth Holmes' conviction upheld by US appeals court*. ABC News. https://abcnews.go.com/Business/theranos-founder-elizabeth-holmes-conviction-upheld-us-appeals/story?id=119135714
22. Knolle, S. (2024, June 9). *AI copyright issues "on shifting sands" but legal protections are coming, experts tell PGA produced by conference*. The Wrap. https://www.thewrap.com/ai-copyright-issues-pga-produced-by-conference-producer/
23. Leswing, K. (2024, June 18). *Nvidia passes Microsoft in market cap to become most valuable public company*. CNBC. https://www.cnbc.com/2024/06/18/nvidia-passes-microsoft-in-market-cap-is-most-valuable-public-company.html
24. Heaven, W. D. (2024, July 10). What is AI? *MIT Technology Review*. https://www.technologyreview.com/2024/07/10/1094475/what-is-artificial-intelligence-ai-definitive-guide/
25. Stern, J. (2023, March 26). AI is like… nuclear weapons? *Atlantic*. https://www.theatlantic.com/technology/archive/2023/03/ai-gpt4-technology-analogy/673509/
26. Kahn, J. (2023, June 30). The atom bomb or the airliner? The battle over A.I. analogies will determine the fate of A.I. regulation. *Fortune*. https://fortune.com/2023/06/30/airliner-or-atom-bomb-microsoft-brad-smith-ai-regulation-europe/
27. Philbrick, I. P., & Wright-Piersani, T. (2023, June 10). *New York Times*. https://www.nytimes.com/2023/06/10/upshot/artificial-intelligence-nuclear-weapons-quiz.html

28. Matthews, D. (2023, June 29). *AI is supposedly the new nuclear weapons—but how similar are they, really?* Vox. https://www.vox.com/future-perfect/2023/6/29/23762219/ai-artificial-intelligence-new-nuclear-weapons-future#
29 Kristof, N. (2024, July 17). A.I. may save us or may construct viruses to kill us. *New York Times*. https://www.nytimes.com/2024/07/27/opinion/ai-advances-risks.html
30. Brooks, D. (2024, July). Many people fear A.I. They shouldn't. *New York Times*. https://www.nytimes.com/interactive/2024/07/31/opinion/ai-fears.html
31. World Intellectual Property Organization. (n.d.). *Frequently asked questions: AI and IP policy.* Retrieved February 24, 2025, from https://www.wipo.int/about-ip/en/artificial_intelligence/faq.html
32. Ibid.
33. Monahan, J. (n.d.). *Artificial intelligence, explained*. Heinz College, Carnegie Mellon University. Retrieved February 24, 2025, from https://www.heinz.cmu.edu/media/2023/July/artificial-intelligence-explained
34. Ibid.
35. EU Artificial Intelligence Act. (2025, February 2). *Article 3: Definitions.* https://artificialintelligenceact.eu/article/3/
36. Corbyn, Z. (2024, June 29). AI scientist Ray Kurzweil: "We are going to expand intelligence a millionfold by 2045." *Guardian*. https://www.theguardian.com/technology/article/2024/jun/29/ray-kurzweil-google-ai-the-singularity-is-nearer
37. Ibid.
38. Ibid.
39. Ibid.
40. O'Neil, C. (2016). *Weapons of math destruction: How big data increases inequality and threatens democracy*. Crown Books.
41. Buolamwini, J. (2016, November). *How I'm fighting bias in algorithms*. TED. https://www.ted.com/talks/joy_buolamwini_how_i_m_fighting_bias_in_algorithms?language=en
42. Algorithmic Justice League. (n.d.). *Our mission.* Retrieved February 24, 2025, from https://www.ajl.org/about
43. Mosley, T. (Host). (2023, November 28). "If you have a face, you have a place in the conversation about AI," expert says. *Fresh Air*. https://www.npr.org/2023/11/28/1215529902/unmasking-ai-facial-recognition-technology-joy-buolamwini
44. Ibid.
45. Castro-Pagán, C. (2024, July 13). Workday loses bid to toss bias claims over AI hiring tools. *Bloomberg Law*. https://news.bloomberglaw.com/litigation/workday-loses-bid-to-toss-bias-claims-over-ai-screening-tools
46. Rhoades, G. (2024, June 3). *Ghosts in the machine: How past and present biases haunt algorithmic tenant screening system*. American Bar Association. https://www.americanbar.org/groups/crsj/publications/human_rights_magazine_home/technology-and-the-law/algorithmic-tenant-screening-systems/
47. Santarino, A., & Piffaré, R. T. (2024, July 18). An algorithm told police she was safe. Then her husband killed her. *New York Times*. https://www.nytimes.com/interactive/2024/07/18/technology/spain-domestic-violence-viogen-algorithm.html
48. MacMillon, D., & Rowland, C. (2024, April 1). Assisted living managers say an algorithm prevented hiring enough staff. *Washington Post*. https://www.washingtonpost.com/business/2024/04/01/assisted-living-algorithm-staffing-lawsuits-brookdale/
49. Ibid.

50. Gregory, A. (2024, March 29). Sepsis blood test combined with AI could offer early detection tool. *Guardian*. https://www.theguardian.com/society/2024/mar/29/sepsis-blood-test-combined-with-ai-could-offer-early-detection-tool
51. Associated Press. (2024, March 29). New York City to test AI-enabled gun scanners in subway system. *Guardian*. https://www.theguardian.com/technology/2024/mar/29/new-york-city-subway-ai-gun-scanners
52. Marchant, J., & Nature Magazine. (2023, October 17). AI reads ancient scroll charred by Mount Vesuvius in tech first. *Scientific American*. https://www.scientificamerican.com/article/ai-reads-ancient-scroll-charred-by-mount-vesuvius-in-tech-first/
53. Ibid.
54. Sample, I. (2024, February 5). AI helps scholars read scroll buried when Vesuvius erupted in AD79. *Guardian*. https://www.theguardian.com/science/2024/feb/05/ai-helps-scholars-read-scroll-buried-when-vesuvius-erupted-in-ad79
55. Ibid.
56. Ibid.
57. Ibid.
58. Orwell, G. (1949). *1984*. Secker & Warburg.
59. Gramlich, J. (2022, September 14). *As more states legalize the practice, 19% of U.S. adults say they have bet money on sports in the past year*. Pew Research Center. https://www.pewresearch.org/short-reads/2022/09/14/as-more-states-legalize-the-practice-19-of-u-s-adults-say-they-have-bet-money-on-sports-in-the-past-year
60. Statista Research Department. (2024, October 29). *Gambling in the United States—statistics & facts*. https://www.statista.com/topics/1368/gambling/
61. Parry, W., & the Associated Press. (2024, February 24). So many Americans were gambling on the Super Bowl online that transactions hit nearly 15,000 per second. *Fortune*. https://fortune.com/2024/02/12/super-bowl-sports-gambling-online-betting/
62. Casaletto, L. (2023, August 29). *Ontario to ban gambling commercials featuring athletes, celebrities*. City News. https://toronto.citynews.ca/2023/08/29/ontario-online-betting-commercials-banned-igaming-agco/
63. Austen, I. (2024, March 4). A call to end gambling ads as athletes and celebrities are shut out of them. *New York Times*. https://www.nytimes.com/2024/03/09/world/canada/canada-sports-gambling-advertising.html
64. Ibid.
65. Ibid.
66. Ibid.
67. Wertheim, J. (2024, June 30). Technology has fueled a sports betting boom and a spike in problem gambling, addiction therapist warns. *60 Minutes*. https://www.cbsnews.com/news/technology-fuels-sports-betting-boom-and-problem-gambling-spike-addiction-therapist-warns-60-minutes-transcript/
68. Ibid.
69. Ibid.
70. Ibid.
71. Ibid.
72. Ibid.
73. Crisis Text Line. (n.d.). *About us*. Retrieved February 24, 2025, from https://www.crisistextline.org/about-us/

74. Ibid.
75. Crisis Text Line. (n.d.). *Data philosophy.* Retrieved February 24, 2025, from https://www.crisistextline.org/data-philosophy/
76. Resnick, B. (2018, June 9). *How data scientists are using AI for suicide prevention*. Vox. https://www.vox.com/science-and-health/2018/6/8/17441452/suicide-prevention-anthony-bourdain-crisis-text-line-data-science
77. Ibid.
78. Levine, A. S. (2022, January 28). Suicide hotline shares data with for-profit spinoff, raising ethical questions. *Politico*. https://www.politico.com/news/2022/01/28/suicide-hotline-silicon-valley-privacy-debates-00002617
79. Ibid.
80. Ibid.
81. Hicks, J., & Lawler, R. (2022, February 1). Crisis Text Line stops sharing conversation data with AI company. *Verge*. https://www.theverge.com/2022/1/31/22906979/crisis-text-line-loris-ai-epic-privacy-mental-health
82. Crisis Text Line. (2022, January 31). *An update on data privacy, our community and our service.* https://www.crisistextline.org/blog/2022/01/31/an-update-on-data-privacy-our-community-and-our-service/
83. US Department of Housing and Urban Development. (2023, December). *The 2023 Annual Homelessness Assessment Report (AHAR) to Congress.* https://www.huduser.gov/portal/sites/default/files/pdf/2023-AHAR-Part-1.pdf
84. Dickey, M. R., Fitzpatrick, A., & Feng, A. (2024, January 4). California tops nation in percentage of unsheltered homelessness. *Axios*. https://www.axios.com/local/san-francisco/2024/01/04/california-homeless-statistics-high
85. US Department of Housing and Urban Development. (2023, December). *The 2023 Annual Homelessness Assessment Report (AHAR) to Congress.* https://www.huduser.gov/portal/sites/default/files/pdf/2023-AHAR-Part-1.pdf
86. Dickey, M. R., Fitzpatrick, A., & Feng, A. (2024, January 4). California tops nation in percentage of unsheltered homelessness. *Axios*. https://www.axios.com/local/san-francisco/2024/01/04/california-homeless-statistics-high
87. Ibid.
88. Feathers, T. (2024, March 25). Revealed: A California city is training AI to spot homeless encampments. *Guardian*. https://www.theguardian.com/technology/2024/mar/25/san-jose-homelessness-ai-detection
89. Pew Research Center. (2023, February 10). *Half of Americans or more aware of common uses of AI, but fewer can identify AI's role in all six examples*. https://www.pewresearch.org/science/2023/02/15/public-awareness-of-artificial-intelligence-in-everyday-activities/ps_2023-02-15_ai-awareness_00-01-png/
90. Tyson, A., & Kikuchi, E. (2023, August 28). *Growing public concern about the role of artificial intelligence in daily life*. Pew Research Center. https://www.pewresearch.org/short-reads/2023/08/28/growing-public-concern-about-the-role-of-artificial-intelligence-in-daily-life/
91. Faverio, M., & Tyson, A. (2023, November 21). *What the data says about Americans' views of artificial intelligence*. Pew Research Center. https://www.pewresearch.org/short-reads/2023/11/21/what-the-data-says-about-americans-views-of-artificial-intelligence/

92. Tyson, A., & Kennedy, B. (2024, March 26). *Many Americans think generative AI programs should credit the sources they rely on*. Pew Research Center. https://www.pewresearch.org/short-reads/2024/03/26/many-americans-think-generative-ai-programs-should-credit-the-sources-they-rely-on
93. Orsi, J. (2024, June 27). *We asked people about using AI to make the news. They're anxious and annoyed*. Poynter. https://www.poynter.org/ethics-trust/2024/we-asked-people-about-using-ai-to-make-the-news-theyre-anxious-and-annoyed/
94. Ibid.
95. Asilomar AI Principles. (n.d.). *Open letter.* Retrieved February 24, 2025, from https://futureoflife.org/open-letter/ai-principles/
96. Ibid.
97. RenAIssance Foundation. (n.d.). *The call.* Retrieved February 24, 2025, from https://www.romecall.org/the-call/
98. RenAIssance Foundation. (2023, January 17). *The Abrahamic commitment to the Rome Call for AI Ethics.* https://www.romecall.org/the-abrahamic-commitment-to-the-rome-call-for-ai-ethics-10th-january-2023/
99. RenAIssance Foundation. (2024, June 25). *AI ethics for peace: World religions commit to the Rome Call.* https://www.romecall.org/ai-ethics-for-peace-world-religions-commit-to-the-rome-call/
100. Ibid.
101. Ibid.
102. Ibid.
103. AI Ethics for Peace. (2024, July 12). *L'Osservatore Romano*. https://www.osservatoreromano.va/en/news/2024-07/ing-028/ai-ethics-for-peace.html
104. Ibid.
105. Ibid.
106. Ibid.
107. Peters, G., & Woolley, J. T. (2023, October 30). *Fact sheet: President Biden issues executive order on safe, secure, and trustworthy artificial intelligence*. American Presidency Project. https://www.presidency.ucsb.edu/documents/fact-sheet-president-biden-issues-executive-order-safe-secure-and-trustworthy-artificial
108. Ibid.
109. Johnson, K. (2023, November 1). Joe Biden wants US government algorithms tested for potential harm against citizens. *Wired*. https://www.wired.com/story/joe-biden-wants-us-government-algorithms-tested-for-potential-harm-against-citizens/
110. Ibid.
111. *Bletchley Declaration by countries attending the AI safety summit.* (n.d.). Retrieved February 24, 2025, from https://www.gov.uk/government/publications/ai-safety-summit-2023-the-bletchley-declaration/the-bletchley-declaration-by-countries-attending-the-ai-safety-summit-1-2-november-2023
112. Ibid.
113. Ibid.
114. Tarka, J. T., & Shapiro, P. (2024, February 5). EU reaches agreement on AI act text: What this means for employers. *National Law Review*. https://www.natlawreview.com/article/eu-reaches-agreement-ai-act-text-what-means-employers

115. Beverley-Smith, H., & Perowne, C. (2024, February 6). *EU Artificial Intelligence Act—final form legislation endorsed by member states*. JDSupra. https://www.jdsupra.com/legalnews/eu-artificial-intelligence-act-final-3315649/
116. Volpicelli, G. (2024, February 2). EU countries strike deal on landmark AI rulebook. *Politico*. https://www.politico.eu/article/eu-countries-strike-deal-ai-law-act-technology/
117. European Commission. (2024, August 1). *AI Act enters into force*. https://commission.europa.eu/news/ai-act-enters-force-2024-08-01_en
118. Stokel-Walker, C. (2024, February 7). Europe's new AI rules could go global—here's what that will mean. *Scientific American*. https://www.scientificamerican.com/article/europes-new-ai-rules-could-go-global-heres-what-that-will-mean/
119. European Union. (n.d.). *AI Act*. Retrieved February 24, 2025, from https://digital-strategy.ec.europa.eu/en/policies/regulatory-framework-ai
120. European Commission. (2024, November 14). *Commission publishes first draft of General-Purpose Artificial Intelligence Code of Practice*. https://digital-strategy.ec.europa.eu/en/news/commission-publishes-first-draft-general-purpose-artificial-intelligence-code-practice
121. Ibid.
122. Ibid.
123. Allyn, B. (2024, September 29). California Gov. Newsom vetoes AI safety bill that divided Silicon Valley. *Morning Edition*. https://www.npr.org/2024/09/20/nx-s1-5119792/newsom-ai-bill-california-sb1047-tech
124. Korte, L., & White, J. B. (2024, September 29). Gavin Newsom vetoes sweeping AI safety bill, siding with Silicon Valley. *Politico*. https://www.politico.com/news/2024/09/29/gavin-veto-ai-safety-bill-00181583
125. *Governor Newsom signs bills to crack down on sexually explicit deepfakes & require AI watermarking*. (2024, September 19). Governor Gavin Newsom. https://www.gov.ca.gov/2024/09/19/governor-newsom-signs-bills-to-crack-down-on-sexually-explicit-deepfakes-require-ai-watermarking/
126. United Nations. (2024, September). *Governing AI for humanity*. https://www.un.org/sites/un2.un.org/files/governing_ai_for_humanity_final_report_en.pdf
127. United Nations. (n.d.). *Ensuring effective governance of AI at all levels*. Retrieved February 24, 2025, from https://www.un.org/tr/node/89386
128. Stokel-Walker, C. (2024, February 2). Act now on AI before it's too late, says UNESCO's AI lead. *Fast Company*. https://www.fastcompany.com/91022817/act-now-on-ai-before-its-too-late-says-unescos-ai-lead
129. Zhang, B., Shi, H., & Wang, H. (2023, June 26). Machine learning and AI in cancer prognosis, prediction, and treatment selection: A critical approach. *Journal of Multidisciplinary Healthcare,* 16, 1779–1791. https://pmc.ncbi.nlm.nih.gov/articles/PMC10312208/
130. Roose, K. (2023, May 30). A.I. poses "risk of extinction," industry leaders warn. *New York Times*. https://www.nytimes.com/2023/05/30/technology/ai-threat-warning.html
131. American Library Association. (n.d.). *Artificial intelligence*. Retrieved February 24, 2025, from https://www.ala.org/tools/future/trends/artificialintelligence
132. American Library Association. (2023, October 11–12). *Generative AI and libraries*. E-forum. https://www.ala.org/core/generative-ai-and-libraries. [Page no longer active.]
133. Panuncial, D. (2024, January 21). Using AI responsibly: Librarians on understanding and harnessing artificial intelligence. *American Libraries*. https://americanlibrariesmagazine.org/blogs/the-scoop/using-ai-responsibly/
134. Ibid.

135. American Library Association. (2022, April 7). *New from ACRL—"The Rise of AI: Implications and Applications of Artificial Intelligence in Academic Libraries."* https://www.ala.org/news/member-news/2022/04/new-acrl-rise-ai-implications-and-applications-artificial-intelligence-academic
136. Coffey, L. (2024, January 17). University libarians see urgent need for AI ethics. *Inside Higher Education*. https://www.insidehighered.com/news/quick-takes/2024/01/17/university-librarians-see-urgent-need-ai-ethics
137. Montana State University Library. (n.d.). *Responsible AI in libraries and archives.* Retrieved February 24, 2025, from https://www.lib.montana.edu/responsible-ai/#overview
138. Mitchell, D. (2023, July 9). *New handbook highlights ways to develop tech ethically*. Voice of America. https://www.voanews.com/a/new-handbook-highlights-ways-to-develop-tech-ethically-/7171672.html
139. Institute for Technology, Ethics, and Culture (ITEC). (n.d.). *Markkula Center for Applied Ethics, Santa Clara University.* Retrieved February 24, 2025, from https://www.scu.edu/institute-for-technology-ethics-and-culture/
140. Mitchell, D. (2023, July 9). *New handbook highlights ways to develop tech ethically*. Voice of America. https://www.voanews.com/a/new-handbook-highlights-ways-to-develop-tech-ethically-/7171672.html
141. Institute for Technology, Ethics, and Culture (ITEC). (n.d.). *Ethics in the age of AI.* Markkula Center for Applied Ethics, Santa Clara University. Retrieved February 24, 2025, from https://www.scu.edu/institute-for-technology-ethics-and-culture/ethics-in-the-age-of-ai/
142. Cary, A. C. (2024, February 5–6). *American Archivist editor: Generative AI in academic publishing.* Society of American Archivists Council Meeting. https://www2.archivists.org/sites/all/files/0224-V-G-AAE.pdf
143. Ibid.
144. Mangan, D. (2023, June 22). *Judge sanctions lawyers for brief written by A.I. with fake citations*. CNBC. https://www.cnbc.com/2023/06/22/judge-sanctions-lawyers-whose-ai-written-filing-contained-fake-citations.html
145. Cary, A. C. (2024, February 5–6). *American Archivist editor: Generative AI in academic publishing.* Society of American Archivists Council Meeting. https://www2.archivists.org/sites/all/files/0224-V-G-AAE.pdf
146. Ibid.
147. National Institutes of Standards and Technology. (2024, February 15). *NIST researchers suggest historical precedent for ethical AI research.* https://www.nist.gov/news-events/news/2024/02/nist-researchers-suggest-historical-precedent-ethical-ai-research
148. Ibid.
149. US Department of Health and Human Services. (n.d.). *The Belmont report*. Retrieved February 24, 2025, from https://www.hhs.gov/ohrp/regulations-and-policy/belmont-report/index.html
150. National Institutes of Standards and Technology. (2024, February 15). *NIST researchers suggest historical precedent for ethical AI research.* https://www.nist.gov/news-events/news/2024/02/nist-researchers-suggest-historical-precedent-ethical-ai-research
151. Ibid.
152. Ibid.
153. Iasevoli, B. (2018, August 17). How one district is preparing students for an AI-powered world. *Education Week*. https://www.edweek.org/teaching-learning/how-one-district-is-preparing-students-for-an-ai-powered-world/2018/08
154. Ibid.

155. Ibid.
156. Kingson, J. A. (2023, November 6). How states are guiding schools to think about AI. *Axios*. https://www.axios.com/2023/11/06/ai-schools-education-teachers-students-states
157. Dusseault, B., & Lee, J. (2023, October). *AI is already disrupting education, but only 13 states are offering guidance for schools*. Center on Reinventing Public Education. https://crpe.org/ai-disrupt-ed-13-states/
158. Ibid.
159. Hancock, L. (2024, February 15). *From ethics to outsmarting Chat GPT, state unveils resource for AI in Ohio education*. Cleveland.com. https://www.cleveland.com/news/2024/02/from-ethics-to-outsmarting-chat-gpt-state-unveils-resource-for-ai-in-ohio-education.html
160. Ohio's AI in Education Coalition. (2024, November 20). *AI strategy.* https://innovateohio.gov/wps/wcm/connect/gov/7484d716-5c6f-47d2-a04a-7914e11bc44c/AI+In+Education+Strategy.pdf
161. Innovate Ohio. (n.d.). *Spotlight: Ohio's AI in Education Coalition.* Retrieved February 24, 2025, from https://innovateohio.gov/aistrategy
162. Hancock, L. (2024, February 15). *From ethics to outsmarting Chat GPT, state unveils resource for AI in Ohio education*. Cleveland.com. https://www.cleveland.com/news/2024/02/from-ethics-to-outsmarting-chat-gpt-state-unveils-resource-for-ai-in-ohio-education.html
163. Dusseault, B., & Lee, J. (2023, October). *AI is already disrupting education, but only 13 states are offering guidance for schools*. Center on Reinventing Public Education. https://crpe.org/ai-disrupt-ed-13-states/
164. Ibid.
165. Ibid.
166. Hoel, E. (2024, March 29). A.I.-generated garbage is polluting our culture. *New York Times*. https://www.nytimes.com/2024/03/29/opinion/ai-internet-x-youtube.html
167. Liang, W., Izzo, Z., Zhang, Y. Lepp, H., Cao, H., Zhao, X., Chen, L., Ye, H.,Liu, S. Huang, Z., McFarland, D. A., & Zou, J. Y. (2024). Monitoring AI-modified content at scale: A case study on the impact of ChatGPT on AI conference peer reviews. *Proceedings of the 41st International Conference on Machine Learning*, Vienna Austria. https://arxiv.org/pdf/2403.07183
168. Ibid.
169. Hoel, E. (2024, March 29). A.I.-generated garbage is polluting our culture. *New York Times*. https://www.nytimes.com/2024/03/29/opinion/ai-internet-x-youtube.html
170. Brewin, K. (2024, April 4). Why I wrote an AI transparency statement for my book, and think other authors should too. *Guardian*. https://www.theguardian.com/books/2024/apr/04/why-i-wrote-an-ai-transparency-statement-for-my-book-and-think-other-authors-should-too
171. Ibid.
172. Ibid.
173. Klein, A. (2024, July 24). *Students weigh ethics of using AI for college applications*. Government Technology. https://www.govtech.com/education/k-12/students-weigh-ethics-of-using-ai-for-college-applications
174. Ibid.
175. Ibid.
176. *Task Force on Law and Artificial Intelligence.* (n.d.). American Bar Association. Retrieved February 24, 2025, from https://www.americanbar.org/groups/centers_commissions/center-for-innovation/artificial-intelligence/
177. Ibid.

178. Greene, K. K., Theofanos, M. F., Watson, C., Andrews, A., & Barron, E. (2024, February). Avoiding past mistakes in unethical human subjects research: Moving from artificial intelligence principles to practice. *Computer, 57*, 53–63. https://www.computer.org/csdl/magazine/co/2024/02/10417823/1Ua1Drdj4bK
179. Markkula Center for Applied Ethics, Santa Clara University. (2021, November 8). *A framework for ethical decision making.* https://www.scu.edu/ethics/ethics-resources/a-framework-for-ethical-decision-making/
180. Horowitz, J. (2024, February 9). The friar who became the Vatican's go-to guy on A.I. *New York Times*. https://www.nytimes.com/2024/02/09/world/europe/italy-artificial-intelligence-ethics.html
181. Ibid.
182. Benanti, P. (2023, June 23). *Algorethics: Opportunities and challenges in the age of AI*. Distinguished Lecture Series, Gregorian University Foundation. https://gregorianfoundation.org/lecture2023/
183. Horowitz, J. (2024, February 9). The friar who became the Vatican's go-to guy on A.I. *New York Times*. https://www.nytimes.com/2024/02/09/world/europe/italy-artificial-intelligence-ethics.html
184. Ibid.
185. Nolin, E. (2024, March 5). Generative AI: Its innovation, implications and use in the Christian world. *Presbyterian Outlook*. https://pres-outlook.org/2024/03/generative-ai-its-innovation-implications-and-use-in-the-christian-world/
186. Rousselle, C. (2024, June 17). *AI lab at Christian university aims to bring morality and ethics to artificial intelligence*. Fox News. https://www.foxnews.com/lifestyle/new-ai-lab-christian-university-aims-bring-morality-ethics-artificial-intelligence
187. Ibid.
188. Ibid.
189. Lebrija, L. (2024, June 4). *AI & the Church summit: Navigating the ethical frontier.* Virginia Theological Seminary. https://vts.edu/deans-commentary/ai-the-church-summit-navigating-the-ethical-frontier/
190. *Executive Summary: AI & the Church summit: A confluence of faith and technology*. (n.d.). TryTank Research Institute. Retrieved February 24, 2025, from https://www.trytank.org/aigathering.html
191. Bayan. (2024, April 30). *Islam and artificial intelligence: Ethical horizons in a digital age*. PR Newswire. https://www.prnewswire.com/news-releases/islam-and-artificial-intelligence-ethical-horizons-in-a-digital-age-302131175.html
192. Ibid.
193. Ibid.
194. Koppel, M. (2024, March 4). What artificial intelligence has in store for Judaism. *Mosaic*. https://mosaicmagazine.com/essay/politics-current-affairs/2024/03/what-artificial-intelligence-has-in-store-for-judaism/
195. Ibid.
196. Shurpin, Y. (n.d.). *What is the Jewish view on AI and intellectual property?* Chabad. Retrieved February 24, 2025, from https://www.chabad.org/library/article_cdo/aid/5941663/jewish/What-Is-the-Jewish-View-on-AI-and-Intellectual-Property.htm
197. Hongladarom, S. (2021, January 6). What Buddhism can do for AI ethics. *MIT Technology Review*. https://www.technologyreview.com/2021/01/06/1015779/what-buddhism-can-do-ai-ethics/
198. Ibid.
199. Ibid.

200. Bhatia, M. (2023, August 23). *Navigating the intersection between AI, automation and religion—3 essential reads*. Conversation. https://theconversation.com/navigating-the-intersection-between-ai-automation-and-religion-3-essential-reads-211587
201. Ibid.
202. Robinson, F. S. (2024, May 3). Artificial intelligence explains there's no God. *Medium*. https://franksrobinson.medium.com/artificial-intelligence-explains-theres-no-god-f91b6a4a4c8b
203. Ibid.
204. Ibid.
205. Mineo, L. (2024, March 14). It may be neither higher nor intelligence: Religious scholars examine value, limits of AI. *Harvard Gazette*. https://news.harvard.edu/gazette/story/2024/03/it-may-be-neither-higher-nor-intelligence/
206. Ibid.
207. Ibid.
208. Swartzenberg, F. (2024, March 7). *Philosophy, ethics, and the pursuit of "responsible" artificial intelligence*. Rochester Institute of Technology. https://www.rit.edu/news/philosophy-ethics-and-pursuit-responsible-artificial-intelligence
209. Ibid.
210. Daly, P. (2024, July 17). *Tech sector examines the risks and rewards of AI. How two paths converged at Northeastern's London campus*. Northeastern Global News. https://news.northeastern.edu/2024/07/17/women-in-ai-ethics/
211. Ibid.
212. Ibid.
213. United Nations. (n.d.). *Greenwashing—the deceptive tactics behind environmental claims.* Retrieved February 24, 2025, from https://www.un.org/en/climatechange/science/climate-issues/greenwashing
214. Swartzenberg, F. (2024, March 7). *Philosophy, ethics, and the pursuit of "responsible" artificial intelligence*. Rochester Institute of Technology. https://www.rit.edu/news/philosophy-ethics-and-pursuit-responsible-artificial-intelligence
215. Distributed AI Research Institute. (n.d.). *Home page.* Retrieved February 24, 2025, from https://www.dair-institute.org
216. *Address of His Holiness Pope Francis*. (2024, June 14). G7 Session on Artificial Intelligence. Vatican. https://www.vatican.va/content/francesco/en/speeches/2024/june/documents/20240614-g7-intelligenza-artificiale.html
217. Ibid.
218. Ibid.
219. Moreno, J. E. (2023, December 30). Boom in A.I. prompts a test of copyright law. *New York Times*. https://www.nytimes.com/2023/12/30/business/media/copyright-law-ai-media.html
220. Morrone, M. (2023, December 1). Copyright law will shape how we use generative AI. *Axios*. https://www.axios.com/2023/12/01/ai-copyright-law-biden-executive-order
221. Moreno, J. E. (2023, December 30). Boom in A.I. prompts a test of copyright law. *New York Times*. https://www.nytimes.com/2023/12/30/business/media/copyright-law-ai-media.html
222. Chan, K. (2024, June 10). *Facebook owner Meta seeks to train AI model on European data as it faces privacy concerns*. AP News. https://apnews.com/article/meta-facebook-instagram-whatsapp-ai-artificial-intelligence-8d6cb3424ee410c641d0acdb154601f5

223. Sawers, P. (2024, June 14). *Meta pauses plans to train AI using European users' data, bowing to regulatory pressure*. TechCrunch. https://techcrunch.com/2024/06/14/meta-pauses-plans-to-train-ai-using-european-users-data-bowing-to-regulatory-pressure/
224. Matney, L. (2018, May 2). *Facebook is using your Instagram photos to train its image recognition AI*. TechCrunch. https://techcrunch.com/2018/05/02/facebook-is-using-your-instagram-photos-to-train-its-image-recognition-ai/
225. Solov, D. J., & Hartzog, W. (2024, July 3). *The great scrape: The clash between scraping and privacy*. SSRN. https://papers.ssrn.com/sol3/papers.cfm?abstract_id=4884485
226. *Naruto v. Slater*, No. 16-15469 (9th Cir. 2018). Justia. https://law.justia.com/cases/federal/appellate-courts/ca9/16-15469/16-15469-2018-04-23.html
227. Small, Z. (2023, August 21). As fight over A.I. artwork unfolds, judge rejects copyright claim. *New York Times*. https://www.nytimes.com/2023/08/21/arts/design/copyright-ai-artwork.html
228. AI Comic Books. (n.d.). *Zarya of the Dawn by Kristina Kashtanova*. Retrieved February 24, 2025, from https://aicomicbooks.com/book/zarya-of-the-dawn-by-kristina-kashtanova-download-now
229. United States Copyright Office. (2023, February 21). *Letter*. https://www.copyright.gov/docs/zarya-of-the-dawn.pdf
230. Ibid.
231. Brittain, B. (2022, August 5). *U.S. appeals court says artificial intelligence can't be patent inventor*. Reuters. https://www.reuters.com/legal/litigation/us-appeals-court-says-artificial-intelligence-cant-be-patent-inventor-2022-08-05
232. Brittain, B. (2023, April 24). *US Supreme Court rejects computer scientist's lawsuit over AI-generated inventions*. Reuters. https://www.reuters.com/legal/us-supreme-court-rejects-computer-scientists-lawsuit-over-ai-generated-2023-04-24/
233. Ibid.
234. *AI cannot patent inventions, UK Supreme Court confirms*. (2023,December 20). BBC. https://www.bbc.com/news/technology-67772177
235. Wu, H. (2024, April 14). *China's first AI-generated image copyright infringement case*. Lexology. https://www.lexology.com/library/detail.aspx?g=15fef69a-cd76-42de-957c-7368f2d9dc68
236. Copyright registration guidance: Works containing material generated by artificial intelligence. (2023, March 16). *Federal Register*. https://www.federalregister.gov/documents/2023/03/16/2023-05321/copyright-registration-guidance-works-containing-material-generated-by-artificial-intelligence
237. Ibid.
238. Ibid.
239. Inventorship guidance for AI-assisted inventions. (2024, February 13). *Federal Register*. https://www.federalregister.gov/documents/2024/02/13/2024-02623/inventorship-guidance-for-ai-assisted-inventions
240. Ibid.
241. Levin, M. (2024, January 31). *The economy and ethics of AI training data*. Marketplace. https://www.marketplace.org/2024/01/31/the-economy-and-ethics-of-ai-training-data/
242. Brittain, B. (2024, June 7). *Google defeats class action over AI training data for now*. Reuters. https://www.reuters.com/legal/transactional/google-defeats-class-action-over-ai-training-data-now-2024-06-06/ https://www.reuters.com/legal/transactional/google-defeats-class-action-over-ai-training-data-now-2024-06-06/
243. Levin, M. (2024, January 31). *The economy and ethics of AI training data*. Marketplace. https://www.marketplace.org/2024/01/31/the-economy-and-ethics-of-ai-training-data/

244. Veltman, C. (2023, July 17). Thousands of authors urge AI companies to stop using work without permission. *Morning Edition*. https://www.npr.org/2023/07/17/1187523435/thousands-of-authors-urge-ai-companies-to-stop-using-work-without-permission
245. Vaughan, P. (2024, February 13). *The New York Times' AI copyright lawsuit shows that forgiveness might not be better than permission.* Conversation. https://theconversation.com/the-new-york-times-ai-copyright-lawsuit-shows-that-forgiveness-might-not-be-better-than-permission-222904
246. Burgess, M., & Rogers, R. (2024, October 12). How to stop your data from being used to train AI. *Wired*. https://www.wired.com/story/how-to-stop-your-data-from-being-used-to-train-ai/
247. Generative AI is a marvel. Is it also built on theft? (2024, April 14). *Economist*. https://www.economist.com/business/2024/04/14/generative-ai-is-a-marvel-is-it-also-built-on-theft
248. O'Leary, D. (2024, June 3). Generative AI is creating a copyright crisis for artists. *Mind Matters*. https://mindmatters.ai/2024/06/generative-ai-is-creating-a-copyright-crisis-for-artists/
249. Fairly Trained. (n.d.). *About*. Retrieved February 24, 2025, from https://www.fairlytrained.org/about
250. Ibid.
251. Fairly Trained. (n.d.). *FAQs*. Retrieved February 24, 2025, from https://www.fairlytrained.org/faqs
252. Fleck, A. (2024, March 18). *Trust in Boeing plummets after safety incidents*. Statista. https://www.statista.com/chart/31928/net-trust-among-us-familiar-with-boeing-brand/
253. Sharma, K., Schoorman, F. D., & Balligner, G. A. (2022, April 2022). How can it be made right again? A review of trust repair research. *Southern Management Association,* 49(1). https://journals.sagepub.com/doi/full/10.1177/01492063221089897
254. McClain, C. (2024, March 26). *Americans' use of ChatGPT is ticking up, but few trust its election information*. Pew Research Center. https://www.pewresearch.org/short-reads/2024/03/26/americans-use-of-chatgpt-is-ticking-up-but-few-trust-its-election-information/
255. Walker, J. (2024, June 19). *McDonald's drive-thru AI is McCanceled*. Kotaku. https://kotaku.com/mcdonalds-drive-thru-ai-failure-tiktok-1851548668
256. IBM. (2023, September 1). *What are AI hallucinations?* https://www.ibm.com/topics/ai-hallucinations
257. Ibid.
258. Broad, W. J. (2024, December 23). How hallucinatory A.I. helps science dream up big breakthroughs. *New York Times*. https://www.nytimes.com/2024/12/23/science/ai-hallucinations-science.html
259. Dick, P. K. (1953, September). The variable man. *Space Science Fiction*. https://www.gutenberg.org/files/32154/32154-h/32154-h.htm
260. Weiss, D. C. (2023, June 26). Lawyers who "doubled down" and defended ChatGPT's fake cases must pay $5K, judge says. *ABA Journal*. https://www.abajournal.com/web/article/lawyers-who-doubled-down-and-defended-chatgpts-fake-cases-must-pay-5k-judge-says
261. Ibid.
262. Ibid.
263. *Rule 5.1: Responsibilities of partners, managers, and supervisory lawyers.* (2019, April 17). American Bar Association. https://www.americanbar.org/groups/professional_responsibility/publications/model_rules_of_professional_conduct/rule_5_1_responsibilities_of_a_partner_or_supervisory_lawyer/
264. Ibid.
265. *Rule 1.1: Competence.* (n.d.). American Bar Association. Retrieved February 24, 2025, from https://www.americanbar.org/groups/professional_responsibility/publications/model_rules_of_professional_conduct/rule_1_1_competence

266. *Rule 1.1 Competence—comment.* (n.d.). American Bar Association. Retrieved February 24, 2025, from https://www.americanbar.org/groups/professional_responsibility/publications/model_rules_of_professional_conduct/rule_1_1_competence/comment_on_rule_1_1/
267. American Library Association. (2021). *ALA code of ethics.* https://www.ala.org/tools/ethics.
268. *Rule 1.1 Competence—comment.* (n.d.). American Bar Association. Retrieved February 24, 2025, from https://www.americanbar.org/groups/professional_responsibility/publications/model_rules_of_professional_conduct/rule_1_1_competence/comment_on_rule_1_1/
269. Ibid.
270. Magesh, V., Surani, F., Dahl, M., Suzgun, M., Manning, C. D., & Ho, D. E. (2024, June 6). *Hallucination-free? Assessing the reliability of leading AI legal research tools*. Stanford University. https://dho.stanford.edu/wp-content/uploads/Legal_RAG_Hallucinations.pdf
271. Ibid.

Chapter 12

Emerging Technologies

Introduction: Tech Gamblers and Game-Changers

Writer and film critic Chris Nashawaty's 2024 book *The Future Was Now* tells the story of eight science fiction films that debuted in the summer of 1982. Nashawaty asserts that these movies—*E.T.: The Extra-Terrestrial*, *Blade Runner*, *Poltergeist*, and *Star Trek II: The Wrath of Khan*, among others—changed the trajectory of science fiction and cinema. Talking about these films with NPR *Fresh Air* host Tonya Mosley in a July 2024 interview, Nashawaty recounted that Disney's 1979 Sci-Fi film foray *The Black Hole* was not the Star Wars–esque financial or critical success the studio had hoped for. Not giving up on the genre though, in 1982 Disney released *Tron*, a daring video-gaming-themed film "about a computer hacker who is abducted into the digital world."[1] *Fresh Air* interviewer Mosley asks, "What did Disney learn from 'The Black Hole' that then maybe helped them with the success of 'Tron'?" Nashawaty replies: "Yeah. I mean, I think it learned that it has to gamble in order to stay alive, and, yes, 'Black Hole' had been sort of an unsuccessful gamble, or at least a push, but they knew that this is the way they had to go in order to stay relevant and to stay in business."[2]

Nashawaty's observation that Disney learned it needed to gamble and take their best shot at making a bold new technology-focused movie like *Tron*, "in order to stay relevant and to stay in business," has great applicability to tech companies today. With self-driving vehicles, companion robots, drones, augmented reality, AI chatbots, and more, technology companies are gambling on the success of new technologies and the creation of new business models. Akin to AI, these other digital-age emerging technologies are influencing societies and information centers. Just like AI, too, these can be used for good and ill purposes, on a continuum of moral potentialities. AI may currently be the A-list star of the emerging technologies stage show, garnering the vast majority of ink, bytes, and airtime, but up-and-comers like the internet of things (IoT), drones, robots, and self-driving vehicles are also worthy of recognition and critical review.

For some people, a unifying element of most, if not all, internet-era emerging technologies is apprehension: fear as to whether humans will be able to retain dominion over them—whether the "machines" will flip the existing paradigm and take control over their "human overlords." Others are more optimistic. Pew Research Center (PRC) captured an illuminating array of comments from experts, reported in its 2023 report, "As AI Spreads, Experts Predict the Best and Worst Changes in Digital Life by 2035." The article's tagline distills the report's glass-half-empty, glass-half-full, frequently nuanced findings: "They have deep concerns about people's and society's well-being. But they also expect great benefits in health care, scientific advances and education."[3]

One of the enduring legacies of atomic and hydrogen bomb technologies of the mid-twentieth century was, and remains, the wholly relatable fear that these world-ending weapons may proliferate beyond the handful of nations that currently possess them. Beyond science fiction writers, however, few worried, until recent times, that these weapons themselves might one day become sentient or capable of being deployed by nonhuman intelligences (i.e., what we now would refer to as AI). Today, such "AI-gone-rogue" fears have become mainstream journalistic fodder for newspaper op-eds and broadcast media moral panic exposés.

As an interesting computing trivia fact, 1983's hugely successful *WarGames* movie depicted Matthew Broderick as a teenage video-games-seeking hacker who ultimately must stop an AI-like military supercomputer from starting World War III. (Spoiler alert: he succeeds.) In a 2008 *Wired* magazine article on the 25th anniversary of this computer-geek-coming-of-age film that became part of the "cult lore of Silicon Valley," Google cofounder Sergey Brin observed that "it was a key movie of a generation, especially for those of us who got into computing."[4]

Differences of opinion on the moral nature, uses, and capabilities of digital-age technologies are a feature of real-world discussions and cultural works of fiction. One such debate is incisively captured in the following brief exchange, as two characters in the Star Wars universe's popular science fiction television show *The Mandalorian* (2019–) share their divergent, life-informed viewpoints of the robot-like machines called "droids":

> "Droids are not good or bad. They are neutral reflections of those who imprint them."—Kuill.
>
> "I have seen otherwise."—The Mandalorian (Din Djarin)[5]

To the "Pantheon of Feared and Welcomed Technologies," already inhabited by IoT and AI, can be added killer and companion/helper robots, lethal and nonlethal drones, and autonomous/self-driving vehicles (sometimes unpredictable and deadly). All of these technologies have proponents heralding "good news" and uses of these technologies that are touted to bring benefits to individuals and communities, some of which are historically disadvantaged and at-risk. Similarly, these technologies also have opponents, raising concerns in the vein of those voiced about IoT and AI. In this chapter, we examine some key pros and cons of these emergent technologies, as well as technology related to extended reality (XR), encompassing augmented reality (AR), virtual reality (VR), and mixed reality (MR). We also identify real-life ways these technologies are already being wielded by information professionals to do good and make positive contributions to societies.

Internet of Things

The internet of things, or IoT, is one of those terms of art that has a certain air of coolness to it while at the same time exuding an elusive, enigmatic quality. So what exactly *is* IoT? A few sources are helpful in fleshing out its meaning and DNA. ZDNet's definition is particularly accessible, noting in a 2020 article titled "What Is the IoT?" that IoT "refers to the billions of physical devices around the world that are now connected to the internet, all collecting and sharing data."[6] This source also observes that a key characteristic of IoT is "digital intelligence." Namely, IoT is made up of smart devices that can "communicate real-time data," as opposed to dumb terminals and tools that exist in solitary silence. Another aspect is that these devices communicate with networks without human intervention or action.

Kevin Ashton is credited with inaugurating the phrase in 1999. As ZDNet reported, Ashton's term, in his own words, is meant to convey that "IoT integrates the interconnectedness of human culture—our 'things'—with the interconnectedness of our digital information system."[7] Despite this definition directly from the term's progenitor, IoT still remains ambiguous to many. Bill Geary's 2022 *Forbes* article "The Internet of Things Demystified" adroitly calls out both IoT's inherent opaqueness and self-anointed technocrats who generate unclear and insular names: "The tech industry establishes a lot of jargon that helps differentiate one technology from another. Those terms are helpful to IT professionals but often serve to confuse everyone else. The Internet of Things (IoT) is a term that creates confusion."[8] In a plainspoken description, Geary succinctly states that "IoT is nothing more than connecting things, collecting information from them, analyzing it and acting upon it accordingly: connect, collect, analyze and act."

The same ZDNet article provides a good rule of thumb for determining examples of IoT devices: "Pretty much any physical object can be transformed into an IoT device if it can be connected to the internet to be controlled or communicate information."[9] One important distinction, the piece points out, is that the "term IoT is mainly used for devices that wouldn't usually be generally expected to have an internet connection, and that can communicate with the network independently of human action." Examples include items like smart thermostats, wearable fitness monitors, internet-connected stuffed animals, and designer coffee makers.

Stakeholders Impacted by IoT

Nearly all members of the public, except for communities that eschew electricity and technology, are potentially impacted by the rapidly expanding numbers of IoT devices and equipment. Beyond every member of the tech-impacted world, other principal stakeholders include Big Tech companies, industries, and transnational conglomerates that manufacture, market, and sell IoT. A host of third parties with links to the aforementioned IoT makers and owners have significant IoT interests as well. These IoT beneficiaries include banks, insurers, lawyers, advertisers, celebrity spokespersons, social media influencers and endorsers, suppliers, distributors, and delivery entities. Governments—local, state/provincial, regional, national—have myriad stakes in IoT devices. Chief among these interests is national security because of the data collection and transmission capabilities of IoT. For example, in February 2024 Gina Raimondo, US secretary of commerce, shared national security and privacy fears about the potential for China to access the biometric data of Americans who drive EVs (electric vehicles) that include China-made sensors.[10]

Specific communities of IoT users also have interests in IoT. For example, people with ability challenges and medical conditions may benefit from the health-monitoring and data-reporting abilities of IoT equipment. Historically disadvantaged communities may experience heightened surveillance and profiling via IoT technologies. On the flip side, first responders such as law enforcement have interests in the various safety and security functions IoT can perform. People affiliated with education, spanning preschool through higher education, have IoT interests, including students, teachers and faculty, staff and administrators, and parents and guardians. This list is by no means exhaustive but offers some tangible examples of IoT stakeholders at the micro and macro levels.

Benefits/Upsides of IoT

IoT devices present an array of positives and beneficial uses. For one, they have the potential to advance knowledge and promote innovation, thereby, arguably, advancing societies. IoT also has many applications for promoting public welfare. For instance, IoT equipment can be used to support first responders like law enforcement, emergency medical technicians, and firefighters. IoT can similarly support vital societal interests like the structural and systemic integrity of critical infrastructure, public transit, disaster preparedness and response, and consumer safety.

Additionally, the inherent data collection and use capabilities of IoT mean that these devices can enable individual preferences to be learned. This, in turn, can facilitate the creation and delivery of personalized service for individuals and groups. (A downside of this knowledge when applied to groups, though, is that it can result in stereotypic profiling and confirmation biases.) Applying personalized knowledge to medical contexts can enable "precision medicine" that is targeted to the singular needs of a particular patient. IoT devices already are yielding specific public health applications and uses: for instance, periodic monitoring of insulin levels for persons with diabetes, as well as data on blood pressures, oxygen levels, and heart rates for people managing heart conditions. In an important sense, IoT devices, hence, can be seen as giving people more agency and autonomy over the monitoring of their own bodies, by affording access to real-time, unfiltered health data on themselves. On the other hand, because third parties may have access to this health data, the privacy interests of users may be compromised, as is discussed more in the next section.

Risks/Downsides of IoT

Probably the most serious and pervasive risk of IoT can be summarized by one word: privacy. This book earlier looked in detail at issues of privacy, data collection and use, cybertracking, confidentiality breaches, and surveillance. All of those issues, and more, are front and center when talking about the internet of things. Because IoT devices are able to transmit information without humans directing those transmissions, information can potentially be sent that a user would not want to be transmitted. Additionally, IoT equipment, by its very nature, can enable data collection and use by the makers and owners of the device being used. Third parties, who often are licensed by IoT manufacturers to collect and use that data, may also be involved. Moreover, bad actors, such as hackers and other cybercriminals, may be able to gain access to user data and use that data for illicit purposes.

As discussed in the privacy chapter, in many parts of the United States at present, consumers have little to no control over their personal data. Efforts to pass comprehensive privacy and data collection and use laws at the federal level have been unsuccessful. Some states, such as California and Virginia, have successfully signed privacy and data collection legislation. The full implications of these laws on their own respective states, as well as on the nation, have yet to be determined. Deep-pocket lobbyists and interest groups, for example, are waging multipronged efforts to hamstring and dilute the mandated requirements of these state-based privacy and data-collection reforms.

Privacy breaches that can occur as a result of IoT data transmission or hacking can enable potential unethical uses of data. Discriminatory treatment may also be perpetrated because of privacy lapses. For example, data may be collected and used that influences decisions related to employment, housing, and financial matters. Security and safety concerns must also be considered, particularly with regard to minors. Children, for example, may be endangered by IoT toys and other devices that transmit personal information, such as names, ages, and addresses.

The intrinsic digital nature of IoT means that there is a heavy reliance on electricity and network infrastructure to store data that is communicated between devices and systems. This in turn raises the issue of greater carbon impacts and ramifications for climate change and environments. Another related downside of IoT is that it fuels digital divides and social inequities. People who do not have access to steady supplies of electricity and who experience financial scarcity are at a disadvantage when compared with those who have abundant access to the necessary utilities and revenue sources that undergird IoT.

Understanding and Managing IoT Trade-Offs

With a better understanding now of the major stakeholders and the upsides and downsides of IoT devices, we turn to the issue of what users can do to make more informed decisions about IoT and the interests that are implicated, like privacy and confidentiality. The March 2022 the Conversation article, "Smart Devices Spy on You," handily lays out the challenges that users face:

> As a user, it is important to make an informed decision by understanding the trade-offs between privacy and comfort when buying, installing and using an internet-connected device. This is not always easy. Studies have shown that, for example, owners of smart home personal assistants have an incomplete understanding of what data the devices collect, where the data is stored and who can access it.[11]

Rather than just calling out the problem, the authors provide measurable and achievable recommendations for what the average consumer can do to take better control of their privacy autonomy: (1) understand your vulnerabilities and (2) limit the privacy damage. The former, understanding your vulnerabilities, is centered on awareness and education. The latter, limit the privacy damage, is focused on taking action: proactive action that, using a football reference, is a mix of effective offense and defense. The article, too, provides a number of references for organizations that offer guidance to consumers. For example, the Federal Trade Commission maintains a consumer advice site called "Securing Your Internet-Connected Devices at Home."[12]

Articles in mainstream publications can also provide useful tips and ideas for "understanding vulnerabilities" and "limiting privacy damage." A 2023 *Forbes* magazine article, "14 Essential Steps to Take to Secure Household IoT Devices," explains what users can do to protect themselves from malicious hackers. With easy-to-understand and important-to-implement strategies, 14 members of a Forbes Technology Council exhort consumers "to leverage best cybersecurity practices to defend each of their IoT devices."[13]

Information professionals and centers have a crucial role to play in aiding patrons and customers in becoming more educated and prepared to make the best possible IoT and privacy-related choices for themselves and their families. Discussion questions at the end of this chapter provide a number of learning exercises and reflective opportunities for thinking about concrete steps that organizations can take to guide users and actively engage communities on their rights and options.

Drones

Unmanned aircraft systems (UASs) or unmanned aerial vehicles (UAVs) are the technical classifications of the emerging technology more commonly referred to as drones. In the United States, definitions of drones vary. This is important to know as these definitions carry varying legal implications and potential liabilities. Some features are common among drones, though: first and foremost, they are aircraft. They are operated remotely by a human who is neither inside nor on the vehicle. In terms of size, they can be as small as a human hand and as a large as commercial aircraft.

Laws addressing drones vary from state to state and country to country. This is understandable given the dynamic state of play with this emergent technology as governing bodies scramble to "get up to speed" with its novelty and potential risks and rewards. The range of applications of drones (known and as yet unknown) further contributes to diverging, often competing, views on how to use and regulate them. Updated sites, such as a March 2023 web page "Current Unmanned Aircraft State Law Landscape" of the National Conference of State Legislatures, are helpful for staying abreast of drone developments at the state level. This site also provides an excellent summary of the fundamental features currently involving drones:

> Unmanned aircraft systems (UAS), commonly called drones, have a host of applications including law enforcement, land surveillance, wildlife tracking, search and rescue operations, disaster response, border patrol and recreational use. Drones have become a part of our daily lives, especially among drone hobbyists. Today, over 1.1 million recreational drones are registered with the Federal Aviation Administration.
>
> State legislatures continue to debate if and how drone technology should be regulated, considering the benefits of their use, privacy concerns and their potential economic impact. Since 2013, at least 44 states have enacted laws addressing drones and an additional three states have adopted resolutions. Common issues addressed in the legislation include defining what a UAS is, how they can be used by law enforcement or other state agencies, how they can be used by the general public and regulations for their use in hunting game.[14]

The US government's Federal Aviation Administration (FAA) provides useful information on drones. As the FAA's home page for drones currently communicates, one of the principal goals, if not the most important, is integrating drones into the "national airspace."[15] As with other emerging technologies examined in this chapter, a natural tension exists between encouraging uses of drones and regulating their use in order to ensure public safety and national security. Drone operators must secure licenses to operate UAS. More than one million Americans are currently licensed drone operators.

Drones can be divided into two categories: lethal and nonlethal. UASs have gained traction in the consciousness of the US public in the first quarter of the twenty-first century in four key ways:

- Military and security purposes: Following the terrorist attacks of September 11, 2001, drone usage was ramped up by the presidential administrations of George W. Bush (2001–2009) and Barack Obama (2009–2017) in various overseas military campaigns, such as Iraq and Afghanistan. US presidents since that time have continued to utilize drones to advance military objectives. After the February 2022 invasion of Ukraine by Russian military forces, drones have been heavily used by both sides in pursuit of larger strategic advantages. In 2024, Houthi rebels in Yemen have used attack drones with deadly effectiveness in hijacking, damaging and sinking tankers, and dramatically impacting global trade and shipping routes in the Red Sea.
- Recreational uses: The public became more aware that drones could be operated for nonlethal purposes as hobbyists increased their use of recreational drones around 2015. Often, recreational drones are used to provide bird's-eye views of both developed and undeveloped settings, raising concerns about privacy, surveillance, and property rights. Drones have also become popular gifts.
- Cultural representations: Hollywood has played a role in drone awareness raising through military-themed films. Some of these films include *Good Kill* (2014), starring Ethan Hawke as a military drone pilot taking on the Taliban, and *Eye in the Sky* (2015), depicting Helen Mirren as a US colonel overseeing drone strikes in Kenya. The characters in these films wrestle with the ethics and ramifications of their actions and decisions.
- Commercial uses: Large companies like Amazon have generated media attention in the past decade because of their desire to use drones for delivering a wide variety of goods to homes and businesses.

These uses and representations of drones collectively encompass the messy, mixed feelings many people have about how drones can and are used, as well as the ways they may be used in the future. Ethical issues like privacy, surveillance, public safety, national security, and military uses make drones a complex legal and policy matter. A quote by scholar US Army Lt. Col. Paul Lushenko during a March 2022 Cornell University panel discussion on drone warfare tellingly conveys the opposing natures and contributions of drones: "Drones can aid, they can watch, and they can kill."[16] Interestingly, the panel moderator, Sarah Krebs, also a Cornell scholar, commenced the discussion by stating that "we're here to make sense of this evolving technology." Making sense of drone warfare is an admirable and understandable goal but is not easily or readily achieved, given the array of competing interests, values, and stakeholders.

The following are some examples of ways that drones are providing positive benefits to societies. Many of these examples also includes aspects of information, data, and computing.

Drones and Medical Care

The University of British Columbia (UBC) reported in March 2022 on a study in Rwanda that found "drones get blood to patients faster—and reduce waste."[17] People not living near medical facilities where they can get blood transfusions expediently are at greater medical risk. A collaborative study between UBC researchers and research partners reviewed medical outcomes from more than 12,000 deliveries of blood over a two-year period. This study, published in the April 2022 issue of *Lancet Global Health*, concluded that "using drones instead of ambulances to deliver emergency blood products has resulted in blood arriving for transfusions 79 minutes faster on average. It also reduced the number of wasted blood units by two-thirds."[18] This story illustrates how drones can literally save lives by getting medical care to patients more quickly than human-driven ambulances.

Drones and Public Safety

Over the past few decades, Australia has experienced significant increases in catastrophic wildfires that are believed to be a result of climate change. Robert Irwin, the son of the late Steve Irwin (1962–2006), who was popularly known as "The Crocodile Hunter," won a Wildlife Photographer of the Year Award, reported by the BBC in 2021, for a photo he took with his drone showing the advancing line of a bushfire in Queensland.[19] Photos and videos taken by drones can be used to not only provide breaking news but also aid in disaster preparedness and response, historical recordkeeping, and conservation efforts. They are also safer to deploy than putting humans at risk to secure visual data.

Drones, sometimes in tandem with robots (discussed later in this chapter), are performing important roles in search and rescue, public safety, and building maintenance and inspection. A February 2024 *New York Times* article discussed New York City Mayor Eric Adams's plan to pair drones with lifeguards patrolling beaches for the five boroughs. Drones can be used to communicate with swimmers in trouble, locate them, drop flotation devices to them, and spot sharks. The New York area has experienced recent severe lifeguard shortages, so drones working with lifeguards may expand safety capabilities. The piece mentions, too, that drones are already being used to inspect buildings and assist with fire emergencies. Illustrating how ethics interests can bump up against one another, a surveillance advocate also raised concerns that beach drones could potentially be utilized to surveil beachgoers.[20]

In a June 2024 article, the *Washington Post* reported that Australia was experiencing a sharp upticks in shark attacks. With more than 100,000 beaches throughout the country, drones and AI are increasingly being deployed to endeavor to keep beachgoers safe. The article reports that "Australia now has the world's largest coastal drone-surveillance operation," though "many areas remain unmonitored."[21]

Drones and Conservation

Video monitoring and footage from drones are proving instrumental and invaluable for conservation of fauna and flora. A November 2023 *Axios* story described a partnership between a biotech start-up ("aiming to bring back the woolly mammoth" and the Tasmanian tiger) and an elephant conservation group in Kenya utilizing "drones and AI to track African elephants and study their behavior."[22] A 2020 article on the website DroneDJ dramatically demonstrated how the counting of Adelie penguins in Antarctica during annual population surveys can be accomplished by drones in record time: what took humans "two full days" to complete via human-flown helicopters was whittled down "to just three hours" with drones.[23] Penguin populations are experiencing sharp declines, so data from surveys like this is critical for better understanding why these drop-offs are happening and what can be done to aid and protect penguin colonies. In addition to performing these surveys faster, the drones and the innovative AI algorithm they employ have broader applications beyond the most southern continent. As the article notes, though the POPCORN algorithm was designed to specifically work with penguins, the scientists' aim is for it to eventually be usable in working with a variety of animals. Tasks the algorithm may someday carry out include counting livestock, assisting with wildlife management, and identifying animal populations before and after natural disasters.

A school librarian in the Blue Ridge Mountains of rural Virginia has used drones from a company called Wing, a subsidiary of Google/Alphabet, to have items delivered to her home. When the COVID-19 pandemic caused nationwide lockdowns of all nonessential businesses in 2020, which included the Montgomery County School District library where she worked, Kelly Passek had a—pardon the pun—novel idea to save the summer reading program: using Wing's drones "to deliver library books by drone to encourage more kids to read and still observe social distancing."[24] A 2020 CBS News story, available on YouTube in a brief video, chronicles this stirring example of ingenuity, collaboration, leveraging technology to promote access to information, and servant leadership to the community's students and their families: "Virginia Librarian Uses Drones to Deliver Books."[25]

Robots

While the subject of drones is quite new to films, robots in movies have a much longer history. A fairly robust number of internet-accessible resources provide robots-in-film information on everything from mundane trivia to more serious scholarship. Indeed, Carnegie Mellon University (CMU), one of the world's premier research institutions for robots and robotics, maintains a Robot Hall of Fame, complete with a federally registered trademark symbol: "The Robot Hall of Fame recognizes excellence in robotics technology worldwide and honors the fictional and real robots that have inspired and made breakthrough accomplishments in robotics. The Robot Hall of Fame was created by Carnegie Mellon University in April 2003 to call attention to the increasing contributions of robots to society."[26]

One of the first robots of the silver screen, a robot called Maria, appears in Fritz Lang's 1927 influential silent black-and-white capitalism-satirizing masterpiece *Metropolis*. The Robot Hall of Fame identifies the significance of Maria: "The robot MARIA stands apart as one of the only female robotic images of early science fiction."[27] Notably, harking back to prior discussion of copyright law and the public domain, the US copyright protection period for this classic movie ended on January 1, 2023, and it entered the US public domain.[28] In contrast, a 2023 CultureSlate article points out that the film will not enter the European Union's public domain until 2046, highlighting how much copyright protection periods can vary around the world.[29]

As mentioned, CMU is recognized as a global robotics leader. Chronicling and making the history of robotics available for study falls squarely within the bailiwick and mission of information centers. To that end, the Robotics Project at CMU is a collaborative initiative to preserve and make accessible the history and future of robotics. Its web page provides more details:

> The Robotics Project, launched in 2019, is a partnership between the University Libraries and the School of Computer Science to create a home for robotics's past, present, and future. Our intent is to capture a wide breadth of stories about the research, impact, and experiences of the robotics community across the university and around the world, creating a global destination for scholars and students thinking about the history of robotics. We preserve the collections in the University Archives and share them through educational and public programming.
>
> Working in consultation with leading archives and museums, we're rethinking how archivists collect and preserve robotics, and creating a community of practice for information professionals working with technology-driven material.[30]

Before moving into exploration of some of the ethical issues involving robots, let's define what the term means and some of the key elements of robots. Chapter 1, "What Is a Robot," in Oxford Press's *Robotics: A Very Short Introduction* (2012) "discusses three definitions of 'a robot': an artificial device that can sense its environment and purposefully act on or in that environment; an embodied artificial intelligence; or a machine that can autonomously carry out useful work."[31] *Merriam-Webster Dictionary*'s definition of a robot is "a machine that resembles a living creature in being capable of moving independently (as by walking or rolling on wheels) and performing complex actions (such as grasping and moving objects)."[32] *Britannica* further uncovers important aspects of a robot and introduces the related term *robotics*: "Any automatically operated machine that replaces human effort, though it may not resemble human beings in appearance or perform functions in a humanlike manner. By extension, robotics is the engineering discipline dealing with the design, construction, and operation of robots."[33]

If a robot were added to the Yellow Brick Road–trodding quartet journeying to the Emerald City in search of its heart's desire from the much-ballyhooed Wizard, that robot would probably be seeking an *ethical code*. Whether robots can in fact be imbued with an ethics-oriented nature and a moral set of values is the Holy Grail for engineers and scientists searching for that answer. Indeed, a 2023 *Nature* article, "How Robots Can Learn to Follow a Moral Code," explains that "many computer scientists are investigating whether autonomous systems can be taught to make ethical choices, or to promote behaviour that aligns with human values."[34]

As with the previous section identifying lethal and nonlethal uses of drones, robots are also categorized by similar categories. *Killer robots* is a term used for robots deployed for military-related purposes. Robots that perform more benign, service-oriented tasks are referred to as helper robots, care robots, and companion robots.

Killer Robots

Probably no other film has infused images of robot lethality in the global public's minds more than *The Terminator* and its five sequels. Released in 1984, *The Terminator* portrayed bodybuilder-cum-actor Arnold Schwarzenegger (eventually governor of California from 2003 to 2011) as a humanoid robot killing machine, sent back in time to snuff out a young human boy before his older self can become the human resistance's rallying leader in a war against self-aware machines. Though *The Terminator* films are fiction, robots that can kill are real and will increasingly serve as weapons of war. A significant corpus of killer-robot-related information, discourse, and scholarship is already available. In addition to scholarly papers on this subject, mainstream media has also provided commentary. One example, a September 2023 the *Hill* article, "Why the Pentagon's 'Killer Robots' Are Spurring Major Concerns," spells out some of the challenges and worries:

> Critics call self-operating weapons "killer robots" or "slaughterbots" because they are powered by artificial intelligence (AI) and can technically operate independently to take out targets without human help. . . .
>
> But there are no international treaties governing the use of these weapons, and human rights groups are uneasy about Washington's ethical guidelines on AI-powered systems and whether they will offer any protection against an array of humanitarian concerns.[35]

The United Nations and Killer Robots

The United Nations, individual nations, and activist groups are some of the stakeholders that have expressed differing stances on robots used for military purposes. A 2021 *New York Times* article noted "a push to ban" killer robots by some, but a lack of agreement by nations on whether to institute a treaty against such weapons.[36] The article points out that killer robots are also "more technically known as Lethal Autonomous Weapons Systems" (LAWS).

The Convention on Certain Conventional Weapons (CCW) is a UN treaty that was adopted in 1980, entered into force in 1983, and is viewed as "a key international humanitarian law instrument."[37] Because it has a human-rights orientation, the CCW has been and continues to be considered as a means for banning or restricting killer robots by incorporating killer robots within this treaty framework. The full name of the CCW is "The Convention on Prohibitions or Restrictions on the Use of Certain Conventional Weapons Which May Be Deemed to Be Excessively Injurious or to Have Indiscriminate Effects."

What is the UN's stance on killer robots and what other UN efforts are being considered for addressing LAWS? Looking back at the past few years sheds some light. Every five years, the CCW holds a review meeting. At the 2021 meeting, the CCW "debated the question of banning autonomous weapons."[38] A December 2021 piece on the Conversation web site, "UN Fails to Agree on 'Killer Robot' Ban as Nations Pour Billions Into Autonomous Weapons Research," points out that consensus on banning LAWS could not be reached. The Conversation article further identified the LAWS arms race that is already well underway: "Militaries around the world are investing heavily in autonomous weapons research and development. The U.S. alone budgeted US$18 billion for autonomous weapons between 2016 and 2020."[39]

In November 2023, the UN's so-called First Committee (Disarmament and International Security) "approved a new draft resolution on lethal autonomous weapons systems (document A/C.1/78/L.56)."[40] Ultimately, this "draft resolution as a whole was approved by a recorded vote of 164 in favour to 5 against... with 8 abstentions."[41] Human Rights Watch (HRW), a nongovernmental organization (NGO) in New York City, explained in a January 2024 post that on "December 22, 2023, 152 countries voted in favor of the General Assembly resolution on the dangers of lethal autonomous weapons systems, while four voted no, and 11 abstained. General Assembly Resolution 78/241 acknowledges the 'serious challenges and concerns' raised by 'new technological applications in the military domain, including those related to artificial intelligence and autonomy in weapons systems.'"[42] The four countries that voted no on the resolution were, in alphabetical order, Belarus, India, Mali, and the Russian Federation. The United States voted yes.[43]

Groups like HRW continue to urge the UN to institute a binding treaty on killer robots. Researchers also have an important role to play in getting the word out about these weapons. A January 2024 article in the *Harvard Gazette*, "'Killer Robots' Are Coming, and U.N. Is Worried," interviewed Bonnie Docherty. Docherty is a senior researcher within HRW's Arms Division, a lecturer at Harvard Law School's International Human Rights Clinic, and attended some of the UN meetings highlighted earlier. Docherty effectively identifies the ethical and legal issues (e.g., lack of human judgment in distinguishing between soldiers and humans) surrounding killer robots, key stakeholder nations and their positions (e.g., Russia opposed to any treaty; the US supporting nonbinding rules, etc.), and impediments to and prospects for a treaty. Here Docherty discusses some of the ethical challenges:

> [*Harvard Gazette*] What are the ethical concerns posed by killer robots?
>
> [Bonnie Docherty] The ethical concerns are very serious. Delegating life-and-death decisions to machines crosses a red line for many people. It would dehumanize violence and boil down humans to numerical values.
>
> There's also a serious risk of algorithmic bias, where discriminating against people based on race, gender, disability, and so forth is possible because machines may be intentionally programmed to look for certain criteria or may unintentionally become biased. There's ample evidence that artificial intelligence can become biased. We in the human-rights community are very concerned about this being used in machines that are designed to kill.[44]

Advocacy Organizations and Civil Watchdogs

In addition to HRW, other advocacy groups have coalesced to speak out about the threats of killer robots and lobby for restrictions on their use and proliferation, as groups have done in past years on the dangers of nuclear weapons and landmines. The International Committee for Robot Arms Control (ICRAC) is an NGO that a 2014 *Business Insider* article indicates was formed in 2009 by academics and scientists concerned about the threats posed by LAWS.[45] ICRAC states that "we are an international committee of experts in robotics technology, artificial intelligence, robot ethics, international relations, international security, arms control, international humanitarian law, human rights law, and public campaigns, concerned about the pressing dangers that military robots pose to peace and international security and to civilians in war."[46]

Interestingly, in light of renewed concerns voiced by the United States and other nations in February 2024 about potential Russian orbital nuclear weapons,[47] ICRAC's mission statement includes a bullet point addressing the need for discussion that "should consider," among other concerns, "The prohibition of the development, deployment, and use of robot space weapons."[48] The nonprofit group Arms Control Association (founded in 1971) reported in a March 2024 post, "Keeping Outer Space Nuclear Weapons Free" by Daryl G. Kimball, that

> the White House confirmed on Feb. 15 that U.S. intelligence uncovered evidence that Russia is developing an ASAT weapon that "would be a violation of the Outer Space Treaty, to which more than 130 countries have signed up to, including Russia." Russian President Vladimir Putin issued a nondenial denial, claiming on Feb. 20 that Russia remains "categorically against… the placement of nuclear weapons in space."[49]

Analysis of ICRAC's web page and other sources indicates that the group has not had an active internet presence since 2020. However, ICRAC's web page provides access to podcasts, such as one titled "Interviews on Social Responsibility in Robotics and AI,"[50] and ICRAC publications, like the paper "What Makes Human Control Over Weapons Systems 'Meaningful'?"[51]

Stop Killer Robots is another nonprofit group lobbying for restrictions on AI-imbued killer robots. Its web page states, "Our campaign calls for new international law on autonomy in weapons systems."[52] Stop Killer Robots is wary of decisions being made by automated systems. Examination of its web page in early 2024 shows sections on

- Automated decision research that was conducted in 2023, some of which involves "monitoring for reports of autonomous weapons use from other conflicts around the globe"[53]
- Information on concerns about "reports of the use of automated targeting systems in the Gaza Strip" by Israel[54]
- An October 2023 interactive multimedia exhibit called *Automated by Design*.[55]

A brief 2021 YouTube video, "The Stop Killer Robots Campaign," explicates the group's fears of uncontrolled robots *and* hopes for humanity in this transcripted snippet from the narrator's voiceover: "Our aim is to preserve humanity—by ensuring it is people, not robots, that are in control."[56]

Killer Robots for Nonmilitary Activities

Killer robots also have applications for nonmilitary purposes. An April 2021 *Guardian* article, "Killer Farm Robot Dispatches Weeds With Electric Bolts," describes killer robots being used for agricultural purposes—in this instance, to remove weeds on a UK farm. The manufacters see these robots as "part of a fourth agricultural revolution... bringing automation and big data into farming to produce more while harming the environment less."[57] Interestingly, farmers advised the robot makers that "killing weeds was the top priority" because of growing resistance to pesticides' efficacy and the need to reduce the amount of chemicals used for weed control.[58]

Security Robots

Security-related robots are another category of robots. These fall somewhere between killer robots and the next category that we explore—helper, care, and companion robots. Security robots aren't as lethal as LAWS (yet), but they also aren't cuddly and submissive, like robot cat companions (discussed later) or the help-oriented robots that, say, pass out food to residents in nursing home dining rooms. Keeping with our cinematic theme in identifying pop cultural representations of robots, 1987's *RoboCop* (and its two sequels and 2014 reboot) would probably be most people's pick of a film franchise depicting a security-themed robot (though the protagonist is actually a human cop transformed involuntarily into a cyborg).

Anecdotally, security robots haven't garnered as much press coverage as killer robots. However, these devices, like their more lethal cousins, raise many ethical considerations and questions. Scholarship is developing in this area as well, such as the Association for Computing Machinery (ACM) Digital Library's open-access article "Would I Feel More Secure With a Robot? Understanding Perceptions of Security Robots in Public Spaces" in the *Proceedings of the ACM on Human-Computer Interaction*. The abstract for this paper effectively distills the pluses and minuses of security robots, as well as gaps in design thinking and community engagement:

> Robots are increasingly being deployed as security agents helping law enforcement in spaces such as streets, parks, or shopping malls. Unfortunately, the deployment of security robots is not without problems and controversies. For example, the New York Police Department canceled its contract with Boston Dynamics in response to backlash from their use of Digidog, an autonomous robotic dog, which sparked fears in the public. However, it is unclear to what extent affected communities have been involved in the design and deployment process of robots. This is problematic because, without input from community members in the processes of design and deployment, security robots are likely to not satisfy the concerns or safety needs of real communities.[59]

The ACM paper's abstract references a thought-provoking canine security robot example, which we now further explore. This case study sheds light, too, on the broader use of security robots in all kinds of communities. Boston Dynamics is a start-up company based, as one might expect, in the Boston, Massachusetts, metro area. It was founded in 1992, at the cusp of the internet age. As *Robotics 24/7* describes, "Boston Dynamics began as a spin-off from the Massachusetts Institute of Technology, where National Academy of Engineering member Marc Raibert and his colleagues first developed robots that ran and maneuvered like animals."[60]

In 2019, Boston Dynamics unveiled a signature commercial product of its decades-long labors: a canine robot. Most likely as a wink and a nod to the name of the dog in the iconic Dick

Figure 12.1 Security robot in a Maryland Lowes store. (© 2023 Kip Currier)

and Jane reading books used in US schools from about 1940 to 1970, the designers bestowed their *novo canid* robot with the moniker Spot (despite nary a spot, dot, speckle, or freckle). How much *does* that robot doggie in the window cost? In 2020 and 2021, reputable sources, such as CNBC,[61] IEEE,[62] and Ars Technica,[63] reported that Spot was fetching a price of approximately $74,500. An upgraded Spot will set one back more. The *New York Daily News* reported in November 2022 that the Los Angeles Police Department was considering adding Spot to "its arsenal": "The price tag for just one Spot with several advanced features—like a robot arm and an infrared camera—is $277,917."[64]

In 2021, the New York Police Department set off a firestorm when the Spot robot dog that it had purchased joined a group of police officers as they searched for and found a suspect with a gun at a Manhattan public housing building. Context is especially key in fully understanding and appreciating why the inclusion of a robot dog on this legally permissible law enforcement activity was so upsetting to so many. The Black Lives Matter movement was and remains a grassroots response, many years in the making, to police brutality against African Americans and controversial deaths suffered by Black Americans. The death of 46-year-old African American George Floyd on May 25, 2020, at the hands of white police officer Derek Chauvin and several other officers in Minneapolis, Minnesota, catalyzed weeks of protest calling for law enforcement overhauls, oversight, and, in some places, disbandment. Against this backdrop of ongoing protests and debates about policing, the appearance of a robot dog in a historically disadvantaged New York setting less than a year after the murder of George Floyd struck an exposed nerve. As the *New York Times* described in an April 2021 article:

> It was what came out of the building next that really grabbed attention while feeding into a far-reaching debate about policing in New York: a 70-pound robotic dog outfitted with lights, cameras and what the police describe as artificial intelligence.
>
> The four-legged device had only gone into and out of the building's lobby without playing an active role in the operation, the police said. Still, its mere presence at a public housing building ignited a fierce backlash, with many people condemning it as a stark example of police power and misplaced priorities even as calls to address both roil the United States.[65]

A complicating factor in any analysis of the pluses and minuses of robot dogs is that they *can* save lives. The *Guardian* reported in March 2024 that state police in Massachusetts sent a Boston Dynamics robot dog into a home where an individual was barricaded, along with two robots of a different kind that are deployed for bomb-related issues. The barricaded person shot the robot dog three times and disabled its communication abilities. The police were able to arrest the individual after using tear gas. A statement from the state police described the tangible ways that the robot dog provided aid that, in this instance, kept law enforcement from being injured or killed. The robot was able to provide crucial "room clearance and situational awareness capabilities" that enabled the police to be able to refrain from "inserting human operators, and may have prevented a police officer from being involved in an exchange of gunfire."[66] Notably, Boston Dynamics stated that this incident was the first time one of their robot dogs had been shot. The company stated that they were happy to know that there were no other casualties experienced other than the robot. They added that this story demonstrates the ways that robots "can be used to save lives."[67]

Security and law enforcement robots like Spot the robot dog trigger numerous ethical questions:

- How should public safety and policing interests be weighed against issues of privacy, surveillance, and data collection when determining whether to use security robots? Does this calculus change at all when considering the use of security robots in communities of color and historically disenfranchised and disadvantaged communities?
- Who decides whether and when to use funds to acquire, maintain, and deploy security robots?
- What ethics codes, guidelines, and guardrails should be used, updated, or revised by communities in making decisions about security robots?
- What roles, if any, do information professionals and centers have with regard to security robots, in general? What roles, if any, do information professionals have pertaining to specific issues, such as the deployment of security robots in communities where information centers are located?

Helper, Care, Companion, Social Robots

Varying kinds of helper-related robots have long been a staple of pop culture and science fiction television show and movies. From Rosie the Robot maid (debuting in 1962) in *The Jetsons* cartoon to the beloved "Danger, Danger, Will Robinson" robot (first appearing in 1965) in the *Lost in Space* TV shows and film and the trash-collecting, lovelorn robot Wall-E in Pixar's 2008 hit flick, robotic helpmates have captured hearts and imaginations. In the twenty-first century, kindness-centered categories of robots are increasingly moving from small and silver-screen depictions to real-world kitchens, health-care venues, and third places like libraries.

As with killer robots, scholarship exists on helper, care, companion, and social robots. For example, researchers are looking at the use of these kinds of robots in residential nursing home settings. A 2023 *JMIR Aging* article, "Robots in Assisted Living Facilities: Scoping Review," examined "the literature on robots used in assisted living facilities and identify gaps in the literature to guide future research."[68] So-called care robots are one suggested "technological intervention" for alleviating staffing shortfalls in assisted living venues and helping with residents' needs for care. The COVID-19 pandemic was noted as contributing to an already present "shortage of professional caregivers." Care robots may be able to assuage other acts of God, as well as economic, labor, and demographic changes that may occur in the future. Although care robots offer some potential pluses, the researchers observe that "concerns about efficacy, ethics, and best practices in the applications of robotic technologies in care settings remain."[69]

Japan has been a longtime leader in research and innovations on helper, care, and companion robots. A 2023 *MIT Technology Review* piece provides a look "inside Japan's long experiment in automating elder care." This article provides insightful analysis that can be instructive for other nations on the complex, nuanced relationships that Japanese citizens have with robots. Researcher James Wright indicates that his Japan-based ethnographic fieldwork at a nursing home conducting trials on three distinct types of robots "has focused on this disconnect between the promise of care robots and their actual introduction and use."[70]

NPR has featured a number of programs involving ethics issues and helper and care robots. In May 2022, NPR Radio's *1A* program asked, "Are robots the solution to understaffed nursing homes?"[71] In addition to looking at issues such as how the COVID pandemic impacted staffing in care facilities and the "chronic loneliness among patients," and how robots might alleviate both of those challenges, the program also examined ethical concerns involving "data privacy and the importance of human-to-human contact."[72] A January 2024 NPR Radio segment, "Preserving Our Humanity in the Age of Robots," interviewed science writer Eve Herold about her new book.[73] Herold's 2024 book *Robots and the People Who Love Them: Holding on to Our Humanity in an Age of Social Robots* explores ways that social robots can aid children on the autism spectrum and patients with dementia and memory impairment, but also exacerbate loneliness and addictive tendencies in others.

Public libraries are investing in robot animals, too, with cats proving especially popular (see Figure 12.2). In October 2023, Oregon Public Broadcasting reported on three robot kitties that can be checked out by patrons at Eugene Public Library. Librarians interviewed for the story explained that although robot cats have been shown to be helpful for those experiencing memory impairment or loneliness, anyone can check them out from the library. In fact, the

Figure 12.2 Robotic cats. (University of Salford, Creative Commons License CC BY 2.0)

Eugene librarians found the cats to be so much fun and stress relieving that they bought two for the staff's own use, as described by a librarian:

> There's research that shows cats purring is therapeutic. So we have a kind of quiet room that we have placed one of our cats. Staff have actually really enjoyed that—maybe you had a stressful part of your day or something's happened and you just need a moment, and you can come hang out with one of our cats that are here that are not circulating.[74]

Access by staff to robot animals is an intriguing and practicable wellness strategy, especially in light of growing awareness of the impacts of stress and burnout on service providers. Researchers Karen E. Fisher and Lauren Alexa Gambrill at the University of Washington Information School have been studying the impacts of trauma on library staff. The significance, rationale, and need for this Institute of Museum and Library Studies–funded research is articulated in their 2022 paper, "Superheroes Need Help Too: Understanding Peri- and Post-Traumatic Symptoms Amongst Library Staff as Essential Frontline Workers": "Trauma-informed care for patrons is a growing movement. However, scant research examines the nature of workplace trauma experienced by library staff as essential frontline workers in today's complex communities."[75] One of the aims of their work is "to identify trauma-informed care tools, policy and procedure that libraries can implement." Additionally, they aim to "create curricula for library students and staff." It will be interesting to see how robot animals may be able to play a broader role in societies in promoting student and staff wellness, as well as for public health in general.

Manistee County Library in Michigan is another public library that has accessioned robot animals into its library of things. While Eugene Public Library opted to focus exclusively on robot cats, Manistee took a diversified approach, acquiring a robot cat, a robot dog, and a robot bird. Its robot animal collection is specifically geared to addressing the needs of senior community members. The different robot animals are intended to promote distinguishable activities and outcomes, as explained by the interim library director, Julie Cirone: "The bird is designed to motivate a person who uses a walker. It will chirp and sing to encourage walking."[76]

Western Michigan University has been studying whether robot animals can alleviate stress in students. In 2019 the university library and the School of Communication partnered "to investigate the potential therapeutic impact of students interacting with robot pets during finals week."[77] Universities, nursing homes, hospitals, and other venues periodically bring in live therapy animals that people can interact with for fun and to reduce stress and loneliness. However, live animals can present certain risks and downsides; these include possible injuries to people as well as to the therapy animals if play is too rough; noise; allergic reactions from hair, fur, or feathers; and hygienic issues emanating from occasional messes and animal waste. Hence, robot therapy animals offer some distinct advantages to their real-life counterparts.

Every action has ripple effects though, and one potential downside to reliance on robot animals may be diminished adoptions of shelter animals in need of "forever homes." Examination of the web page of the American Society for the Prevention of Cruelty to Animals in February 2024 revealed, for instance, that about "6.3 million companion animals enter U.S. animal shelters nationwide every year. Of those, approximately 3.1 million are dogs and 3.2 million are cats.... Each year, approximately 920,000 shelter animals are euthanized (390,000 dogs and 530,000 cats)."[78] Time will tell as to how real-life and robot animals will coexist with their respective human companions and creators. The choices that humans make will impact the futures of all organic *and* inorganic creatures great and small.

Personal Delivery Devices

Personal delivery devices (PDDs) are another category of robots. The principal purpose of these mechanized devices is telegraphed in their name: transport and delivery of a range of items. Edible and potable consumable goods, medicines, and books are some of the things that are conveyed. In recent years, start-up companies have been conducting beta tests of PDDs in select US cities. University campuses are also popular sites for some PDDs.

A 2022 Robot Report piece noted that Kiwibot, a PDD company, "was established in 2017, and in the same year launched its first pilot program at the University of California-Berkeley campus."[79] In September 2021, a TribLive article covered a pilot partnership between the City of Pittsburgh and Kiwibot. The article reported that residents in parts of the Pittsburgh area neighborhoods of Bloomfield, Garfield, and Lawrenceville would potentially be seeing nonhuman deliveries of medicines, food items, and books. Local businesses collaborated with the city and Kiwibot. One of those organizations was the Carnegie Library of Pittsburgh, which provided free deliveries of books to people residing within the library's area of service.[80]

It is noteworthy to point out that an earlier April 2021 *Pittsburgh Post-Gazette* article highlighted concerns of some residents of the neighborhoods where Kiwibots were to be deployed.[81] Some of those concerns were about sharing sidewalk space with the PDDs and liability that could result from collisions with the devices. Americans with Disabilities Act issues are a concern as well: a 2019 TribLive piece informed readers that "in 2019, Starship Technologies launched a fleet of automated food delivery robots in Oakland, though they were taken off the streets amid concerns they were blocking sidewalk access for people in wheelchairs and others who may have limited mobility."[82]

To make it possible for PDDs to legally operate in Pennsylvania communities, in 2020 the Commonwealth of Pennsylvania legislature amended an existing statute (Section 102 of Title 75 of the Pennsylvania Consolidated Statutes) to enable PDDs to be included within the definition of a pedestrian:

> "Pedestrian." [A natural person afoot.] Any of the following:
>
> (1) An individual afoot.
> (2) An individual with a mobility-related disability on a self-propelled wheelchair or an electrical mobility device operated by and designated for the exclusive use of an individual with a mobility-related disability.
> (3) A personal delivery device.[83]

As a result of this statutory change, "The robots are permitted to operate in any pedestrian area, including sidewalks, on the berm or shoulder of a roadway or on a roadway posted at 25 mph or less."[84] The City of Pittsburgh touted the benefits of this emerging technology. Mayor Bill Peduto's chief of staff, Dan Gilman, explained that in additon to embracing these kinds of new technologies, the city wanted to ensure that the technologies address "greater societal needs"—for example, to "close gaps in access to food, medicine and libraries, while at the same time supporting local businesses, and doing so in partnership with the communities they serve."[85]

This case study implicates a number of ethical issues:

- Whether individuals and communities had meaningful input into decisions by the City of Pittsburgh to permit Kiwibots to be deployed in select Pittsburgh neighborhoods.
- Who chose the neighborhoods where Kiwibots were tested? Why were those neighborhoods selected? If applicable, which other neighborhoods were considered and why were they ultimately not chosen as testing sites?
- What alternate plans were put in place if PDDs were stolen or unable to make their deliveries of foods and medicines?
- In the neighborhoods where PDDs were tested, was information provided about where people could register complaints or concerns about PDDs? If so, where and in what formats was the information made accessible?
- What safeguards did entities like Carnegie Library of Pittsburgh consider before deciding to join the pilot project in some of its service areas?

- What factors should libraries and other information centers weigh in considering whether to use PDDs to provide access to materials?
- What rationales were used by the Pennsylvania legislators to warrant amending the statute to treat PDDs as pedestrians?
- Do PPDs possess qualities that are necessary to be deemed pedestrians? Pedestrians have historically been recognized to be human beings. Human beings have sentience and can (generally) exercise agency in their capacity as pedestrians.
- What happens to the data that is collected by PDDs? How is that data protected?
- What data is accessible on incidents between humans and PDDs?
- In what ways, if any, is the use of PDDs influencing employment rates of human deliverers? Are PDDs displacing humans from delivery jobs?

Perusal of Kiwibot's web page in February 2024 indicated that their PDDs are located on more than a couple dozen college and university campuses, including Gonzaga University (Spokane, Washington), Howard University (Washington, DC), Marquette University (Milwaukee, Wisconsin), New Mexico State University (Las Cruces), and Tulane University (New Orleans, Louisiana).[86] Positioning and deployment of PDDs on higher education campuses also raises ethical questions:

- Which administrative bodies and officials are making the decisions to permit PDDs on these campuses?
- Do individuals, groups, and communities have opportunities to publicly comment on the use of PDDs on these campuses?
- Is Kiwibot making decisions about where to station its PDDs based on demographic data? If yes, what kinds of data and where is that data coming from?
- Is Kiwibot targeting historically disadvantaged communities or vulnerable populations of college students?
- What kinds of local businesses is Kiwibot partnering with? What kinds of local businesses is Kiwibot not partnering with?

Robots and Information Professionals

Having examined the varying kinds of robots in the previous sections, some of the categories have a more readily apparent link to information centers. The easiest through line from "robot types" to the information professions are two categories: companion robots and delivery robots. Public libraries, like Eugene Public Library, are already making robot animals available to users and staff. Additionally, public libraries, like Carnegie Library of Pittsburgh, have experimented with robots, such as PDDs, that can transport physical library items. I did not locate instances of information centers that are presently utilizing security robots, though it is likely only a matter of time until examples appear. However, libraries *are* located in communities where these robots are already patrolling. For example, security robot maker Knightscope has supplied Portland, Oregon, with several robots that are patrolling the downtown area, arguably in the metropolitan vicinity of Portland's Multnomah County Central Library and other libraries, as reported by *Willamette Week* in a December 2023 piece: "The 5-foot-5, 420-pound droid, named Rob, is made by a Mountain View, Calif.-based company called Knightscope Inc. Rob is at least the third Knightscope bot to go on duty in the metro area."[87] Another robot patrols areas around a hospital and a hotel. A Knightscope press release noted that that robot was named Robby and had "been scouting for vandals since September 2022."[88]

A quick internet search handily reveals examples of academic libraries that are collecting and making killer-robot-related books, films, peer-reviewed articles, and other resources accessible to patrons. The following sampling of items also shows the utility of these information objects as authoritative sources that offer different perspectives on this topic across a period of years:

- University of Minnesota–Duluth's Kathryn A. Martin Library: *Killer Robots: Legality and Ethicality of Autonomous Weapons* (2009) by Armin Krishnan (book)
- Stanford University Libraries: *Losing Humanity: The Case Against Killer Robots* (2012) by Bonnie Docherty (book)
- University of Wisconsin–Oshkosh: *The Truth About Killer Robots* (2018), directed by Maxim Pozdorovkin (film, available to current UW Oshkosh students, faculty, and staff)
- MIT D-Space repository: "Wrestling With Killer Robots: The Benefits and Challenges of Artificial Intelligence for National Security" (2021) by Erik Lin-Greenberg (article, *MIT Case Studies in Social and Ethical Responsibilities of Computing Journal*)

Information professionals are in the business of stewarding and curating information, in addition to facilitating access. They review, assess, select, categorize, annotate, program, teach, guide, ask, answer, fail to answer or abstain from answering where applicable, make findable, advocate, lead, follow, partner, engage, and stay current with information on general and specific topics. As demonstrated, the topic of killer robots is generating considerable attention, debate, scholarship, and legal, ethical, and policy decision-making. Hence, from these standpoints alone and collectively, it behooves information professionals to have at least a basic understanding and awareness of this category of robots, in addition to the more benign varieties. It also is not disrespectful to killer robots and those affiliated with them to state that most of us likely hope that the former will continue to be less pertinent to the daily lives of information professionals and their customers than, for example, robot tabbies and book-delivering robots. But—just in case—information professionals acutely understand the importance of being informed and ready to lead the way to the information that individuals and societies need, including that which illuminates the killer kind of machine and enables people to make up their own minds on where they stand.

Robots and Drones Partnering With Humans

Some technology experts see an encouraging future of humans and robotic devices working together. Daniela Rus and Gregory Mone's 2024 book *The Heart and the Chip: Our Bright Future With Robots* chronicles collaborative examples, such as robotic drones being used to study whale populations and robots that can reduce physical wear and tear on human workers in warehouses or freezing-cold storage lockers. They describe ways, too, that robots may be able to remotely enable humans to experience faraway lands, even the smells of, say, fresh croissants wafting from a Parisian *boulangerie*. Robotic arms and appliances can also aid older and physically challenged people in retrieving items from higher shelves or picking up heavy objects. In all of these cases the authors assert there is a central theme: robots can extend the reach of humans. In their own words, "These examples of how we can pair the heart with the chip to extend our perceptual reach range from the whimsical to the profound. And the potential for other applications is vast."[89]

This chapter has illustrated some of the ways that robots and information professionals and users are already interacting and expanding the boundaries of service. Information centers have a long history of innovative, human-centered services, and we can imagine the kinds of fantastic and life-enhancing human and robot collaborations that are just over the horizon.

Autonomous/Self-Driving Vehicles

Imagine, for a moment, the following scenario: It's morning. You're walking to your car to drive to the museum where you work as the new executive director. The commute from home will take approximately forty-five to sixty minutes, depending on traffic. You're scheduled to give a 15-minute presentation on "AI Ethics and Indigenous Museum Collections" via Zoom to the twenty on-site and virtual-attending members of the museum's board of trustees shortly after the workday starts. You want to use the commuting time to review your PowerPoint slides and practice your talk before arriving at work. You get into the back of the car, so you'll be able to recline and look at your tablet screen more comfortably. With a voice command from you, and a biometric scan of your irises and face to verify your identity from the visually undetectable sensors within the car's nanotech windshield, you turn the car on. Then, you tell the car to "play calming music" (preprogrammed to "Ocean Bliss" or "Forest Reverie") and drive to your (also preprogrammed) workplace. Traffic is lighter than usual today. Exactly 42 minutes and 11 seconds after departing from your driveway, you and your car pull safely into your reserved parking space. Because of the time you saved by not having to concentrate on the task of driving, you were able to rehearse your talk, respond to urgent emails from two employees and the museum's general counsel, and center yourself. You smile as you confidently walk into the boardroom, ready to "knock the board's socks off" with your presentation.

That technology-centered scenario—no-hands-necessary, worry-free, safety-first, AI-driven transportation—has been the primary vision and value proposition of self-driving (also referred to as autonomous) vehicles for decades. Despite the promotion of self-driving cars and trucks by some Big Tech companies and smaller start-ups as a close-at-hand, realizable goal, the reality, unfortunately, has yet to be achieved.

Autonomous Vehicle Risks and Controversies

Digital-age testing of self-driving vehicles has been occurring in the United States for more than a decade. *IEEE Spectrum* observed in a September 2014 article that "the world's first self-driving test

Figure 12.3 Argo company's autonomous vehicle on test drive in Pittsburgh, Pennsylvania. Argo went out of business in October 2022. (© 2020 Kip Currier)

took place on May 1, 2012 in Las Vegas."[90] Since that time, testing has been happening in more than a dozen cities throughout the United States. Indeed, over the past decade, I have observed testing of self-driving cars (most of them Uber test cars) at least twenty times in a number of Pittsburgh neighborhoods (see Figure 12.3). On all of these occasions, humans were observed to be sitting in the front seats of these test cars. This is important to note because, in many cities where testing has been occurring, it has been a requirement that at least one person be present in a test car while it is driving, to serve as a fall-back safety mechanism. Should a self-driving car malfunction for some reason and present an imminent danger, the thinking behind having a human, or humans, in the test car is to enable human beings to reassume control of the vehicle.

It is necessary to note that having a human testing monitor in a self-driving vehicle is no guarantee that an injury cannot occur at the hands of an autonomous vehicle. Indeed, the first fatality that occurred as a result of a self-driving car had a human monitor in that car. As a November 2019 article recounted in Tempe, Arizona, 49-year-old Elaine Herzberg was killed when she was hit by an Uber autonomous vehicle while she was walking her bicycle across a street in an area not within a designated crosswalk.[91] A report on the incident stated that the Uber SUV's human operator, Rafaela Vasquez, was viewing "an episode of *The Voice* while operating the test vehicle."[92] As a result of this fatality, Uber suspended all testing in Tempe, Pittsburgh, San Francisco, and Toronto. Roughly a year and a half later, a National Transportation Safety Board investigation revealed that the Uber test car "did not recognize [Herzberg] as a jaywalking pedestrian and its braking system was not designed to avoid an imminent collision."[93] Ultimately, Vasquez, the Uber human backup driver, "pleaded guilty to one count of endangerment and was sentenced to three years of probation" on July 28, 2023.[94]

Another significant autonomous vehicles milestone occurred in October 2020: the launch of the world's first so-called robotaxi in Phoenix, Arizona. An April 2022 *Nikkei Asia* story describes this landmark event: "The term 'robotaxi' has been around for years, and companies worldwide are testing autonomous ride-hailing services with a safety driver behind the steering wheel. But Waymo, owned by Google's parent company Alphabet, was the first to provide commercial driverless taxis to the public."[95] After the Phoenix debut of Waymo robotaxis, San Francisco followed suit in 2022 with service by Cruise, a General Motors–supported effort: "For the past decade, San Francisco and Phoenix—the first, America's second-densest city, the second a sprawling metropolis—have served as laboratories for the companies vying to commercialize the burgeoning technology."[96] A September 2023 *San Francisco Standard* piece calls San Francisco the "unofficial birthplace of the robotaxi revolution, with hundreds of completely autonomous vehicles crawling around the streets," Cruise and Waymo the main companies represented.[97]

Unfortunately, robotaxi service in San Francisco has been plagued by various setbacks, and a calamitous high-profile incident. On October 2, 2023, a pedestrian walking downtown was hit by a car driven by a human driver. A nearby Cruise robotaxi ran over the woman, dragging her 20 feet, where she remained trapped underneath the vehicle until she was extricated.[98] As a result, two days later Cruise halted all robotaxi service in San Francisco.[99] Investigations ensued. The California Department of Motor Vehicles suspended Cruise's permits when it found that Cruise had hidden video of the incident.[100] Recriminations ensued against Cruise, as well as the City of San Francisco for enabling Cruise to operate robotaxis in its neighborhoods.

This incident harmed not only the pedestrian who was dragged and injured but also other interests: namely, public trust. As the *Austin American-Statesman* related in November 2023, "After that incident, Cruise has been working to 'rebuild public trust.'"[101] On November 14, 2023, Cruise posted an update:

> On October 26, we announced a pause of all our driverless operations while we take time to examine our processes, systems, and tools and improve how we operate. In the coming days, we are also pausing our supervised and manual AV operations in the U.S., affecting roughly 70 vehicles. This orderly pause is a further step to rebuild public trust while we undergo a full safety review. We will continue to operate our vehicles in closed course training environments and maintain an active simulation program in order to stay focused on advancing AV technology.
>
> Cruise is dedicated to rebuilding trust and operating at the highest standards of safety. We are committed to keeping our customers, regulators, and the public informed throughout this process.[102]

In the wake of incidents like the Tempe and San Francisco respective pedestrian death and injury, difficult questions regarding the technology are emerging:

- Was self-driving vehicle technology oversold? Was it capable of doing what its proponents claimed it could do or would soon be able to do?
- Was public safety accorded lower weight, in essence sacrificed, because of interests that were given greater weight, like commercial, for-profit interests?

- What models and ethical theories, if any, are relied on by policy makers who render decisions about emerging technologies like self-driving vehicles?
- What laws should be enacted and enforced to ensure that individuals are protected from technologies that are unsafe and not thoroughly tested? Who determines when a minimum threshold of safety has been reached?
- What are the responsibilities—ethically and legally—and civil and criminal penalties for people, companies, and governments that make legal and policy decisions with injurious, even lethal, impacts on citizenries and communities?
- What avenues do members of the public and communities have (or should have) for not only sharing their voices in decision-making processes but also challenging decisions that are made about emerging technologies?

As the hard-bitten Alaskan chief-of-police character played by Jodie Foster in the 2024 prestige HBO drama *True Detective: North Country* repeatedly poses to her law enforcement colleagues, it's vital to "ask the right [expletive] questions." An October 6, 2023, *Axios* article peels back the onion on one of the most sobering questions about self-driving vehicles, asking whether driverless cars need to "drive more like humans."[103] Big Tech leaders and champions such as Elon Musk, technologists, venture capitalists, and countless others have touted the supposed superior driving capabilities of autonomous vehicles for years. But maybe self-driving cars *won't* be better drivers than flesh-and-blood humans. *Axios* recounts that the San Francisco pedestrian incident "is raising questions about whether autonomous vehicles can handle the unexpected as well as, or better than, humans."[104] The article goes on to note that the ability of autonomous vehicles to *predict* is one of the most challenging things for them to do because they "lack the instincts of a human driver." The "big picture" takeaway from *Axios* is that "the promise that driverless cars will be safer than human drivers is almost the entire selling point of the technology." The right question that follows then is this: "would a human driver have behaved differently?"

The testing and use of robotaxis and autonomous vehicles proceeds with a mix of advances and retreats. A March 1, 2024, *Los Angeles Times* article reported that state regulators had approved Waymo's expansion of its driverless taxi services into Southern California's Los Angeles County and Northern California's San Mateo County, though a specific launch date was not determined. The article further noted that "the decision by the California Public Utilities Commission will open the streets of America's second-largest city to a fleet of autonomous vehicles—even as self-driving cars continue to be the subject of safety concerns and some public criticism."[105]

In the meantime, humans may soon be sharing highways and byways with AI trucks. As a October 2023 TribLive headline telegraphs, "Self-Driving Vehicle Testing in Pittsburgh Area Shifting Gears to Focus on Trucking Industry." The article explains why focus may shift from self-driving cars to self-driving trucks: "Companies and investors are looking at vehicles in more predictable scenarios—think trucks on the interstate rather than in city traffic—to advance self-driving technology."[106] As one observer stated, "We've learned a lot in the last five years about the technology and about regulation and the public's willingness to accept autonomous vehicles."[107] Why might autonomous trucks be more efficacious than vehicles in cities? "Highways have fewer variables than urban roads, where design quirks, pedestrians, bicycles and unpredictable driver behavior—including motorists making the so-called Pittsburgh Left—can throw a wrench in the works. Highways are more predictable. Big rigs don't navigate tight urban centers."[108] Our ethics and values tell us that we must continue to consider what the right questions, and the right answers, are as societies navigate the promise and perils of new technologies and their evolving applications.

Autonomous Vehicles and Information Centers

Self-driving vehicles, if they can be realized, portend several intriguing opportunities for information centers. Libraries, as New York Public Library Director Anthony W. Marx has said, are in the information access business. Some libraries are in cooperative systems, or what are sometimes referred to as "library consortia," where physical resources are shared among brick-and-mortar locations and are able to be borrowed by patrons in different communities and cities. Library A may not have a physical copy of a particular book in its collection, for example, but a copy of that book can be borrowed from Library B and transported to Library A, where a patron can then check out Library B's copy of that book. This kind of physical resource sharing enables information centers to augment collection capabilities beyond their own physical locations. In this way, too, they can further the important goal of facilitating access to information.

Information centers expend lots of money each year on transportation vehicles that deliver those shared materials. Those transportation vehicles are driven by humans. Hence, should autonomous vehicles ever realize their envisioned promise of safe, dependable nonhuman transportation, it is possible to imagine scenarios in which driverless vehicles could be used to convey materials. This could save money for libraries, archives, and other information organizations. At the same time, though, it would also likely displace at least some number of real people from driving transport vehicles and earning money from doing so.

Another way that autonomous vehicles could promote access to information is specifically people-centered. In other words, self-driving vehicles could be used by users to physically travel to information centers, rather than, say, by bus or other means of transit. Moreover, one could imagine a library perhaps purchasing an autonomous van or bus-like vehicle that could pick up users and take them to and from the library. Such services might be free or provided at low cost. City governments or other municipal entities and for-profit and nonprofit institutions might also partner with information centers in making these kinds of services available. In turn, these transit possibilities might increase library, archives, and museum access to lower-income people and resource-scarce communities.

ALA and Self-Driving Cars

As with prior technologies we have examined, ALA's website has a helpful section on self-driving cars.[109] This is another technology for which ALA provides information as part of its Library of the Future initiative.[110] The site includes a beneficial overview of this emerging technology and articulates five distinct ways that autonomous vehicles "matter," and why they matter, vis-à-vis libraries and information centers:

> Self-driving cars could solve one of the biggest problems in urban design—parking. If self-driving cars improve transportation efficiency to the point that residents choose not to own automobiles, the amounts of parking needed in cities and communities could drop significantly. This could be an especially important change for libraries' physical locations, where considerations for parking availability can challenge facilities planning, affect patrons' visits, or even raise concerns for safety.[111]

Self-driving cars may, thus, influence libraries as physical spaces. One way they may do so is to reduce the footprints (i.e., the amounts of land and space) needed by libraries. Reduced footprints can be seen as potentially having both positive and negative implications for libraries, as noted in the preceding excerpt.

The site's "Why It Matters" section further identifies ways that these vehicles may have impacts on libraries: "The design and function of self-driving cars' interiors could radically change users' experiences.... Libraries could become interested in self-driving technology as a means of improving materials delivery, outreach services, or transportation between location.... Like other advances in artificial intelligence, self-driving technologies could have a significant impact on labor and the workforce."[112] A final point in this section effectively shows the unknowns that still exist regarding this emerging technology. On the one hand, self-driving vehicles have the potential to "help increase socialization in communities." On the other hand, they "just as likely might be the further isolation of communities."

Immersive Technologies: Virtual Reality, Augmented Reality, and More

"San Junipero," the Emmy and BAFTA Award–winning season 3 (2016) episode of the *Black Mirror* anthology series, turns on its ear the usual conceit of the stand-alone *Twilight Zone*–inspired "technology morality plays." As a 2016 AV Club review remarks, "Throughout its three seasons, *Black Mirror* has been consistent on a single idea: technology is a trap.... In Charlie Brooker's malignant worldview, that app you just downloaded is always going to bring out the worst in you.... If new technology is an opportunity to change the world, there's no guarantee we won't be changing for the worse."[113] Spoiler alert: The place referred to as San Junipero is actually a virtual reality (VR) for visitors still alive and those who have died and had their consciousnesses uploaded. As viewers, we vicariously feel the bittersweet life experiences of Kelly and Yorkie, two of the still living, as they're faced with making individual choices about whether or not to be permanently uploaded to San Junipero's idyllic surfside VR afterlife. The heart of "San Junipero" is the "technology-isn't-a-trap" free-will choice it offers, namely, To-Permanently-San Junipero or Not-to-Permanently-San-Junipero, with none of *Black Mirror*'s trademark dystopic twists. AV Club captures its VR-tech-enabled allure: "The delightful freedoms of San Junipero, a virtual reality experience that allows users to enjoy the same perfect city through different eras, are exactly what they appear to be. No one is plotting against anyone. There are no strings, no horrible reveals. Just this once, things turn out okay."[114]

As the fictional "San Junipero" episode depicts, advanced digital-age technologies can make it possible for people to immerse themselves in realities that feel, look, smell, sound, and, in all the ways that matter, seem real. Though the real world isn't yet at the level of technical prowess seen in San Junipero's near-future alternate reality (which also shows characters getting around in self-driving vehicles), this world *is* fast developing life-changing immersive reality capabilities and applications.

Let's define immersive technology and its related but distinct variations. "The Future of Immersive" on *Medium* explains that "immersive technology describes a range of technologies that facilitate immersion in a simulated environment. These experiences are achieved via screens or headsets. Some forms of immersive tech provide access to entirely simulated worlds. However, others project digital objects over the physical world."[115] ALA defines virtual reality as "the computer-generated simulation of images or whole environments that can be experienced using special electronic equipment."[116] Some well-known movie examples depicting VR include *Tron* (1982), *The Matrix* (1999), and *Ready Player One* (2018). *Oxford Dictionary* defines augmented reality (AR) as "vision technologies that superimpose a computer-generated object on an image of a real-world scene."[117] The Franklin Institute in Philadelphia, Pennsylvania, offers a widely recognized AR example: "Perhaps the most famous example of AR technology is the mobile game *Pokémon Go*, which was released in 2016 and quickly became an inescapable sensation. In the game, players locate and capture Pokemon characters that pop up in the real world—on your sidewalk, in a fountain, even in your own bathroom."[118] We will examine ethical issues of the *Pokémon Go* mobile app game later in this section. The ability of smartphone users to overlay, say, animal heads on their faces is another kind of popular use of AR, blending overlaid computer images with real people.

Per Microsoft, mixed reality (MR) "is a blend of physical and digital worlds, unlocking natural and intuitive 3D human, computer, and environmental interactions."[119] Microsoft also notes its provenance: "The term 'mixed reality' was introduced in a 1994 paper by Paul Milgram and Fumio Kishino, 'A Taxonomy of Mixed Reality Visual Displays.'"[120] The paper was published by the Institute of Electronics, Information and Communication Engineers (IEICE).[121] Hewlett-Packard fleshes out key aspects of extended reality (XR):

> In a nutshell, XR is "reality-plus" tech using any kind of display. XR is VR plus AR.
>
> XR stands for "extended reality," an umbrella term that covers VR, AR, and MR. All XR tech takes the human-to-PC screen interface and modifies it, either by 1) immersing you in the virtual environment (VR), 2) adds to, or augments, the user's surroundings (AR), or 3) both of those (MR).
>
> The term XR has been around for decades. It first popped up in the 1960s when Charles Wyckoff filed a patent for his silver-halide "XR" film, intended for photographing extremely bright light events, such as nuclear explosions.[122]

Forbes, in 2019, observed that XR "is an emerging umbrella term for all the immersive technologies."[123] The *Forbes* article, "What Is Extended Reality Technology? A Simple Explanation for Anyone," also explains that XR is "used to describe immersive technologies that can merge the physical and virtual worlds."

To summarize and taxonomically organize these terms then, immersive technologies is the broad, overarching, general term from which varying "reality" technologies radiate or extend. (In a similar organizing way, intellectual property [IP] is the broad term encompassing the four types of IP: copyright, patents, trademarks, trade secrets.) XR is a way of expressing the concept of "immersive technologies" via a clever two-syllable acronym. Like immersive technologies, XR enfolds the different kinds of existing immersive tech (e.g., VR, AR, MR) "plus those that are still to be created."[124]

Legal and Ethical Considerations of Immersive Technologies

In March 2022, the Markkula Center for Applied Ethics hosted a virtual webinar titled "Virtual Reality, Real Virtues, and Augmented Norms and Laws."[125] The speakers, Erick Ramirez, associate professor of philosophy at Santa Clara University and human rights attorney Brittan Heller, provided illuminating insights on "some of the novel ethical and legal issues associated with the widening adoption of VR and AR technology, and new norms and laws that might develop in response to the ecosystem that some call the 'metaverse.'"[126] Ramirez's presentation pointed out some of the immersive technology ethical issues currently facing societies. These include the need for stronger consumer protections, revising and updating institutional review board requirements governing the conduct of experiments, and acknowledging the limitations of these technologies.

Heller, who advises the UN on the intersection of technologies with human rights, emphasized that VR and AR differ greatly from social media. For example, VR and AR capture much more information and data about a person. She evocatively describes AR and VR technology as akin to a polygraph with six cameras: watching, sensing, recording, collecting, and transmitting data about its users. Heller explains that this practice, "biometric psychography," "is a new concept for a novel type of bodily centered information that can reveal intimate details about users' likes, dislikes, preferences, and interests."[127] Heller's 2020 paper "Watching Androids Dream of Electric Sheep: Immersive Technology, Biometric Psychography, and the Law" compellingly unpacks some of the risks posed by collection and use of this data:

> Pupillometry offers the perfect illustration of this risk. Many people are surprised by how much can be revealed through evaluating the motions of the eye.... Some researchers have found that autism in some young children can be gauged by irregular eye motion patterns. Other serious ailments, like schizophrenia, Parkinson's disease, ADHD, and concussions can also be diagnosed though [*sic*] eye tracking.[128]

One of the other noteworthy aspects of these immersive technologies is their ability to track users' involuntary physical responses in addition to voluntary responses. Hence, Heller stressed that conceptions of personally identifiable information (PII) need to be very different from social media and that existing biometric laws will not offer sufficient protections. Informed consent laws, too, need to be strengthened to better plug loopholes in existing laws. For example, informed consent regulations need to be tailored to address not only an individual who is electing to use immersive technologies but also people nearby who may not have provided consent. PII and biometric data from nearby people may intentionally or inadvertently be "swept up" when data is collected. Thus, it is important to be ethically mindful and legally compliant with securing "bystander consent."

Heller and Ramirez suggested a number of immersive tech areas and questions that remain to be considered and managed. All of these have legal, ethical, and policy dimensions:

- Need to define best practices
- Need to create industry standards and compliance frameworks for adhering to standards
- Need to define VR/AR technology for all populations, such as making immersive technologies inclusive for those with ability challenges
- Recognition that AR/VR makers must treat children differently than adults: What are the developmental impacts on children? Do spatial perceptions of children using AR/VR differ from adults? What are the dissociation challenges for children using this tech?
- Need for more research on therapeutic applications of AR/VR
- Need for more research to determine whether the alleged benefits of AR/VR are in fact clinically proven; for example, these technologies are sometimes called "empathy machines" because, it is asserted, users can attain more empathy by using this tech and engaging virtually with the lived experiences of others, "walking in their shoes," and seeing through their eyes—such assertions have yet to be empirically confirmed, however
- Necessity of determining whether there should be any limits on what people can embody within immersive environments
- Consideration of ways that immersive tech can be used to manipulate people's physical and emotional states
- Examination of ways that misinformation and disinformation are present in VR/AR and how to address that

The panelists also covered additional challenges presented by immersive technologies:

- Thinking about how the so-called terrible people may utilize immersive tech in ways that cause harm and injury
- No standardized terminology or taxonomy currently existing for immersive tech; as the panelists suggested, "Imagine that the way to make a 911 emergency call was different *everywhere*." This is an area in which information professionals are naturally suited to significantly contribute recognized expertise in information classification and organization
- What responsibilities will AR/VR designers and manufacturers bear for impacts on users?

- What are the unintended impacts of VR/AR technology?
- Keeping compassion in mind and thinking about all people who may use or be impacted by VR/AR technologies
- What existing or new regulatory frameworks and laws may be needed
- Courage to say no to a product that has or may have foreseeable harms

Heller and Ramirez identified a number of positive and promising uses of AR/VR:

- Use of immersive tech for trainings, such as medical training, learning how to operate a submersible, and so on
- Artistic representation
- Medical uses, such as pain mitigation
- Therapeutic purposes, such as counseling
- Experiential upshots of AR/VR use, such as the sense of wonder one may experience from simulating flying through the air or swimming with a pod of whales

The panel's moderator, Irina Raicu, the director of Santa Clara University's Internet Ethics Programs at the Ethics Center, noted the importance of "grace" in looking at immersive technologies from multiple angles. As she opined, "There are going to be missteps but VR/AR also holds promise."[129]

Controversial Uses of Immersive Technology

The previous section identifies many important issues, as well as some positive and negative aspects of AR and VR technologies. The following are some examples of immersive tech that raise a number of ethical concerns, many of which intersect with prior areas examined in this book.

VR Headsets

VR headsets raise privacy, surveillance, and data collection and use concerns. These concerns build on points raised by Brittan Heller regarding biometric psychographic data. A 2022 *Washington Post* piece noted that "companies are already gorging on our data."[130] Moreover, a privacy expert with the Electronic Frontier Foundation points out that "one of the potential problems with virtual reality is that we still haven't answered many of the privacy problems we encounter in normal reality."[131] Stakeholders impacted and potentially impacted by VR and AR tech include not only employees who may be using immersive tech in their jobs now and in the future, but also parents, children, families, and anyone and anything within those immersive technology-touched homes and businesses.

The *Daily Targum*, the official student newspaper of Rutgers University, reported in February 2022 that "AR/VR headset users are at risk of revealing sensitive information via speech":

> Rutgers researchers are currently studying the security vulnerabilities of augmented reality and virtual reality (AR/VR) headsets, which could allow hackers to steal users' sensitive information....
>
> Yingying Chen, a professor in the Department of Electrical and Computer Engineering, led the creation of a tool called "Face-Mic" to demonstrate how hackers could infer sensitive information by making use of users' speech-associated facial movements while wearing AR/VR headsets, according to the release.[132]

The researcher also highlighted security issues and deanonymization issues that could occur as a result of this data collection and identification of users. Cybertracking and profiling of these users could result, too.

Pokémon Go

In 2016, an AR game called *Pokémon Go* took the world by storm. Per Statista, in the United States alone, *Pokémon Go* users reached a peak of 45 million in July 2016.[133] The game became such a cultural phenomenon in the United States that anyone who was alive then (and more than a few years old) was most likely either playing the game or knew someone who was. With the virality of the game, though, came ethical controversies.

Some brief background sets the stage for examining the controversies, beginning with the genesis of the game's name. *Pokémon* is a Japanese-language portmanteau term derived by combining portions of two English-language words: *pocket* and *monster*. As the BBC describes, two Japanese men invented Pokémon: Satoshi Tajiri and his illustrator friend Ken Sugimori.[134] From the video gaming magazine *Game Freak* that he started and worked on with friends, Tajiri eventually moved to "making his own video games, instead of writing about them." In 1989, Tajiri and friends morphed *Game Freak* the magazine into Game Freak the video gaming company. The BBC explains how Pokémon was born: "In the early 1990s, inspired by his childhood exploring forests and finding bugs and tadpoles, Satoshi came up with the idea for Pocket Monsters (or as it's better known Pokémon), and pitched it to Nintendo."

In 1996, the first Pocket Monsters video games were released. Tajiri worked "with legendary game maker Shigeru Miyamoto (the man behind Mario and The Legend of Zelda) to make Pocket Monsters a reality."[135] That name was later shortened to Pokémon (after an early name Capsule Monsters was also jettisoned). From the success of the Pokémon video games came a lucrative portfolio of derivative works, including trading cards (i.e., Pokémon Trading Card Game), *manga* (Japanese comic books), and *anime* (Japanese cartoons). One of the pocket monsters, a yellow-hued creature named Pikachu, became the famous face of Pokémon. As Den of Geek's 2021 "How Pikachu Became Pokémon's Mascot" reveals, "Pikachu has served as the face of the Pokémon franchise in a way that makes it easy to imagine someone instantly identifying the electric creature even if they'd never played a *Pokémon* game."[136]

Then, in 2016, another offshoot of the Pokémon empire materialized: an AR mobile gaming application that could be accessed and played on people's phones *anywhere*. *Britannica* explains: "The mobile application *Pokémon GO* was a blockbuster success upon its release in 2016. Developed by Niantic Labs, a former division of Google, the game used GPS data and the mobile device's camera to create an 'augmented reality' that allowed users to capture Pokémon from real-life locations."[137] In a nutshell, *Pokémon Go* players could be anywhere with a smart device and see AR creatures overlaid on whichever environment where they were located. What the designers of the game either hadn't thought about, or had dismissed if they had foreseen it, was the sensitivity around having AR gaming images superimposed on solemn and sacred spaces. A July 12, 2016, *Washington Post* headline unforgettably captures one of these controversies: "Holocaust Museum to Visitors: Please Stop Catching Pokémon Here."[138]

Why were the game users going into this Washington, DC, museum, in furtherance of *Pokémon Go*'s objectives? The game is purposefully designed for players to acquire various items at so-called PokéStops (as well as at gyms), including stops at landmarks and monuments. As the game's maker Niantic describes, "Visit PokéStops and Gyms around the world to gather Poké Balls, Potions, Eggs, and other items. You'll find these near great places that encourage exploration, exercise, or social interactions with others. When you visit PokéStops and Gyms, spin the Photo Disc to generate items for your inventory."[139] Now, imagine the designers intentionally including three PokéStops within parts of the Holocaust Museum. On top of that, consider all of the other parties, including legal counsel and corporate executives at all levels, who, in essence, greenlit these design aspects of *Pokémon Go*.

A Holocaust Museum official sums up the concerns and controversy: "'Playing the game is not appropriate in the museum, which is a memorial to the victims of Nazism,' Andrew Hollinger, the museum's communications director, told The Post. 'We are trying to find out if we can get the museum excluded from the game.'"[140] Sadly, the burden is on the museum officials to take action to have their venue removed from the game's design interface. Niantic provides information on how to submit a request for a PokéStop or Gym to be modified or removed, which the company then reviews.[141] Similar ethics controversies have also occurred at places like Arlington National Cemetery (ANC), where the nation's military veterans and members of their families are buried. A July 12, 2016, Twitter (now designated X) post from ANC communicates its position on the matter: "We do not consider playing 'Pokemon Go' to be appropriate decorum on the grounds of ANC. We ask all visitors to refrain from such activity."[142]

In one sentence, the *Washington Post* further encapsulates a design dilemma of AR apps and the ethical issues and stakeholder interests that are involved: "The Holocaust Museum's plight highlights how apps that layer a digital world on top of the real one can create awkward situations, especially since the owners of the physical locations often cannot weigh in on how their spaces are being used."[143] The ethical issues and sensitivities may in some cases be additionally compounded by legal issues: namely, the property interests of those whose private land is (allegedly) trespassed on by those who are playing AR games like *Pokémon Go*. For example, in September 2023, a *Pokémon Go* user submitted to Niantic, at the request of a private property owner, a request for removal. The property owner's garden, next to a gym included in the game's app, had been damaged by people who visited while playing the game. The request was accompanied by screenshots of the property owner's garden.[144]

In addition to being violative of people's private property rights to control access, entry, and egress to and from their land (barring an exception like a legal easement or other right of way), inclusion of private property venues in AR games implicates other legal interests. One of these is the right to quiet enjoyment of one's property, which is also a standard term included in most leasehold/rental contracts. Landowners, too, may in certain instances be liable for injuries and deaths that occur on their property, particularly if there are known hazards or foreseeable consequences that may cause harm. Moreover, even if one does not own the property at issue, but rather is a lessee or tenant, the legal rights of people who are renting may be deleteriously impacted as well, such as, again, the right to quiet enjoyment. Ethical issues and concerns also may arise. These include, but are not limited to, privacy, cybertracking, data collection and use, security, and safety.

A September 2016 Reuters news story's headline, "'Get Off My Lawn': Pokemon Go Tests Global Property Laws," demonstrably conveys the human and legal tensions aroused by some immersive technologies.

> The global popularity of the game and other video games that put digital "characters" into real places—from private homes to shops, parks and even monuments and museums—has fueled debate on land rights, the legal boundaries of private property, and what constitutes trespass.
>
> Experts say the inter-section between virtual reality and property law is not clear and nobody really knows what the rights of property owners are when digital characters or structures appear on their land.[145]

The article points out that Niantic has strategically included a statement at the start of each *Pokémon Go* that instructs users "not to trespass on private property." This could potentially protect the company from liability for trespassing by game users. Legal theories like willful blindness (also called conscious avoidance) might still, though, render a company legally responsible for actions it could have or should have foreseen but chose not to. The Reuters article also triggers thought-provoking questions that are yet to be answered:

- What if an AR game adversely affects a property value?
- What if an AR game with violent content creates safety concerns?
- What are the responsibilities of law enforcement and other government officials to assist landholders whose private properties, or their private properties adjacent to public properties, are impacted by AR game overlays?

Apple Vision Pro

On February 2, 2024, Apple launched sales of its new tech innovation, Apple Vision Pro. As an January 8, 2024, Apple press release described in part, "Vision Pro is a revolutionary spatial computer that transforms how people work, collaborate, connect, relive memories, and enjoy entertainment. Vision Pro seamlessly blends digital content with the physical world and unlocks powerful spatial experiences in visionOS, controlled by the most natural and intuitive inputs possible—a user's eyes, hands, and voice."[146] Soon after sales of the devices began, the *Guardian*, as well as other media outlets, reported alarming instances of people using the devices in unsafe settings.[147] While it was noted that some may have been using the equipment as stunts to garner attention, other users were observed wearing the visors while driving, raising serious public safety concerns and admonitions, including from Pete Buttigieg, the US secretary of transportation.

A groundbreaking immersive technology developer has voiced his fears about the power of these unchecked technologies in a March 2022 VentureBeat article: "XR Pioneer Calls for Metaverse Regulation": "Louis Rosenberg, the XR pioneer who developed the first interactive augmented reality system in 1992 at the Air Force Research Laboratory, believes we'll face the same concerns with the metaverse that we face now with social media (but potentially much, much worse)."[148] Rosenberg urges that action be taken *now*, before we lose the ability to influence the roles and regulation of immersive tech in our societies, as has largely happened—or, perhaps more aptly, has been allowed to happen through US legislative indifference, inaction, and inertia—vis-à-vis social media: "'Twenty years ago, everybody was very excited about the possibilities of social media to bring people together, to democratize the world. We saw it as a utopia. But over two decades later, we now feel like social media is creating a dystopia,' [Rosenberg] said."[149]

Beneficial Uses of Immersive Technology

Immersive technologies catalyze a range of concerns, as demonstrated by the examples identified in the previous sections. However, these technologies are also already being used in ways that have many positive influences. AR/VR, as one example, can be used to promote educational access and equity, as well as student learning. *School Library Journal*, in October 2023, shared the VR/AR experiences of students of a resource-scarce school in rural Maine:

> Piscataquis County in central Maine is the state's poorest and least populated, but for the roughly 400 students at rural Piscataquis Community Elementary School, they've visited parts of the world most students their age can only dream about.
>
> That's because Piscataquis is home to one of the only fully immersive 360-degree classrooms in New England. Nicknamed the "Pirate Portal" after the school's mascot, the learning environment is equipped with multiple projectors that create a seamless panoramic experience, allowing students to swim with sharks and explore distant planets.[150]

Efforts like this have the upshot of promoting access to technologies and information, particularly for communities that are historically underserved. Immersive tech can be utilized to assist people with job searching, employment retraining, and career preparation. A March 2022 *Library Journal* article reported on the launch of a job training program at Georgia's Clayton County Library System (CCLS). The initiative is believed to be the first of its kind at a US public library. Patrons who participate in the program are able to use virtual reality (VR) headsets to have the simulated experience of working in jobs like "plumbing, commercial and residential HVAC, and even solar panel installation."[151] The article shared the story of a 70-year-old property owner who wanted to learn about plumbing so she could take care of those kinds of issues on her own. Firsthand observations by CCLS Director Rosalind Lett underscore the ways that information centers and emerging technologies like AR and VR can make positive differences in people's lives, especially when they are experiencing tough circumstances:

> Lett hopes that people go to the library and learn about new innovations and technology, instead of paying a lot of money for various classes. . . . "We just kind of want them to know that the library is the place for them to come. It's our job. We're not the first responders out there, but we are the second responders, and we're the ones to help them kind of put their life back together once life is beating them down."[152]

Museums and Immersive Tech

Museums are at the forefront of cultural heritage institutions that are experimenting with and incorporating immersive technologies. XR offers seemingly endless ways to reimagine a host of museum-related roles. One is with regard to the ways that physical and virtual collections are presented and made accessible to users. Users can, for example, interact with museum objects. They can also interface with, say, holographic representations of living and deceased persons.

Indeed, these kinds of possibilities are already being offered in some museums, as well as being prepared for future museum settings and even home access. The following example tips more toward AI but is illustrative of intended hybrid emerging technologies. William Shatner is an actor best known for playing the *Star Trek* TV and film series' twenty-third-century captain of the starship *Enterprise*, James T. Kirk. He revealed in March 2021, as he was turning 90 years old, that over the course of five days he had recorded answers to various questions that users would be able to pose in an interactive video made by a California company named StoryFile. Reuters explained in an interview with the company's cofounder, Stephen Smith, that beginning in May 2021 users would be able to "ask questions of the Shatner video, and artificial intelligence will scan through transcripts of his remarks to deliver the best answer."[153] Exploration of the StoryFile website yielded samples of questions for Shatner that were already formulated:

- When did you record this interview?
- Where did you grow up?
- Does it bother you to be known mostly for *Star Trek*?
- Do you have a favorite type of acting role?
- Do you believe life has meaning?

As described by StoryFile's founder, one can also enter questions in a search box and the AI will provide what it reasons is "the best answer." To try out the interface, I typed in the questions, "What is your favorite food?" (eliciting a quite detailed response involving "a last meal" of sushi, or great Italian pasta, or grilled wagyu steak, or Mediterranean food, or a creation of his wife's), "What is your favorite Star Trek episode?" (Answer: 1967's "The City on the Edge of Forever"), and "What is your best job advice?" ("Work long, work hard, and eventually you will be successful").

Three-dimensional cameras were used to record the actor's answers to 650 wide-ranging questions, with the aim of enabling a holographic Shatner to be able to transport into future people's homes. One can also imagine the immersive technology applications of this kind of interface for museums, archives, libraries, and many other organizations. Public and private individuals could also use this type of hybrid technology for a wide number of purposes. One purpose is to record oral histories; another is to facilitate genealogical research. Ethical issues can also be imagined, such as the likelihood of misinformation, as even well-intentioned purposes will never lead to "perfect memory" of events. Additionally, these kinds of technologies that meld recorded histories and immersive technologies could enable disinformation: purposefully false or misleading information. A number of authoritative scholarly books and articles are critically examining the intersection of museums and immersive technologies. These resources also address ethical issues that bump up against this research area. Some examples of these scholarly works are the following:

- *Augmented Reality in Tourism, Museums and Heritage: A New Technology to Inform and Entertain* (2021)[154]
- *Emerging Technologies and Museums: Mediating Difficult Heritage* (2022)[155]
- *Museums and Technologies of Presence* (2023)[156]

Historical Records, Memory, and Underserved Communities

Documentation of textual and visual historical records and preservation of oral histories are vital missions of cultural heritage institutions, such as archives, museums, and libraries. As highlighted with the StoryFile and Shatner example, emerging technologies are already being used to document and preserve history and memory. In the Hill District, a historically Black neighborhood of Pittsburgh, immersive tech is playing an instrumental, innovative part in supporting those missions. In February 2024, PittWire reported an inspiring example of how VR is shedding light on the rich histories and stories of past and present peoples with connections to the Hill District. The Time Traveling Project, created by Tim Huang, assistant professor in the University of Pittsburgh School of Computing and Information, employs immersive visual technologies to link the photography of Charles "Teenie" Harris with local historical information. Harris's photographs documented twentieth-century African American individuals and communities in Pittsburgh. Huang converted Harris's still photos into short videos that can then be viewed through VR headsets while users listen to "oral histories from people alive during the era when the images were taken."[157]

Huang and his collaborators' motivations and goals are centered on giving participants a feeling of belonging. The project's efforts are also focused on filling in gaps in the historical record, and giving voice to the often invisible and unknown lived experiences of historically disenfranchised individuals and communities. Another aim is that the project can serve as a kind of "time capsule" that preserves and interprets the significance of both this historic neighbohood community and the groundbreaking photography of Harris. As Huang explained, "I want my work to help underserved populations gain representation. The story of the Hill District and the people who live there is absent from mainstream media. We are recreating the experience of what it was once like to live there and documenting people's stories."[158]

VR and Veterans

The Veterans Administration (VA) began utilizing immersive technologies in several VA facilities in August 2019. Four years later, by August 2023, the program known as VA Immersive is available in all 50 US states, American Samoa, Guam, and Puerto Rico:

> VA Immersive is defining a new reality in Veteran health care by leading the nation in the clinical implementation of immersive technology and uncovering new ways to embrace employees' whole health. By engaging the senses of sight sound, and touch, immersive technology cultivates a new level of engagement to each Veteran's health care experience.
>
> VA has now deployed more than 1,200 virtual reality (VR) headsets across more than 160 VA medical centers and outpatient clinics in all 50 U.S. states and Puerto Rico—advancing VA's commitment to connecting Veterans to the soonest and best care. The program also engages with industry, academia, and other Government agencies to co-develop content and engineer devices that are beneficial for our Nation's Veterans.[159]

The VA Immersive website provides resources and information, including the "Introductory Guide to Immersive Technology,"[160] the "Immersive Playbook"[161] "to share resources, lessons learned, successes, and failures" with the public, and the January 2024 "Literature Review"[162] on virtual reality use in health care. The literature review has sections on benefits and barriers to VR use in health care. It also contains separate sections on the use of VR for rehabilitation, mental health, peer support, pulmonary rehabilitation, clinical education, training, patient education, and palliative care. Each of these subsections provides its own discussion of benefits and barriers.

What are some of the kinds of issues that VA immersive technologies are used for? VR is used to alleviate and manage symptoms of posttraumatic stress disorder (PTSD), anxiety, and depression as well as to facilitate pain management and rehabilitation, among other uses. Medications are absolutely an important part of health care, but one of the pluses of immersive technologies is that they offer nonpharmaceutical options to treatment and rehabilitation.

Videos posted to YouTube by the VA enable veterans to talk about ways that VR can help them. A 2023 video, "New Horizons in Health Episode 2: The VA Immersive Experience," is one example.[163] In this video, a veteran named Dean talks about the "phantom limb" pain he experiences from the loss of his left leg when he removes his prosthetic leg and goes to sleep at night. Through participating in a 4-week VR beta test simulation, in which Dean "has to kick things and do things with both feet," like kicking at a ball on a VR screen, he finds that his pain is lessened and he is able to fall asleep sooner. Dean also poignantly explains another upshot of his VR experience:

> The way that VR was set up, it did a mirror image of my good foot, my right foot. When I look into the screen and watch the video game, it looked like I had two real legs. And I had discovered early on that you want to do it wearing shorts. And I would take my shoes and sock off on my right foot so that when it did the mirror image of my right foot, I saw a left foot. And that that's actually a pretty cool thing. I've been without a leg for 18 years now. So to see a left foot was just—it was exciting.[164]

The positive effects of his beta test participation, Dean noticed, lasted "for six months, maybe even a full year after the trial."[165] He further says that it has been over a year and a half since the beta test, that the "phantom limb thing" is starting to come back, and "it's annoying at night again." So, with a grin, he says, "In fact, I told the people at the trial that when you're ready to do another beta test, give me a call because I'm ready. I'm ready."

Firsthand stories like Dean's are powerful and persuasive examples of the ways that immersive technologies are helping people right now. It is exciting to imagine how and where these technologies can be applied in the future to ease suffering and aid people in living their best lives, particularly those who are experiencing adversity and losses of many kinds.

ALA and VR

As with previous technologies, ALA has a section of its website providing information on immersive technologies, as part of its Library of the Future initiative.[166] In addition to offering an overview of the origins and recent developments of VR-related technology, a "Why It Matters" section identifies some innovative ways that this topic is relevant to information professionals and centers:

- Exposure to emerging tech. Information professionals and centers have historically been at the vanguard in introducing and helping to acclimate individuals and communities to new technologies.
- Storytelling, which has long been a staple of children's activities provided by libraries. Immersive storytelling, as ALA describes, "will become a more frequent tool for conveying information, especially as major media outlets invest in in-house virtual reality production."[167]
- Opportunities for VR and education. Information professionals play important roles in educating users, particularly with regard to collection-related materials and programming. VR can augment these kinds of educational experiences and promote innovations in information center educational offerings.

The Future and Here-and-Now of Immersive Tech

Like AI, IoT, robots, drones, and self-driving vehicles, the story of how XR will continue to evolve is unknown and unwritten. Like all technologies, emerging technologies can be used to do good things and bad things, and all kind of things that lie between both of those poles, just like their creators. The following Electronic Frontier Foundation excerpt about XR technologies, written with a perspective informed by the acute COVID pandemic period of 2020–2022, serves as a cautionary warning to the demonstrated as well as the yet-to-be-proven promises of XR technologies:

> Extended reality technologies (XR), which include Virtual Reality (VR) and Augmented Reality (AR), are rapidly maturing and becoming more prevalent to a wider audience, especially as the pandemic drives more people to virtual activities. This technology provides the promise to entertain and educate, to connect and enhance our lives, and even to help advocate for our rights. But it also raises the risk of eroding them online. The headsets and devices can gather

deeply personal information about you and the world around you, and the XR services often store this data within their servers. And by introducing artificial intelligence to XR, the privacy and security hazards are augmented, too, as the devices and services gather and analyze data at an unprecedented level.[168]

Summary

Like AI, other twenty-first-century emerging technologies discussed in this chapter implicate myriad ethical issues, necessitating thoughtful consideration of risks and benefits. Internet of things (IoT) smart devices can, for example, make our lives easier and more informed, by means of the tasks they are able to perform and the real-time data they can provide at our fingertips. However, they also carry potential safety risks and privacy intrusions because of the data they collect, store, and communicate to third parties. Bad actors, as an example, may carry out cyberattacks to use illegally obtained data for malicious purposes. Similarly, drones and robots can be deployed in ways that may help or cause hurt. Drones can deliver library books. Robotic animal companions can be made available in a library or senior center to comfort people who are lonely or cognitively impaired. Security robots can be utilized for purposes like public safety and threat monitoring but may impinge on civil liberties by targeting historically marginalized communities. The reliability of autonomous vehicles is still an open question, but professional associations like ALA are already considering the ethical ramifications of issues like being able to use self-driving vans that can transport materials, while weighing the downsides of diminished job opportunities for the humans who could be displaced by such technology. Immersive technologies, like AR and VR, can offer new kinds of gaming that amaze and delight, but which may cause emotional and psychological pain to some individuals and groups due to game design insensitivities and unforeseen harms. At the same time, immersive tech is already making many innovative experiences possible, including novel treatment regimens for veterans experiencing PTSD and vocational simulation experiences for those wanting to explore a new career before committing to a lifestyle change. Information professionals can bring their unique skill sets to bear in raising greater awareness of the advantages and disadvantages presented by emerging tech, while also advocating for regulatory safeguards and implementing ethical guardrails in their own institutions to mitigate potential harms.

Questions for Reflection, Exploration, Action, Discussion (READ)

1. A 2022 the Conversation article, "Smart Devices Spy on You—2 Computer Scientists Explain How the Internet of Things Can Violate Your Privacy," encourages people to limit "privacy damage" by IoT data-collecting-technologies. Brainstorm some innovative ways that information professionals can educate users on steps that can be taken to limit potential privacy breaches and data misuses by the companies that make their smart devices.
2. Identifying risks and benefits, as well as stakeholders, is an important component of ethical analysis and decision-making. This chapter highlighted a number of drone and helper/care/companion

robot examples. Drawing from the examples presented, as well as other examples you know or can find, compose a bulleted list of (1) three risks of drones and helper/care/companion robots and (2) three benefits of drones and helper/care/companion robots.

3. Having thought about and identified some of the risks and benefits of drones and robots, let's now focus on applying your insights to libraries and information centers. Imagine you are the director of a public library or an academic library (choose whichever of the two you want). Your board wants you to consider implementing a robot cat program for your public library's patrons or a drone delivery program for your academic library's students and faculty. Identify the biggest benefit and the biggest risk of the proposal. Then, state your decision on whether to proceed with the robot cat program or the drone delivery program for your institution. Explain the rationale(s) for your decision.
4. PDDs are potential delivery mechanisms for library books and archival materials. A SWOT analysis is a strategic planning tool that identifies strengths, weaknesses, opportunities, and threats. Imagine a library considering whether to agree to a local tech company's request for the library to participate in a beta test of a PDD. A popular independent bookstore near the library has already agreed to participate in the PDD beta test. The PDD will be used to deliver library books within a one mile radius of the library's main branch for a one-month duration. Identify two strengths, two weaknesses, two opportunities, and two threats involved with the library participating in this beta test.
5. Public libraries are increasingly incorporating VR technologies that patrons can use for job searches, career planning, and learning. List three to six safeguards a library should consider when deciding whether to use VR technology.
6. Service providers can do more to include individuals and communities in the formulation of ethics guidelines and policy making on emerging technologies. Make a list of some concrete steps information centers can take to seek and include the voices and perspectives of diverse users in ethics and policy decisions involving new tech.
7. Values are strongly linked to ethics-centered adoption and assessment of new technologies. Articulate three to six values that can serve as ethics guardrails and assessment metrics for organizations using emerging technologies.
8. Create a best practices tool kit of ethical decision-making approaches, codes, guidelines, and resources that can be used to make policy decisions about the kinds of disruptive emerging technologies profiled in this chapter.

Notes

1. Lisberger, S. (Director). (1982). *Tron*. IMDb. https://www.imdb.com/title/tt0084827/
2. Mosley, T. (Host). (2024, July 31). From "E.T." to "Blade Runner," how the summer of 1982 changed cinema forever. *Fresh Air*. NPR. https://www.npr.org/2024/07/31/g-s1-14578/from-e-t-to-blade-runner-how-the-summer-of-1982-changed-cinema-forever
3. Anderson, J., & Rainie, L. (2023, June 21). *As AI spreads, experts predict the best and worst changes in digital life by 2035*. Pew Research Center. https://www.pewresearch.org/internet/2023/06/21/as-ai-spreads-experts-predict-the-best-and-worst-changes-in-digital-life-by-2035/
4. Brown, S. (2008, July 21). *WarGames*: A look back at the film that turned geeks and phreaks into stars. *Wired*. https://www.wired.com/2008/07/ff-wargames/

5. Chow, D. (Director). (2019, December 18). Chapter 7: The reckoning. *The Mandalorian* (Season 1, Episode 7). https://www.imdb.com/title/tt9121544/
6. Ranger, S. (2020, February 3). *What is the IoT? Everything you need to know about the internet of things right now*. ZDNET. https://www.zdnet.com/article/what-is-the-internet-of-things-everything-you-need-to-know-about-the-iot-right-now/
7. Ibid.
8. Geary, B. (2023, October 12). The internet of things demystified: Connect, collect, analyze and act. *Forbes*. https://www.forbes.com/sites/forbestechcouncil/2023/10/12/the-internet-of-things-demystified-connect-collect-analyze-and-act/?sh=672fe0a3eef6
9. Ranger, S. (2020, February 3). *What is the IoT? Everything you need to know about the internet of things right now. ZDNET*. https://www.zdnet.com/article/what-is-the-internet-of-things-everything-you-need-to-know-about-the-iot-right-now/
10. Marshall, A., & Knight, W. (2024, February 29). The White House warns cars made in China could unleash chaos on US highways. *Wired*. https://www.wired.com/story/china-cars-national-security-threat-investigation/
11. Yus, R., & Pappachan, P. (2022, March 14). *Smart devices spy on you—2 computer scientists explain how the internet of things can violate your privacy*. Conversation. https://theconversation.com/smart-devices-spy-on-you-2-computer-scientists-explain-how-the-internet-of-things-can-violate-your-privacy-174579
12. Federal Trade Commission, Consumer Advice. (n.d.). *Securing your internet-connected devices at home.* Retrieved February 26, 2025, from https://consumer.ftc.gov/articles/securing-your-internet-connected-devices-home
13. Expert Panel. (2023, February 22). 14 Essential steps to take to secure household IoT devices. *Forbes*. https://www.forbes.com/sites/forbestechcouncil/2023/02/22/14-essential-steps-to-take-to-secure-household-iot-devices
14. National Conference of State Legislatures. (2023, March 27). *Current unmanned aircraft state law landscape*. https://www.ncsl.org/transportation/current-unmanned-aircraft-state-law-landscape
15. Federal Aviation Administration. (n.d.). *Drones*. Retrieved February 26, 2025, from https://www.faa.gov/uas
16. Hanchett, J. (2022, March 17). Panel: Drone warfare is increasingly sophisticated, deadly. *Cornell Chronicle*. https://news.cornell.edu/stories/2022/03/panel-drone-warfare-increasingly-sophisticated-deadly
17. UBC Faculty of Medicine. (2022, March 21). *Drones get blood to patients faster—and reduce waste.* University of British Columbia. https://www.med.ubc.ca/news/drones-get-blood-to-patients-faster-and-reduce-waste/
18. Nisingizwe, M. P., Ndishimye, P., Swaibu, K., Nshimiyimana, L., Karame, P., Dushimiyimana, V., Musabyimana, J. P., Musanabaganwa, C., Nsanzimana, S., & Law, M. R. (2022). Effect of unmanned aerial vehicle (drone) delivery on blood product delivery time and wastage in Rwanda: A retrospective, cross-sectional study and time series analysis. *Lancet Global Health,* 10(4), e564–e569. https://doi.org/10.1016/s2214-109x(22)00048-1
19. *Robert Irwin: Steve Irwin's son wins award for bushfire image.* (2021, February 10). BBC News. https://www.bbc.com/news/entertainment-arts-56008271
20. Rubinstein, D., & Kilgannon, C. (2024, February 20). N.Y.P.D. drones carrying rafts could join lifeguards in beach rescues. *New York Times*. https://www.nytimes.com/2024/02/20/nyregion/nyc-lifeguard-drones-beaches.html
21. Miller, M. (2024, June 13). Shark attacks are rising in Australia. Technology is helping stop them. *Washington Post*. https://www.washingtonpost.com/world/2024/06/13/shark-attack-prevention-drones-ai-australia/

22. Snyder, A. (2023, November 9). AI and drones are helping scientists study secrets of elephant behavior. *Axios*. https://www.axios.com/2023/11/09/ai-drones-elephants-colossal
23. Spires, J. (2020, November 12). *Drones now used to count Antarctic penguin colonies in 3 hours*. DroneDJ. https://dronedj.com/2020/11/12/drones-now-used-to-count-antarctic-penguin-colonies-in-3-hours/
24. Lerman, R. (2020, June 10). Google-backed drones will drop library books so kids in Virginia can do their summer reading. *Washington Post*. https://www.washingtonpost.com/technology/2020/06/10/wing-library-book-drone-delivery/
25. CBS News. (2020, June 16). *Virginia librarian uses drones to deliver books*. YouTube. https://www.youtube.com/watch?v=79qxPvHLPRI
26. Robot Hall of Fame. (n.d.). *Welcome to the Robot Hall of Fame*. Retrieved February 26, 2025, from http://www.robothalloffame.org/
27. Robot Hall of Fame. (2006). *2006 inductees*. http://www.robothalloffame.org/inductees/06inductees/maria.html
28. Wexler, E. (2022, December 29). These works are now in the public domain. *Smithsonian Magazine*. https://www.smithsonianmag.com/arts-culture/the-top-ten-works-entering-the-public-domain-2023-180981370/
29. Parnell, L. (2023). *"Metropolis" enters the public domain and what it means for this classic film*. CultureSlate. https://www.cultureslate.com/explained/metropolis-enters-the-public-domain-and-what-it-means-for-this-classic-film
30. Carnegie Mellon University Libraries. (n.d.). *The robotics project*. Retrieved February 26, 2025, from https://library.cmu.edu/university-archives/initiatives/robotics-project
31. Winfield, A. (2012). What is a robot? In *Robotics: A very short introduction*, 1–19. https://doi.org/10.1093/actrade/9780199695980.003.0001
32. Merriam-Webster. (n.d.). *Robot*. Retrieved February 26, 2025, from https://www.merriam-webster.com/dictionary/robot
33. Britannica. (n.d.). *Robot*. Retrieved February 26, 2025, from https://www.britannica.com/technology/robot-technology
34. Savage, N. (2023). How robots can learn to follow a moral code. *Nature*. https://doi.org/10.1038/d41586-023-03258-1
35. Dress, B. (2023, September 28). Why the Pentagon's "killer robots" are spurring major concerns. *Hill*. https://thehill.com/policy/defense/4225909-why-the-pentagons-killer-robots-are-spurring-major-concerns/
36. Satariano, A., Cumming-Bruce, N., & Gladstone, R. (2021, December 17). Killer robots aren't science fiction. A push to ban them is growing. *New York Times*. https://www.nytimes.com/2021/12/17/world/robot-drone-ban.html
37. United Nations. (n.d.). *The Convention on Certain Conventional Weapons*. Retrieved February 26, 2025, from https://disarmament.unoda.org/the-convention-on-certain-conventional-weapons
38. Dawes, J. (2021, December 20). *UN fails to agree on "killer robot" ban as nations pour billions into autonomous weapons research*. Conversation. https://theconversation.com/un-fails-to-agree-on-killer-robot-ban-as-nations-pour-billions-into-autonomous-weapons-research-173616
39. Ibid.
40. United Nations. (2023, November 1). *First committee approves new resolution on lethal autonomous weapons, as speaker warns "An algorithm must not be in full control of decisions involving killing."* https://press.un.org/en/2023/gadis3731.doc.htm
41. Ibid.

42. Human Rights Watch. (2024, January 3). *Killer robots: UN vote should spur action on treaty.* https://www.hrw.org/news/2024/01/03/killer-robots-un-vote-should-spur-action-treaty
43. *United Nations, General Assembly Plenary meeting (NY237376A).* (2023, December 22). https://reachingcriticalwill.org/images/documents/Disarmament-fora/1com/1com23/votes/L56-GA.pdf
44. Mineo, L. (2024, January 12). "Killer robots" are coming, and U.N. is worried. *Harvard Gazette*. https://news.harvard.edu/gazette/story/2024/01/killer-robots-are-coming-and-u-n-is-worried/
45. Bender, J., & Ingersoll, G. (2014, January 28). Automated killbots could replace soldiers. *Business Insider*. https://www.businessinsider.com/automated-killbots-could-replace-soldiers-2014-1
46. International Committee for Robot Arms Control. (n.d.). *About ICRAC.* Retrieved February 26, 2025, from https://www.icrac.net/about-icrac/
47. Reuters. (2024, February 20). *US believes Russia developing space-based nuclear weapon, says source.* https://www.reuters.com/world/us-believes-russia-developing-space-based-nuclear-weapon-says-source-2024-02-20/
48. International Committee for Robot Arms Control. (n.d.). *About ICRAC*. Retrieved February 26, 2025, from https://www.icrac.net/about-icrac/
49. Kimball, D. G. (2024, March). *Keeping outer space nuclear weapons free*. Arms Control Association. https://www.armscontrol.org/act/2024-03/focus/keeping-outer-space-nuclear-weapons-free
50. Sharkey, N. (Host). (2015, May 4). *Interviews on social responsibility in robotics and AI*. ICRAC. https://www.icrac.net/podcast-6-interviews-on-social-responsibility-in-robotics-and-ai/
51. International Committee for Robot Arms Control. (2019). *What makes human control over weapons systems "meaningful"?* https://www.icrac.net/wp-content/uploads/2019/08/Amoroso-Tamburrini_Human-Control_ICRAC-WP4.pdf
52. Stop Killer Robots. (n.d.). *Home page*. Retrieved February 26, 2025, from https://www.stopkillerrobots.org
53. Stop Killer Robots. (2023, December 21). *2023 automated decision research year in summary.* https://www.stopkillerrobots.org/news/2023-automated-decision-research-2023-year-in-summary/
54. Stop Killer Robots. (2023, December 1). *Use of automated targeting system in Gaza.* https://www.stopkillerrobots.org/news/use-of-automated-targeting-system-in-gaza/
55. Stop Killer Robots. (2023, November 25). *Exhibition launch: Automated by design.* https://www.stopkillerrobots.org/news/automated-by-design/
56. *The Stop Killer Robots campaign.* (n.d.). You Tube. Retrieved February 26, 2025, from https://www.youtube.com/watch?v=-PVmPbFJm9E
57. Carrington, D. (2021, April 29). Killer farm robot dispatches weeds with electric bolts. *Guardian*. https://www.theguardian.com/environment/2021/apr/29/killer-farm-robot-dispatches-weeds-with-electric-bolts
58. Ibid.
59. Marcu, G., Lin, I., Williams, B., Lionel, R., & Schaub, F. (2023). "Would I feel more secure with a robot?": Understanding perceptions of security robots in public spaces. *Proceedings of the ACM on Human-Computer Interaction,* 7(CSCW2), 1–34. https://doi.org/10.1145/3610171
60. *Boston Dynamics Inc. profile.* (n.d.). Robotics247. Retrieved February 26, 2025, from https://www.robotics247.com/company/boston_dynamics
61. Stieg, C. (2020, June 22). *This $75,000 Boston Dynamics robot "dog" is for sale—take a look*. CNBC. https://www.cnbc.com/2020/06/22/75000-boston-dynamics-robot-dog-for-sale-take-a-look.html
62. Ackerman, E. (2020, June 16). Boston Dynamics' Spot robot dog now available for $74,500. *IEEE Spectrum*. https://spectrum.ieee.org/boston-dynamics-spot-robot-dog-now-available

63. Amadeo, R. (2021, February 3). *Boston Dynamics' robot dog gets an arm attachment, self-charging capabilities*. Ars Technica. https://arstechnica.com/gadgets/2021/02/boston-dynamics-robot-dog-gets-an-arm-attachment-self-charging-capabilities/
64. Cain, J. (2022, November 8). LAPD could soon add Boston Dynamics' $300K robot dog to its arsenal. *Daily News*. https://www.dailynews.com/2022/11/08/lapd-could-soon-add-boston-dynamics-300k-robot-dog-to-its-arsenal/
65. Zaveri, M. (2021, April 14). N.Y.P.D.'s robot dog returns to work, touching off a backlash. *New York Times*. https://www.nytimes.com/2021/04/14/nyregion/robot-dog-nypd.html
66. Associated Press. (2024, March 28). Massachusetts: Robotic police dog shot and credited with averting bloodshed. *Guardian.* https://www.theguardian.com/us-news/2024/mar/28/massachusetts-robotic-police-dog-shot
67. Ibid.
68. Trainum, K., Tunis, R., Xie, B., & Hauser, E. (2023). Robots in assisted living facilities: Scoping review. *JMIR Aging,* 6, e42652. https://doi.org/10.2196/42652
69. Ibid.
70. Wright, J. (2023, January 9). Inside Japan's long experiment in automating elder care. *MIT Technology Review*. https://www.technologyreview.com/2023/01/09/1065135/japan-automating-eldercare-robots/
71. *Are robots the solution to understaffed nursing homes?* (2022, May 24). NPR. https://www.npr.org/2022/05/24/1100985010/are-robots-the-solution-to-understaffed-nursing-homes
72. Ibid.
73. Barber, R. G., Carlson, R., & Ramirez, R. (2024, January 10). Preserving our humanity in the age of robots. *Short Wave.* NPR. https://www.npr.org/2024/01/10/1198909122/social-robots-ai-lonely-humanity
74. McDonald, R. (2023, October 28). *This robot cat offers purrfect companionship for Eugene public library patrons to check out*. OPB. https://www.opb.org/article/2023/10/28/eugene-public-library-robotic-cats
75. Fisher, K. E., & Gambrill, L. A. (2022, September 7). *Superheroes need help too: Understanding peri- and post-traumatic symptoms amongst library staff as essential frontline workers*. https://repository.ifla.org/items/72cc01bb-3ca3-462d-9382-91eb46f6f468
76. Fraley, S. (2023, December 14). Manistee library has robot pets to check out. *Manistee News Advocate*. https://www.manisteenews.com/news/article/manistee-library-robot-pets-check-18550095.php
77. Western Michigan University, School of Communication. (n.d.). *Robot petting zoo.* Retrieved February 26, 2025, from https://wmich.edu/communication/robot-zoo
78. ASPCA. (n.d.). *Pet statistics.* Retrieved February 26, 2025, from https://www.aspca.org/helping-people-pets/shelter-intake-and-surrender/pet-statistics
79. Wessling, B. (2022, February 28). *Kiwibot brings in $7.5 million in Pre-Series A funding.* The Robot Report. https://www.therobotreport.com/kiwibot-brings-in-7-5-million-in-pre-series-a-funding/
80. Felton, J. (2021, September 21). *Robots to deliver books, medicine, food to Pittsburgh neighborhoods*. TribLIVE. https://triblive.com/local/robots-to-deliver-books-medicine-food-to-pittsburgh-neighborhoods/
81. Trombola, N. (2021, April 26). Bloomfield residents raise concerns about sharing sidewalk space with delivery robots. *Pittsburgh Post-Gazette*. https://www.post-gazette.com/local/city/2021/04/26/kiwibot-personal-delivery-robots-PDD-pilot-program-bloomfield-pittsburgh-liability-concerns-DOMI/stories/202104220188
82. Martines, J. (2019, October 24). *Pitt suspends delivery robots after wheelchair user reports safety hazard*. TribLIVE. https://triblive.com/local/pittsburgh-allegheny/pitt-suspends-delivery-robots-after-wheelchair-user-reports-safety-hazard/

83. Pennsylvania General Assembly. (2020). *2020 Act 106.* https://www.legis.state.pa.us/cfdocs/Legis/LI/uconsCheck.cfm?txtType=HTM&yr=2020&sessInd=0&smthLwInd=0&act=106
84. Ibid.
85. Felton, J. (2021, September 21). *Robots to deliver books, medicine, food to Pittsburgh neighborhoods*. TribLIVE. https://triblive.com/local/robots-to-deliver-books-medicine-food-to-pittsburgh-neighborhoods/
86. *Kiwibot*. (n.d.). Retrieved February 15, 2024, from https://www.kiwibot.com/campuses
87. Effinger, A. (2023, December 18). *Security robots, made by a money-losing company, are on the rise in Portland*. https://www.wweek.com/news/2023/12/18/security-robots-made-by-a-money-losing-company-are-on-the-rise-in-portland/
88. Ibid.
89. Rus, D., & Mone, G. (2024, March 5). The robots are coming. And that's a good thing. *MIT Technology Review*. https://www.technologyreview.com/2024/03/05/1087646/the-robots-are-coming-and-thats-a-good-thing-2/
90. Harris, M. (2014, September 10). How Google's autonomous car passed the first U.S. state self-driving test. *IEEE Spectrum*. https://spectrum.ieee.org/how-googles-autonomous-car-passed-the-first-us-state-selfdriving-test
91. Gonzales, R. (2019, November 7). *Feds say self-driving Uber SUV did not recognize jaywalking pedestrian in fatal crash*. NPR. https://www.npr.org/2019/11/07/777438412/feds-say-self-driving-uber-suv-did-not-recognize-jaywalking-pedestrian-in-fatal-
92. Ibid.
93. Ibid.
94. Vanek, C. (2023, July 28). Arizona driver in fatal autonomous Uber crash in 2018 pleads guilty, sentenced to probation. *Arizona Republic*. https://www.azcentral.com/story/news/local/tempe/2023/07/28/rafaela-vasquez-pleads-guilty-in-in-fatal-uber-self-driving-crash-killed-pedestrian-elaine-herzberg/70488361007
95. Yu, Y. (2022, April 28). What it's like to ride first and only public robotaxi in U.S. *Nikkei Asia*. https://asia.nikkei.com/Business/Technology/What-it-s-like-to-ride-first-and-only-public-robotaxi-in-U.S
96. Cano, R. (2023, October 17). Driverless cars debuted in two US cities. Only SF has problems. *San Francisco Chronicle*. https://www.sfchronicle.com/bayarea/article/robotaxi-driverless-cars-18364714.php#
97. Lindqwister, L. (2023, September 16). A brief history of robotaxis in San Francisco. *San Francisco Standard*. https://sfstandard.com/2023/09/16/san-franciscos-robotaxi-revolution-explained/
98. Ingram, D. (2023, October 3). *Self-driving car runs over pedestrian hit by human driver in San Francisco*. NBC News. https://www.nbcnews.com/tech/tech-news/driver-hits-pedestrian-pushing-path-self-driving-car-san-francisco-rcna118603
99. Farivar, C. (2023, October 27). Cruise immediately halts all robotaxis nationwide, seeks to "rebuild trust." *Forbes*. https://www.forbes.com/sites/cyrusfarivar/2023/10/26/cruise-immediately-halts-all-robotaxis-nationwide-seeks-to-rebuild-trust
100. Council, S. (2023, October 24). *Calif. DMV says Cruise hid video of SF crash, halts driverless permits*. SFGate. https://www.sfgate.com/tech/article/cruise-driverless-permits-suspended-dmv-18445296.php
101. Carlson, K. (2023, November 16). Self-driving car company Cruise pauses even more operations as it aims to rebuild trust. *Austin American-Statesman*. https://www.statesman.com/story/business/technology/2023/11/16/cruise-self-driving-car-austin-tx-autonomous-vehicle-company-pause-more-operations-rebuild-trust/71592824007/
102. Cruise. (2023, November 14). *Important updates from Cruise.* https://getcruise.com/news/blog/2023/important-updates-from-cruise/

103. Muller, J. (2023, October 6). Driverless cars may need to drive more like humans. Axios. https://www.axios.com/2023/10/06/cruise-driverless-car-pedestrian-accident-san-francisco
104. Ibid.
105. Solis, N., Uranga, R., & Garcia, K. (2024, March 1). Waymo is cleared to launch robotaxi service in Los Angeles, sparking new protests. *Los Angeles Times*. https://www.latimes.com/california/story/2024-03-01/waymo-cleared-to-launch-robotaxi-service-in-los-angeles
106. Ritenbaugh, S. (2023, October 1). *Self-driving vehicle testing in Pittsburgh area shifting gears to focus on trucking industry*. TribLIVE. https://triblive.com/local/regional/self-driving-vehicle-testing-in-pittsburgh-area-shifting-gears-to-focus-on-trucking-industry/
107. Ibid.
108. Ibid.
109. American Library Association. (n.d.). *Self-driving cars*. Retrieved February 27, 2025, from https://www.ala.org/tools/future/trends/selfdriving
110. American Library Association. (n.d.). *Center for the Future of Libraries*. Retrieved February 27, 2025, from https://www.ala.org/future
111. American Library Association. (n.d.). *Self-driving cars.* Retrieved February 27, 2025, from https://www.ala.org/tools/future/trends/selfdriving
112. Ibid.
113. Handlen, Z. (2016). *Black Mirror finds love (and a great episode) in a hopeful place*. AV Club. https://www.avclub.com/black-mirror-finds-love-and-a-great-episode-in-a-hope-1798189275
114. Ibid.
115. Sid. (2022, September 12). The future of immersive. *Medium*. https://medium.com/@sidpix/the-future-of-immersive-3ca0be5b7c3a
116. American Library Association. (n.d.). *Virtual reality.* Retrieved February 27, 2025, from https://www.ala.org/future/trends/virtualreality
117. Oxford Reference. (n.d.). *Augmented reality.* Retrieved February 27, 2025, from https://www.oxfordreference.com/display/10.1093/oi/authority.20110803095434380
118. Franklin Institute. (n.d.). *What is augmented reality?* https://fi.edu/en/what-is-augmented-reality
119. *What is mixed reality?* (2023, January 24). Microsoft Learn. https://learn.microsoft.com/en-us/windows/mixed-reality/discover/mixed-reality
120. Ibid.
121. Milgram, P., & Kishino, F. (1994, December 1). A taxonomy of mixed reality visual displays. *IEICE Transactions on Information Systems, E77-D* (12), 1321–1319. https://search.ieice.org/bin/summary.php?id=e77-d_12_1321
122. Gerencer, T. (2021, April 3). *What is extended reality (XR) and how is it changing the future?* Hewlett-Packard. https://www.hp.com/us-en/shop/tech-takes/what-is-xr-changing-world
123. Marr, B. (2019, August 12). What is extended reality technology? A simple explanation for anyone. *Forbes*. https://www.forbes.com/sites/bernardmarr/2019/08/12/what-is-extended-reality-technology-a-simple-explanation-for-anyone/?sh=5cf51ae57249
124. Ibid.
125. Markkula Center for Applied Ethics. (2022, March 15). *Virtual reality, real virtues, and augmented norms and laws.* YouTube. https://www.youtube.com/watch?v=jQzSGrW2TkQ
126. Markkula Center for Applied Ethics. (n.d.). *IT, ethics, and law lecture series.* Retrieved February 27, 2025, from https://www.scu.edu/ethics/events/it-ethics-and-law-lecture-series/

127. Kentbye. (2021,April 8). *Defining "biometric psychography" to fill gaps in privacy law to cover XR data: Brittan Heller's human rights perspectives*. Voices of VR. https://voicesofvr.com/988-defining-biometric-psychography-to-fill-gaps-in-privacy-law-to-cover-xr-data-brittan-hellers-human-rights-perspectives/
128. Heller, B. (2021). Watching androids dream of electric sheep: Immersive technology, biometric psychography, and the law. *Vanderbilt Journal of Entertainment & Technology Law,* 23(1), 12–2020. https://scholarship.law.vanderbilt.edu/cgi/viewcontent.cgi?article=1000&context=jetlaw
129. Markkula Center for Applied Ethics. (2022, December 5). *Determining users' ages online: Ethical, legal, and technical considerations.* YouTube. https://www.youtube.com/watch?v=DPqX2I98eL0
130. Hunter, T. (2022, January 13). Surveillance will follow us into "the metaverse," and our bodies could be its new data source. *Washington Post.* https://www.washingtonpost.com/technology/2022/01/13/privacy-vr-metaverse/#
131. Ibid.
132. Marques, M. (2022, February 24). AR/VR headset users are at risk of revealing sensitive information via speech, new Rutgers study finds. *Daily Targum*. https://dailytargum.com/article/2022/02/ar-vr-headset-users-are-at-risk-of-revealing-sensitive-information-via
133. Statista. (2016). *U.S. Pokémon GO user count 2016.* https://www.statista.com/statistics/589213/pokemon-go-user-number-us/
134. BBC. (2021, February 28). *Pokémon at 25: A history—from Pocket Monsters, to TCG and Pokémon GO.* CBBC Newsround. https://www.bbc.co.uk/newsround/56167405
135. Ibid.
136. Byrd, M. (2021, April 28). *How Pikachu became Pokémon's mascot.* Den of Geek. https://www.denofgeek.com/games/pikachu-pokemon-mascot-history-nintendo/
137. Brittanica. (n.d.). *Pokémon GO.* https://www.britannica.com/topic/Pokemon-GO
138. Peterson, A. (2016, July 12). Holocaust Museum to visitors: Please stop catching Pokémon here. *Washington Post*. https://www.washingtonpost.com/news/the-switch/wp/2016/07/12/holocaust-museum-to-visitors-please-stop-catching-pokemon-here/
139. Pokémon GO Help Center. (n.d.). *Visiting PokéStops and Gyms*. Retrieved February 27, 2025, from https://niantic.helpshift.com/hc/en/6-pokemon-go/faq/117-visiting-pokestops-and-gyms/
140. Peterson, A. (2016, July 12). Holocaust Museum to visitors: Please stop catching Pokémon here. *Washington Post*. https://www.washingtonpost.com/news/the-switch/wp/2016/07/12/holocaust-museum-to-visitors-please-stop-catching-pokemon-here/
141. Niantic Support. (n.d.). *Request modification or removal of a PokéStop or Gym.* Retrieved February 27, 2025, from https://niantic.helpshift.com/hc/en/5-report-an-issue-with-a-pokestop-gym/faq/1008-request-modification-or-removal-of-a-pokestop-or-gym/?contact=1&p=web
142. Arlington National Cemetery. (2016, July 12). *Twitter post.* https://twitter.com/ArlingtonNatl/status/752959640816582656
143. Peterson, A. (2016, July 12). Holocaust Museum to visitors: Please stop catching Pokémon here. *Washington Post*. https://www.washingtonpost.com/news/the-switch/wp/2016/07/12/holocaust-museum-to-visitors-please-stop-catching-pokemon-here/
144. RiselkaOwO-PGO. (2023, September 18). *Gym on private property*. Niantic Wayfarer. https://web.archive.org/web/20230923233308/https://community.wayfarer.nianticlabs.com/discussion/46719/gym-on-private-property
145. Taylor, L., & Hayden, S. (2016, September 22). *Get off my lawn! Pokemon Go tests global property laws*. Reuters. https://www.reuters.com/article/technology/get-off-my-lawn-pokemon-go-tests-global-property-laws-idUSKCN11S1GY/

146. Apple. (2024, January 8). *Apple Vision Pro available in the U.S. on February 2*. Apple Newsroom. https://www.apple.com/newsroom/2024/01/apple-vision-pro-available-in-the-us-on-february-2/
147. Reed, B. (2024, February 7). Safety concerns after people filmed using Apple Vision Pro headset—video report. *Guardian*. https://www.theguardian.com/technology/video/2024/feb/07/safety-concerns-after-people-filmed-using-apple-vision-pro-headset-video-report
148. Witt, T. (2022, March 19). *XR pioneer calls for metaverse regulation*. VentureBeat. https://venturebeat.com/datadecisionmakers/xr-pioneer-calls-for-metaverse-regulation/
149. Ibid.
150. Bauld, A. (n.d.). From job exploration to geography, AR and VR lift student learning. *School Library Journal*. https://www.slj.com/story/From-Job-Exploration-to-Geography-AR-and-VR-Lift-Student-Learning
151. Shoenberger, E. (2022). VR job training launched at Clayton County Library System. *Library Journal*. https://www.libraryjournal.com/story/vr-job-training-launched-at-clayton-county-library-system
152. Ibid.
153. Reuters. (2021, March 26). *Ask him anything: William Shatner's life story to live on through AI*. https://www.reuters.com/technology/ask-him-anything-william-shatners-life-story-live-through-ai-2021-03-26/
154. Geroimenko, V., & Springerlink Online Service. (2021). *Augmented reality in tourism, museums and heritage: A new technology to inform and entertain*. Springer International.
155. Stylianou-Lambert, T., Bounia, A., & Heraclidou, A. (2022). *Emerging technologies and museums: Mediating difficult heritage*. Berghahn Books.
156. Shehade, M., & Stylianou-Lambert, T. (2023). *Museums and technologies of presence*. Taylor & Francis.
157. Faina, N. (2024, February 22). *A Pitt professor is using virtual reality to connect to the Hill District's past*. Pittwire. https://www.pitt.edu/pittwire/features-articles/virtual-reality-hill-district
158. Ibid.
159. US Department of Veterans Affairs, Office of Healthcare Innovation and Learning. (n.d.). *VA Immersive defining a new reality in health care.* Retrieved February 27, 2025, from https://www.innovation.va.gov/hil/views/immersive/immersive.html
160. US Department of Veterans Affairs, VA Immersive. (n.d.). *Introductory guide to immersive technology.* Retrieved February 27, 2025, from https://www.innovation.va.gov/hil/assets/documents/va-immersive-introduction-guide-summer-2023-public.pdf
161. US Department of Veterans Affairs, VA Immersive. (2023). *VA Immersive playbook.* Retrieved February 27, 2025, from https://www.innovation.va.gov/hil/assets/documents/va-immersive-playbook-summer-2023-public.pdf
162. US Department of Veterans Affairs. (2024, June 3). *Immersive virtual reality in health care literature compendium.* https://www.innovation.va.gov/hil/assets/documents/va-immersive-lit-review-2024.pdf
163. Veterans Health Administration. (2023, October 25). *New horizons in health, episode 2: The VA immersive experience.* YouTube. https://www.youtube.com/watch?v=PsNgRS3JRgg
164. Ibid.
165. Ibid.
166. American Library Association. (n.d.). *Virtual reality.* Retrieved February 27, 2025, from https://www.ala.org/future/trends/virtualreality
167. Ibid.
168. Electronic Frontier Foundation. (n.d.). *Extended reality (XR).* Retrieved February 27, 2025, from https://www.eff.org/issues/xr

Part 5

Conclusion

Chapter 13

Codas and Continuances

Introduction: Order and Chaos

An overarching aim of this book has been to critically investigate key issues and developments at the intersection of ethics, information, and technology. Some of these issues fall squarely within the area of professional ethics and character, like misrepresentation, plagiarism, transparency, and accountability. Others sit within the category encompassing principles that align with democratic values and ideals: access, privacy, intellectual freedom, freedom of expression, free and independent presses, truth. The antithetical counterparts to these ideals (i.e., drawing from "a literary device called the 'topsy-turvy motif,' that portrays a world upturned," a la the Superman canon's Bizarro World inversions)[1] are often associated with autocracies and dictatorships: lack of access; surveillance; censorship; restrictions on what can be read, written, and said; falsehoods and lies.

Another category of issues derives from largely Western legal principles, laws, and jurisprudence. This grouping includes intellectual property (IP) and its four subareas: copyright, patents, trademarks, and trade secrets. Yet legal systems like IP are viewed by some as artifacts of colonial systems that have maintained systemic hegemony and exploitation of peoples who, through no fault of their own, were born into or reside in geographical locations outside of the physical spaces where such ownership systems originated and remain. Amid these colonial institutions that have not only domestic influence but also extraterritorial reach and impacts, people within and outside of those structures have shone light on some of the inherent inequities of these foundational relics. As a result of greater awareness of some of the critiques of, for example, Western-style copyright legal systems, new upstart movements like open source, open access, Creative Commons, open data, open educational resources (OER), open government, and open science have been midwifed, uplifted and raised in their formative early years, and nurtured toward maturation, growth, and evolution. Indeed, in August 2024, Creative Commons (CC) openly discussed its own internal process of rethinking how it should and will move forward with the changes that are transpiring in the ongoing AI age. In a post that has since been removed, CC's CEO Anna Tumadóttir noted that "the world has changed a lot" just in the past few years since the organization launched a 2021–2025 strategic plan and that a "reset" was needed. In addition to "thinking about AI a lot," CC gathered input via roundtables, conference workshops, and public forums. This effort resulted in its 2025–2028 Creative Commons Strategy. Most interestingly, the plan does not specifically address AI; rather, it does so implicitly when it speaks of "increasing concentrations of power online," "monopolization of power... amplified exponentially through technological breakthroughs," and control of the internet and ways that people can connect online "by the few who disproportionately benefit from the many in ways that deepen existing inequities."[2] The "antidote" to this concerning diagnosis, CC says, is "a strong and resilient open infrastructure of sharing, enabling a healthy and thriving creative commons powered by reciprocity and community in the public interest."[3]

Just as AI has catalyzed introspective rethinking and repositioning as to how and why CC exists, so too are AI and generative AI posing hard questions and soul-searching about systems and frameworks for creators, publishers, third-party commercial entities, licensors, geographical regions, users, and other stakeholder groups. For some, copyright systems represent order, reliability, remuneration, and revenue. For others, these systems are a manifestation of the IP have-nots' lack of agency, representing a kind of continuing chaos, uncertainty, indebtedness, and scarcity. This reality triggers iterative questions with profoundly complex and moral implications:

- Who benefits from order?
- Who benefits from chaos?

Ethically, these are questions to which we must regularly return to ensure that we repeatedly assess the costs and benefits of our actions as human beings with agency who are creating new technologies that can so consequentially influence the lives of our fellow humans and earthly island home. Recognizing this great power and responsibility as shepherds of such constructive and destructive forces, it is incumbent on us to be mindful of the importance of intellectual humility and our own limits as flawed beings striving to be more perfect. Faith that we can do good and use technologies like AI for good is a comforting touchstone and strengthening belief. Yet we would also be well advised to remember that, as some theologians and futurists tell us, the opposite of faith is not doubt but rather certainty. Bob Johansen, a futurist with the Institute for the Future (TITF), observes that "faith is a lot like clarity. The future will reward clarity, but punish certainty. Certainty is a lot like extreme belief."[4]

As of 2025, AI technologies continue to assume increasing importance and prevalence in many societies. Newspaper headlines and broadcast news stories now include AI developments every day. Some people—the AI-aware—know of these burgeoning developments and keep up with them diligently, even eagerly. Many, though, still do not have AI on their everyday radars and are seemingly unknowing, or perhaps heedless, of what is already here and yet to come. What then are the ethical responsibilities, if any, of information professionals to put their efforts toward raising awareness of the AI revolution that is all around us but not yet recognized by all?

Moreover, how should the AI revolution be characterized and communicated? Should AI be welcomed as a new arrival to our shores, ready to enter into our homes, neighborhoods, and larger communities? Are AI technologies a new tool, ready to make our lives better, easier, and more productive, like the horse, the plow, or the digitally enhanced twenty-first-century tractor? Or, rather, is AI tech akin to an invasive species: stealthily entering our ecosystems and replacing humans with a new, more advanced non-life-form that can already perform so many of the things that humans thought could only be done by people? Just as humans are inclined to "other" people (i.e., to fear and distrust those who are not "us"), is there a tendency to "other" new technologies?

The 1988 film *Alien Nation*, and a subsequent eponymous TV series of the 1990s, explored the tensions between citizens of Earth and a quarter of a million asylum-seeking alien refugees from the planet Tencton, referred to by Earthers as Newcomers. Just as AI has a number of advantages over humans, so too, allegorically, do the Tenctonese adapt to new environments, learn fast, and outlive humans. How real-life humans will adjust to AI-technology "Newcomers" is an unfolding story whose plot and twists and turns are yet to be written and definitively known. Ethical questions, dilemmas, and issues, though, will undoubtedly be an integral element of this real-world narrative.

This concluding chapter includes some additional synthesizing observations on the issues we've examined in prior chapters. Ethics-related questions and areas for further research are identified. Concluding reminders and some best-practice suggestions for how we can all be better prepared to face and navigate ethical issues and challenges are also highlighted.

Technology Autonomy and Algorithmic Justice

Humans tend to fall into three camps when it comes to how they relate to new technologies (as well as how they handle change):

- early adopters and evangelists
- wait-and-see, potentially persuadable skeptics and undecideds
- never-gonna-use-it rejectors, resisters, and abstainers

Early adopters are already using AI tech, experimenting with different AI tools, networking with other early adopters, sharing their experiences with AI tech users and nonusers, and incorporating it into areas of their professional and personal lives. These people lean in to using new tech and are inclined to view it as having more upsides than downsides. Those who are not fully on board with AI (yet, if ever) may be taking in more information about it before deciding how they ultimately feel or are holding off on getting on board with it until a certain tipping point of adoption and product penetration has been reached. Weighing the pros and cons, or the costs and benefits, and employing a utilitarian approach as they make up their minds on how they ultimately feel about AI technologies are approaches that are emblematic of those who are undecided and somewhat skeptical.

Societies will likely always have some individuals and groups who reject, resist, and abstain from new technologies like AI. In the United States and Canada, Old Order Amish and Old Order Mennonites eschew most technologies and technological devices, such as electricity, motor vehicles, television, radios, and computers. Yet some technologies are used or modified by these groups, depending on the strictness or leniency of those in authority of certain areas. For example, I am acquainted with Amish families in Western Pennsylvania who have elected to maintain a traditional cord-connected phone, housed in a separate shed away from the main dwelling areas, where phone calls can be made and received and voicemail messages can be retrieved. Additionally, some Amish may ask non-Amish people (generally referred to as "English") to purchase items for them via digital sites like Amazon or hire drivers to transport them to jobs, stores, or hospitals. Other Amish and Mennonite groups, though, may take more stringent approaches and eschew virtually any kind of technology. The salient point is that these groups are setting their own rules and norms for how technologies fit or do not fit within their belief systems and ways of living. In essence, they have developed and practice a kind of customizable technological autonomy. For some members, their tech autonomy has the capacity to adapt and evolve and incorporate certain uses and application of technologies, while for others, their decision is to remain static and orthodox and shun technologies.

As AI becomes more embedded in all aspects of societal life, though, how should societies ethically address the choices of those who decline to accept AI tech or outright reject its place in their lives? I recall working in a public library in a Midwest community at the turn of the millennium when decisions were made by the state legislature to close unemployment offices and require people to apply for unemployment benefits by going online. Significant numbers of the unemployed were on the side of the digital divide *without* reliable and readily available computer access. Yet these individuals were likely never asked what processes would best work for them to apply for benefit support, reinforcing their marginalization and lack of agency by those in authority. In a similar vein, in what ways may AI and new technologies potentially disenfranchise people and communities that don't have a say in where, when, and how these technologies are adopted in ways that have direct impacts on their lives?

Thankfully, through the work of researchers like Cathy O'Neil, Joy Buolamwini, and others, we have evidence that such practices are occurring. How such practices can and will be dealt with is another matter altogether. One way is through legal actions. In August 2024, the US Department of Justice (DOJ) brought a federal lawsuit against a group of landlords and a company called RealPage. In the complaint, the DOJ alleged that RealPage's AI algorithms enabled landlords to collude in inflating rents for housing, per DOJ's August 23, 2024, press release, "Justice Department Sues RealPage for Algorithmic Pricing Scheme That Harms Millions of American Renters," with the subtitle "RealPage's Pricing Algorithm Violates Antitrust Laws":

> The Justice Department, together with the Attorneys General of North Carolina, California, Colorado, Connecticut, Minnesota, Oregon, Tennessee, and Washington, filed a civil antitrust lawsuit today against RealPage Inc. for its unlawful scheme to decrease competition among landlords in apartment pricing and to monopolize the market for commercial revenue management software that landlords use to price apartments. RealPage's alleged conduct deprives renters of the benefits of competition on apartment leasing terms and harms millions of Americans. The lawsuit was filed today in the U.S. District Court for the Middle District of North Carolina and alleges that RealPage violated Sections 1 and 2 of the Sherman Act.
>
> The complaint alleges that RealPage contracts with competing landlords who agree to share with RealPage nonpublic, competitively sensitive information about their apartment rental rates and other lease terms to train and run RealPage's algorithmic pricing software. This software then generates recommendations, including on apartment rental pricing and other terms, for participating landlords based on their and their rivals' competitively sensitive information. The

> complaint further alleges that in a free market, these landlords would otherwise be competing independently to attract renters based on pricing, discounts, concessions, lease terms, and other dimensions of apartment leasing. RealPage also uses this scheme and its substantial data trove to maintain a monopoly in the market for commercial revenue management software. The complaint seeks to end RealPage's illegal conduct and restore competition for the benefit of renters in states across the country.[5]

Just as societies make policies and laws that impact issues like bodily autonomy, resulting in conflicts and polarization, so too may decisions that are already being made, and that will continue to be made, about AI and other technologies potentially contribute to conflicts and polarization between tech haves and have-nots. Whether and how some individuals and groups can exercise technology autonomy and achieve algorithmic justice for ethical improprieties and legal breaches, in the face of majority populations that have adopted technologies without the input and consent of minority populations, will continue to be challenging. Yet these issues also implicate ethical considerations and dilemmas that are neither easily reconcilable nor morally excusable.

Perfect Imperfections, Trade-Offs, and Acceptable Risks

We may not like it at times, but most of us accept that humans—you, me, we—are not perfect. John Legend's 2013 song "All of Me" full-throatedly embraces all the "perfect imperfections" of the recipient of his ardor. What expectations do we have, though, about our analog and digital tools? The truth is that we don't celebrate or unconditionally love the "perfect imperfections" of tools and technologies. At the same time, however, we accept a certain degree of fallibility associated with tools, devices, and machines. We know that cars occasionally get flat tires and cracked windshields. Hair dryers go on the fritz, and wearable fitness trackers malfunction. We don't realistically expect tools like our toasters, lawnmowers, dishwashers, smartphones, computers—and even spacecraft that aren't ours (e.g., Boeing Starliners and SpaceX Starships)—to *never* break down. Returning to scientific principles like order and chaos, discussed earlier, another inescapable scientific principle we must pay heed to is entropy, defined by Merriam-Webster as "a process of degradation or running down or a trend to disorder."[6]

That things—all organic and inorganic things—break down is a universal law and truth. The unavoidable, inherent fallibility of our machines, devices, tools, and technologies gives rise to numerous queries in search of cogent responses:

- What is the line we will accept on AI and other emerging tech tools that *will* inevitably break, make mistakes, and not work as they were/are advertised?
- What remedies do we have, or should we have, for tools that break down and that may cause foreseeable or unforeseeable harms?
- What are the ethical, legal, and policy rationales for such remedies?
- Do we expect more performance reliability from AI tools than we do from other kinds of non-AI tools? If so, why? If not, why not?

- In what ways are AI tools and generative AI products different from, say, a smart coffee maker, fridge, or laptop?
- What are the acceptable risks we as users are willing to click on and agree to in the terms and conditions of a generative AI app?
- What are the trade-offs we are willing to accept to have tools and technologies in our lives, communities, countries, and world?

Autonomous vehicles are another emerging technology that is grappling with issues of acceptable risk. At what threshold will self-driving cars, taxis, and trucks be considered "safe enough"? Will policy makers and legislators decide at some near future time that the risks of autonomous vehicles malfunctioning outweigh the downsides and uncertainties of driverless transportation? Robots used for military and law enforcement purposes are also subject to ongoing concerns, scenario planning, and decision-making on acceptable risks and trade-offs. During the July 2024 "AI Ethics for Peace" conference in Hiroshima, Dr. Jeremy Wertheimer, a member of the Yeshiva University Academic Committee, Broad Institute of MIT, and Harvard and former vice president of engineering at Google, voiced a number of thought-provoking observations that relate to the issue of acceptable risks and trade-offs. Wertheimer said that a right that does not get as much attention as other rights is the right to technology. He asserted that there is much good that AI technologies can do and that there are opportunity costs to not realizing AI technology. For example, with regard to the education of children, an important objective of the United Nations, Wertheimer noted that AI tutoring systems offer significant possibilities for helping children to learn. Perhaps his most persuasive point, though, was during his discussion of the trade-offs in deciding what pharmaceutical medicines to approve for use. Wertheimer recounted that there are some medications for which scientists and researchers still do not understand the reasons why they are effective in treating patients, despite decades of research. Yet decisions are made to bring those medicines to market because the good they can do now outweighs the unknown. This example underscores the incredibly complex ethics and policy decisions that must be critically considered and responsibly made at the nexus of data, technology, and human subjects protections.

Other Models, Approaches, and Frameworks for Ethical Decision-Making

In various chapters, we've considered the usefulness of ethical reasoning and decision-making models to aid us in facing ethical challenges, making the best decisions possible, and reflecting on our decisions and the outcomes of those decisions. This kind of deliberative, metacognitive (i.e., thinking about our own thinking) process can hone our ethics competencies. Even more importantly, though, it can help us to be more holistically cognizant of the people and elements that can shape how we approach questions of what to do or not do in different situations. We can work on being more aware of our blind spots. We can be more discerning of the lenses and innate and learned biases through which we view the world, as well as the assumptions that we sometimes make about the world that may benefit from being challenged with more information and varied voices. Being more conscious of different ethical approaches can also expand our available options.

"A Framework for Ethical Decision Making" from the Markkula Center for Applied Ethics, described at the outset of this book, is a model I have found particularly useful. However, other models, frameworks, and approaches are available for examination, experimentation, and implementation. We can benefit from looking at these other schema and seeing what aspects may be of use to us, which are not, and which ones we may choose to modify. For example, the University of Kansas has a nine-step ethical decision-making model. Step 9 of that model sets out an interesting recommended action that may prove practical to apply as a preparatory step when facing ethical challenges and dilemmas, particularly when confronting contentious issues and stakeholders, such as, say, public hearings on proposed book ban challenges: "Decide on the proper ethical action and be prepared to deal with opposing arguments."[7] This University of Kansas example illustrates the utility of periodically exploring other approaches to ethics matters.

Literacies

Literacy, the ability to read, is a bedrock principle of education and learning. It's also foundational to many aspects of societies, including libraries and other information centers. Access to information and technology fosters literacy that can be applied to other areas of our professional and personal lives. Indeed, the access chapter of this book underscores the essential link between access and varying kinds of literacies.

Information professionals, such as librarians, archivists, and cultural heritage practitioners, have long understood and championed literacies and access to information and technologies. On this point, Tony Marx, president of New York Public Library, has described libraries as being in the "access to information business."[8] For these reasons, library leadership and staff have been at the vanguard of promoting the freedom to read, think, and express oneself through decades-long advocacy efforts like American Library Association's (ALA) annual Banned Books Week initiative. In more recent years, since the advent of the internet in the 1990s, information professionals of many kinds have spearheaded digital literacy and media literacy through school, public, academic, and special libraries. Users of all kinds can benefit from additional literacies that are germane to the nexus of ethics, information, and technologies, such as the following:

- ethics literacy
- copyright literacy
- data literacy
- AI literacy

Information professionals have crucial roles to play in helping users and nonusers of their organizations augment their own suites of literacies that can facilitate their information needs and the ability to better negotiate and traverse potential ethics pitfalls and pathways.

Ethics Codes and Codes of Conduct

This book has also critically evaluated various ethics codes and codes of conduct of information centers and other types of organizations. Opinions vary as to the usefulness and efficacy of ethics codes. However, I take the position that ethics codes and codes of conduct serve a variety of valuable purposes, while also recognizing that they have limitations. Chief among these limitations is the element of enforceability or "teeth." Much criticism has, for example, been leveled at the US Supreme Court for the unveiling of its nonbinding self-generated code of ethics in November 2023.

All organizations, including information centers, are strongly advised of the need to think about regularly assessing and, as necessary, updating their ethics codes to consider whether and how to address new issues arising, such as emerging technologies like AI. Moreover, institutions can inclusively maximize diverse perspectives and voices by involving members from a variety of communities and populations in efforts such as the initial generation of ethics codes as well as periodic revisiting and updating of these codes. Just as strategic plans are living documents that benefit from routine review and revisions as warranted and desired, so too should ethics codes be viewed as snapshot principles in time that need to be regularly reviewed and reconsidered in light of changes and novel developments.

Listening to and Learning From Experts

Writers of books like *The Death of Expertise: The Campaign Against Established Knowledge* (2017) by Tom Nichols and Andrew Keen's *The Cult of the Amateur: How Today's Internet Is Killing Our Culture* (2007) are representative of thought leaders who have expressed concerns over the past couple of decades about the ways that expertise, proficiency, and competence have been undermined by technologies like the internet and social media, as well as by political movements that cast doubt on the importance of expertness and knowledge. Certainly, aspects of internet and social media cultures that have collapsed barriers to entry by persons who, pre-internet, would not have had digital platforms like YouTube, Facebook, Amazon, or TikTok to express their unfiltered viewpoints and participate in knowledge economies, can be seen as a net positive in terms of democratic ideals like freedom of expression and equity of access. However, expertise also, arguably, has a critical role to play in our societies. Experience, particularly lived experience, that is, "knowledge based on someone's perspective, personal identities, and history, beyond their professional or educational experience," per the United States' Office of Human Services Policy, is also often integral to expertise.[9] Lived-experience expertise also enriches organizations and societies by including individuals from communities that have traditionally been marginalized.

People with expertise frequently have experiences and viewpoints that can add richness, depth, and breadth to issues like those involving ethics, information, and technologies. My "Keep Notes" smartphone app is replete with collected notes and information that are useful for purposes like the writing of this book, teaching graduate courses, and conducting research, in addition to self-edification. Some examples of notes that are relevant to the kinds of issues examined in this book include the following select sampling:

- Veteran newsperson Steve Brill's assertion that "AI becomes a force multiplier"[10]
- MSNBC program host Ali Velshi's suggestion of "triangulating your information" by checking at least three sources for your news, in the way that cell towers triangulate locations by pinging different cell towers[11]
- 2021 Nobel Peace Prize winner Maria Ressa's experiential observations from her courageous media work in the Philippines that "disinformation is like cocaine" and that "by design social media divides and radicalizes"; Ressa also states that "cynicism and hopelessness are the tools of a tyrant" and asks "what are you willing to sacrifice for the truth?"[12]

Perspectives, arguments, and evidence-based information from people with varying types of expertise and lived experiences can be invaluable in myriad settings, as well as when we are dealing with complicated ethics-based issues like the ones focused on in this book.

Advocacy

Another theme of this book is the vital role that advocacy plays in influencing ethics, laws, and policies. In virtually every chapter we see examples of civil watchdog groups, like Electronic Frontier Foundation, professional associations such as ALA, and everyday private and public figures who are expressing their voices, views, and advocacy positions.

Information professionals as individuals and members of professional organizations and workplaces can consider ways they can get involved in advocacy efforts. Some organizations, particularly news media entities that are concerned about conflicts of interest, appearances of impropriety, and the lines between subjectivity and objectivity of their employees and the organization itself, may specify limits on what activities individuals can engage in within the confines of their employment. Such restrictions can bump up against important constitutional principles like freedom of expression and association and the right to petition one's government, so it is advisable to get clarity on how to proceed with individual advocacy desires before embarking into potentially problematic or nebulous employment areas.

"Move Fast and Break Things" or "If You Have Doubt, Don't Do It"?

Mark Zuckerberg, cofounder and CEO of Facebook (now under the corporate umbrella name Meta) famously, or infamously, ascribed to a philosophy of "Move Fast and Break Things" in the initial years of the World Wide Web. As explained in a 2023 *Medium* article, "When Companies Move Fast, They Do More Than Break Things," the mantra "referred to a culture in which technical excellence turning an idea into executable code quickly was the name of the game."[13] The "Move Fast" ethos, fueled by all manner of tech companies, venture capitalists, boards of trustees, and shareholders, certainly ushered in untold examples of technological innovation. It also made its adherents fabulously wealthy and, in many cases, very powerful, especially in political terms. As has been detailed at various places in this book, too, this style of tech company governance has caused various types of harms and in some cases, arguably, contributed to losses of life. Most notably, social media has been linked to tragic examples of bullying and cyberbullying that contributed to the victims of these acts taking their own lives.

Despite high-profile incidents involving Facebook, such as Cambridge Analytica's procurement of data from the Facebook accounts of millions users without their consent, first revealed in late 2015, as well as other tech company controversies, some commentators suggested that these events had somehow chastened tech companies and given them pause to rethink their ways, lest the US Congress institute strict regulatory frameworks that were anathema to tech leaders and their shareholders. In January 2019, Hemant Taneja wrote a piece for *Harvard Business Review* titled "The Era of 'Move Fast and Break Things' Is Over" suggesting "that in the wake of the Facebook scandal, the public is less tolerant of tech startups that ignore the societal ramifications of their innovations."[14] He asserted that "'Minimum viable products' must be replaced by 'minimum virtuous products'—new offerings that test for the effect on stakeholders and build in guards against potential harms."

However, notwithstanding Taneja's claim that the "Move Fast" era was at an end, difficult and impactful questions remain in the mid-2020s and call that assertion into question:

- Are tech companies testing for the effects of products, like AI chatbots and other tools, on stakeholder groups?
- Have tech companies built in sufficient "guards against potential harms" caused by their products?
- Have tech companies learned from tragic examples like the suicide victims of cyberbullying?
- Have these companies and their leaders abandoned the mantra of "Move Fast and Break Things"?
- What ethical guardrails and guidelines do tech companies follow, amid continuing modus operandi of "Move Fast and Break Things"?
- What can the average citizen, including information professionals, proactively do to influence how tech companies operate and do business, given what we have learned from the "Move Fast and Break Things" era?

Concerns about the "Move Fast" philosophy are amplified for many when we consider the exponentially increasing powers and societal influences of existing AI, as well as the perhaps "Omega-level" future capabilities of AI 2.0, 3.0, and AI-progeny technologies that are yet unknown. Indeed, examples abound that the "Move Fast" way of operating continues, though not for all tech companies in all instances. Former AI experts working at Google, for example, have been so concerned by Google's handling of AI development at times that they have departed and started their own companies. One example is Anthropic, whose website proclaims "AI research and products that put safety at the frontier."[15] Microsoft has endeavored to lead on AI ethics in a number of ways. As a reminder, Microsoft is one of the original signatories of the Rome Call for AI Ethics in 2020. It also maintains efforts that are focused on responsible AI practices[16] and its AI for Good Lab.[17]

In February 2024, Google CEO Sundar Pichai was mobilized to do damage control from an imbroglio involving image results from its Gemini AI tool that offended its users. As NPR reported, Google halted the tool's ability to generate images after some of those images went viral on social media. The posts included Gemini-created images that portrayed "America's Founding Fathers as black, the Pope as a woman and a Nazi-era German solider with dark skin."[18] Pichai subsequently sent an email to its employees stating that some of Gemini's responses to user queries had "offended our users and shown bias." He acknowledged that that was "completely unacceptable and we got it wrong." Pichai pledged that the tool would be fixed and relaunched, with process steps like "updated product guidelines."[19] Had Google moved too fast in making its newest chatbot available to the public?

In looking at the controversy in its totality, Google's design teams had been working to ameliorate prior known issues with AI-generated bias against people of color. For example, Google had been trying to remedy the issue of AI-generated images that "perpetuate bias and stereotype" by using datasets that predominantly have images of Caucasians or that depict professions in stereotypic ways.[20] Google had endeavored to sidestep one kind of public relations controversy but had found itself in another.

To his and Google's credit, the tech giant did take responsibility for the missteps and offense created by its Gemini tool. Of note as well, harkening back to this previous chapter's discussion of fallibility of humans *and* machines, Pichai acknowledged, "No AI is perfect, especially at this emerging stage of the industry's development."[21] He further noted that there is a high bar for what is expected of the company and that they would continue to "keep at it for however long it takes."

Returning to the "AI Ethics for Peace" gathering in Japan in July 2023, one of the Jewish scholars shared his thoughts about the responsibilities the world faces and the respective roles of the world's major religions in shaping the use of AI for the good of people. He recounted the writings and teachings of a Jewish theologian-intellectual from centuries ago who urged caution when facing a decision fraught with ethical consequences. The English translation of that ancient learned scholar's Hebrew admonition is "If you have doubt, don't do it." The difficult, consequential questions remain:

- How shall we proceed with emerging technologies?
- Do we countenance a "Move Fast and Break Things" approach for the potential greater good of innovation and innovation's offshoots?
- Do we follow the long-ago admonitory dictum "If you have doubt, don't do it," for fear of calamitous effects from potentially risky or ill-advised pursuits?
- Do we find some kind of middle ground?
- Who shall decide and when, where, and how such rules or guidelines are applied and enforced?

Technology for Good, Technology for Bad

As mentioned, Microsoft's AI for Good Lab is an example of the tech industry's efforts to leverage AI in ways that benefit society. As part of my participation in the "AI & the Church Summit" in Seattle in mid-August 2024, I had the opportunity to visit Microsoft headquarters in suburban Seattle for a half day. A half dozen Microsoft administrators spoke with us about ways that the company is striving to make Microsoft AI technologies accessible throughout the world in ways that improve people's lives. One presenter, Juan M. Lavista Ferres, corporate vice president and chief data scientist, recounted a compelling example demonstrating how AI tech can be used to perform time-sensitive surgeries in places like the Global South to save the vision of premature babies before the 24- to 48-hour window of operability closes and these infants become blind for the rest of their lives.

Another intriguing example of technology being used for good comes from Lisa Schirch, Richard G. Starmann Sr. Professor of the Practice of Peace Studies at the University of Notre Dame. As a presenter at the "AI Ethics for Peace" conference in July 2024 in Hiroshima, Schirch talked about ways that AI-equipped robots are being used to resolve conflicts around the globe. Sometimes, she recounted, AI robots are even preferred over humans in negotiations with parties that are in conflict with another.

Amy and Ryan Green, the creators of a video game called *That Dragon, Cancer*, shared their remarkably inspiring story at the 2024 "AI & The Church Summit." When their one-year-old son Joel was diagnosed with terminal cancer, the Greens deployed their considerable technological skills to create "an immersive adventure game that retells Joel Green's 4-year fight against cancer through about two hours of poetic, imaginative gameplay."[22] Unfortunately, Joel did not survive his heroic battle against cancer, passing on before the game's development was completed. In the aftermath of their young son's passing, his parents continue to share the story of their experience as a way to help others and show the powerful impacts that gaming and immersive storytelling can have on people's lives. A 2016 *New York Times* article about the Greens' experience, "This Video Game Will Break Your Heart," explained that

> "That Dragon, Cancer" is not the first memoir to arrive in the form of a video game, but it is probably the most ambitious.
>
> "I feel like 'That Dragon, Cancer' is one of those breakthrough moments," said John Sharp, a professor of games at the New School and the author of "Works of Game," a book about the intersection of games and art. "I think we may look back at this game and see this as a touch point, a moment of rethinking what people can do with interactivity."[23]

While all of the prior examples rather persuasively and poignantly show the good that technologies can catalyze, it is key to be ever mindful of the bad things for which technologies can also be used. Another of the valuable contributions to ethical decision-making is the Markkula Center for Applied Ethics' "Ethical Toolkit for Engineering/Design Practice." Markkula Center notes that the tools it offers "represent concrete ways of implementing ethical reflection, deliberation, and judgment into tech industry engineering and design workflows."[24] The tool kit includes seven tools, such as ethical risk sweeping, ethical pre- and postmortems, and so on. For the purposes of this section on the capability of technologies being used for bad purposes, it is useful to briefly unpack the sixth tool, which covers thinking about "terrible people" and criminal behavior. A number of helpful questions for consideration are provided to think through during the design stage. For example, one question involves considering who might steal, destroy, or weaponize what has been created. Another includes weighing who may utilize the technology "with alarming stupidity/irrationality."[25] In summation, though many of us would prefer not to have to think about how bad actors may use technologies, ethics mandates that we do so, to safeguard public safety, well-being, and the mitigation of harms.

One final cautionary story derives from Father Paolo Benanti, AI ethics advisor to the late Pope Francis, discussed in the earlier AI chapter of this book. During his opening remarks at the July 2024 "AI Ethics for Peace" conference in Japan, Father Benanti spoke about the work of the late twentieth-century Italian philosopher Umberto Eco, perhaps best known for his 1980 novel *The Name of the Rose*. Eco described the tendency of people to "construct an enemy" upon whom they can focus their fears.[26] Father Benanti observed that technologies like generative AI deep fakes could be used by malevolent forces as a kind of tireless vehicle for constructing enemies, inflaming conflicts, and fomenting fear of the "other." For this reason, it is incumbent that we be vigilant to the bad aims that technologies can be used to carry out, while uplifting the many good deeds that technologies can also enable.

Ethics Backpack

The last section of this concluding chapter is a collective mix of items to include in a metaphorical "ethics backpack." This backpack holds some concluding reminders, utilitarian tips and tools, and inspirational calls to both action and reflection that we can carry with us. Prepared and fortified for the challenges that may lie ahead of each of us, we can be reasonably assured that we have the knowledge and resources we need to embark on our own paths of exploration, discovery, and, hopefully, enlightenment into ethics-rich environments of peaks, valleys, and rapids we may need to traverse.

Proactiveness, Rather Than Reactiveness

This first item for our backpack is the reminder that we can often do much more than we think we can. Being proactive is motivating and empowering. Groups have already coalesced around the United States and other parts of the world, made up of people who want to think about positive ways to use technologies like AI. At colleges and universities these groups are asking probing, multi-faceted questions about how to responsibly use AI and create guidelines for higher education stakeholders. One such group at Arizona State University is proactively taking a leadership role on addressing "ethical questions about the role of AI in higher ed."[27] In an August 27, 2024, article, ASU News reported that "as artificial intelligence becomes more ubiquitous in our everyday lives, the AI and Ethics Workgroup at Arizona State University's Lincoln Center for Applied Ethics is working to establish ethical guidelines and frameworks for the deployment of AI technologies."[28] The article explains that the working group will be interdisciplinary. To that end, the cochairs are Sarah Florini, an associate professor of film and media studies, and Nicholas Proferes, as associate professor of social and behavioral sciences. The group's aim is to assist people in maneuvering amid complicated ethical issues that are occurring due to AI's accelerating developments. An initial white paper produced by the group addresses AI tools in college and university settings.

The workgroup's cochairs beneficially provide their responses to questions like, "What can educators and institutions start doing today to instill more responsible, ethical adoption of AI-related technologies?"[29] Sarah Florini advises people to "Get involved and participate in the conversations surrounding these technologies."[30] She notes that everyone needs to be involved in the processes that will influence how AI is incorporated into higher education because there are lots of stakeholders with diverse ideas about what to do. Florini further suggests that we need to look critically at these technologies and assess how this tech aligns with values. Nicholas Proferes, like Florini, recommends talking with one's community members and all stakeholder that may be impacted by the technologies and decisions. He also stresses looking at core values, identifying what the end goals are for using the tech, and creating policies and procedures that will support those objectives. Florini observes that technology development can often feel like an exercise in which "ethics is treated as a checkbox, rather than an ongoing process of reflection and consideration."[31]

The bottom line that can be applied to other higher education environments as well as other types of organizations: think about ways that enable you to get involved in research and teaching discussions, brainstorming sessions, policy formation groups, service efforts, and other gatherings that can impact issues involving ethics, information, and technologies.

Continuity of Purpose

A reminder for us as we engage with doing the often challenging work of ethical reasoning and decision-making involving ethics and technologies is that these are *ongoing* efforts and labors. Hard work can also be good work. One of the features that attracted me to the "AI & the Church Summit" is that it was not a one-and-done event. Rather, attendees were sorted into ecumenical cohorts by AI, using our responses to various questions beforehand about our learning styles, interests, and the like. Cohorts will determine what actions they want to take over the next year, such as writings, action, advocacy, teaching, outreach, and so forth. Then, attendees are invited to convene in Minneapolis for another summit in September 2025. Other subgroups, such as an AI for Good group, may also work on additional projects. This example of a framework encompassing AI and faith communities reminds us that having a continuity of purpose is important for getting work done, contributing to the growth of movements, and cultivating a kind of spoke-and-wheel schema to advance missions, visions, and goals.

Grayness of Ethics

One of the hallmarks of ethics is that as a discipline and approach it tends to be more gray in nature than black and white. Answers, or rather responses, to ethics questions and dilemmas are not always clear-cut. Ways to approach ethics issues and queries can turn on the changing of a single fact or the application of a different ethical theory. Lawyers are famously lampooned for the reply, "It depends," when asked a question, but this trope is actually factually accurate. Law and ethics questions frequently defy easy answers, and answers depend on myriad factors, circumstances, and out-and-out uncertainties; some of the most consequential uncertainties of trials, for example, are the makeup of the jury and who the judge is.

Keeping center of mind that ethics questions may not be easily answerable doesn't mean we must accept a defeatist attitude when engaging with ethical conundra; rather, we enter an ethical inquiry with the expectation that a "right" or "perfect" answer is likely not possible. Equipped with helpful tools like an ethical framework or a trusted person with whom to consult, we can arrive at the best possible ethical response. Even if, upon reflection, the course of action chosen was not the one we would have hoped for, the evaluative steps of postdecision introspection can aid us in making better ethical decisions in the future.

Getting Comfortable With Uncomfortable

On a winter day in January 2018, I was preparing to work out with a longtime trainer at a fairly new CrossFit gym called Mecka that had opened in suburban Pittsburgh. The gym where we had trained for almost a decade had closed and the trainer decided we would train at this location. In addition to training, I also decided on my own to join several rowing classes. These classes, taught by a 50-something rowing expert with experience training elite rowing athletes at the collegiate level, were among the most challenging physical training exercises I had ever done. For sixty intensive minutes, without rest periods, we rowed and intermittently hoisted kettlebells, pulled levers down and toward us on machines, and exerted ourselves at a breakneck pace. Soon after having joined Mecka, a quick stop in the men's restroom one morning revealed a sign on the wall; this changeable sign enabled the gym staff to press letters into the soft center to communicate inspiring phrases and motivational quotes. The words on the sign this particular day immediately grabbed my attention: "GET COMFORTABLE WITH UNCOMFORTABLE."

Those words hit home. Most of us, myself included, do not seek out or readily embrace that which is "uncomfortable." However—and I remember thinking this from the very moment of reading those words—getting comfortable with uncomfortable is sometimes exactly what we need to lean into. Of course, there are situations where uncomfortableness is not something to accept or get comfortable with. Examples include situations that are abusive, inappropriate, or nonconsensual or morally and legally indefensible. Yet there are also many circumstances in life when we are confronted with ideas that challenge our beliefs or ways of operating, and it can serve us well to consider sitting with our discomfort to see what insights, lessons, or epiphanies may be revealed over time.

Identifying the Essential Questions

"Will technologies like AI replace humans?" This question—and derivations of this question, such as "Will technologies like AI replace artists and writers?"—seems to be one of the greatest fears that people have in the face of new technological innovations. Much is being written now about whether this or that sector of worker, creator, professional, or even clergyperson will or will not be replaced by a machine. The truth is that we do not know. No one knows the future. What futurists and AI technologies have in common is not the ability to predict but rather the ability to identify *probabilities*. It is probable that AI technologies will displace people within some industries and sectors, but no one yet knows the magnitude of what these changes may in fact be.

In the midst of such looming uncertainties and potentially paradigmatic changes, what we as humans can do is focus on identifying the essential questions:

- If AI is able to complete tasks and venture into creative realms like art, authorship, and musicality, what will this mean for human creators?
- What can humans bring to the creative process that machines alone cannot?

Some creators are already collaborating with technologies to create new kinds of art and culture, as this book has identified. These activities are spawning profound ethical and legal challenges for people and societies. We've also explored some ways that religion and faith communities are engaging with AI tech to think about ways of enhancing ministries and service. For example, if AI technologies can be used to perform tasks that are more menial in nature (e.g., keeping prayer lists updated, keeping track of contributions, making payroll, etc.), faith leaders and staff can target their energies toward "high touch" activities, like visiting ill or bereaved persons in hospitals, assisted living facilities, and their homes.

What *is* clear from these examples is that ethical inquiry will continue to be a necessary component of our daily lives. A number of faith and thought leaders are already engaging their congregants and wider communities in these kinds of probing discussions about AI and other technologies. For example, in 2023 the Rev. Doyt L. Conn Jr. at Seattle's Epiphany Episcopal Church delivered a series of AI sermons on issues like "AI and the Human Identity"[32] and "The Formation of the Human Heart."[33] Such leaders and fellow spiritual travelers continue to remind us to contemplate essential questions like the following, given all that AI technologies can and likely will do in the future:

- What makes us human?
- What are the inimitable—that which cannot be imitated—qualities that make human beings unique in all of creation?

One of those qualities may be empathy. *Merriam-Webster's Dictionary* defines this vital ability: "The action of understanding, being aware of, being sensitive to, and vicariously experiencing the feelings, thoughts, and experience of another."[34] The interjection of the element of "empathy" into the equation of ethics, information, and technology initiates some of the most potentially consequential and revelatory questions confronting many foundational cornerstones and building blocks of our human societies, institutions, and world:

- Can a machine feel empathy?
- Will AI be able to be sensitive and truly responsive to the feelings and thoughts of a flesh-and-blood person?
- Can technologies understand the lived experiences of a human?

No one yet knows. Science may someday tell us, but that day has not yet arrived. Unsurprisingly, great and contemplative minds have been and are engaging with these kinds of deeply probative questions. As science fiction author Arthur C. Clarke observed in *The Light of Other Days*, "What we need is a machine that will let us see the other guy's point of view."[35] Alas, we don't yet have such a machine. Whether such a device will be possible to create one day is one of the Holy Grails of science and technology.

Academic researchers, technologists, philosophers, theologians, and others are actively engaging with these types of questions. As just one example, a 2022 webinar offered by the Markkula Center for Applied Ethics examined the intersection of AR/VR, ethics, and law.[36] Presenters for the webinar have written thought-provoking articles on issues related to these questions and issues, such as Brittan Heller's "Reimagining Reality: Human Rights and Immersive Technology"[37] and Erick Ramirez's "It's Dangerous to Think Virtual Reality Is an Empathy Machine."[38]

Moreover, if we do someday create machines that can feel and be sensitive to the thoughts and experiences of a human being, in ways that are not purely performative or probabilistic because of a human-engineered subroutine or coded algorithm, what will be our responsibilities to such machines? What ethical principles will guide how we treat our technological devices and tools? We are just beginning to culturally and philosophically explore such incredibly complex questions through thoughtful films like *After Yang* (2022) and *The Creator* (2023) that explore the relationships between humans and AI humanoids. Returning to and adding to the prior question raised by Rev. Conn Jr. about what makes us human in a world of machines that are likely to exceed mortal abilities, what can the ways that we treat our technological devices and machines, now and in the future, also tell us about what it means to be human? Thinking about and choosing to act as ethically and morally responsible as possible on the approaches and answers to these kinds of essential questions should continue to be our guiding North Star, informing and shedding clarifying light on what it means to be human.

Notes

1. Martin, F. (2022, April 21). *How imagined "bizarro worlds" invite us into the real worlds of ancient Israel and Egypt*. Stroum Center for Jewish Studies, University of Washington. https://jewishstudies.washington.edu/israel-hebrew/bizarro-worlds-ancient-israel-egypt-topsy-turvy-bible/
2. Creative Commons. (n.d.). *Creative Commons 2025–2028 strategy*. Retrieved February 27, 2025, from https://drive.google.com/file/d/1PCQvwqcfB0bRQh2Txx4Ta-L7fYoZQBJ3/view?pli=1
3. Ibid.
4. Johansen, B. (2018). *The new leadership literacies*. Penguin Random House.
5. *Justice Department sues RealPage for algorithmic pricing scheme that harms millions of American renters*. (2024, August 23). US Department of Justice. https://www.justice.gov/opa/pr/justice-department-sues-realpage-algorithmic-pricing-scheme-harms-millions-american-renters
6. Merriam-Webster. (n.d.). *Entropy*. Retrieved February 27, 2025, from https://www.merriam-webster.com/dictionary/entropy

7. May, D. R. (n.d.). *Steps of the ethical decision-making process*. EESE Faculty Development Workshop. University of Kansas Initiative on Ethics Education in Science and Engineering. https://crenshaw consultingassociates.com/wp-content/uploads/2019/01/EESE_EthicalDecisionmakingFramework.pdf
8. *NYPL CEO: We're in the "access to information" business*. (2020, April 23). Yahoo Finance. https://finance.yahoo.com/video/nypl-ceo-access-information-business-133620962.html
9. US Department of Health and Human Services. (2022). *What is lived experience?* https://aspe.hhs.gov/sites/default/files/documents/5840f2f3645ae485c268a2784e1132c5/What-Is-Lived-Experience.pdf
10. Brill, S. (2024). *The death of truth: How social media and the internet gave snake oil salesmen and demagogues the weapons they needed to destroy trust and polarize the world—and what we can do*. Penguin Random House.
11. Velshi, A. (2024). *Small acts of courage matter: A legacy of endurance and the fight for democracy*. Macmillan.
12. Ressa, M. (2022). *How to stand up to a dictator: The fight for our future*. HarperCollins.
13. Dans, E. (2023, November 8). When companies move fast, they do more than break things. *Medium*. https://medium.com/enrique-dans/when-companies-move-fast-they-do-more-than-break-things-770740e248c2
14. Taneja, H. (2019, January 22). The era of "move fast and break things" is over. *Harvard Business Review*. https://hbr.org/2019/01/the-era-of-move-fast-and-break-things-is-over
15. Anthropic. (n.d.). *Home page*. Retrieved February 27, 2025, from https://www.anthropic.com/
16. Microsoft. (n.d.). *Empowering responsible AI practices*. Retrieved February 27, 2025, from https://www.microsoft.com/en-us/ai/responsible-ai
17. Microsoft (n.d.). *AI for good lab*. Retrieved February 27, 2025, from https://www.microsoft.com/en-us/research/group/ai-for-good-research-lab/
18. Allyn, B. (2024, February 28). *Google CEO Pichai says Gemini's AI image results "offended our users."* NPR. https://www.npr.org/2024/02/28/1234532775/google-gemini-offended-users-images-race
19. Ibid.
20. Ibid.
21. Ibid.
22. That Dragon, Cancer. (n.d.). Retrieved February 27, 2025, from http://www.thatdragoncancer.com/#home
23. Suellentrop, C. (2016, February 5). This video game will break your heart. *New York Times*. https://www.nytimes.com/2016/02/06/arts/that-dragon-cancer-video-game-will-break-your-heart.html
24. Vallor, S. (2018, June 22). *An ethical toolkit for engineering/design practice*. Markkula Center for Applied Ethics, Santa Clara University. https://www.scu.edu/ethics-in-technology-practice/ethical-toolkit/
25. Ibid.
26. Eco, U. (1980). *The name of the rose*. Harcourt.
27. Fitzgerald, K. (2024, August 27). *ASU workgroup addresses ethical questions about the use of AI in higher ed.* ASU News. https://news.asu.edu/20240827-science-and-technology-asu-workgroup-addresses-ethical-questions-about-use-ai-higher-ed
28. Ibid.
29. Ibid.
30. Ibid.
31. Ibid.
32. Conn, Rev. D. L., Jr. (2023, May 7). *AI and the human identity—the royal priesthood.* Epiphany Seattle. https://epiphanyseattle.org/sermon/sermon-series-1-ai-and-the-human-identity-the-royal-priesthood/

33. Conn, Rev. D. L., Jr. (2023, May 21). *The formation of the human heart*. Epiphany Seattle. https://epiphanyseattle.org/sermon/sermon-series-3-the-formation-of-the-human-heart/
34. Merriam-Webster. (n.d.). *Empathy*. Retrieved February 27, 2025, from https://www.merriam-webster.com/dictionary/empathy
35. Clarke, A. C., & Baxter, S. (2000). *The light of other days*. Macmillan.
36. Raicu, I. (2022, March 7). *An upcoming webinar on AR/VR, ethics, and law*. https://www.scu.edu/ethics/internet-ethics-blog/an-upcoming-webinar-on-arvr-ethics-and-law/
37. Heller, B. (2020). *Reimagining reality: Human rights and immersive technology*. SSRN. https://papers.ssrn.com/sol3/papers.cfm?abstract_id=4563877
38. Ramirez, E. (2018, October 26). *It's dangerous to think virtual reality is an empathy machine*. Aeon. https://aeon.co/ideas/its-dangerous-to-think-virtual-reality-is-an-empathy-machine

Index

About the Author

Kip Currier, PhD, JD, MLIS, is Assistant Professor in the School of Computing and Information at the University of Pittsburgh, where he teaches technology ethics and intellectual property. He regularly blogs on ethics at https://kipcurrierethics.blogspot.com and on IP at https://kipcurriercopyright.blogspot.com.